LIT

*Active Leadership in Education
Enterprise and Engagement*

An Introduction to

Metallurgy

Second Edition
SI Units

Alan Cottrell

Jesus College, Cambridge

Edward Arnold

*Printed and bound in Great Britain
at The Pitman Press, Bath*

Preface

Because of their great practical value and scientific interest, metals lie at the cross-roads of many scientific and technological disciplines. Chemists are interested in the oxidation and reduction of metals, the catalytic properties of metals and the laws by which metals combine together to form alloys. Chemical engineers apply their general principles of chemical processing to the production of pure metals from ores. Solid-state physicists are fascinated by the electronic and atomic structures of metals and by the ways in which these structures determine the characteristic properties of metals and alloys. Mechanical engineers are interested in the plastic working of metals, structural engineers in the mechanical performance of metals in practical use, and electrical engineers in all the special electrical and magnetic properties obtainable from metallic materials.

The contributions to the science and technology of metals made by people in these fields are immensely valuable. Nevertheless, and quite naturally, each of them sees only his own side of the subject. The essential task of the *metallurgist* is to complement and coordinate the work of these specialists by acting as a general practitioner over the whole field. To take an example, if we wish to make an intelligent choice of a steel for a nuclear reactor pressure vessel we must know about the chemistry of steelmaking, about the rolling and welding of steel, about corrosion, about brittle fracture at low temperature and creep deformation at high temperature, about effects of damage to the atomic structure of metals by nuclear radiation, about commercial and economic factors, and about the interrelations of all these and many other things with one another. An equally wide spectrum of knowledge is required in other problems; for example, to develop a new turbine alloy or to diagnose the cause of failure in a broken aircraft undercarriage. A metallurgist, however much he may specialize in practice, must be able, when required, to take this wide, all-embracing view.

It was with such thoughts in mind that I considered an invitation to prepare a new edition of my book *Theoretical Structural Metallurgy*. When that book was written, the pure science of metals was still quite new and there seemed a good case for bringing it to the attention of metallurgists in an elementary but specialized book. The position is now different. This science has become well established and triumphant. The need now, it seems to me, is to re-assert the unity of all metallurgical knowledge and to link up this new science with the more traditional aspects of the subject. A new edition under the old title could not do this. I was thus led to attempt, instead, a complete survey of the whole metallurgical field. What I have tried to do particularly is to show metallurgy as a single applied science. This has meant developing the science as a continuous thread running through the subject, from atomic theory, through thermodynamics, reaction kinetics and crystal physics, to elasticity and plasticity, but stopping at all suitable places to show how the characteristic features of metals, alloys, and the processes of practical metallurgy, such as extraction, refining, casting, working, and heat-treatment, grow out of this science. In choosing this pattern I have also been influenced by the feeling that the most intellectually exciting thing to do in metallurgy today is to apply the newly-discovered science to the invention and development of new metallurgical processes and materials.

Naturally, I have not been able to go very far into any one aspect of the subject. The book has been aimed, however, at those who are at the beginning of their metallurgical studies and so I hope that it will be forgiven its admitted lack of depth, for the sake of what I hope is a broad view.

Christ's College, Cambridge   ALAN COTTRELL
1966

Acknowledgements

Much of this book has grown out of the more elementary parts of the courses taught in the Department of Metallurgy, University of Cambridge, and I am grateful to many friends there for their advice and help. I would additionally like to thank Mr G. C. Smith, Dr R. B. Nicholson, Dr B. Ralph, Dr S. Ranganathan and Dr J. R. Griffiths for kindly making available the photographs used in this book.

Acknowledgement is made for kind permission to reproduce the following diagrams: Figs. 15.6 and 15.14 from DESCH, C. H. (1944) *Metallography* 6th ed, Longmans, Green, London; Figs. 17.8 and 17.9 from CHALMERS, B. (1949) *Prog. Metal Phys.* **1**, 77 (Pergamon Press, Oxford); Figs. 17.10 and 17.12 from CHALMERS, B. (1953) *Prog. Metal Phys.* **4**, 205 (Pergamon Press, Oxford); Fig. 19.5 from the Clarendon Press, Oxford; Fig. 19.12 from SMITH, C. S. (1948) *Trans. Am. Inst. Min. metall. Engrs.*, **175**, 15; Fig. 19.14 from The Institute of Physics and The Physical Society.

Preface to the Second Edition

Since the first Edition was written the scientific community has largely yielded to the pressures brought upon it to conform to the SI system of units. I have therefore accepted the inevitable and converted the text, in this second Edition, also to these units. For those who nevertheless find other systems, such as c.g.s., more congenial, I have included some conversion factors in a table at the front of the book.

I have also taken the opportunity to up-date some of the information in the text.

Jesus College, Cambridge
1974

ALAN COTTRELL

Contents

List of Plates

Units, Conversion Factors, Physical Constants

The basic SI units are *metre* (m) for length, *kilogramme* (kg) for mass, *second* (s) for time, degree *kelvin* (K) for temperature, *mole* (mol) for amount of substance, *ampere* (A) for electric current and *candela* (cd) for luminous intensity.

1 m = 39·37 inches = 3·281 feet = 10^{10} angstroms (Å).

1 kg = 2·205 pound mass (lb) = $9·84 \times 10^{-4}$ long tons = 10^{-3} metric tons (tonne) = $6·03 \times 10^{26}$ atomic mass units (u).

273·15 K = 0° C.

1 mol = $6·03 \times 10^{23}$ particles (Avogadro's number).

1 coulomb = 1 amp.sec. (A s) = 3×10^9 electrostatic units (esu).

1 joule (J) = 1 watt.sec. = 10^7 erg = 0·239 calories = $6·26 \times 10^{18}$ electron-volts (eV) = $9·478 \times 10^{-4}$ British Thermal Units (BTU).

1 eV per particle = 96 500 J mol^{-1} = 23 069 cal.mol^{-1}.

1 MeV = 10^6 eV.

1 u = 931 MeV.

1 Faraday = 96 500 coulombs.

1 Newton (N) = 10^5 dynes = 0·225 pound force.

1 N m^{-2} = 10 dyne.cm^{-2} = $1·02 \times 10^{-7}$ kg.mm^{-2} = $1·45 \times 10^{-4}$ pound in $^{-2}$ (psi) = $9·87 \times 10^{-6}$ atmospheres.

1 tesla = 10^4 gauss (magnetic fields).

c = velocity of light = $2·997 \times 10^8$ m s^{-1}.

e = electron charge = $1·6 \times 10^{-19}$ coulomb = $4·8 \times 10^{-10}$ esu.

m = electron mass = $9·1 \times 10^{-31}$ kg.

h = Planck constant = $6·624 \times 10^{-34}$ J s.

R = gas constant = 8·312 J mol^{-1} K^{-1}.

k = Boltzmann's constant = $1·38 \times 10^{-23}$ J K^{-1}.

Chapter 1

Prologue

1.1 The art and science of metals

Metallurgy is the art and science of making metals and alloys in forms and with properties suitable for practical use. Most people know it only as an ancient and mysterious art. Certainly it had a place in ancient history, having brought us out of the Stone Age into the Bronze Age and then into the Iron Age. The seemingly miraculous conversion of dull earths into shining metals was the very essence of alchemical mystery; and there was no science of metals to bring rationality and enlightenment into the mediaeval world of secret formulae for tempering metals and blending alloys.

Some of this air of mystery still lingers over metallurgy to this day. No space ship in science fiction is respectable without its own secret 'wonder metal'. This mystery may be a legacy from the past but it is also an unconscious acknowledgement of the many striking achievements of the modern metallurgist in producing new metals and alloys for jet engines, nuclear reactors, electronic circuits and other advanced pieces of engineering. These successes have not been achieved by old dark magic, however, but by the enlightened and logical application of scientific principles. Metallurgy is now a disciplined applied science based on a clear understanding of the structures and properties of metals and alloys. The mystery of the modern wonder metals is due to no more than the simple fact that this science is mostly too new to have filtered down, yet, into the more elementary levels of scientific education.

1.2 Chemical metallurgy

The least remote part of the subject is *chemical metallurgy*. This deals with all the chemical properties of metals, including the uniting of different metals with one another to form alloys, but a very large part of it is concerned with the oxidation-reduction reactions of metals, for two main practical reasons. First, most metals occur in nature as oxides, sulphides, chlorides, carbonates, etc., and the critical step in converting these *ores* into metals, i.e. in *extraction metallurgy*, is a process of chemical reduction. The basic chemical reactions involved are often simple; the scientific challenge in this part of the subject is to achieve these reactions on a massive scale economically. Second, when the finished piece of metal goes into service afterwards and is exposed to the environment, these same chemical reactions tend to occur spontaneously in reverse. The metal thus reverts from the metallic state to the oxidized state, i.e. it rusts or corrodes. The main tasks of the chemical metallurgist are thus to get metals into the metallic state and then to keep them there.

The origins of extraction metallurgy go back into pre-history. The first discoveries must have been made accidentally in camp fires and hearths where stones of easily reducible metallic ores would have been converted to metal by the heat and reducing flames. Copper, lead and tin were amongst the first metals to be made by such a *smelting* process, over 5000 years ago. At a very early age the alloy *bronze*, usually about 10 parts of copper to one of tin, was made by smelting mixed ores of the two metals together and was much prized because of its great hardness and because, when melted, it could be *cast* easily into intricate shapes by letting it solidify in shaped holes in clay or sand moulds. Early *brasses* were similarly made by smelting mixed copper and zinc ores. The modern method of making alloys by mixing metals was developed later.

Iron ores are also easily reduced but the high melting point of the metal prevented iron from being produced in a liquid form. Instead, a pasty porous mass of *sponge iron* mixed up with *slag* (a crude glass containing un-reduced oxides and silicates), was produced and this had to be compacted, while hot and soft, by *beating* or *forging* it down with hammers, so making something rather like *wrought iron*. The need for higher temperatures to achieve greater outputs led to the gradual evolution of the early iron-making hearth into the *blast furnace*, with an air blast directed into the hot zone above the hearth and a tall enclosed stack above, down which the ore and charcoal fuel travelled.

A great advance occurred in the fourteenth century. Temperatures became high enough to produce liquid iron. The blast furnace could then be operated continuously, being periodically 'tapped' to run out the pool of molten iron at the bottom, and this greatly increased its output. The liquid *pig iron* produced in this way contained about 4 wt per cent dissolved carbon, picked up from the furnace fuel. This carbon greatly lowered the melting point and so made the metal easy to re-melt and cast into moulds. This *cast iron* was, however, brittle due to the carbon, which forms a brittle iron carbide, and other impurities, and so could not be used for the same pur-poses as forged sponge iron. The problem of converting pig iron to a ductile form by refining away the carbon was solved by *Cort* in the eighteenth century with his *puddling process* for making wrought iron. These two forms of iron, wrought and cast, remained the staple ferrous constructional materials until the later part of the nineteenth century.

The delicate carbon control required to make *mild steel* (about 0·25 wt per cent carbon) was beyond the scope of the metallurgy of those days. Admittedly, a type of *tool steel* for swords and cutting tools, which contained about 1 wt per cent carbon and which could be hardened by *quenching*, red-hot, into cold water, was made from very early times by the *cementation* process in which forged sponge iron was heated in charcoal; and in 1740 *Huntsman* made tool steel by melting irons of different carbon contents in a crucible, which was the foundation of the Sheffield cutlery industry. But the discovery that cheap low-carbon steel could be made on a large scale for constructional uses did not come until the mid-nineteenth century, when *Bessemer* invented his *converter* process. This was followed a few years later by the *open-hearth* steelmaking process and the modern age of steel was then begun.

Electricity plays a large part in many modern extraction processes. The decisive step was the *Hall-Héroult process* for the commercial production of *aluminium*, announced in 1886. Many other metals such as magnesium, sodium and calcium are also produced electrically, and these metals in turn are now used to produce the 'modern' metals such as titanium, zirconium, uranium and niobium.

The science of extraction metallurgy has developed rapidly in recent years, with the application of *thermodynamics* and the theory of *reaction kinetics* to its problems. The thermodynamics of metallurgical reactions is now well established but there are many opportunities for further advances, both scientific and technological, in the study and control of reaction rates. Many of the newest extraction processes, such as oxygen steel-making, flash roasting, spray refining and the zinc blast furnace process, depend critically on reaction kinetics.

1.3   Mechanical metallurgy

Metallurgy is a branch of a wider subject, known as *materials science and engineering*, which deals with all materials, e.g. metals, ceramics, glasses, organic plastics and polymers, wood and stone. The reason why metallurgy stands by itself as such a large and self-contained subject is, of course, the outstanding importance of metals as constructional materials. Society as we know it would be quite impossible without metals. The production of metals and metallic goods represents about one-fifth of the gross national product in a modern industrial country.

Metals owe their importance to their unique mechanical properties, the combination of *high strength* with the ability to change shape *plastically* (*ductility* and *malleability*). This plasticity enables them to be shaped, e.g. into motor car bodies, tin cans and girders, by processes of *mechanical working* such as pressing, drawing, rolling and forging. Even more important, this same plasticity gives strong metals their extraordinary *toughness*, the ability to endure all the knocks and shocks of long rough service without breaking or crumbling away.

Mechanical metallurgy deals with all these aspects of the subject; in particular with mechanical working, the testing of mechanical properties, the relations between these properties and engineering design and selection of materials, and the performance of metals in service. It is the oldest part of metallurgy. The earliest known metals, copper, silver and gold, were found *native*, as metallic nuggets. Meteorites were a source of iron-nickel alloys. All these naturally occurring metals are malleable and, from very early days, they were shaped into ornaments, tools and weapons by hammering. The forging of metals became widely established once extraction metallurgy began to provide copper, bronze, sponge iron and other metals in larger amounts. The Romans made extensive use of lead sheet and pipe in water supply systems. *Coining*, the indenting of a design on a metal surface with a punch and die, developed at an early age. The advantages of mechanical working at various temperatures were also realized; *cold working*, because it increased the hardness and strength of metals such as copper and iron; *hot working*, particularly of sponge iron, because metals proved to be much softer and more malleable at high temperatures and also because they could be joined by *pressure welding* if hammered together while hot.

The mechanical working of metals remained a largely hand-craft industry, typified by the blacksmith's forge, for many centuries. The need for larger forgings and the advent of water power eventually led to the power-operated hammer and the forging press. A major development was the rolling mill, which first came into wide use in the eighteenth century. Other processes, such as drawing, machining and extrusion, also gradually evolved. Many new processes have been developed recently, including the cold-forging of steel, using high-pressure lubricants, and explosive forming in which the metal is driven against a die by the force of an explosion. Hydrostatic forming, in which the metal is worked while under great hydrostatic pressure to prevent cracking, seems likely to open up a whole new phase of mechanical metallurgy by enabling the more brittle metals and alloys to be worked.

The science of mechanical metallurgy consists of three main, closely related parts. First, the basic mechanical properties have to be explained from an atomic theory of metals, analogous to the kinetic theory of gases. Here mechanical metallurgy unites with *physical metallurgy*. Then, starting from these basic properties, the behaviour of metals in mechanical working operations has to be understood and controlled. The attack on this problem has produced a new branch of applied mechanics, the *theory of plasticity*. Thirdly, again in terms of basic properties, the mechanical behaviour of metals in service has to be understood and improved, to avoid failures due to plastic collapse, brittle cracking, metal fatigue, etc., and to provide a rational basis for engineering design and the efficient and safe use of materials. This is a very active part of the subject at present.

1.4 Physical metallurgy

Few things in nature seem more inanimate than a piece of metal. The casual observer sees only his own reflection in its bright, still surface and nothing of its world within. This internal world is, however, a place of ceaseless activity. Electrons dash from end to end at immense speeds. The atoms themselves also move and exchange places, even when the metal is completely solid. Changes of temperature can cause the atoms to rearrange themselves suddenly into a radically different pattern of organization. In a quenched steel this can happen in a few micro-seconds, even at temperatures far below room temperature. Plastic deformation occurs through the passage of faults, called *dislocations*, which move at great speed through the metal and cause large-scale slippages between enormous masses of atoms. The traffic of dislocations can become very dense. Huge traffic jams build up, which bring the dislocations to rest and make the metal hard. When this *work-hardened* metal is heated (*annealing*) it rids itself of these dislocations in a wave of reorganization of the entire atomic pattern (*recrystallization*). Completely new atomic patterns can be produced by alloying and these in turn can be changed by heat-treatment. For example, when an aluminium alloy is rested at room temperature, after quenching, its alloy atoms move through the solid to congregate together in small clusters, like water droplets in a mist, and these clusters make the metal hard by getting in the way of dislocations (*precipitation hardening*).

The study of all effects such as these belongs to the province of physical metallurgy, the part of the subject that deals with the structures of metals and alloys with the aim of designing and producing those structures that give the best properties. Physical metallurgy has obvious links with mechanical metallurgy; but it also has close links with chemical metallurgy, particularly in connection with the casting of metals, the formation of alloys, corrosion, and the many effects of impurities on the structures and properties of metals and alloys. It is the newest part of metallurgy although the processes of quenching and tempering, work-hardening, annealing and alloying were discovered and used, in a completely empirical way, in ancient times. Imaginative attempts to construct a theory of metals, including the essential idea that solid metals might be *crystalline*, i.e. have their atoms arranged in orderly patterns, were made in the seventeenth and eighteenth centuries. However, there was no way to put these ideas to experimental test in those days and most scientists chose instead to work in fields such as mechanics, astronomy, electricity and chemistry where progress was easier. So the classical pattern of the history of science developed.

The decisive break-through for physical metallurgy was the development by *Sorby*, in the second half of the nineteenth century, of the *metallographic* technique for observing the structures of metals and alloys with a reflecting optical microscope. The great barrier of the surface mirror was at last penetrated, by a process of polishing and chemical etching, to reveal the internal structure beneath. What was then seen was the *grain* structure of metals, an assemblage of minute interlocked crystals. Great changes in this microstructure were seen to be brought about by alloying, working and heat-treatment. Ideas about the nature of these changes sharpened rapidly once it became possible to put them to the test of observation. At about the same time, the theory of thermodynamics was clarifying the understanding of what happens when different substances are blended together and this provided a scientific basis for the study of alloying. The combination of systematic alloy research with optical microscopy opened most of the doors into physical metallurgy. The effects of carbon in steel could be largely understood, as well as the processes of quench-hardening and tempering; the structures and properties of the older alloys, such as the brasses and bronzes, could be rationalized; and a method was at last available for the systematic development of alloys deliberately designed to have certain properties.

The metallurgical microscope is still the most useful general-purpose instrument available to the physical metallurgist. It could not, of course, give direct proof of the crystalline atomic arrangements in metals, although it left little room for doubt. The direct proof had to wait until the discovery of the *x-ray diffraction* method, the application of which ushered in a second major phase of physical metallurgy in the 1920s. Single crystals of metals were also grown at about this time and their mechanical properties threw much light on the processes of plastic deformation.

The next big advances were theoretical. By the early 1930s the quantum theory of electrons and atoms had become powerful enough to provide a real theory of the metallic state, which could explain what a metal really is and how it conducts electricity. The forces holding metal atoms together could then be understood and a start was made on the theory of alloys. Corrosion was seen (and demonstrated experimentally) to be as much an electrical process as a chemical one and the mobility of atoms in metals was explained in terms of certain well-defined imperfections in the crystal structure (dislocations and vacancies). Theoretical metallurgy was forced along still further in the years immediately after the second world war by the need to develop metals and alloys that would resist high temperatures for jet engines, or which would resist damaging nuclear radiations in nuclear reactors; and by the demands for special materials to be used in the electrical industry. Still more recently, experiment has taken the lead once more, due to the development of extremely powerful electron microscope and field-ion microscope techniques, which allow the structure of metals to be observed right down to the atomic scale of magnitude. The study of dislocations and atomic structures in metals has now become mainly an experimentalist's science.

The innumerable advances that have taken place in the basic science of metals in recent years have left the considerable problem of digesting them all and converting them into corresponding advances in the applied science.

Nevertheless, we can now see clearly how to design the microstructures of metals and alloys so as to develop the basic properties to best effect. Some of the proposed new microstructures are very different from the traditional ones and there is a great technological challenge to realize these commercially on a large scale. As regards the basic science there are still many areas where fundamental problems remain, particularly to do with the theory of alloys, with liquid metals and with the more complex mechanical properties such as metal fatigue.

1.5   Metallurgical science and industry

Metallurgy today is an applied science. Its fascination lies in the challenge of using science to give mankind the best engineering materials that the laws of nature and the Earth's natural resources will allow. This simple fact is often overlooked or forgotten. This is partly due to the fact that for thousands of years industrial metallurgy was an empirical art in which the 'right way' of doing things was learnt by hard experience. However, that is history and the science is now with us. But in building up this science academic metallurgy has sometimes seemed to be a pure science, unrelated to the industry. Certainly, the explanation of properties of metals in terms of microstructure and the development of the theory of alloys can stand in their own right as contributions to pure science. But means must not be confused with ends. The long-term aim of this work has been to provide a scientific base for further practical advances. A good example is provided by the history of superconductivity. For years a branch of pure physics, superconductivity immediately became a subject of intense metallurgical interest once the possibility of making useful superconductive devices was clearly realised.

The applied science of metallurgy links the science of metals to the metal industries. This link is only maintained and strengthened by deliberate care and attention, for there is always a tendency for the scientific and industrial sides to drift apart. It is as natural for the research man to commit himself totally to his scientific problems as it is for the industrialist to commit himself totally to his production problems, but this all too often has the result that each has little time for the other. This tendency must, however, be resisted at all costs; for without the practical purpose the science must eventually become pointless and trivial, and without the scientific stimulus the industry will stagnate technically and survive only by cheap labour and clever accountancy.

The difficulty of this problem should not be underestimated. The qualities that help to make a good research man—an ability to fix attention on a single scientific problem to the exclusion of all else and to suspend judgement patiently until the facts are quite clear—are not well suited to a member of a design or production team where breadth of knowledge, quick response and good intuitive judgement are all-important. Few people will ever be able to contribute fully to both the scientific and practical sides of the subject, at least not at the same stages of their career. Furthermore, the qualities that make a good experiment—choice of special conditions and experimental materials to display the critical effects as clearly and simply as possible, rigorous control of all unwanted variables—often lead the experiment far away from the industrial problem it may have been designed to illuminate.

For these reasons the research metallurgist cannot help becoming separated from his more practical colleagues. But he should never give up the challenge of making his work as directly relevant to theirs as it possibly can be, without sacrificing the principles of good science. The research metallurgist must seek and extract his problems very precisely from the heart of industry itself, but they must be good problems scientifically. The ability to do this, and pleasure in doing it, is something quite foreign to the mainstream of pure scientific research; and the inculcation of this ability and attitude is perhaps the main justification for the teaching of metallurgy as a separate academic discipline.

The industrial metallurgist also has his challenges. He must remain aware of the science and yet has to solve his urgent problems by the quickest route, which may often be largely empirical because there is no time to stop and fill in the missing scientific background. Intuitive judgement, an ability to conceive and try quick *ad hoc* solutions and to move on without further ado, if these are successful, are essential here. However, a good background of analytical science is equally important for narrowing the choice of possible empirical routes, for coordinating all the innumerable scraps of information into a coherent picture and for placing value and emphasis on the various parts of the programme, to show where the ground may be covered quickly and where one must tread carefully. This kind of metallurgy also is applied science and demands great analytical powers.

In the chapters that follow, we shall try to develop a unified view of both the scientific and industrial aspects. Even in an introductory course there is a great deal of science and a large number of facts to be learnt. The science cannot take wing without the facts, but a long preliminary recitation of the facts of industrial metallurgy, without the science, can only be stupefying. To try to overcome this problem we shall work through the science, starting at the atomic nucleus and building up gradually to the complex structures of industrial metals, to provide a continuous backbone to the subject, but on the way we shall stop at all suitable places to show how this science relates to the characteristic features of industrial metallurgy. By continually crossing to and fro between the science of metals and industrial practice, we shall thus try to view metallurgy as the applied science linking these two sides.

Chapter 2

The Atomic Nucleus

2.1 Composition of nuclei

The general picture of an atom is now familiar. At the centre is the *nucleus*, 10^{-15} to 10^{-14} m radius, and round it swarm the *electrons*, elementary particles of about 10^{-15} m radius. The electrons move about rapidly (10^6 m s^{-1}) in a region of about 10^{-10} m ($=1$ Ångström) radius centred on the nucleus. The mass of the electron, at rest, is 9.1×10^{-31} kg, and each *nucleon*, i.e. each elementary particle in the nucleus, is about 1840 times heavier, so that practically all the mass of an atom is in the nucleus. The nucleon can exist in two forms; as a *neutron*, which is electrically uncharged, or as a *proton*, which carries a positive charge ($e = 1.6 \times 10^{-19}$ coulomb) equal and opposite to the negative electrical charge carried by the electron.

The *atomic number* Z and the *neutron number* N are the numbers of protons and neutrons in the nucleus respectively. The chemical properties of an atom depend on the atomic number and each chemical element has its own particular value of Z (cf. Table 3.1, p. 21). The *atomic mass number* is $A = Z + N$. In their natural state most chemical elements are mixtures of *isotopes*, i.e. mixtures of nuclei with the same Z but different N.

The *atomic weight* (Table 3.1) is roughly equal to the atomic mass number. The *chemical scale* of atomic masses assigns the standard value 16 to the naturally occurring mixture of oxygen isotopes. The *physical scale* assigns 16 to the most common isotope (i.e. $A = 16$) of oxygen and atomic weights on this scale are 1·000275 times those on the chemical scale. If an atomic weight of an element is measured out in kg, this sample contains 10^3 mol of atoms of the element. The *atomic mass unit* (u) is one-sixteenth of the mass of the oxygen-16 atom and 1 u $= 1.66 \times 10^{-27}$ kg. The reciprocal of this is related to *Avogadro's number*, 6.03×10^{23}, the number of particles in 1 mol. The masses of the proton, neutron and electron are respectively 1·00758, 1·00897 and 0·00055 u.

The simplest nucleus is that of hydrogen, a single proton ($A = Z = 1$). Its isotopes are deuterium ($N = 1$, $Z = 1$), i.e. heavy hydrogen, and tritium

($N = 2$, $Z = 1$). The next element is helium ($Z = 2$) which commonly has $N = 2$. The various chemical elements follow in sequence, up to heavy elements such as uranium, each element having one proton and usually one or two neutrons more than its predecessor. Thus the main isotope of uranium has $A = 238$ and $Z = 92$. This is conveniently written as $^{238}_{92}\text{U}$, using a notation useful for describing nuclear reactions. For example, the reaction

$$^1_1\text{H} + {}^7_3\text{Li} \rightarrow {}^4_2\text{He} + {}^4_2\text{He}$$ **2.1**

means that a proton and a lithium-7 nucleus have united and then split into two helium-4 nuclei, i.e. into two *alpha particles*.

2.2 Nuclear binding

The masses of nuclei and nuclear particles can be measured accurately in a *mass spectrometer*. In general a nucleus weighs slightly less than the sum of the individual weights of its nucleons, the difference being called the *mass defect*. It is due to the release from the nucleus, when it was made, of the *binding energy* which comes from the attractive *nuclear force* that binds the nucleons to one another as a stable dense cluster of particles. The connection between the loss of mass and the release of the binding energy is provided by *Einstein's* famous formula

$$E = mc^2$$ **2.2**

which shows that energy E is mass m multiplied by the square of the velocity of light ($c = 3 \times 10^8 \text{ m s}^{-1}$). The proportionality factor c^2 is so large that the loss of mass in ordinary chemical reactions is negligible. Thus, when 1 kg of coal is burned, about $33 \cdot 5 \times 10^6$ joule (J) of energy are released. The loss of mass is then only $33 \cdot 5 \times 10^6/(3 \times 10^8)^2 = 3 \cdot 7 \times 10^{-10}$ kg. Nuclear binding energies are so large, however, that their equivalent masses are appreciable. Such energies are usually measured in MeV; here 1 MeV $= 10^6$ eV and the eV is the *electron-volt*, the work done when an electron moves through an electrical field of one volt. We note the following conversion factors:

1 eV $= 1 \cdot 6 \times 10^{-19}$ joule
1 eV per particle $= 96\,500$ joule per mole of particles
1 u $= 931$ MeV.

(A more complete set of conversion factors is given at the front of the book.)
As an example, consider an atom of oxygen-16. We regard it as equivalent to eight hydrogen atoms, each of mass $1 \cdot 00813$ u, and eight neutrons. The total mass of the individual particles is thus $8(1 \cdot 00813 + 1 \cdot 00897) = 16 \cdot 13680$. The mass of oxygen-16 is 16 ($= A$). Hence the mass defect is $0 \cdot 13680$. This is equivalent to a total binding energy of $0 \cdot 1368 \times 931 = 127 \cdot 4$ MeV, i.e. to a binding energy per nucleon of $127 \cdot 4/16 = 7 \cdot 96$ MeV.
Fig. 2.1 shows that most elements have binding energies of about 8 MeV

per nucleon. This value is fairly constant because the nuclear force is a short-range force. It is extremely strong at distances between nucleons of order 10^{-15} m, far stronger than any electrical force, but falls off so sharply at larger distances that the binding of a nucleon is due almost entirely to the forces from its immediate neighbours in the nucleus. Consistent with this, experiment has shown that the volume of a nucleus is directly proportional to the number of nucleons present.

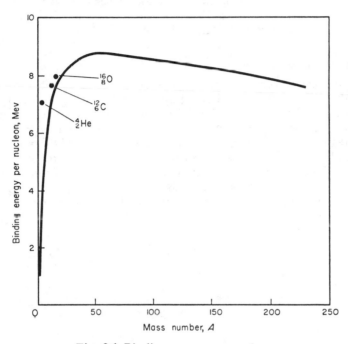

Fig. 2.1 Binding energy per nucleon

The stabilities of the various nuclei depend on several effects:

(1) In small nuclei most of the nucleons are near the surface and their binding is reduced by the absence of neighbours on one side. This effect, analogous to surface tension in water droplets, leads to a release of nuclear energy when small nuclei *fuse* together to form large ones.

(2) Very large nuclei are unstable because of the electrostatic repulsion between their protons. The greatest binding occurs in nuclei of medium size, e.g. iron ($A = 56$). Whereas small nuclei have $N \simeq Z$, the large ones have relatively fewer protons, i.e. $N \simeq 1.5Z$. The largest nuclei, e.g. uranium and the trans-uranic elements, tend to break up by *fission*. A typical fission reaction is

$$^{235}_{92}\text{U} + n \rightarrow {}^{140}_{54}\text{Xe} + {}^{94}_{38}\text{Sr} + 2n \qquad \textbf{2.3}$$

where n = neutron. The energy released, about 200 MeV per fission, is of course the basis of nuclear power.

(3) *Even-even* nuclei (i.e. Z = even, N = even) are more stable than even-odd, odd-even and odd-odd nuclei. Pairing of protons and of neutrons gives increased stability. The most abundant nuclei in nature are even-even. Thus natural uranium contains 99·3 per cent of $^{238}_{92}U$ and only 0·7 per cent of $^{235}_{92}U$. The even-odd character of uranium-235 makes it more fissionable than uranium-238, which is important for nuclear power. Amongst the lighter nuclei, those that can be regarded as clusters of α-particles are particularly stable, e.g. $^{12}_{6}C$, $^{16}_{8}O$.

2.3   Radioactive changes

Changes can occur spontaneously in unstable nuclei and can also be induced more generally by irradiating substances with nuclear particles and energetic radiations. Fission is an infrequent spontaneous process in uranium but it occurs readily in nuclear reactors due to the entry of neutrons, released by previous fissions, into the uranium nuclei. The neutron, being electrically neutral and so able to penetrate the electrostatic field round a nucleus, is particularly effective in promoting nuclear changes.

Most radioactive changes lead to the emission of α-particles, β-particles (i.e. electrons, or *positrons* which are the positively charged *anti-particles* of electrons), and γ-rays (high-frequency electromagnetic waves). Typical α-particle emissions are

$$^{239}_{94}Pu \rightarrow {}^{235}_{92}U + {}^{4}_{2}He \qquad\qquad 2.4$$

$$n + {}^{10}_{5}B \rightarrow {}^{7}_{3}Li + {}^{4}_{2}He \qquad\qquad 2.5$$

the latter being induced by a neutron. Electrons are emitted from nuclei with surplus neutrons and are produced by the reaction

$$\text{neutron} \rightarrow \text{proton} + \text{electron} \qquad\qquad 2.6$$

which conserves the total electrical charge since the proton and electron have equal and opposite charges. An example is

$$^{131}_{53}I \rightarrow {}^{131}_{54}Xe + e^- + \nu \qquad\qquad 2.7$$

where e^- is the electron and ν is a *neutrino*, an uncharged particle of very small mass which carries away some of the energy released. Nuclei with insufficient neutrons commonly decay by the process,

$$\text{proton} \rightarrow \text{neutron} + \text{positron} \qquad\qquad 2.8$$

which also conserves charge. An example is

$$^{13}_{7}N \rightarrow {}^{13}_{6}C + e^+ + \nu \qquad\qquad 2.9$$

A γ-ray is usually emitted when a nucleus is initially in a state of energetic internal vibration, i.e. in an *excited* state, and then drops into a state of lesser vibration and gives away the surplus vibrational energy as a pulse of electromagnetic radiation, i.e. as a *photon* of γ-rays. This radiation is identical to radio waves, light waves and x-rays, except for its short wavelength.

The precise instant when an unstable nucleus will make a spontaneous change is quite unpredictable. Only the *probability* that the change will occur during a given period of time can be known. This leads to the *law of radioactive decay*. At time t there are N nuclei still unchanged and at an infinitesimally later time $t + dt$ there are only $N(1 - \lambda\, dt)$, where λ is the *decay constant* of the process. Then $dN/dt = -\lambda N$ which integrates to give

$$N = N_0 e^{-\lambda t}$$ 2.10

where $N = N_0$ at $t = 0$. The *half-life*, equal to $0.693/\lambda$, is the time to halve the number still unchanged. In ten half-lives the radioactivity drops by about one thousandfold. Nuclei vary enormously in their half-lives; for example, $^{12}_{5}B$, 0.03 sec; $^{131}_{53}I$, 8 days, $^{60}_{27}Co$, 5.3 years; $^{14}_{6}C$, 5600 years.

The unit of radioactivity is the *curie*, 3.7×10^{10} nuclear changes per second. This is approximately the rate of disintegration of 10^{-3} kg of radium, on which the unit is based.

2.4   Uses of radioactivity

Radioactive isotopes, usually made by neutron irradiation of substances in reactors, have many uses. Radioisotopic atoms behave, apart from their radioactivity, almost exactly like ordinary atoms of the element, but they can readily be detected and identified from their half-life and from the energy and nature of their radiation. The addition of a trace of a radioisotope to a substance thus enables the substance to be located precisely and its movements followed. This has innumerable applications. In metallurgy it has been widely used to study the distributions of impurities and added substances in metals—for example, the segregation of phosphorous inside cast steel ingots—and also to locate the sources of impurities and foreign inclusions picked up by metals during melting and casting.

Usually in such work β and γ emitter isotopes are used because these radiations, if of fairly high energy (e.g. 0.1 to 4 MeV), can penetrate thick pieces of metal and other solids. The range of α-particles is generally much shorter and few will penetrate even an 0.05 mm thick aluminium sheet. The radiations can be detected and measured by various instruments, e.g. ionization chambers, Geiger counters and scintillation counters. In *autoradiography* a polished flat surface of the specimen to be examined is held against a photographic plate. The radiations emitted by the radioisotopic atoms in the surface layers of the specimen blacken the photographic plate locally and so produce an image of the distribution of the element in the specimen. This technique is useful for studying the fine-scale distribution of substances in metals and also for determining the migrations of atoms in solids (*diffusion*).

Another application is chemical analysis by *radioactivation*. This is particularly useful for determining traces of impurity in nominally pure substances. The specimen is irradiated, usually in a nuclear reactor, to activate the impurity and its radioactivity then compared against a standard, usually after chemical separation. When several impurities are activated their separate contributions can usually be distinguished by analysing their time-decay curves in relation to their known half-lives and by determining the nature and energy of the radiations.

Finally, there is *γ-radiography*, useful for detecting cracks, holes and other faults inside thick pieces of metal. The γ-rays from cobalt-60, because of their high energy, about 1·3 MeV, can penetrate about 15 cm of steel.

2.5   Materials in nuclear reactors

The development of nuclear power has led to special metallurgical problems concerning the use of materials in nuclear reactors. First, there are problems due to the particular materials that must be used. These materials must have certain nuclear properties, e.g. low absorption of neutrons in structural materials, and those chosen for their nuclear properties, e.g. U, graphite, Mg, Zr, Be, Nb, often have disadvantages in other directions; they may be scarce, technologically undeveloped and may have poor engineering properties.

Second, there is the *radiation damage* done to such materials by the high-energy radiations and nuclear particles which pass through them. There are three main types of radiation damage:

(1) Breaking of the chemical bonds holding the atoms together. This can be very severe in some organic substances but metals are immune (cf. § 4.6).

(2) *Knock-on* damage in which atoms inside the material are knocked from their sites by the impacts of nuclear particles and of other atoms already knocked on. In solids this produces atomic-sized holes, called *vacancies*, from which the atoms have been ejected, and intruder atoms, called *interstitials*, where the ejected atoms have come to rest in the material (cf. § 19.10).

(3) *Transmutation* effects, in which atoms of various chemical elements are created and become impurities in the material. Particularly harmful are *noble gas* elements; i.e. krypton and xenon, produced by fission in uranium and plutonium, and helium produced in various substances (e.g. B, Li, Be, Mg) by α-emitter reactions, usually promoted by neutrons. At high temperatures these atoms can move about inside a solid and collect together to form compressed gas pockets. The expansion of these pockets causes the material to swell and crack.

The chance that a given atomic nucleus will be struck by a particle passing through a material depends on the 'target area' which the nucleus presents to the bombardment. From the diameters of nuclei we expect such target areas, called *cross-sections*, to be based on the order of magnitude of 10^{-28} m². This in fact is the *barn*, the unit of nuclear cross-section. Consider a material with a cross-section σ m² exposed to a *flux* of particles ϕ m⁻² s⁻¹, over a time t. The proportion of its nuclei which suffer collisions is then $\sigma \phi t$. The fission cross-section for natural uranium exposed to slow neutrons (kinetic energy $\simeq 0.025$ eV) is about 4 barns. Hence, in 1 month ($\simeq 2.6 \times 10^6$ sec) in a power reactor (e.g. $\phi = 10^{17}$ m⁻² s⁻¹) about 1 atom in 10^4 in natural uranium undergoes fission. Each fission produces 2 atoms, about one-tenth of which are noble gases.

The nucleus is, of course, very small compared with the electronic part of the atom. Atoms in solids are spaced at a distance of a few Ångströms, so that the geometrical cross-section of an atom in a solid is about 10^9 barns. Hence, for a nucleus with a 1 barn cross-section, the bombarding particle would on average pass through 10^9 atoms before striking a nucleus, i.e. its *mean free path* inside the material between successive nuclear collisions would be about 0·3 m.

The magnitude of the nuclear cross-section varies with the nature of the target nucleus, the type and energy of the bombarding particle or radiation and the nature of the nuclear process in question. For neutron irradiation there are three main processes; *fission* (limited mainly to heavy nuclei such as uranium), *absorption* of the neutron into the nucleus (which one generally tries to avoid as far as possible in nuclear reactors since it wastes neutrons) and *scattering*, i.e. the neutron bounces off the nucleus. The absorption cross-section is small for neutron energies below about 10 eV (hence the need to slow down, i.e. *moderate*, the neutrons in certain reactors), is very large in some materials for neutron energies from 10 eV to 0·1 MeV, and is fairly small for energies above 0·1 MeV. The cross-section for neutron scattering is usually a few barns.

The scattering of energetic particles (e.g. *fast* neutrons with kinetic energies of order 1 MeV) produces knock-on damage. Consider a head-on collision in which the incident particle, of mass M_1 and speed v_1, bounces back along its path of flight with speed v_1^*, having given the target atom, of mass M_2, a forward speed v_2. At its simplest this is like a collision between two elastic spheres and we then have conservation of momentum and energy, i.e.

$$M_1 v_1 = M_2 v_2 - M_1 v_1^* \qquad\qquad \textbf{2.11}$$

$$\tfrac{1}{2} M_1 v_1^2 = \tfrac{1}{2} M_2 v_2^2 + \tfrac{1}{2} M_1 v_1^{*2} \qquad\qquad \textbf{2.12}$$

Writing $E = \tfrac{1}{2} M_1 v_1^2$ and $E_M = \tfrac{1}{2} M_2 v_2^2$ for the kinetic energies, and eliminating v_1^*, we obtain

$$E_M = \frac{4 M_1 M_2}{(M_1 + M_2)^2} E \qquad\qquad \textbf{2.13}$$

where E_M is of course the maximum energy the target can acquire for any angle of elastic collision with this bombarding particle. Some values of E_M for a 2 MeV neutron are, in MeV: H, 2·0; C, 0·56; Al, 0·28; U, 0·033. These are far greater than the minimum ($\simeq 25$ eV) for a knock-on collision and in fact the struck atom will itself become a bombarding particle in the surrounding material and produce a cascade of *secondary knock-ons* before brought to rest (cf. § 19.10).

The fission process in uranium is very violent since the two *fission fragment* nuclei produced by a fission share about 200 MeV of kinetic energy between them. The knock-on damage is then correspondingly severe; e.g. every atom in natural uranium is knocked out of its site about once a week in an average power reactor. It is remarkable that the material can withstand such damage and still largely keep its original shape and properties. We shall see later that metals are particularly able to heal effects of radiation damage in themselves (§ 19.10).

2.6   Origin and abundance of the elements

Astrophysical research has provided estimates of the distribution of the various chemical elements throughout the universe. Taking silicon as a standard, the relative numbers of atoms of the most abundant are as in Table 2.1.

TABLE 2.1.   ABUNDANCE OF ELEMENTS IN THE UNIVERSE

H	He	O	N	C	Fe	Si
12,000	2800	16	8	3	2.6	1
Mg	S	Ni	Al	Ca	Na	Cl
0·89	0·33	0·21	0·09	0·07	0·045	0·025

By comparison, most of the other elements are rare, even familiar ones; e.g. Cu, 7×10^{-4}; Au, $1 \cdot 5 \times 10^{-6}$.

We see that hydrogen is by far the most abundant element in the universe. It is fairly certain that all other elements have been created from hydrogen by nuclear fusion in stars, the energy so released being mainly responsible for maintaining the temperature of stars. The barrier to such processes is the long-range electrostatic repulsive force which tends to keep nuclei apart at distances too great for the strongly attractive short-range nuclear force to take over. This barrier can be overcome at very high temperatures, where the nuclei have enough kinetic energy to penetrate one another's electrostatic fields and get within reach of their nuclear forces. Stellar reactions are thus mainly *thermonuclear*.

The formation of a star begins by the collecting together of interstellar hydrogen into a cloud. These atoms are pulled together by their gravitational attraction and, as the cloud shrinks, the temperature rises due to the conversion of gravitational energy into heat. Eventually temperatures are reached where protons can fuse together (and eject positrons) to make nuclei of deuterium and other elements. At the temperature of the centre of the sun, about 15 million degrees, the main reactions are $^1_1H + ^1_1H \rightarrow ^2_1H + e^+$; $^2_1H + ^1_1H \rightarrow ^3_2He$; and $^3_2He + ^3_2He \rightarrow ^4_2He + ^1_1H + ^1_1H$. In stars a little hotter than this, helium-4 is produced more rapidly from hydrogen by a sequence of reactions which use the carbon-12 nucleus as a catalyst. This nucleus changes, by absorbing protons, into an unstable nitrogen-16 nucleus and this then breaks down to carbon-12 again by emitting an α-particle. The temperature at this stage is raised to about 3×10^7 °C, when the hydrogen at the centre collapses inwards and becomes hotter. Above about 10^8 °C α-particles begin to unite to form carbon-12 and oxygen-16; at 2×10^8 °C elements such as Ne, Mg, Si and Ca begin to form, largely because α-particles released in other reactions have enough nuclear energy to penetrate these heavier nuclei. At still higher temperatures, over 10^9 °C, iron and its neighbouring elements are created. It is difficult to form elements much heavier than iron because the growth of large nuclei absorbs energy instead of releasing it (cf. Fig. 2.1). The heavy elements may be created from supernovae, immense stellar explosions which throw out neutrons, protons and nuclei far into space where they may meet other nuclei and unite with them.

TABLE 2.2.   ABUNDANT ELEMENTS IN EARTH'S CRUST
(weight per cent)

O	Si	Al	Fe	Ca	Na	K
49·2	25·67	7·50	4·71	3·39	2·63	2·40

Mg	H	Ti	Cl	P	Mn	C
1·93	0·87	0·58	0·19	0·11	0·08	0·08

The distribution of elements in the earth's crust is very different from that in the universe, as Table 2.2 shows. There is almost no helium and little hydrogen, the earth's gravitational field being too weak to hold these light elements as chemically uncombined gases. The other differences are due to chemical and density effects. Most of the earth's iron remains chemically uncombined and, being a fairly heavy metal, has sunk into the centre leaving only a relatively small amount in the crust as iron oxide. By contrast, the elements aluminium, magnesium, calcium and silicon, being light and chemically active, have combined with oxygen to form silicates (sand, granite) and alumino-silicates (clay). These make up most of the crust.

FURTHER READING

BORN, M. (1951). *Atomic Physics*. Blackie, London.

HALLIDAY, D. (1960). *Introductory Nuclear Physics*. John Wiley, New York and London.

STEPHENSON, R. (1954). *Introduction to Nuclear Engineering*. McGraw-Hill, New York and Maidenhead.

Chapter 3

Atomic Structure

3.1 Electropositive and electronegative elements

Three-quarters of the chemical elements are *metals*. We recognize them by their characteristic properties. They conduct electricity extremely well. They combine with acids to form salts and with oxygen to form basic oxides and hydroxides. Many are strong, ductile and tough. They can often dissolve in one another, even in the solid state, and so form *alloys*. We shall see that all these properties stem from one general feature of all metals—easily removable electrons in their atoms. However, only the chemical properties follow in an immediately obvious way from this, for only these are direct properties of single atoms. All the others, including the mechanical properties that make metals so useful, are properties of *large groups* of atoms.

Electrons are electrostatically attracted to protons. A nucleus of charge $+Ze$ attracts Z electrons to itself to form an electrostatically neutral atom and it is in this way that the atomic number Z determines the chemical individuality of an atom. An atom can, however, exist with either fewer or more electrons than its atomic number requires. It is then a *positive* or *negative ion*. In very hot flames electrons 'evaporate' from their atoms and the flame gas then becomes a *plasma* of positive ions and free electrons. Ionization is manifest when salts such as sodium chloride are dissolved in water. The charges on the separately dissolved sodium and chlorine ions make the solution an electrical conductor. A voltage applied across immersed electrodes produces an electrolytic drift of negative Cl^- ions (*anions*) to the positive electrode (*anode*) and of positive Na^+ ions (*cations*) to the negative *cathode*.

Work has to be done to pull an electron off an atom against the electrical forces. The work to remove one electron from an initially neutral atom in its most stable state is called the *first ionization energy I* of the atom. Fig. 3.1 gives values. Metals have lower values than other elements; cf. 4–5 eV for alkali metals, 6–9 eV for most other metals, 13·6 eV for hydrogen, 10–17 eV for halogen elements and 12–24 eV for the noble gases.

Easy dissociation into positive ions and free electrons is the central property of a metal. Metals are *electropositive* or *electron-donor* elements, by contrast with *electronegative* or *electron-acceptor* elements such as oxygen, sulphur and the halogens. A *neutral* electron-acceptor atom attracts an electron and releases energy when it captures it and becomes a negative ion. Values of this *electron affinity* in eV are:

F	Cl	Br	I	O	S	H
4·13	3·72	3·49	3·14	3·07	2·8	0·72

The chemical attraction between metals and electronegative elements follows obviously from this. Consider sodium and chlorine. An energy of

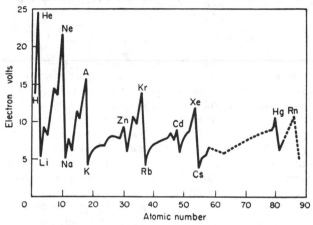

Fig. 3.1 First ionization energies of elements

5·14 eV is needed to take an electron off a neutral sodium atom; 3·72 eV of this is returned when the electron is captured by a distant neutral chlorine atom. This gives two oppositely charged ions and the electrostatic energy released when these come together more than compensates for the (5·14 − 3·72) eV still outstanding, so that the net energy balance is favourable. The electrostatic force between these ions, at a spacing r between their centres, is

$$f = e^2/4\pi\varepsilon_0 r^2 \qquad\qquad \textbf{3.1}$$

with the factor $4\pi\varepsilon_0$ ($\varepsilon_0 = 8\cdot85 \times 10^{-12}$) introduced in order to give correct results when SI units are used but which has no physical significance other than through the conventions on which these units are based. The work done by this electrostatic force in bringing the ions to this spacing, starting from infinite spacing, is

$$\int_r^\infty f \, dr = \frac{1}{4\pi\varepsilon_0} \int_r^\infty \frac{e^2}{r^2} \, dr = \frac{e^2}{4\pi\varepsilon_0 r} \qquad\qquad \textbf{3.2}$$

This is positive since the ions attract. We define the *electrostatic potential energy* of the ions as *zero* at infinite spacing. It is then $-e^2/4\pi\varepsilon_0 r$ at the spacing

r since the ions have released $+ e^2/4\pi\varepsilon_0 r$. If $r = 2\cdot5 \times 10^{-10}$ m, which is typical of atomic sizes, then $e^2/4\pi\varepsilon_0 r = (1\cdot6 \times 10^{-19})^2/4\pi \times 8\cdot85 \times 10^{-12} \times 2\cdot5 \times 10^{-10} = 9\cdot2 \times 10^{-19}$ J $= 5\cdot75$ eV $= 555\,000$ joule per mol of ion-pairs. This roughly indicates the bond strength. Divalent compounds such as MgO and CaO are much more strongly bonded because here the metal gives the non-metal two electrons so that the electrostatic energy is raised fourfold.

The ionization of neutral metal atoms to positive ions is called *oxidation* (even when oxygen is not the electron-acceptor) because metals so often occur naturally as oxides. Similarly the reverse process is referred to as *reduction*, because this is the key step in converting metallic ores to metals.

3.2 The periodic table

Fig. 3.1 shows that the ionization energy varies periodically through the elements. The *periodic table*, Table 3.1, shows that chemical properties have the same periodicity, chemically similar elements lying in vertical columns. As the atomic number increases along successive horizontal rows the chemical properties recur in cycles; first a cycle of two elements (H and He), then two cycles of eight (Li to Ne; Na to A) and three of eighteen.

Ignoring small irregularities, we see that the ionization energy increases along each horizontal row of the table and then drops sharply at the alkali metal which starts each new row. The increase is easily understood in terms of the increasing nuclear charge $+Ze$ which binds the electrons more strongly in the atom. The sharp drop, which gives the chemical periodicity, shows that as the number of electrons in the atom builds up, with increasing Z, the added electrons cannot all continue to be accepted by the atom on exactly the same basis. At certain stages of atom building all the places for strongly bound electrons in the atom are evidently filled up. This occurs in the *noble gas elements* (He, Ne, A, Kr, Xe, Rn). The next electron added, at the alkali metal which follows each noble gas in the table, finds no room available amongst the strongly bound electrons and so is only weakly bound to the atom and can easily be ionized. Along each horizontal row, following an alkali metal, the increasing nuclear charge gradually strengthens its grip on the weakly bound electrons until these also become strongly bound in the elements at the end of the row.

The experimental facts show that a kind of 'shell' structure must exist in atoms, each shell consisting of a group of *states of motion* in which there is room for only a limited number of electrons. The electrons in a given shell all move on average at about the same distance from the nucleus and so enjoy the electrostatic force from the nucleus to about the same extent. As the nuclear charge increases, each shell is pulled in more closely and the electrons in it become more bound to the nucleus. To balance the increased nuclear charge, however, more electrons have to be added, for a neutral atom, and these gradually fill up the shell. Electrons added beyond

TABLE 3.1. THE PERIODIC TABLE OF THE ELEMENTS

KEY

H = chemical symbol
1 = atomic number
1·008 = atomic weight

H 1 1·008																	**He** 2 4·003
Li 3 6·939	**Be** 4 9·012											**B** 5 10·81	**C** 6 12·01	**N** 7 14·01	**O** 8 16·00	**F** 9 19·00	**Ne** 10 20·18
Na 11 22·99	**Mg** 12 24·31											**Al** 13 26·98	**Si** 14 28·09	**P** 15 30·97	**S** 16 32·06	**Cl** 17 35·45	**A** 18 39·95
K 19 39·10	**Ca** 20 40·08	**Sc** 21 44·96	**Ti** 22 47·90	**V** 23 50·94	**Cr** 24 52·00	**Mn** 25 54·94	**Fe** 26 55·85	**Co** 27 58·93	**Ni** 28 58·71	**Cu** 29 63·54	**Zn** 30 65·37	**Ga** 31 69·72	**Ge** 32 72·59	**As** 33 74·92	**Se** 34 78·96	**Br** 35 79·91	**Kr** 36 83·80
Rb 37 85·47	**Sr** 38 87·62	**Y** 39 88·90	**Zr** 40 91·22	**Nb(Cb)** 41 92·91	**Mo** 42 95·94	**Tc** 43 (97)	**Ru** 44 101·1	**Rh** 45 102·9	**Pd** 46 106·4	**Ag** 47 107·9	**Cd** 48 112·4	**In** 49 114·8	**Sn** 50 118·7	**Sb** 51 121·8	**Te** 52 127·6	**I** 53 126·9	**Xe** 54 131·3
Cs 55 132·9	**Ba** 56 137·3	**La*** 57 138·9	**Hf** 72 178·5	**Ta** 73 180·9	**W** 74 183·9	**Re** 75 186·2	**Os** 76 190·2	**Ir** 77 192·2	**Pt** 78 195·1	**Au** 79 197·0	**Hg** 80 200·6	**Tl** 81 204·4	**Pb** 82 207·2	**Bi** 83 209·0	**Po** 84 (210)	**At** 85 (210)	**Rn** 86 (222)
Fr 87 (223)	**Ra** 88 226·0	**Ac** 89 (227)	**Th** 90 232·0	**Pa** 91 (231)	**U** 92 238·0	†											

* The *rare earths* occur here (**La, Ce, Pr, Nd, Pm, Sm, Eu, Gd, Tb, Dy, Ho, Er, Tm, Yb, Lu**).
† The *transuranic* elements follow here: e.g. Np(93), Pu(94), Am(95), Cm(96), Bk(97), Cf(98), Es(99), Fm(100), Md(101), No(102).

this stage must then go into the next shell further out from the nucleus. Being further out, they are only weakly attached to the atom and can easily be removed. As the nuclear charge continues to increase, so this new shell in turn becomes strongly bound and fills up·with electrons.

3.3 Quantum mechanics

Classical physics is quite incapable of explaining this behaviour. It cannot even explain the stable existence of atoms, since the moving electrons should spiral into the nucleus, losing their energy by emitting electromagnetic radiation. Some *non-classical* effect evidently prevents atoms from shrinking to less than about 10^{-10} m radius. There is also the problem of the sharp spectral lines. When an electron alters its state of motion a pulse or *quantum* of electromagnetic radiation (i.e. a *photon* of light or x-rays) is absorbed or emitted according as the electron increases or decreases its energy. If the change of energy is ΔE, the frequency of the radiation is ν, where

$$\Delta E = h\nu \qquad\qquad 3.3$$

h being *Planck's* constant, $6\cdot624 \times 10^{-34}$ J s. Classical physics predicts an infinite number of orbits for the electron, differing infinitesimally in their energies. Thus, all values of ν should be observed in the spectrum. The fact that only *discrete* spectral lines are observed at a few special values of ν, instead of a continuous spectrum, shows again that some non-classical effect controls the motion of electrons. Only certain states of motion are allowed; these are called *stationary states* because they do not change with time (*Bohr* theory of the atom).

Through the work of *Bohr, de Broglie, Schrödinger, Heisenberg, Dirac, Born* and *Pauli* a new system of mechanics, *quantum mechanics* or *wave mechanics*, was created which could deal satisfactorily with the motion of electrons. A moving particle behaves in certain ways like a wave. This is not explained by the theory but has to be accepted as a fact of nature. Based on this starting point, however, the theory is able to account quantitatively for an enormous range of physical and chemical behaviour. Quantum mechanics applies to all particles and radiation quanta, whether these are photons, electrons, protons, neutrons, atoms, molecules or even billiard balls. Classical mechanics is a simplified form of quantum mechanics, just as geometrical optics is a simplified form of wave optics; it holds well enough for large things, e.g. billiard balls on a table, but gives wrong answers for very small things, e.g. electrons in an atom.

Quantum mechanics is often regarded as a difficult subject. The main trouble is due to the depth with which a false idea about mechanics is ingrained in our outlook. Familiarity with the world of large things leads us to take for granted the idea that a particle, whether a billiard ball or an electron, has at any given instant of time a perfectly precise position and momentum. But experiment proves that this idea is false; it only *seems* to be true in the world of large things because here our experimental techniques are not delicate enough to prove it wrong.

Consider a beam; say of electrons or light rays. If it is a beam of *particles* it will produce recoil effects when it hits atoms and other particles, which can be seen in a cloud chamber; and it will produce point scintillations where it strikes a fluorescent screen. If it is a beam of *waves*, it will produce an interference pattern when it goes through a diffraction grating (a crystal serves as such a grating for wavelengths $\lambda \simeq 1$ Å; see Chapter 17, p. 283). In complete conflict with classical ideas, one and the same beam is found in fact to behave as *particles* when tested by the particle detector or as *waves* when tested by the wave detector.

What does 'electron waves' mean? To examine this we repeat the wave diffraction experiment but now reduce the intensity of a beam of electrons until only one electron at a time enters the grating. It goes through and is then seen to hit the screen as a *particle*, producing a sharply defined point impact, not a wave pattern. Its wavelike behaviour in fact turns out to be connected with *where* it hits the screen. This hit may take place anywhere, but usually it occurs at places where the diffracted electron beams of the earlier experiment were strongest; similarly the 'dark' regions of the diffraction pattern are avoided by the particle.

Such experiments display the whole basis of wave mechanics. First, a particle cannot have both a precise position and, at the same time, a precise velocity; there is a certain indeterminacy in its motion. Second, this indeterminate motion is not completely haphazard but is controlled by underlying laws which happen to be the same mathematically as those of wave motion. These are empirical facts. The first destroys the classical theory. The second, even though we understand neither the indeterminacy nor the resemblance to wave behaviour, enables a new set of laws of motion to be worked out, complete, exact and able to predict every property that can be reached by experiment, in which we deal no longer with the exact values of position and velocity but only with the *probability* that their values lie in various ranges.

For a particle with *momentum mv* (m = mass, v = velocity) and *energy* E ($= \frac{1}{2} mv^2$) we use a *wavelength* λ and *frequency* v given by

$$\lambda = h/mv \qquad\qquad\qquad \textbf{3.4}$$

$$v = E/h \qquad\qquad\qquad \textbf{3.5}$$

where h is Planck's constant. We can regard these as empirical formulae, established by experiment, although the theory of relativity gives some theoretical support for them.

Consider an electron ($m = 9 \cdot 1 \times 10^{-31}$ kg) in an atom. Typically $v \simeq 10^6$ m s^{-1}, so that $\lambda \simeq 7$ Å. Because this is larger than the atomic diameter, the electron does not follow a well-defined orbit in the atom. It is best to picture it as equivalent to an electrical 'cloud', a smeared distribution of electrical charge round the nucleus. Of course, if a stream of electrons goes through a cloud chamber, for example, along a track say 0·01 cm wide, it will look then like a sharply defined beam, just as a searchlight beam which is broad compared to the wavelength of light also looks sharply defined.

For dealing with the stationary states of atoms we do not need to bring *time* into the wave equations. We know from ordinary experience that time-independent waves are *standing waves* (e.g. the smooth hump of a water wave flowing steadily over a weir). Our problem then is to work out the patterns of standing waves which may exist in the atom. Suppose we have such a wave; e.g. a simple hump, the height or *amplitude* of which is $\psi(x,y,z)$ at a point of space with coordinates x, y, z. We write ψ as $\psi(x,y,z)$ to indicate that in general ψ is a function of x, y and z. What is the significance of ψ? It must tell us the *probability*, i.e. proportion of time, that the electron spends near the point x,y,z. We have to take the probability as proportional to ψ^2, however, not ψ itself. We see this partly by analogy with wave optics, where the amount of light is proportional to the square of the wave amplitude, and partly from the fact that, whereas probability obviously cannot ever be negative, the wave function ψ must be able to be both positive and negative in order to show interference effects when different waves are superposed. We say then that $\psi^2\, d\tau$, where $\psi = \psi(x,y,z)$, is the probability of the electron being in a small element of volume $d\tau$ situated at the point x,y,z. This definition assumes that we have *normalized* ψ, i.e. chosen the scale of measurement of ψ in such a way that the probability of this electron being somewhere in the whole of space is *unity*; thus

$$\int \psi^2\, d\tau = 1 \qquad\qquad 3.6$$

For *time-dependent* processes, e.g. the flight of an electron through an apparatus, we cannot use ψ in such a simple form as this. It becomes necessary to use *complex* wave functions, $\psi = f + ig$ and $\psi^* = f - ig$, where f and g are ordinary functions and $i = \sqrt{-1}$. The probability is then equal to $\psi\psi^*\, d\tau$, i.e. $(f^2 + g^2)\, d\tau$. This reduces to the simple form, sufficient for describing stationary states, when $g = 0$.

3.4 The Schrödinger equation

The simplest form of standing wave in one dimension (x) is $\psi = A \sin(2\pi x/\lambda)$. We differentiate this twice with respect to x and then replace λ^2 by $h^2/2mE$ to obtain a wave equation,

$$\frac{d^2\psi}{dx^2} + \frac{8\pi^2 mE}{h^2}\, \psi = 0 \qquad\qquad 3.7$$

Although not yet sufficiently general to describe atomic structure, this equation shows some of the main features of quantum mechanics. Suppose that an electron is confined, by impenetrable boundary walls at $x = 0$ and $x = L$, to move up and down a limited region of length L. Within this region it can move about freely but it is turned back at the boundaries as if these were perfect mirrors. The solution $\psi = A \sin (2\pi x/\lambda)$ of eqn. 3.7 then has to satisfy the *boundary conditions* $\psi = 0$ at $x = 0$ and $x = L$, since the electron probability must be zero, by definition, beyond the walls. The boundary condition at $x = 0$ is immediately satisfied since $\sin 0 = 0$. That at $x = L$ is satisfied only if $2\pi L/\lambda = n\pi$, where $n = 1, 2, 3 \ldots$ etc. Hence, the electron cannot exist in *any* stationary state but only one for which

$$\lambda = 2L/n \qquad (n = 1, 2, 3 \ldots \text{etc.}) \qquad 3.8$$

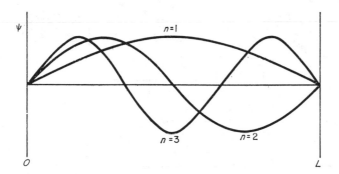

Fig. 3.2 Wave functions of an electron in a box

Only certain *discrete quantum states* are thus allowed to the electron, governed by a *quantum number*, n. Corresponding to these the energy, $E = h^2/2m\lambda^2$, of the electron is restricted to certain *energy levels*

$$E = n^2h^2/8mL^2 \qquad 3.9$$

The existence of similar sharp energy levels in atoms is, of course, the basis of discrete atomic spectra.

Fig. 3.2 shows a few ψ patterns of long wavelength. The longest wave, $\lambda = 2L$, corresponds to the *ground state* in which the electron has lowest energy and has *no* nodes (except at the boundaries). All the other quantum states have nodes in them, across which ψ changes sign. Nodes are a characteristic feature of standing wave patterns. In two-dimensional waves, e.g. vibrations of the skin of a drum, the nodes lie along *lines*; in three-dimensions, e.g. standing sound waves in a room, they lie in *surfaces*. The more nodes there are within an electronic ψ pattern, the higher is the kinetic energy of the electron, since the more sharply does the ψ curve have to bend to and fro to fit its nodes into a given length, and so the shorter is the wavelength. We notice that if the walls in Fig. 3.2 are moved together the energy of every electronic state, including the ground state, is increased. A similar effect forms the basis of the stability of atoms. The electrostatic force from the nucleus tries to pull the electrons close to the nucleus but this confinement raises the kinetic energy of the electrons. A balance is struck between the electrostatic energy and the kinetic energy at a radius of order 1 Å.

We have now to generalize eqn. 3.7. First, to take account of the potential energy V which an electron may possess due to a field of force acting on it: in eqn. 3.7 we introduced E as the kinetic energy $\frac{1}{2}mv^2$, but in this particular case where $V = 0$ we could equally well regard E as the *total* energy (kinetic plus potential). We shall now in fact take E as the total energy. In the derivation of eqn. 3.7, however, the energy came in, through λ, as kinetic energy.

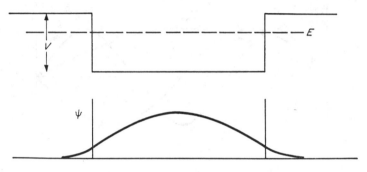

Fig. 3.3 Electron in a potential energy well

Hence we must now replace E by $E - V$ in the equation. This gives us the *Schrödinger equation* of wave mechanics in its one-dimensional, time-independent form,

$$\frac{d^2\psi}{dx^2} + \frac{8\pi^2 m}{h^2}(E - V)\psi = 0 \qquad\qquad \textbf{3.10}$$

To show an important feature of this equation, we consider in Fig. 3.3 an elaboration of the problem of Fig. 3.2. We now have an electron of total energy E moving in a potential which is zero inside a certain region and is V, where $V > E$, outside it. The electron is thus in a *bound state* in this potential energy 'box'. Since the kinetic energy $E - V$ can never be negative in classical mechanics, the electron could not penetrate the walls of its box to the slightest extent. In quantum mechanics, however, a certain penetration of potential energy barriers is possible. Inside the box, $E - V$ is positive and the solution of eqn. 3.10 is a sine wave, as in the previous problem. Outside the box, where $E - V$ is negative, eqn. 3.10 still has a solution, of the form $\psi = Ae^{-bx}$ where $b = \sqrt{8\pi^2 m(V - E)}/h$. The value of A has to be chosen so that this part of the solution joins smoothly on to the ends of the sine curve at the edges of the box, as shown for the ground state in Fig. 3.3. It follows that ψ is not zero at the edges of the box and that the electron has some chance of appearing outside the box. Because of the exponential form of ψ in this region, however, the probability is very small when $V - E$ is large and it decreases rapidly with increasing distance x beyond the edges of the box.

If the potential barrier is narrow, as in Fig. 3.4, the exponential 'tail' of the ψ curve can reach the far side of the barrier and ψ can then increase again in the low V region beyond the barrier. There is now a good chance of finding the electron on the far side of the barrier, i.e. it is possible for electrons to *escape* through potential barriers even where, according to classical mechanics, this would be impossible because $V > E$. This is the *tunnel effect*. The rate of escape, which can be calculated from time-dependent wave mechanics, falls off rapidly with increasing height and

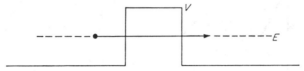

Fig. 3.4 The tunnel effect

width of the barrier. The tunnel effect is important for the escape of α-particles from unstable atomic nuclei and has practical applications in electronic devices.

The three-dimensional form of the Schrödinger equation,

$$\frac{\partial^2 \psi}{\partial x^2} + \frac{\partial^2 \psi}{\partial y^2} + \frac{\partial^2 \psi}{\partial z^2} + \frac{8\pi^2 m}{h^2}(E - V)\psi = 0 \qquad 3.11$$

is a straightforward generalization of eqn. 3.10. The partial differentials imply differentiation with respect to one coordinate when the other two are held constant.

3.5 The hydrogen atom

To describe the hydrogen atom we substitute $V = -e^2/4\pi\varepsilon_0 r$, which represents the electrostatic field of the nucleus, into eqn. 3.11. Since $e^2/4\pi\varepsilon_0 r$ is the same in all directions out of the nucleus, i.e. has *spherical symmetry*, at least some of the solutions of eqn. 3.11 must also have spherical symmetry. We consider then those solutions for which $\psi = \psi(r)$ only; thus $\partial\psi/\partial x = (\partial\psi/\partial r) \times (\partial r/\partial x)$. Making use of this, together with the standard formulae $r^2 = x^2 + y^2 + z^2$, $\partial r/\partial x = x/r$, $\partial(uv) = u\,\partial v + v\,\partial u$ etc., we can bring eqn. 3.11 into the form

$$\frac{d^2\psi}{dr^2} + \frac{2}{r}\frac{d\psi}{dr} + \frac{8\pi^2 m}{h^2}\left(E + \frac{e^2}{4\pi\varepsilon_0 r}\right)\psi = 0 \qquad 3.12$$

We find by substitution that the simplest solution of this is the quantum state

$$\psi(r) = Ae^{-r/a} \qquad\qquad 3.13$$

where

$$a = h^2/4\pi^2me^2 \doteq 0.53 \text{ Å} \qquad\qquad 3.14$$

and $A(=\sqrt{1/\pi a^3})$ is the normalization factor. This solution occurs at the energy level

$$E = 2\pi^2me^4/h^2 = -13.6 \text{ eV} \qquad\qquad 3.15$$

which agrees with the *measured* ionization energy of the hydrogen atom in its ground state. We can express the probability of where the electron is in

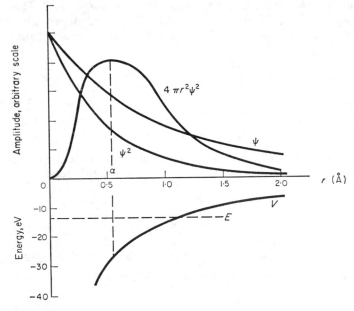

Fig. 3.5 Wave function and energy of the ground state of the hydrogen atom

the atom in two ways. First, as the *electron density* at any point, which is simply ψ^2. Second, as the *radial density*, i.e. the probability that the electron is in the region between spheres of radii r and $r + dr$; this is $4\pi r^2\psi^2 \, dr$. We see from Fig. 3.5 that the electron density is greatest at the centre of the atom and decreases smoothly outwards from the centre, never quite reaching zero at any finite distance. The radial density by contrast, because of the $4\pi r^2$ factor, is zero at the centre and reaches a maximum at the radius $r = a$, which is the distance at which the electron is most likely to be found. Hence, although an atom strictly has no definite radius, we can regard a as a measure of the size of the atom.

A hydrogen atom is most stable, i.e. has lowest energy, in this ground state. By absorbing energy, e.g. from a photon of light, it can, however, be raised out of its ground state, at least temporarily, into an *excited* state in which its electron has a wave pattern belonging to a higher energy level. These higher wave patterns, which can be found by further analysis of Schrödinger's equation, have *nodal surfaces* running through them. A pattern with $n - 1$ nodal surfaces is said to have a *principal quantum number n*, where $n = 1, 2, 3 \ldots$ etc., and there are n^2 distinct patterns (i.e. independent solutions of the equation) for each value of n. In the hydrogen atom it happens, because V is spherically symmetrical and varies as r^{-1}, that all the n^2 patterns with the same n have the same energy level E_n,

$$E_n = -2\pi^2 m e^4 / n^2 h^2 \qquad \textbf{3.16}$$

When different wave functions belong to the same energy level, they are said to be *degenerate*. Degeneracy can usually be removed by applying an external electric or magnetic field to the atom; this adds an external contribution to V and causes the degenerate level to *split* into a group or *band* of slightly different levels.

Before describing the wave patterns of excited states it is important to notice that the wave equation is *linear* (i.e. contains only the first power of ψ and its derivatives) and its solutions can be *superposed*. This property of the equation is essential since an interference pattern depends on the simple superposition and algebraic addition of separate waves from the slits of a diffraction grating. Superposition means that if $\psi_1, \psi_2, \psi_3 \ldots$ etc., are all solutions of the equation belonging to the same energy level, then their sum $a_1\psi_1 + a_2\psi_2 + a_3\psi_3 + \cdots$, where $a_1, a_2 \ldots$, are numerical constants, is itself also a solution for that energy level and should be considered as such *on equal merit* with the others. As a result, there is no unique way of describing degenerate wave patterns. For a given n we can construct innumerable wave patterns, but only n^2 of these can be truly independent, i.e. not derivable from one another by linear superposition; all the others are merely these same n^2 patterns disguised in various superposed forms. It is a matter of convenience which n^2 independent patterns we choose to regard as basic.

For an isolated atom we choose the n^2 *most symmetrical* patterns. The nodal surfaces of these are either *spherical*, centred on the nucleus, or they are *planes* or *cones* that pass through the nucleus. A wave pattern or quantum state is denoted by the symbols $ns, np, nd, nf \ldots$, where n is the principal quantum number and $s, p, d, f \ldots$ indicate the number of nodal surfaces $0, 1, 2, 3 \ldots$, respectively, that pass through the nucleus. For example, $5s$ has four spherical nodes; $3p$ has one spherical and one through the centre; $4d$ has one spherical and two through the centre.

The s states are spherically symmetrical. Fig. 3.5 shows the $1s$ and Fig. 3.6 shows the radial density for the $2s$. The others are similar, with more spherical nodes in them. These nodes tend to shift the electron distribution outwards, away from the nucleus, as we would expect from the higher energies of these states.

In the $n = 2$ group there is one $2s$ state and three $2p$ states. The single nodal surface of a $2p$ state is a *plane*, through the nucleus. We obtain such a solution by choosing a function of the type $xf(r)$ for ψ, where $f(r)$ is some function of r, which ensures $\psi = 0$ in the plane $x = 0$. When substituted

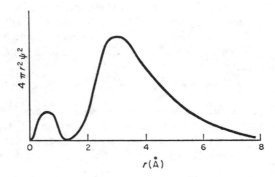

Fig. 3.6 Radial density for the $2s$ state of hydrogen

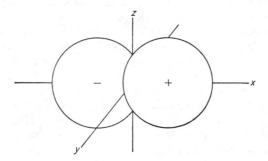

Fig. 3.7 Boundary surface of a $2p$ state

into eqn. 3.11 (with $V = -e^2/r$) this leads to an ordinary differential equation for $f(r)$ which can be solved. Clearly the electron density will be greatest in directions away from the nodal plane; i.e. the state has a 'dumb-bell' shape, along the x axis for the $xf(r)$ state. Fig. 3.7 shows this in a simplified form. We can regard this as a sketch of the 'boundary surfaces' inside which lie most of the electron density. The + and − signs denote the change in the sign of ψ from one side of the nodal plane to the other. There are three mutually perpendicular orientations for a nodal plane and so there are three independent $2p$ states, which we may write as $2p_x$, $2p_y$, $2p_z$ etc., according as x, y or z is the dumb-bell axis. Because of linear superposition it is possible to construct these three states along any set of Cartesian axes, whatever its orientation, in the atom.

The higher p states, e.g. $3p$, $4p$ etc., also occur in groups of three, which we denote as p_x, p_y and p_z, but they contain spherical nodes as well as a nodal plane through the nucleus.

The $n = 3$ group contains one $3s$ state, three $3p$ states, and five $3d$ states. The wave functions of d states are of the forms $xy\phi(r)$ and $(x^2 - y^2)\phi(r)$, where $\phi(r)$ is a function of r only. The d_{xy} type has two nodal planes through the origin, at $x = 0$ and $y = 0$, and appears as in Fig. 3.8. The other two states of this type are $yz\phi(r)$ and $zx\phi(r)$, similarly arranged about the other axes. The $d_{x^2-y^2}$ state has nodal planes at $x = \pm y$ and appears as in Fig. 3.8. Again other states of this type can be based on the other axes. However, these are not all independent since $(x^2 - y^2)\phi(r) + (y^2 - z^2)\phi(r)$

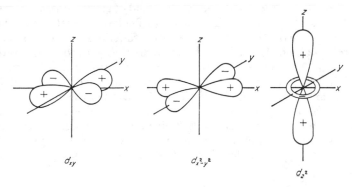

$$d_{xy} \qquad d_{x^2-y^2} \qquad d_{z^2}$$

Fig. 3.8 Boundary surfaces of d states

$+ (z^2 - x^2)\phi(r) = 0$. As a result for the fifth independent d state it is usual to take a linear combination of the $d_{y^2-z^2}$ and $d_{z^2-x^2}$ states, known as a d_{z^2} state, which has conical nodal surfaces as in Fig. 3.8.

In higher energy levels more complicated wave patterns appear. For example, in the $n = 4$ group there are seven $4f$ states as well as five $4d$ (one spherical node), three $4p$ (two spherical nodes) and one $4s$ (three spherical nodes).

3.6   Atomic structures of the elements

The direct effect of increasing the nuclear charge beyond that of hydrogen is easily allowed for replacing e^2 by Ze^2 in the wave equation; it simply pulls the wave functions closer to the nucleus and binds the electrons more firmly. The indirect effects, the mutual interactions between the additional electrons brought into the atom to balance the nuclear charge, are much more profound. As well as the electrostatic repulsive forces between electrons there is also a striking quantum-mechanical effect. Magnetic measurements show

Magnetic measurements show that an electron behaves rather like an electrically charged ball, spinning about its own axis as well as moving about in the atom; and only two states of spin, of equal and opposite angular momentum about the same axis, are available to the electron. These spin properties can be deduced by making the wave equation satisfy the requirements of relativity theory (*Dirac* theory of the electron). The experimental facts also show that not more than one electron with a given spin, or two electrons with opposite spins, are allowed to occupy the same quantum state. This is the *Pauli exclusion principle*, also explainable only by relativistic quantum mechanics. Without it, the heavier atoms would all be extremely small and stable, for their electrons would all crowd close to the nucleus in a highly shrunken $1s$ state.

Because the motion of each electron depends on that of every other one, the wave equation can no longer be solved exactly. Instead, we must pretend that each electron moves in a fixed *average* field of all the others, superposed on the field of the nucleus. This average field is at least approximately spherically symmetrical and so creates wave functions of $s, p, d, f \ldots$ type, similar to hydrogen. However, the potential energy V no longer varies as r^{-1} since the field experienced by a given electron varies, from about Ze^2/r when the electron is very near the nucleus, to about e^2/r when the electron is far out in regions where the $+Ze$ charge of the nucleus is largely *screened* or *shielded* by the $-(Z-1)e$ charge of the $Z-1$ other electrons. Because of this, quantum states with a given n are no longer all degenerate. The energy level is split into a number of levels, one for each of the $s, p, d, f \ldots$ groups of states. Fig. 3.9 shows schematically the order of these energy levels.

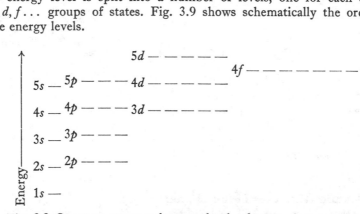

Fig. 3.9 Quantum states and energy levels of many-electron atoms

The s states have the lowest energies because they have no nodes through the nucleus and so their electrons spend relatively more time near the nucleus. The more nodes there are through the nucleus, the less time does an electron in such a state spend near the nucleus and the higher is its energy. The order of energies for a given n is thus $s < p < d < f \ldots$

This effect raises the energy of an nd electron to about that of an $(n + 1)$ s. An nf electron is raised even higher.

We now have the key to the periodic table. We can work through this, atom by atom, increasing the nuclear charge by one proton at a time and placing the corresponding extra electron in that quantum state of lowest energy which, after allowing for the Pauli principle, is not yet full. Starting with hydrogen, with a single $1s$ electron with ionization energy $I = 13.6$ eV, we place the additional electron of helium also in the $1s$ state, with spin opposite to the first. These two electrons equally experience the field $2e$ of the nucleus without much screening of either by the other, since on average they move at equal distances from the nucleus. Hence they are strongly bound (first ionization energy $I = 24.6$ eV). The third electron of lithium has to go into the $2s$ state. The *effective* nuclear charge on it is reduced almost to e by the screening effect of the two $1s$ electrons round the nucleus. It is thus weakly bound ($I = 5.4$ eV), like an electron in a $2s$ state in hydrogen. Thus, metallic properties and a new row of the periodic table start at lithium. The extra electron of beryllium also goes into $2s$, with opposite spin to the first, and these two $2s$ electrons are now more strongly bound ($I = 9.32$ eV) by the increased effective nuclear charge ($\simeq 2e$) acting on them. Next is boron, with one electron in $2p$. This electron is less firmly bound (cf. Fig. 3.9) and the ionization energy is slightly reduced ($I = 8.3$ eV).

After boron we meet a new problem. Does the sixth electron of carbon go into the same $2p$ state, say p_x, as the fifth, with opposite spin, or does it go into one of the other p states? It turns out that the energy of mutual interactions of two such electrons is lowest when the electrons have parallel spins because then on average they keep well away from each other (*Hund's rule*). Hence the sixth electron goes into one of the other p states, say p_y, with a parallel spin. We then go on as far as neon, thus:

Element	B	C	N	O	F	Ne
p_x	↑	↑	↑	↑↓	↑↓	↑↓
p_y		↑	↑	↑	↑↓	↑↓
p_z			↑	↑	↑	↑↓
I (eV)	8.28	11.27	14.55	13.62	17.42	21.56

The arrows denote spin 'up' or spin 'down' directions. We notice the break in the sequence of ionization energies at oxygen; the half-filled configuration, at nitrogen, is particularly stable but the fourth p electron of oxygen is forced to join an existing electron, with opposite spin, in the same state. At neon the $2p$ and $2s$ states are all completely full and the configuration is very stable. Every group of states, whether s, p, d or f, once completely full, forms a spherically symmetrical charge cloud of electrons and this is always a good configuration since a maximum number of electrons then equally enjoy the electrostatic field of the nucleus with a minimum of mutual screening. Because of its spherical symmetry such a filled group of electrons is referred to as a *shell*.

After neon, a new row of the periodic table starts with the alkali metal sodium. The filling of $3s$ and $3p$ is then very similar to that in the previous row. Thus:

Element	Na	Mg	Al	Si	P	S	Cl	A
$3s$ electrons	1	2	2	2	2	2	2	2
$3p$ electrons	0	0	1	2	3	4	5	6
I (eV)	5·14	7·64	5·97	8·15	10·9	10·36	12·9	15·76

Again, a break occurs in the sequence of ionization energies, at sulphur, after the $3p$ group is half filled.

A new effect appears in the elements following argon. The $3d$ states are still empty but for these elements the $4s$ energy level is slightly lower than that of the $3d$, so that the $n = 4$ group begins to fill before the $n = 3$ one is complete. The order of filling is as follows:

	K	Ca	Sc	Ti	V	Cr	Mn	Fe	Co	Ni	Cu	Zn	Ga	Ge	As	Se	Br	Kr
$3d$	0	0	1	2	3	5	5	6	7	8	10	10	10	10	10	10	10	10
$4s$	1	2	2	2	2	1	2	2	2	2	1	2	2	2	2	2	2	2
$4p$	0	0	0	0	0	0	0	0	0	0	0	0	1	2	3	4	5	6

This now begins with an alkali and an alkaline earth metal, potassium and calcium, and then continues, from scandium to copper, with a group of *transition metals* in which the $3d$ states fill up, leaving one or two electrons in $4s$. We notice that at chromium and copper one extra electron is pulled into the d group, at the expense of the s group. This shows again the extra stability of a half-full or full group of quantum states.

Similar behaviour occurs in the later rows of the periodic table. The energy levels of $4d$ relative to $5s$ and $5d$ relative to $6s$ lead to two more groups of transition metals. An additional feature of the last group is that the filling of the $5d$ states is interrupted, while the fourteen $4f$ states fill up, in the *rare earth metals*.

This delayed filling of $(n - 2)f$ states and of $(n - 1)d$ states, after the ns states have started to fill, means that most of the elements in the lower part of the periodic table have one or two easily ionizable s electrons in their outermost group and so are metals.

3.7 Sizes of atoms and ions

As the nuclear charge increases, more electrons go into quantum states with large n, far outside the nucleus, but the wave patterns of all states are pulled closer to the nucleus. These two effects approximately compensate so that heavy atoms are not much larger than light ones. There is some increase, however, and this is reflected in, for example, the smaller ionization energies of the heavier alkali metals:

	Li	Na	K	Rb	Cs
I (eV)	5·39	5·14	4·34	4·16	3·87

TABLE 3.2. ATOMIC AND IONIC RADII, ÅNGSTRÖMS

Notation: (12) = coordination number 12, (8) = coordination number 8, $(+x)$ = cation with charge $+x$, $(-x)$ = anion with charge $-x$, bold type = calculated value.

Element	Type of Bond			
	van der Waals	Covalent	Metallic	Ionic
H	1·2	0·37		2·08(−1)
He	**0·8**			
Li		1·22	1·52(8)	0·7(+1)
Be		0·89	**1·13**(12)	0·34(+2), **0·2**(+3)
B		0·88		**0·2**(+3)
C		0·77		**2·6**(−4), 0·20(+4)
N	1·5	0·74		1·7(−3), 0·16(+3), 0·15(+5)
O	1·4	0·74		1·35(−2), **0·09**(+6)
F	1·35	0·72		1·33(−1), **0·07**(+7)
Ne	1·59			
Na		1·57	1·85(8)	0·98(+1)
Mg		1·37	**1·6**(12)	0·75(+2)
Al		1·25	1·43(12)	0·55(+3)
Si		1·17		1·98(−4), 0·4(+4)
P	1·9	1·10		1·86(−3), 0·44(+3), 0·35(+5)
S	1·85	1·04		1·82(−2), 0·37(+4), 0·30(+6)
Cl	1·80	0·99		1·81(−1), 0·34(+5), **0·26**(+7)
A	1·91			
K		2·02	2·25(8)	1·33(+1)
Ca		1·74	1·96(12)	1·05(+2)
Sc		1·44	1·63(12)	0·83(+3)
Ti		1·32	1·45(12)	0·76(+2), 0·70(+3), 0·64(+4)
V		1·22	1·31(8)	0·88(+2), 0·75(+3), 0·61(+4) **0·59**(+5)
Cr		1·17	1·25(8)	0·89(+2), 0·65(+3), 0·36(+6)
Mn		1·17	**1·3**(12)	0·91(+2), **0·62**(+3), 0·52(+4) **0·46**(+7)
Fe		1·16	1·24(8)	0·83(+2), 0·67(+3)
Co		1·16	1·25(12)	0·82(+2), 0·65(+3)
Ni		1·15	1·25(12)	0·78(+2)
Cu		1·17	1·27(12)	**0·96**(+1), 0·72(+2)
Zn		1·25	1·37(12)	0·83(+2)
Ga		1·25	**1·35**(12)	0·62(+3)
Ge		1·22		**2·72**(−4), 0·65(+2), 0·55(+4)
As	2·0	1·21		1·91(−3), 0·69(+3), **0·47**(+5)
Se	2·0	1·17		1·93(−2), 0·5(+4), 0·35(+6)
Br	1·95	1·14		1·96(−1), 0·47(+5), **0·39**(+7)
Kr	2·01			

TABLE 3.2—(*cont.*)

Element	Type of Bond			
	van der Waals	Covalent	Metallic	Ionic
Rb		2·16	2·49(8)	1·49(+1)
Sr		1·91	2·13(12)	1·18(+2)
Y		1·61	1·81(12)	0·95(+3)
Zr		1·45	1·60(12)	0·80(+4)
Nb		1·34	1·42(8)	0·74(+4), **0·7**(+5)
Mo		1·29	1·36(8)	0·68(+4), 0·65(+6)
Tc			1·35(12)	0·56(+7)
Ru		1·24	1·34(12)	0·65(+4)
Rh		1·25	1·34(12)	0·75(+3), 0·65(+4)
Pd		1·28	1·37(12)	0·80(+2), 0·65(+4)
Ag		1·34	1·44(12)	1·13(+1), 0·89(+2)
Cd		1·41	1·52(12)	0·99(+2)
In		1·50	1·57(12)	0·92(+3)
Sn		1·41	1·58(12)	**2·94**(−4), 1·02(+2), 0·74(+4)
Sb	2·2	1·41	1·61(12)	2·08(−3), 0·90(+3), **0·62**(+5)
Te	2·20	1·37		2·12(−2), 0·89(+4), **0·56**(+6)
I	2·15	1·33		2·20(−1), 0·94(+5), **0·50**(+7)
Xe	2·20			
Cs		2·35	2·63(8)	1·70(+1)
Ba		1·98	2·17(8)	1·38(+2)
La		1·69	1·87(12)	1··15(+3)
Ce		1·65	1·82(12)	1·18(+3), 1·01(+4)
Pr		1·65	1·82(12)	1·16(+3), 1·00(+4)
Nd		1·64	**1·82**(12)	1·15(+3)
Pm			**1·8**(12)	**1·09**(+3)
Sm		1·66	**1·8**(12)	1·13(+3)
Eu		1·85	1·98(8)	1·29(+2), 1·13(+3)
Gd		1·61	**1·8**(12)	1·11(+3)
Tb		1·59	**1·77**(12)	1·09(+3), 0·81(+4)
Dy		1·59	**1·77**(12)	1·07(+3)
Ho		1·58	**1·76**(12)	1·05(+3)
Er		1·57	**1·75**(12)	1·04(+3)
Tm		1·56	**1·74**(12)	1·04(+3)
Yb		1·70	1·93(12)	1·00(+3)
Lu		1·56	**1·74**(12)	0·99(+3)
Hf		1·44	**1·59**(12)	0·86(+4)
Ta		1·34	1·43(8)	0·73(+5)
W		1·30	1·36(8)	0·68(+4), 0·65(+6)
Re		1·28	**1·38**(12)	0·72(+4), 0·56(+7)
Os		1·25	**1·35**(12)	0·88(+4), 0·69(+6)
Ir		1·26	1·35(12)	1·66(+4)
Pt		1·29	1·38(12)	1·06(+2), 0·92(+4)
Au		1·34	1·44(12)	**1·37**(+1), 0·85(+3)
Hg		1·44	**1·55**(12)	1·27(+1), 1·12(+2)

TABLE 3.2—(*cont.*)

Element	van der Waals	Covalent	Metallic	Ionic
			Type of Bond	
Tl		1·55	**1·71**(12)	1·49(+1), 1·05(+3)
Pb		1·54	1·74(12)	2·15(−4), 1·32(+2), 0·84(+4)
Bi		1·52	**1·82**(12)	2·13(−3), 1·20(+3), **0·74**(+5)
Po		1·53	**1·7**(12)	0·67(+6)
At				0·62(+7)
Rn				
Fr				1·80(+1)
Ra				1·42(+2)
Ac				1·18(+3)
Th		1·65	1·80(12)	1·02(+4)
Pa				1·13(+3), 0·98(+4), 0·89(+5)
U		1·42	1·5(8)	0·97(+4), 0·80(+6)
Np				1·10(+3), 0·95(+4)
Pu			**1·63**(12)	1·08(+3), 0·93(+4)
Am				1·07(+3), 0·92(+4)

(PAULING, L. (1947). *J. Am. chem. Soc.*, 69, 542; MOELLER, T. (1952). *Inorganic Chemistry*, John Wiley, New York and London; WYCKOFF, R. W. G. (1948, 1951 and 1958). *Crystal Structures*. Interscience, London.)

It is difficult to define an atomic size precisely because the electron cloud has no sharp boundary. The radius of the maximum in the wave pattern, $4\pi r^2\psi^2$, as in Fig. 3.5, gives one indication. Experimentally, sizes can be found from measured distances between nuclei of neighbouring atoms in molecules and crystals but this also raises problems because the effective radius of a given atom varies considerably with the nature and strength of the interatomic forces involved. For example, the radius of the sodium ion in salts is 0·98 Å whereas the atomic radius in sodium metal is 1·85 Å. Even for the same type of bond, the radius of an atom or ion may change. For example, the radius of the *ferrous* Fe^{2+} ion is 0·83 Å and that of the *ferric* Fe^{3+} ion is 0·67 Å. The atomic radius in a metallic solid depends on the *coordination number* z, i.e. the number of nearest neighbours of the atom. The radii for $z = 8$, 6 and 4 are about 3, 4 and 12 per cent smaller respectively than that for $z = 12$.

Table 3.2 summarizes values of atomic and ionic radii for various types of bonds, as discussed in Chapter 4. We notice the following features:

(1) the trend of atomic sizes follows the periodic table, generally falling along each row of elements and then increasing sharply again at each alkali metal, where the principal quantum number increases;

(2) negative ions are large, having gained weakly-held electrons, and positive ions are small, having lost their outermost electrons;

(3) the stability of a half-filled group can be seen again in the rare-earth elements, where the $4f$ states become occupied; for example the europium atom tends to be larger than its neighbours, and its ion tends to be divalent, because one of the outer electrons is pulled in to complete the half $4f$ group;

(4) the difference between ionic and atomic radius, which is so marked for the alkali metals, is much smaller for copper, silver and gold due to the filling up of the $(n - 1)\,d$ states beneath the ns states; this has a considerable effect on the chemical and physical properties of these two groups of elements, which are described as *open metals* and *full metals* respectively.

FURTHER READING

BORN, M. (1951). *Atomic Physics*. Blackie, London.

FEYNMAN, R. P. (1965). *Lectures on Physics*, Vol. III. Addison-Wesley, London.

HEITLER, W. (1945). *Elementary Wave Mechanics*. Oxford University Press, London.

HUME-ROTHERY, W. and COLES, B. R. (1969). *Atomic Theory for Students of Metallurgy*. Institute of Metals, London.

MOELLER, T. (1952). *Inorganic Chemistry*. John Wiley, New York and London.

MOTT, N. F. (1952). *Elements of Wave Mechanics*. Cambridge University Press, London.

RICE, F. O. and TELLER, E. (1949). *The Structure of Matter*. John Wiley, New York and London.

Chapter 4

Chemical Bonding

4.1 Forces between atoms

The physical meaning of chemical affinity is that an *attractive* force exists between the atoms. There is also a *repulsive* force at close range which is the basis of the 'impenetrability' of matter. The *equilibrium spacing* of a pair of atoms is that at which these two forces are equal. Because force is rate of change of energy with distance, the energy of interaction of the pair reaches its lowest value at the equilibrium spacing. Quantum mechanics deals directly with energy, not force, and so it is usual to discuss problems of chemical bonding in terms of energy.

We bring two atoms together from infinity, having defined their potential energy of interaction as zero at infinity. Little happens and the energy stays near zero until the atoms are within about an atomic spacing of each other. As they move still closer, the attractive force begins to be felt and the potential energy falls, since work is being done *by* the atoms as they move together under this force. When they get really close, the short-range repulsive force then also comes into play and soon dominates the attractive force. The potential energy rises at this stage since the atoms are now being pushed together. This behaviour is shown in Fig. 4.1 which gives the potential energy due to the attractive (curve a), repulsive (curve b) and total (curve c) forces as a function of the distance between nuclei. The potential energy 'well' is deepest at the equilibrium spacing r and its depth D there is equal to the work required to dissociate the atoms completely.

Whether a collection of such atoms will form a gas or *condense* together into a liquid or solid depends both on D (in relation to the temperature and pressure) and on the nature of the forces. Molecular gases can form even when D is large (e.g. H_2, O_2, N_2, CO_2, at room temperature) if the atomic forces show *saturation*, a quantum-mechanical effect in which an atom bonds strongly with one or two neighbours to form a molecule which then has little affinity for other atoms or molecules. The bonding force in every case is always the ordinary electrostatic attraction between opposite charges but it is often so disguised in its action by quantum-mechanical effects that chemical bonding generally seems very different from simple electro-statics.

The small affinity which exists between chemically saturated molecules enables molecular gases to condense at low temperatures into molecular crystals, in which molecules with strong internal bonds are weakly held together by small residual forces. Many organic substances (waxes, oils) are of this type. The weak intermolecular bonds can easily be broken by heat, so vaporising the molecules, but the bonds inside the molecules of the gas can stand much higher temperatures before they break and the gas dissociates into atoms.

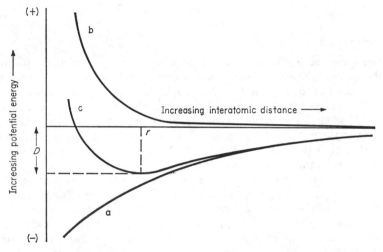

Fig. 4.1 Potential energy between atoms

In many substances saturation does not occur. Large groups of strongly bound atoms can then form and grow indefinitely by the continued addition of more atoms. These can be regarded as giant molecules. Metallic solids and liquids are examples. Here vaporization is small except at temperatures where the vapours consist mostly of single atoms.

4.2 Origin of interatomic forces

We picture a molecule, or liquid or solid, as two or more atomic nuclei embedded in a cloud of negative electrical charge, all held together by electrostatic forces. The strength of the bond depends on the density of the electron cloud at various points between the nuclei and this, of course, is governed by the permissible wave mechanical patterns. The nuclei, being heavy, move slowly compared with the electrons and we can pretend that they are at rest when we determine the wave functions. Even with this

simplification it is usually impossible to solve the wave equation for molecular problems unless we first form some simplifying idea about the type of
bond in each particular case. Four characteristic types are recognised:
van der Waals, covalent, ionic and *metallic*.

The van der Waals force is the only appreciable force exerted between well
separated atoms and molecules. It is a weak attractive force that acts between
all atoms and is responsible for the condensation of noble gases and chemically saturated molecules to liquids and solids at low temperatures. Its
explanation requires time-dependent quantum mechanics because it
involves fluctuations and surges of electronic charge in the atom. When two
atoms approach fairly closely these fluctuations can occur in unison so that
one atom has its electrons slightly nearer the other nucleus whenever this
nucleus happens, through the movements of its own electrons, to be more
exposed in this direction than usual.

In the ionic bond, discussed in § 3.1, electrons pass from an electropositive atom to an electronegative one and the positive and negative ions so
formed are pulled together electrostatically. The repulsive force which
balances this attraction at small distances sets in when the outermost com-
pletely-filled electron shells of the ions begin to *overlap*. Since there is no
sharp boundary to an atom, overlapping strictly occurs at all spacings.
However, electronic clouds fade away exponentially at large distances from
their nuclei and so the effect of overlapping is negligible until the nuclei
approach closely and the dense parts of the clouds meet. Overlap forces are
thus *short-range* forces.

When two shells begin to overlap, an electron originally in one of them
no longer belongs to one atom but is shared by both and belongs to the
molecule as a whole. Its motion and energy are still governed by the
quantum laws and in particular by the Pauli principle. As a result, if
the shells are both full to begin with, the single group of electrons formed
by their union is too large for all to go into quantum states of low energy.
Some must be *promoted* into higher energy levels. This increase in energy
gives the repulsive overlap force. It occurs whenever completely filled
shells overlap, whatever the attractive forces pulling the atoms together.

When overlap occurs between *partly empty* shells of quantum states,
however, the result is very different and leads in general to strong bonds of
the *covalent* or *metallic* type. An important feature of these bonds is that
they can exist between atoms of the same type, between which there can be
little or no ionic bonding. The atoms in, for example, H_2, Cl_2, diamond and
copper are held together by them. The interatomic forces in such cases are
brought about by *valency electrons*, i.e. the electrons in the outlying partly-
empty quantum states, becoming *shared* between the nuclei. Their atomic
wave patterns spread out farthest from their parent nuclei and so overlap
first when the atoms are brought together. The repulsion between the inner
filled shells of electrons only appears when the electron clouds of the atoms
interpenetrate deeply.

4.3 Covalent bonding

The hydrogen molecular ion H_2^+, i.e. one electron and two protons, provides the simplest example of a covalent bond. In Fig. 4.2 we show the protons, 1 and 2, at a fixed distance R and the electron, e, at distances r_1 and r_2 from them. The wave equation of this electron is then obtained directly by substituting

$$V = -\frac{1}{4\pi\varepsilon_0}\left(\frac{e^2}{r_1} + \frac{e^2}{r_2}\right) \qquad\qquad 4.1$$

into eqn. 3.11.

This equation has been solved exactly to obtain the wave function ψ, in Fig. 4.2, and the electron energy E as a function of R. The *total energy* can then be obtained by adding the proton-proton repulsion, $+ e^2/4\pi\varepsilon_0 R$, to E.

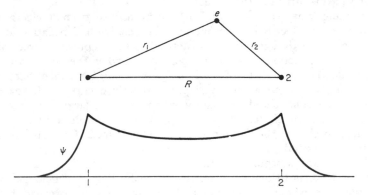

Fig. 4.2 The hydrogen molecular ion, showing the variation of wave amplitude along the line between nuclei

It reaches its minimum, 2·8 eV below the energy of a separate hydrogen atom and proton, at $R = 1·06$ Å, in agreement with experiment. The physical basis of the bond is simple. In the region between the nuclei the electron is electrostatically attracted to both nuclei and this attraction is stronger than the proton-proton repulsion because the electron-proton distance is shorter. Wave mechanics comes in merely to determine the kinetic energy of the electron and to settle how much time the electron spends in the favourable region between the nuclei.

There are, of course, no filled inner electron shells in this molecule to give repulsion at close distances and we might think that by a continuous scaling down of all distances in the molecule the energy could continue to decrease. We recall, however, that when an electron cloud shrinks its kinetic energy increases. If, to avoid this, the nuclei were to move together without the electron cloud also shrinking the $+ e^2/4\pi\varepsilon_0 R$ repulsion would rise disproportionately.

For molecules more complicated than H_2^+ it is usually too difficult to solve the wave equation. Fortunately, this problem can be by-passed and a fair approximation to molecular wave functions got by the simple method, first suggested by *Heitler and London* of *adding together, unaltered wave functions of the separate atoms*. Fig. 4.3 shows this applied to H_2^+. When ψ is of the same sign in both atoms, the overlapping atomic wave functions reinforce each other and ψ^2 is then large between the nuclei, giving bonding. When ψ has opposite signs in the two atoms, however, the atomic wave functions cancel each other by overlapping and produce a nodal plane N between the nuclei; this represents an excited state of the molecule, or an *anti-bonding* molecular wave function.

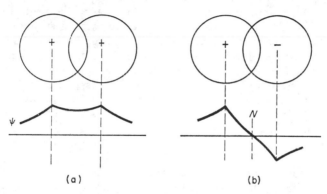

Fig. 4.3 Boundary surfaces (upper diagrams) of atomic wave functions which are superposed to produce approximate molecular wave functions (a) and (b) of the hydrogen molecular ion (lower diagrams). The bonding function (a) should be compared with the exact solution in Fig. 4.2

Mathematically, this amounts to taking two separate atomic wave functions, ψ_1 on nucleus 1 and ψ_2 on nucleus 2, and forming a linear combination of them, $a\psi_1 + b\psi_2$, where a and b are numbers, which we then assume to be an approximate wave function for the whole molecule. For identical nuclei we take $a = \pm b$. This gives two approximate molecular wave functions, σ and σ^*

$$\sigma = A(\psi_1 + \psi_2) \quad \text{and} \quad \sigma^* = B(\psi_1 - \psi_2) \qquad 4.2$$

where the numbers A and B are chosen so that σ and σ^* are normalized (cf. eqn. 3.6). Figs. 4.2 and 4.3(a) show that σ gives a wave function similar to the exact one for H_2^+. *Even though we use undisturbed atomic wave functions there is a heaping up of electronic charge between the nuclei.* This is because the electron density is given by $(\psi_1 + \psi_2)^2$ and this, half-way between the nuclei (i.e. where $\psi_1 \simeq \psi_2$), is *twice* what would be given by superposing *undisturbed atomic electron clouds*. Physically, of course, charge heaps up between nuclei entirely because the electrostatic forces from the nuclei attract the electrons there. Luckily, because charge density increases as the *square* of the total wave amplitude, this heaping up of charge requires us to make little change in the atomic wave functions. This is a mathematical windfall which is exploited to develop approximate theories of molecular bonds without solving the molecular wave equations. We must not be deceived by it, however, into thinking that the heaping up actually occurs because of this mathematical feature of wave functions.

The other wave function σ^* of eqns. 4.2 gives a charge density $(\psi_1 - \psi_2)^2$ that is zero half-way between the nuclei. This represents a wave function with a nodal plane between the nuclei, i.e. an *anti-bonding* function.

The bonding wave function of Fig. 4.2 or 4.3(a) can, of course, accept two electrons of opposite spins. This is the basis of the neutral hydrogen molecule, H_2. This bond (4.7 eV) is not quite twice as strong as the one-electron bond because of the electrostatic repulsion between the electrons.

If another hydrogen atom tries to join this molecule there is no room in the bonding quantum state for the third electron and this has to go into an anti-bonding state similar to that of Fig. 4.3(b). As a result, this atom does not bond itself to the molecule (except weakly through van der Waals forces). Since the Pauli principle, which produces this effect, applies to all quantum states in all atoms and molecules, we thus have the basis of the *saturation* of covalent bonds and of the prevalence of two-electron chemical bonds.

Helium atoms cannot bond covalently because there are no unpaired electrons in them. Both the bonding and anti-bonding molecular wave functions would be completely filled in this case, leading to the *promotion* of electrons and the *closed-shell repulsion* discussed earlier. The additional $2s$ electron of lithium, however, opens up the possibility of new molecular wave functions formed from the $n = 2$ states. There are four of these (one $2s$ and three $2p$), each of which can hold two electrons of opposite spins so that there are always far more bonding states than electrons to fill them. Saturation cannot occur, however many lithium atoms we bring together, and so, instead of a small covalent molecule, we get an indefinitely large molecule formed by adding more and more lithium atoms; i.e. we get *lithium metal*, a huge aggregate of lithium atoms all bonded together by unsaturated covalent bonds.

Beryllium behaves somewhat similarly although now there are two valency electrons per atom so that the atoms are more strongly bound together, as is shown by the higher boiling point of the metal: Be, 3040 K; Li, 1600 K. The number of valency electrons per atom and the bond strength increase still further in boron and carbon but now the number of such electrons begins to be so large, in relation to the number of bonding states, that a restriction is exerted on the number of neighbours to an atom. A notable feature of the carbon atom (four $n = 2$ electrons) is its preference for *tetrahedral* bonds, i.e. bonds with four neighbours which stand out from the atom symmetrically in the directions of a tetrahedron. Examples of this are the diamond crystal and the methane CH_4 molecule.

To explain this, we recall that the classification of atomic wave functions into s, p, d, f types is to a large extent arbitrary and that any linear combination of functions with the same number of nodes is equally valid as an atomic wave function.

It happens that the combinations

$$2s + 2p_x + 2p_y + 2p_z$$
$$2s + 2p_x - 2p_y - 2p_z$$
$$2s - 2p_x + 2p_y - 2p_z \qquad \text{4.3}$$
$$2s - 2p_x - 2p_y + 2p_z$$

give four atomic wave functions in which the electron density is strongly concentrated in four tetrahedral directions out of the atom, which is particularly suitable for making covalent bonds with four symmetrically disposed neighbours. This process of mixing conventional atomic wave functions to make functions suitable for bonding is called *hybridization*.

In the elements beyond carbon, the $n = 2$ states are more than half-full and the number of *unpaired* atomic electrons available for covalent bonds decreases. Hence we get valencies of 3, 2, 1 and 0, respectively, in nitrogen, oxygen, fluorine and neon.

The next row of the periodic table develops somewhat similarly, although here the bond strengths are rather weaker because of the presence of the additional filled shell ($n = 2$) in the atom beneath the valency electrons. Interactions involving d electrons are important in the later rows and give very strong cohesion in the transition metals.

4.4   Transition to ionic bonding

Covalent bonds can form between atoms of different elements. However, these atoms usually have different ionization energies and electron affinities. The covalent bond then does not form symmetrically between them and the electrons tend to be heaped up on the more electronegative partner, so giving some ionic character to the bond. Thus in the sequence $CH_4 \rightarrow NH_3 \rightarrow H_2O \rightarrow HF$ the hydrogen atom loses more and more of its electronic charge to its increasingly electronegative partner. In terms of the approximate molecular wave function $a\psi_1 + b\psi_2$, where $a = b$ for the pure covalent bond, we have $a < b$ when atom 2 is more electronegative than 1 and, in the limit of a pure ionic bond, $a = 0$ and $b = 1$. In general, the presence of some ionic character strengthens an otherwise covalent bond. For example, the diamond-like form of boron nitride (BN) appears to be a stronger solid than diamond itself.

Pauling has prepared a table of electronegativities of the elements from which it is possible to deduce roughly the amount of ionic character in a bond. Table 4.1 gives the table, as modified by more recent work. Most transition metals have values of about 1·6 and rare earths about 1·3. The ionic character of the bond increases with $x_1 - x_2$, where x_1 and x_2 are the electronegativities of the participating atoms. Roughly, $x_1 - x_2 = 1$ gives 15 per cent ionic bond; 2 gives 50 per cent; and 3 gives 85 per cent. For example, $x_1 - x_2 = 3·0 - 2·1 = 0·9$ for HCl, giving about 15 per cent ionic bond. We notice that a given atom can behave very differently according to the nature of its partner; for example, H bonds partly ionically as an electron-acceptor with electropositive metals such as Ca, Li and Na; covalently with B, C, S, and Si; and partly ionically as an electron-donor with F and O.

TABLE 4.1.   ELECTRONEGATIVITY SCALE

F	3·9	Ag, Cu	1·7
O	3·5	Ga, Ti	1·6
N, Cl	3·0	Al, In, Zn, Zr	1·5
Br	2·8	Be, Cd, Tl	1·4
C, S, I	2·5	Sc, Y	1·3
Se	2·4	Mg	1·2
Au, H, P, Te	2·1	Ca, Li, Sr	1·0
As, B	2·0	Ba, Na	0·9
Hg, Sb	1·9	K, Rb	0·8
Bi, Ge, Pb, Si, Sn	1·8	Cs	0·7

(PRITCHARD, H. O. and SKINNER, H. A. (1955). *Chem. Rev.* **55**, 745)

The balance between covalent and ionic bonding depends on several other factors. In ionic *crystals* the ionic character is developed more strongly because each ion has *several* ions of the opposite sign as its immediate neighbours; for example, AgCl forms a mainly covalent molecule in the vapour but an ionic crystal in the solid. A factor which favours covalent character in ionic bonds is *polarizability*. The electronegative ion, having gained an electron or electrons, has a rather large and loosely held outer electron cloud, the shape of which can easily be deformed by nearby charges. The electropositive ion, having lost electrons and become small, provides a strong centre of positive charge which can thus pull the electron cloud of the negative ion partly back towards itself. For example, LiI is fairly covalent because the large I^- ion (cf. Table 3.2) is strongly polarized by the small Li^+ ion. Because of polarization there are chemical similarities between the following *diagonally situated* elements in the periodic table,

$$\text{Li} \quad \text{Be} \quad \text{B} \quad \text{C}$$
$$\text{Na} \quad \text{Mg} \quad \text{Al} \quad \text{Si}$$

For example, the polarizing power of the Al^{3+} ion is *increased*, relative to Be^{2+}, by the greater charge but is *decreased* by the greater ionic radius; the result is a certain chemical similarity between these two metals.

The simple idea of *ionic valency*, i.e. the number of electrons lost or gained to produce an ion with the nearest noble gas electronic structure, works well for elements near noble gases in the periodic table but not for those in the centre of the table. To ionize four electrons off carbon, for example, requires 150 eV and this amount of energy cannot be got back from ionic bonds. Highly charged ions, particularly of metals (e.g. Al^{3+}, Cr^{3+}, Fe^{3+}), are, however, stabilized by an effect which can occur with all ions; the attachment of *ligands* to them. Most ligands are either molecular ions of opposite sign (e.g. SO_4^{2-}) or *neutral polar molecules*, i.e. molecules such as water in which the valency electrons are concentrated mainly at one end (e.g. on the oxygen atom in H_2O). They attach themselves electrostatically in large numbers to the central ion. The simplest example of this is the *hydration* of positive ions by the attachment of water molecules. An example is $Al(OH_2)_6^{3+}$. Here the central aluminium atom is surrounded by six negatively charged oxygen atoms which form partly covalent bonds with it and the triple positive charge is then shared out amongst the twelve hydrogen atoms round the outside. The six water molecules are arranged *symmetrically* about the central ion at the corners of an octahedron.

In many such *coordination compounds* an important part is played by d electrons. Consider for example, $Ti(OH_2)_6^{3+}$. Here again the water molecules are arranged (with oxygen atoms pointing inwards) octahedrally, i.e. along the ends of x, y, z axes through the central ion. From the wave functions of d states in Fig. 3.8 we see that in three of these (d_{xy}, d_{yz}, d_{zx}) an electron avoids the x, y, z axes and so can largely avoid the electrostatic repulsion of the oxygen electrons along these axes; whereas in the remaining d states the electron concentrates in these axes and is repelled by the oxygen electrons. It follows that the octahedral pattern of ligands splits the energy level of the d states into two levels, a low one for the states of d_{xy} type and a higher one for those of $d_{x^2-y^2}$ and d_{z^2} type. This splitting of the energy level can be large, 2 eV or more. The strong absorption of light which accompanies the jumping of a d electron from the lower to the higher level is responsible for the bright colour of transition metal compounds, as used in pigments. In certain cases the higher-energy d electrons may join with those of the ligands to form covalent bonds.

Coordination compounds and complex ions have great value in chemical metallurgy. The volatile *nickel carbonyl* $Ni(CO)_4$ compound provides the basis of the *Mond process* for extraction of nickel (cf. § 8.8). The formation of complex ions often provides a means for getting sparingly soluble substances into solution. The stability of the $AuCl_4^-$ ion is responsible for the ability of *aqua regia* ($HCl + HNO_3$) to dissolve gold. Complexes with the cyanide ion CN^- are important in electrodeposition (cf. § 9.4) and also in the *extraction of gold and silver*.

Gold is chemically unreactive and exists naturally as the metal. Where it occurs in large nuggets or in river gravel it can be separated simply by washing away the lighter minerals, as in *panning*, practised by old-time prospectors. The main production comes, however, from large deposits finely and sparsely distributed in hard quartz (SiO_2) rock, which has to be ground to a fine powder (*slime*). The gold is too fine to be separated by physical methods and too inert for most solvents. It can, however, dissolve as the cyanide ion. The slime is thus treated with dilute KCN solution in large open vats in the presence of air, which produces the reaction

$$4Au + 8KCN + 2H_2O + O_2 \rightarrow 4KAu(CN)_2 + 4KOH \qquad \textbf{4.4}$$

The gold goes into solution and is taken to *extractor boxes* where it is brought into contact with zinc dust and is precipitated,

$$2KAu(CN)_2 + Zn \rightarrow ZnK_2(CN)_4 + 2Au \qquad \textbf{4.5}$$

Excess zinc is then removed from the deposit by dissolution with sulphuric acid.

Native silver is extracted in a similar way. Silver chloride, which may occur as the mineral or may be produced as a result of preliminary chemical treatments of the ore to remove arsenic, is also suitable for cyanide extraction, through the reaction

$$AgCl + 2NaCN \rightarrow NaAg(CN)_2 + NaCl \qquad \textbf{4.6}$$

4.5 Transition to metallic bonding

Consider a large number of carbon (or silicon or germanium) atoms brought together. Each bonds covalently with four neighbours in a tetra-

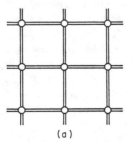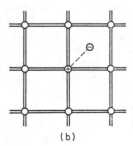

(a)   (b)

Fig. 4.4 (a) Schematic covalent bond structure in diamond-like substances (C, Si, Ge, grey Sn); (b) a 5-valent atom in the structure represented as an electron ⊖ and unit positive ion ⊕

hedral pattern and this is the basis of the diamond crystal structure. The whole aggregate forms a single large covalent molecule with a bond structure as shown schematically in Fig. 4.4(a), where each circle represents an atom and each pair of lines a single electron-pair bond.

The substance is an *electrical insulator*. The rapid flow of electricity through matter takes place by the movement of electrons from atom to atom. The insulating properties of the diamond structure are due not to any immobility of the electrons but to the saturated covalent bond structure. The only way that electrons can move about, while still remaining part of the covalent bond structure, is for two of them in different bonds *to change places*. But this produces no flow of electrical charge through the material, i.e. no conduction.

Consider now the effect of a foreign atom in the material; let us say a phosphorous atom in silicon. By using four of its five outer electrons to bond covalently with four silicon neighbours it can fit into the ordinary bond structure of silicon, as in Fig. 4.4(b). There remains its fifth electron. We may think of this electron as attached to a 'monovalent' positive ion as in a hydrogen or an alkali metal atom. The theory of this 'monovalent' phosphorous atom follows that of Chapter 3, except that the atom does not exist in ordinary space but in a 'space' whose properties depend on the covalent bonds and silicon atoms which fill it. This has two effects. First the electrostatic potential V in the wave equation no longer goes as $(4\pi\varepsilon_0)\ V = -e^2/r$ because the covalently bonded atoms become *polarized* in the electrical field of the positive ion, their electrons being pulled slightly towards it. This polarization effect is allowed for by introducing a *dielectric constant* κ into the analysis, in the form $(4\pi\varepsilon_0)\ V = -\ e^2/\kappa r$. The second effect is that the electron of this 'monovalent' atom has to move through the local electrical fields of the covalent bonds and atoms in the neighbourhood. This can be allowed for in the wave equation by replacing the real mass m of the electron by an *effective mass* m^*.

For silicon, $\kappa = 11.9$ and $m^* \simeq 0.25\ m$. If we repeat the theory of the ground state of the hydrogen atom (§ 3.5), using these values, we obtain a $\simeq 25$ Å and $E \simeq -0.025$ eV. (In practice, the value of E is almost doubled by subsidiary effects.) The important effect of the large κ is that the electron moves far out from its ion; and the ionization energy E required to remove it completely is very small. The heat energy available at room temperature (cf. § 6.6) is in fact sufficient to ionize such atoms. Once the electron is ionized it can move about anywhere in the entire crystal structure. It cannot get captured in a covalent bond since these are all full. Except possibly for getting trapped on another ionized impurity atom, from which it may also escape by ionization, it is a *free* electron, not attached to any particular atom in the structure. If the material is placed between charged electrodes this electron will be attracted to the positive electrode and will move towards it, so providing a unit of electrical current. This material is not a *metal*, however, but an *impurity semi-conductor* because its conductivity disappears at very low temperatures where thermal ionization of the impurity atoms no longer occurs. A metal, by contrast, conducts electricity better at low temperatures than high.

Even without foreign atoms it is possible to obtain some conductivity, called *intrinsic semi-conductivity*, from a covalent structure provided the temperature is high enough to ionize electrons out of the *covalent bonds*.

This ionization energy, E_g, is generally much larger than E and is made up partly of the excitation energy required to promote a valency electron into a non-bonding quantum state and partly of the electrostatic energy then required to pull it away from the remanent positive charge, called a *positive hole*, which it leaves behind. Values of E_g are, in eV: diamond 6·5, silicon 1·1, germanium 0·7.

It is usual to represent all this by means of an *energy band diagram*, as in Fig. 4.5, in which energy is measured vertically and position in the material horizontally. The bands are analogous to energy levels in free atoms. Each band represents the energy levels of a large group of quantum states, *degenerate* (cf. § 3.5) in isolated atoms, but split in the solid or liquid into different, closely grouped energy levels by the interactions between the atoms (cf. Fig. 24.12). The bands are horizontal because in a uniform environment the energy of an electron in a given quantum state is indepen-

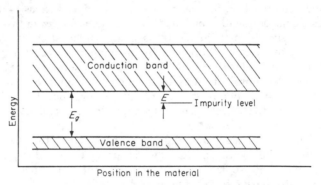

Fig. 4.5 Energy band diagram for a covalent solid

dent of where that electron is in the material. For example, if an electron is taken out of a covalent bond and put into a positive hole in another covalent bond some distance away, its energy is then exactly the same as before; similarly for a free electron moving about in a uniform field of covalently bonded atoms.

We thus have two energy bands, a lower *valence band* which represents the energy levels of electrons in the covalent bonds and an upper *conduction band*, which represents free electrons not covalently bonded. These are separated by the *energy gap E_g*, which we may regard as a *forbidden energy band* in the sense that there are no quantum states in the material with energies within E_g above the valence band. When a foreign donor atom is present, however, e.g. phosphorous in silicon, we know that only a small energy E is required to ionize its extra electron. Hence such an atom can be represented by a localized *impurity energy level* at a depth E below the conduction band, as shown in Fig. 4.5.

When the density of conduction electrons becomes high, a major new effect sets in which makes further ionization very much easier (*N. F. Mott*, 1958, *Nuovo Cimento* (Supplement) **7**, 312). *The material becomes too good an electrical conductor to allow long-range electrostatic fields to exist inside it.* Positive charges then become screened by heaping up of the conduction electron charge cloud round them. Individual conduction electrons are similarly screened by the formation of *positive holes* in the charge cloud immediately round them, as shown in Fig. 4.6. Here the central spike represents the electron in question and the depressed region of the charge distribution near it represents the 'electron hole', formed by the nearby conduction electrons avoiding one another as far as possible. These screening effects of conduction electrons are rather like the dielectric polarization

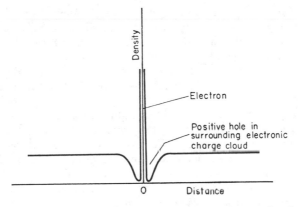

Fig. 4.6 Variation in density of conduction electron charge cloud near a given electron

of the covalent electrons but much more drastic; the e^2/r interaction is no longer replaced by $e^2/\kappa r$ but by the *screened charge* interaction V, where

$$(4\pi\varepsilon_0)\ V = -\frac{e^2}{r}\exp\left(-r/r_0\right) \qquad\qquad 4.7$$

where r_0 is of the order of spacing of the conduction electrons. This V is now a *short-range* interaction which almost vanishes at distances larger than r_0. When r_0 is of the order of an atomic spacing the screened potential becomes too weak to capture electrons at all and the ionization energy vanishes. There are then lots of conduction electrons, even at 0 K, and the substance is a *metal*. The transition from the semi-conducting state to the metallic state is sharp and drastic because, once the screening effect begins to act, further ionization can then occur which leads to more screening and hence to a catastrophic general ionization.

The effect can be produced in a semi-conductor by increasing the impurity content, but in ordinary metals it occurs naturally because E_g is reduced to zero. This is a result of the *unsaturated covalent bond structure* of metals (cf. § 4.3); there are always more bonding states than electrons to fill them. As a result, the energy of unbonding is no longer required when an electron moves out of such a bond since, wherever it goes in the metal, there are always empty bonding states available for it. Every atom can then behave like a donor impurity offering easily removable electrons for ionization and the ionization energy is eliminated by the screening effect. The gap between the valence and conduction bands vanishes and large numbers of valency electrons become free, quite irrespective of temperature (*Mott's* theory of metals).

4.6   The metallic state

We thus picture a piece of metal as a large number of positive ions embedded in a 'cloud' or 'sea' of free electrons. This *free electron* theory of metals, which started with *Drude* (1902) and *Lorentz* (1916), has been most successful in explaining the properties of metals. The extremely good electrical and thermal conductivities of metals are explained immediately by the flow of free electrons through the material. The *optical opacity* and *reflectivity* of metals are also explainable; the free electrons oscillate in the alternating electric field of the incident light beam, absorbing energy at all wavelengths and so making the metal opaque. In turn, the oscillating electrons emit light waves and in this way produce the reflected beam.

Direct evidence for free electrons in metals was obtained by *Tolman and Stewart* (1917). A piece of metal was accelerated sharply so that the free electrons, by their inertia, were thrown towards its rear, thereby producing a detectable pulse of current. Measurements of this showed that the ratio of charge to mass of the particles forming the current was the same, within the experimental uncertainty, as that of electrons.

Since there is no saturation of covalent bonding states in a metal there is almost nothing, apart from geometry, to restrict the number of nearest neighbours which any atom may have. As a result atoms in metals tend to be packed together like close-packed spheres. In pure metals this leads to very simple crystal structures (cf. § 13.1). In certain alloys, e.g. *Laves phases* such as $MgCu_2$ and KNa_2, the difference in sizes of the participating atoms enables even closer packings to be achieved by forming more complex crystal structures. The unsaturated metallic bond is also responsible for the ability of some metals to accept small atoms (e.g. H, N, C) into the *interstitial* spaces between the metal atoms, which is of great practical importance in carbon steel.

The unsaturated nature of the metallic bond also accounts for the *alloying* properties of metals. When two metals such as copper and nickel are mixed together, each atom reacts fairly unspecifically to the other since they are held together by the common free electron cloud to which both have contributed. It is thus possible to make alloys over wide ranges of composition by randomly replacing atoms of one metal by those of another, so forming *substitutional solutions*. The insensitivity of the metallic bond to the particular metal atoms involved is also responsible for the ability of metals to be joined by welding and soldering. In principle, it is sufficient merely to bring clean metal surfaces into contact for them to bond.

The undirected nature of the metallic bond also influences the surface and mechanical properties of metals. If a metal is cut into two pieces along some surface, the bonds which previously crossed that surface transfer themselves to atoms within each of the new surfaces so created. The *surface energy* of a metal is thus rather smaller than might be expected purely from the number of bonds intersected by the cut. As regards mechanical properties, it is easier in metals than most other substances for atoms to slide over one another without breaking bonds. Ultimately, the ductility and toughness of metals are due to this property, although their explanation requires a detailed analysis of the way the atoms move (cf. Chapter 21).

The immunity of metals to *radiation damage by ionization* is also a consequence of the free electron structure. Fast charged particles and ionizing rays can strip electrons off the atoms they pass through on their journey. In metals this matters little because the positive holes so formed can immediately fill up again from the free electron sea, leaving no trace of the damage apart from a little heat. In non-metals, by contrast, where the electrons are localized in bonds, electronic excitation and ionization by radiation can have drastic effects. In ionic solids and glasses the electrons may become trapped at atomic irregularities in the structure and there produce 'colour centres', so that the material becomes coloured and eventually black. In organic molecules the ionization of an electron may destroy a vital bond holding the structure together so that large macromolecules may become broken into pieces.

FURTHER READING

As for Chapter 3; also:

COULSON, C. A. (1961). *Valence*. Oxford University Press, London.

PAULING, L. (1960). *The Nature of the Chemical Bond*. Cornell University Press, New York.

RICE, O. K. (1940). *Electronic Structure and Chemical Binding*. McGraw-Hill, New York and Maidenhead.

SHOCKLEY, W. (1950). *Electrons and Holes in Semi-Conductors*. Van Nostrand, New York and London.

Chapter 5

Heat and Energy

5.1 Thermodynamics

Thermodynamics, the science of energy, is undoubtedly the most important single branch of general science used in metallurgy. The extraction of metals from their ores depends on energies of oxidation. Most metallurgical operations, e.g. smelting, melting, alloying, hot working and annealing, have to be done at high temperatures, e.g. 500 to 2000°C, and so require expensive furnaces and fuels. The development of various equilibrium microstructures in metals and alloys, upon which many important properties depend, is controlled by general thermodynamic laws. In this chapter and the next we shall outline the elements of thermodynamics.

The collection of atoms, molecules, etc. which forms the object of study is called a *thermodynamic system*. It can be *isolated* from most influences by enclosing it in an *adiabatic container*, such as an ideal vacuum flask. If left in isolation for a long time it eventually settles down into a *state of equilibrium* in which its overall structure and properties no longer change with time, even though its atoms are still in motion. Thermodynamics deals with such equilibrium systems.

We have to describe such a system. First, we describe its *content*. The simplest equilibrium systems are *homogeneous*, i.e. uniform everywhere in their composition, structure and properties (apart from temporary fluctuations due to random movements of the atoms, which become small in large samples). The air in a still room is an example. In such cases we need only report the *components* (i.e. the elementary chemical substances; O_2, N_2, H_2O, CO_2, etc. in air) of the system and their amounts. Systems also exist in *heterogeneous equilibrium* and then consist of a number of different parts, each homogeneous in itself; for example, ice and water. These different parts meet at *boundary* surfaces and in principle (and often in practice) can be separated by dissecting them along these boundaries. When a system is completely dissected into homogeneous parts, and all parts of the same kind are collected together into their own homogeneous group, each such group is called a *phase* of the system. The content of the system is then defined by specifying the components and amounts in each phase in the system.

We have also to describe the *thermodynamical state* of the system. It is clearly impossible and unnecessary to specify the position and velocity of every particle. In fact, an equilibrium system can be specified unambiguously by giving the values of a few bulk properties. In the simplest problems two such properties are sufficient, the *pressure P* and *volume V*, and these two variables have a special importance in thermodynamics. It is not sufficient to specify one only; for example, at fixed pressure the volume can vary as the system is heated or cooled.

Consider two isolated systems, each in equilibrium, one with P_1, V_1, the other with P_2, V_2. Bring them together and replace the adiabatic wall between them by a *thermally conducting* wall. Then in general they are no longer in equilibrium. Changes occur in their pressures and volumes and we infer that heat is flowing from the hotter to the colder. Eventually these changes are complete and the two systems are then in *thermal equilibrium*. An important fact is that all systems in thermal equilibrium with any given reference system are also in thermal equilibrium with one another. It can be deduced from this that an equilibrium system possesses a certain property, called its *temperature T*, which is a function of its pressure and volume, i.e. $T = f(P, V)$, called an *equation of state*; and that thermal equilibrium between systems 1, 2, 3 . . . etc. is obtained when $f(P_1, V_1) = f(P_2, V_2) = f(P_3, V_3) = \cdots$ etc. We can then make a temperature scale by graduating some suitable property of a suitable reference system. The best is the volume of a fixed mass of a 'permanent' gas at fixed low pressure. This gives the *absolute temperature scale* (K), on which ice melts at 273·15 K and water boils at 373·15 K (at 1 atm). This scale is, of course, based on the familiar equation of state of a *perfect gas*,

$$PV = RT \qquad \qquad 5.1$$

where R is the *gas constant*, 8·314 J mol^{-1} K^{-1}.

5.2 Internal energy and enthalpy

The equations of thermodynamics are based on quantities that depend only on the thermodynamic state of the system and not upon the sequence of changes by which that state was reached. Such quantities are called *functions of state*; for example $T = f(P, V)$. The *internal energy E* is such a property. For example, however we raise the temperature of a cupful of water from 20°C to 50°C, whether by simple heating or by stirring with paddles, or by other means, the internal energy of the water at 50°C always exceeds that at 20°C by the same amount.

The internal energy is the sum of all the individual kinetic energies of motion and energies of interaction (potential energies) of the particles in the system. When a system is left to itself its internal energy remains constant but when it is brought into interaction with its surroundings it can alter its internal energy by receiving or giving energy. This energy may be transferred either as *work*, through the displacement of an applied force, or as *heat*, through the fine-scale individual interactions of the particles (including photons) in thermal contact. Suppose that a system changes its internal energy from E (before) to $E + dE$ (after) by absorbing a small amount of heat dQ from its surroundings and by doing a small amount of work dW on its surroundings (e.g. pushing the atmosphere back by expanding against atmospheric pressure). Then the *first law* of thermodynamics, i.e. the principle of conservation of energy and heat, requires that

$$dE = dQ - dW \qquad\qquad 5.2$$

Because E is a function of state, dE is a *total differential*. This means that we can simply integrate it,

$$E_2 = E_1 + \int_1^2 dE \qquad\qquad 5.3$$

to find the internal energy E_2 of the system in a state 2 in terms of E_1 in state 1. We cannot however similarly determine the 'heat in the system' from $\int dQ$ and the 'work in the system' from $\int dW$, because dQ and dW are not total differentials and the values of their integrals depend on the path taken from state 1 to state 2. For example, we can warm water to the same temperature with a heater ($dW = 0$) or by stirring with paddles ($dQ = 0$) or by any combination of these.

The integrals $\int dQ$ and $\int dW$ can nevertheless be determined in special types of change. Clearly, if $dQ = 0$ or $dW = 0$ we then have $dW = -dE$ or $dQ = dE$, respectively; since dE is a total differential so also is dW or dQ under these circumstances. The determination of internal energy changes in chemical reactions from the heat absorbed in a *bomb calorimeter* exemplifies the second of these.

In practice, we are mainly interested in changes at *constant pressure* and in which all the work dW is done by the pressure P acting through the volume change dV. Clearly, from the expansion of a system at constant pressure in a cylinder and piston, done slowly without friction, we then have $dW = P\,dV$. Since $dP = 0$ we can write

$$dE + P\,dV = dE + P\,dV + V\,dP = dE + d(PV) = d(E + PV) \qquad 5.4$$

so that in such changes we are dealing with another *function of state*,

$$H = E + PV \qquad\qquad 5.5$$

called the *enthalpy*. We can regard the enthalpy as the sum of the *internal* and *external energies* of the system. The internal energy is, of course, E. The external energy is PV, which we can regard as a potential energy due to the system having pushed its surroundings away to create the volume V for itself, against a constant pressure P. We also notice, from eqns. 5.2 and 5.4, that $dQ = dH$. Hence dQ can be integrated,

$$H = \int_1^2 dQ \qquad 5.6$$

so that we can fairly regard H as the *heat content* of the system. In fact, when a chemical reaction occurs at constant pressure, the actual heat absorbed, i.e. the *heat of reaction*, is ΔH, the change of H in the reaction. Part of this heat goes into E; the rest, $P\Delta V$, into the work done as the system expands by ΔV. A *positive* ΔH means that heat is absorbed, i.e. an *endothermic* reaction; conversely a *negative* ΔH means heat is released, i.e. an *exothermic* reaction. A large negative ΔH indicates of course a strong chemical affinity between the reactants.

5.3 Standard thermodynamic properties

The internal energy, enthalpy and other thermodynamic properties of substances vary with the temperature and pressure. It is thus necessary (a) to define some *standard state* of a substance and give values of E, H, etc. in that state, (b) develop methods and provide the necessary data (specific heats, latent heats, etc.) for calculating thermodynamical properties in other states in terms of those in the standard state. The standard state of a substance is defined as the most stable state of the pure substance at 1 atm pressure and at a specified temperature; i.e. pure solid or pure liquid, or gas at 1 atm pressure. The standard value of a thermodynamic property is indicated by a small superscript $\ominus$, e.g. $H^\ominus$, and refers to 1 mole of the substance. The temperature to which it refers, usually 25°C or 298 K, is indicated by a subscript, e.g. $H^\ominus_{298}$.

The enthalpies of the chemical elements in the standard state at 298 K are conventionally taken as *zero*. Those of compounds can then be found from calorimetrically measured heats of reaction. For example, suppose that a moles of reactant A and b of B, react as follows

$$aA + bB + \Delta H \to cC + dD \qquad 5.7$$

where ΔH is the heat absorbed by the system during the reaction. Let H_A, etc. be the enthalpy per mole of A, etc. Then

$$aH_A + bH_B + \Delta H = cH_C + dH_D \qquad 5.8$$

so that H_A, for example, can be found from a knowledge of ΔH, H_B, H_C, H_D, and the amounts of reactants involved. *Hess's Law* is useful in analysing such reactions; i.e. the overall change of heat content in a reaction is the same whether the reaction takes place in one step or through a number of intermediate reactions.

As a simple example, consider the oxidation of carbon under standard conditions. We have, for 1 mole of C,

$$\underset{\text{(graphite)}}{C} + \underset{\text{(gas)}}{O_2} \rightarrow \underset{\text{(gas)}}{CO_2} + 393\ 500\ J \qquad\qquad 5.9$$

Thus $\Delta H^{\ominus}_{298} = -393\ 500$ joule per mole of CO_2 and the reaction is exothermic. Since $H^{\ominus}_{C}$ and $H^{\ominus}_{O_2}$ are zero we also have $H^{\ominus}_{CO_2} = \Delta H^{\ominus}_{298}$. Now consider

$$C + CO_2 \rightarrow 2CO - 172\ 500\ J \qquad\qquad 5.10$$

This is endothermic, with $\Delta H^{\ominus}_{298} = +172\ 500\ J\ mol^{-1}$ C, and produces two moles of CO. We have

$$-393\ 500 = 2H^{\ominus}_{CO} - 172\ 500 \qquad\qquad 5.11$$

so that $H^{\ominus}_{CO} = -110\ 500\ J\ mol^{-1}$.

To obtain heats of reaction and enthalpies at other temperatures we need to know the *molar heat capacity* C_p, i.e. the specific heat per mole at constant pressure, and also the *latent heats* when there are changes of state. The heat capacity of a substance at room temperature and above can usually be described fairly well by an empirical formula such as

$$C_p = a + bT - cT^{-2} \qquad\qquad 5.12$$

Values of a, b and c, are given in tables of thermochemical data. To raise the temperature of 1 mole from T_1 to T_2 at constant pressure, the heat

$$Q = H_2 - H_1 = \int_1^2 C_p\, dT \qquad\qquad 5.13$$

must be supplied, together with latent heat if any transitions occur.

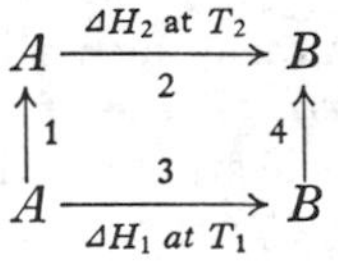

Fig. 5.1 A reaction cycle

To obtain the heat ΔH_2 at T_2 of a reaction $A \rightarrow B$ (A = reactants, B = products) in terms of ΔH_1 at T_1, we consider the cycle in Fig. 5.1. In step 1 the reactants are heated from T_1 to T_2, which requires heat $\int_1^2 C_p^A\, dT$ where C_p^A is the sum of the heat capacities of the reactants. Step 4 similarly absorbs $\int_1^2 C_p^B\, dT$. Hence, equating the heats absorbed in steps 1 + 2 with those absorbed in steps 3 + 4, we have

$$\Delta H_2 = \Delta H_1 + \int_1^2 (C_p^B - C_p^A)\, dT \qquad\qquad 5.14$$

As an example, we shall find the heat of the reaction

$$3C + Fe_2O_3 \rightarrow 2Fe + 3CO \qquad\qquad \textbf{5.15}$$

at 1540°C (1813 K) given that its value at 25°C (298 K) is 485 800 J, and given the following heat capacity data:

	a	b	c
2Fe	51·0	0·0268	0
3CO	82·8	0·0151	0
3C	33·5	0·0326	1 460 000
Fe$_2$O$_3$	103·3	0·0669	1 760 000

We obtain

$$C_p^B - C_p^A = -3\cdot0 - 0\cdot0576T + 3\,220\,000T^{-2} \qquad\qquad \textbf{5.16}$$

and hence

$$\Delta H_2 = 485\,800 - 3(T_2 - T_1) - 0\cdot0576\,(T_2^2 - T_1^2) \\ - 3\,220\,000\,(T_2^{-1} - T_1^{-1}) \qquad \textbf{5.17}$$

Substituting $T_2 = 1813$ and $T_1 = 298$, we obtain $\Delta H = 398\,000$ J at 1540°C.

The effect of *pressure* on enthalpy can usually be neglected (except at high pressures). It is shown in the kinetic theory of gases that the internal energy of a perfect gas is a function only of temperature, not pressure. Since PV is constant at constant temperature, the enthalpy also is a function only of temperature, not pressure. Real gases approximate to this fairly well except at high pressures and low temperatures. Solids and liquids are only slightly compressed by pressures of order 1 atm and their enthalpy is then hardly affected.

5.4 Combustion

The combustion of fuels is important in metallurgy as a means of producing high temperatures. The main oxidizable constituents of common fuels are carbon and hydrogen, which burn exothermically thus:

$$\underset{\text{(graphite)}}{C} + \underset{\text{(gas)}}{O_2} \rightarrow \underset{\text{(gas)}}{CO_2}; \qquad \Delta H_{298}^{\ominus} = -393\,500 \text{ J mol}^{-1}\text{ C} \qquad \textbf{5.18}$$

$$\underset{\text{(gas)}}{H_2} + \underset{\text{(gas)}}{\tfrac{1}{2}O_2} \rightarrow \underset{\text{(water)}}{H_2O}; \qquad \Delta H_{298}^{\ominus} = -285\,700 \text{ J mol}^{-1}\text{ H}_2 \qquad \textbf{5.19}$$

Usually, of course, the H_2O is produced as steam at about 1 atm. Its latent heat of vaporization at room temperature is 2 450 000 J kg^{-1}, so that 1 mol (0·018 kg) requires 44 000 J. Hence, when steam is produced (at 25°C) we replace $\Delta H^{\ominus}$ in eqn. 5.19 by $\Delta H = -241\,700$ J mol^{-1}H$_2$.

The main property of a fuel is its *calorific value*, the heat released by the complete combustion of unit mass. For carbon (1 mol $= 0.012$ kg) this is $393\,500/0.012 \simeq 33\,000\,000$ J kg^{-1}. The precise value varies slightly with the state of the carbon. For carbon in coal it is usual to take the calorific value as $34\,000\,000$ J kg^{-1} C $\simeq 14\,500$ Btu/lb C. For hydrogen burning to water, the calorific value is $142\,000\,000$ J kg^{-1} H$_2$. This is much larger than the carbon value, despite the smaller heat of reaction, because of the low mass of the hydrogen atom. The value of hydrocarbon fuel oils, due to their high calorific values and high thermal storage values, follows from this and makes them particularly suitable for transport.

The calorific value of a coal can either be measured calorimetrically or calculated from the composition. A typical analysis is:

	C	H	O	N	S	H$_2$O	Mineral Ash
Weight fraction	0.79	0.07	0.066	0.014	0.009	0.018	0.033

By forming a weighted average of the heats of combustion of the constituents, in such a coal, the *gross calorific value* of the coal is obtained. The useful practical value, the *net calorific value*, is then obtained by subtracting the heat absorbed in changing the water in the coal, including that formed by the combustion, to steam.

The use of a coal depends on the content of volatile constituents (hydrocarbons and H$_2$O); this commonly ranges from about 40 per cent in *bituminous* coal, which burns with a long flame and is used for making gas, to about 5 per cent in *anthracite*, which burns with a short flame and reaches higher temperatures. *Coke*, extremely important in iron-making and in metallurgy generally, is made by heating a coal containing about 20 per cent volatiles with just sufficient air to raise the temperature to 1000–1100°C by a little burning. The volatile constituents are driven off to be used as fuel gas and hard porous coke is left behind. To support, without crushing down, the great weight of the charge in an iron-making blast furnace, coke must have mechanical strength. Coke is expensive and coal of the right type for making it is very scarce in many parts of the world. This has important effects on iron-making practice; e.g. careful preparation of the furnace charge (by crushing and sintering; cf. § 11.4) to improve the thermal efficiency; injection of other fuels, e.g. gas, oil, powdered coal, in the (heated) air blast into the furnace; use of electrothermal and direct reduction processes (cf. § 11.1).

Oil fuels are important in many metallurgical processes, particularly in steel-making and melting furnaces; they are clean, easy to use, and give intense luminous flames. *Gaseous* fuels are also important; they are easy to use, can be smokeless and allow good control of the air-fuel mixture. The main types are *producer gas*, mainly CO and H$_2$; *town gas (coal gas, coke-oven gas*), mainly H_2 and CH_4; and *natural gas*, mainly CH_4. Producer gas is made by passing air and steam through coke or anthracite at about 1000°C. The steam keeps the temperature constant by counteracting the *exothermic* $C + O_2 \rightarrow CO_2$ and $2C + O_2 \rightarrow 2CO$ reactions with its own *endothermic* $C + H_2O \rightarrow CO + H_2$ and $C + 2H_2O \rightarrow CO_2 + 2H_2$ reactions. *Water gas* is made similarly, except that the air and steam are blown through alternately.

The amounts of oxygen and air required for combustion can be calculated from the reactions involved and from the molecular weights (O_2 32; N_2 28; H_2 2; C 12; S 32). For example, in the reaction $C + O_2 \rightarrow CO_2$, 12 kg C consumes 32 kg O_2 and produces 44 kg CO_2; for ordinary air this involves $4 \times 28 = 112$ kg N_2, i.e. 144 kg air. This is called the *theoretical air* and in practice about 40 per cent more air is required to ensure complete combustion. The air brought into the combustion chamber upstream of the fuel is called *primary air*. As it passes through the fuel it loses oxygen and, on the downstream side of the combustion zone, the shortage of oxygen may cause the endothermic $CO_2 + C \rightarrow 2CO$ reaction to set in. To prevent this, *secondary air* is introduced into the combustion chamber at this point. With a volatile long-flamed coal a third air entry, further downstream, may be required to burn the volatilized hydrocarbons.

The transfer of heat in a combustion chamber at high temperatures occurs mainly by radiation from incandescent flames. In general, flames do not give black-body radiation. Pure CO and H_2 give non-luminous flames because the molecules radiate only in narrow energy bands in the infra-red end of the spectrum. These are 'hard' flames; they are very hot and heat only where they strike. Flames become radiant and luminous when they carry small carbon particles in them, which act as black-body radiators. Sooty flames are 'soft' and relatively cool because of this loss of heat by radiation. They are used for general mild radiant heating.

The highest temperature attainable in a combustion chamber is that of the flame. The *theoretical flame temperature* is calculated by adding the heat of combustion of the flame gas to the initial heat contents of the gas and air entering the flame; and then dividing the sum by the total heat capacity at constant pressure of the products of combustion. Some values in air are: coal gas 1920°C, carbon monoxide 1950°C, hydrogen 2045°C, and acetylene 2330°C. In practice, of course, heat losses usually keep the actual temperatures well below the theoretical values. By using oxygen in place of air to avoid the wasteful heating of nitrogen, much higher temperatures can be reached: e.g. oxygen-hydrogen 2900°C at 1 atm and 3140°C at 20 atm, oxygen-acetylene 3140°C at 1 atm, aluminium powder in oxygen 3000°C, zirconium powder in oxygen 4000°C. To reach still higher temperatures, other methods have to be used, e.g. electric arcs, focused mirrors (solar furnaces), plasma jets, magnetically pinched plasma and explosive shock waves.

Flame temperatures can be usefully increased by *preheating* the fuel and air before combustion. This is particularly effective for gaseous fuels such as producer gas, which have large specific heats, but hydrocarbon fuels (e.g. town gas) cannot be preheated because they would decompose. The *recuperation* method of preheating, shown in Fig. 5.2, is a continuous counterflow process. In principle this is very efficient because the outgoing gases leave at the lowest temperature and the ingoing gases leave at the highest temperature, but it makes severe demands on the tube materials which have to operate at high temperatures in contact with hot reactive gases. The tube wall must be thin, otherwise the drop in temperature across it is large.

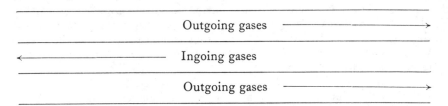

Fig. 5.2 Preheating by recuperation

The alternative is preheating by *regeneration*. Here the gases go through large chambers loosely filled with an open lattice of firebricks. The hot outgoing gas heats its chamber, while a second chamber heats the incoming gas. The flow through the two chambers is then reversed at suitable intervals. This is an intermittent process, but nevertheless can achieve high preheating temperatures since it avoids a temperature drop across walls separating outgoing and ingoing gases.

The most striking examples of regenerators are the large, tower-like *Cowper stoves* which stand alongside iron-making blast furnaces. These are heated by burning the gas (30 per cent CO) from the furnace and in turn they heat the ingoing air blast. Until early in the nineteenth century a cold air blast was used and about 8 kg of coke were used per kg of iron made. In 1828 James Neilson introduced the hot blast. This reduced the coke consumption to 5 kg immediately and to 3 kg by 1840. Some modern furnaces, increased in size and efficiency, now require only 0·6 kg of coke per kg or iron.

5.5 Furnaces

There are of course innumerable different kinds of furnaces, but they can be classified into a few main types. Those heated by *direct combustion* can be grouped according to the separation of the charge and fuel. In the first, charge and fuel are mixed. The simplest example is the *shaft furnace* or *cupola*, used for example for melting foundry iron. Greatly enlarged and sophisticated, the iron-making blast furnace is of the same type. Fig. 5.3 shows a simple shaft furnace. The charge and coke fuel are loaded at the top of the *stack*, a steel cylinder about 7 m high and 1 m radius lined with firebrick, and they slowly descend to the *hearth* against a stream of hot furnace gases coming up from air blown in at nozzles called tuyères near the bottom. Two *tapholes* are periodically opened, the lower one for molten metal, the upper for molten slag. This is a thermally efficient, counterflow furnace which can be operated continuously and is very economical, but the metal produced is very impure.

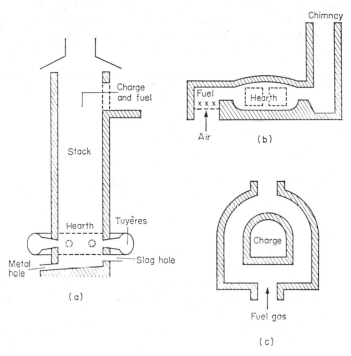

Fig. 5.3 Simple types of (a) shaft, (b) reverberatory, and (c) muffle furnace

In the second group of furnaces the charge is exposed only to the fuel gases, in a *shallow hearth*, as in the simple *reverberatory furnace* of Fig. 5.3. The charge is heated from above by flames which play on its surface. Such furnaces are used when it is necessary to manipulate the charge through the working doors (e.g. the *puddling* process for making wrought iron) or to refine the charge by making additions and taking samples. Modern furnaces may be gas or oil fired and can usually be tilted to run off the finished metal. The *open-hearth* steel-making furnace is an advanced type.

In the third group the charge is isolated from the fuel and its gases by a thermally conducting container. A simple example is a *crucible furnace*, used for melting metals and alloys. The charge is held in a pot, generally of a material with good thermal conductivity, e.g. a graphite-clay mixture, or a metal, covered with a lid and placed in a combustion chamber. Modern crucible furnaces use gas or oil firing and can be tilted for discharging. Another simple furnace in this group is the *muffle*, shown in Fig. 5.3, used for the heat-treatment of metals. Such furnaces are less thermally efficient since the charge is separated from the fuel and are limited by the capabilities of the refractory materials used to isolate the charge, but they protect the charge from contamination with impurities.

Electricity for heating is expensive but clean and adaptable. There are various types of electrical furnace, according to the method of converting electrical energy into heat. Simplest are the *resistance furnaces* in which a heating wire (e.g. nickel-chromium, platinum, molybdenum in hydrogen)

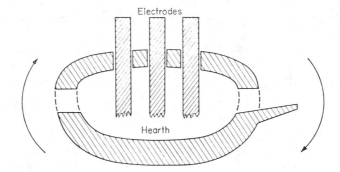

Fig. 5.4 Tilting electric arc furnace

is wrapped round a silica or other refractory tube, or a cage is made of heating elements (e.g. carbon or silicon carbide rods) and the whole is enclosed in a thick thermally insulating jacket.

For large melting and refining operations, particularly steelmaking, tilting electric arc furnaces as shown in Fig. 5.4 are used. Three-phase alternating current (e.g. 100 volts, 30 000 amps) is supplied through three carbon electrodes and arcs are struck from them on to the metal charge beneath. The electrodes are lowered into the furnace as their ends get consumed by oxidation and volatilization.

In *induction* furnaces the metal charge forms the secondary in a transformer circuit, the primary usually being a water-cooled copper tube wound round the container or crucible. Low-frequency induction furnaces are used on a large scale for melting alloys. High-frequency furnaces are used for small-scale laboratory work, for making special magnet and tool steels and for heat-treating metal surfaces.

For melting chemically reactive metals easily contaminated in air (e.g. Ti, Zr, U, Nb, Be), *vacuum furnaces* such as those in Fig. 5.5 are used. In the *cooled-crucible arc furnace* the charge is supported by a thin-walled metal (usually copper) hearth cooled by water. The electrode may be either *consumable* (i.e. made of the metal to be melted), or *non-consumable* (tungsten

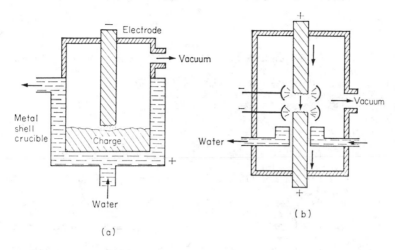

(a)

(b)

Fig. 5.5 Vacuum melting furnaces; (a) cooled-crucible arc, (b) electron-bombardment

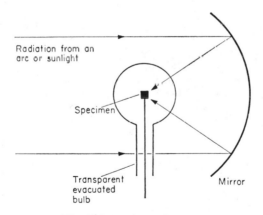

Fig. 5.6 A mirror furnace

or graphite) in which case the metal is fed into the arc through a pipe. In the *electron-bombardment* furnace the tip of a consumable electrode of the metal is melted by a beam of electrons directed on to it from a negative electrode held at about 10 000 V. The molten drops fall on to a similarly heated tip of an ingot electrode, which is slowly pulled down through a cooling jacket.

Mirror furnaces, shown in Fig. 5.6, are occasionally used for heating small specimens to very high temperatures in a vacuum.

FURTHER READING

BICHOWSKY, F. R. and ROSSINI, G. D. (1936). *The Thermochemistry of Chemical Substances*, Reinhold, New York.

DARKEN, L. S. and GURRY, R. W. (1953). *The Physical Chemistry of Metals*. McGraw-Hill, New York and Maidenhead.

FERMI, E. (1937). *Thermodynamics*. Dover Publications, London.

ETHERINGTON, H. and ETHERINGTON, G. (1961). *Modern Furnace Technology*. Griffin, London.

GILCHRIST, J. D. (1963). *Furnaces*. Pergamon Press, Oxford.

KUBASCHEWSKI, O. and EVANS, E. L. (1958). *Metallurgical Thermochemistry*. Pergamon Press, Oxford.

LEWIS, G. N. and RANDALL, M. (1923). *Thermodynamics*. McGraw-Hill, New York and Maidenhead.

STEINER, L. E. (1941). *Introduction to Chemical Thermodynamics*. McGraw-Hill, New York and Maidenhead.

THRING, M. W. (1962). *The Science of Flame and Furnaces*. Chapman and Hall, London.

TRINKS, W. and MAWHINNEY, M. (1953). *Industrial Furnaces*. John Wiley, New York and London.

Chapter 6

Entropy and Free Energy

6.1 Direction of chemical change

Metallurgy involves many problems of chemical change and equilibrium. The central problem in extraction metallurgy is to manipulate chemical and physical changes so as to turn ores into pure metals. What determines the direction of chemical change? Not the *heat of reaction*, even though many spontaneous reactions are highly exothermic; for there are also many spontaneous *endothermic* reactions, e.g. $C + CO_2 \rightarrow 2CO$ at $1000°C$.

The true criterion must involve *force* and *work* rather than *heat*. When a system is out of equilibrium and changing spontaneously, there is a one-way passage of its particles from one place or molecular arrangement to another. This systematic flow can, certainly in principle and often in practice (e.g. in engines and fuel cells), be harnessed to do mechanical work, e.g. by pushing pistons or paddles. *The system can thus do work while it is changing spontaneously.* It can no longer do work when it reaches equilibrium, because then any such flow of particles is balanced by an equal counterflow in the reverse direction, so that the paddles are pushed equally forwards and backwards.

A chemical change thus goes in the direction in which the system can release energy as work. We have then to divide the total energy in the system into two parts:

(1) that which is available to be released as work, called the *free energy*;

(2) that which is not releasable as work, called the *bound energy*.

The system thus changes spontaneously in the direction that releases its free energy and it comes to equilibrium when its remaining free energy is as small as possible.

6.2 Entropy

We need two new *functions of state* for these two sub-divisions of the total energy. As we saw in § 5.2, Q and W cannot be used for this. Consider then a *reversible change*, i.e. one in which all the variables, e.g. P, V, T, change so slowly and gradually that the system passes through a continuous series of equilibrium states. For example, we would expand a gas, not by allowing it to rush into a vacuum, but by letting it slowly push back a frictionless piston against an external pressure only marginally less than the gas pressure. Let ΔQ_a be a measured small amount of heat absorbed by the system during a step in a reversible change when the temperature is T_a. It is found experimentally that the sum of $\Delta Q_a/T_a$ over all steps $a, b, c, \ldots$ of the change, from a given initial state to a given final state, is *practically independent of the path taken between these two states*. This improves into an exact result as the change is made ideally reversible, as the steps taken are made infinitesimally small and when the temperature is measured on the perfect gas scale. We can thus define a new function of state S,

$$S_2 = S_1 + \int_1^2 \frac{dQ}{T} \qquad (reversible \; changes) \qquad \textbf{6.1}$$

called the *entropy* of the system. This is a perfectly definite property of the system, like energy and enthalpy. Since it is proportional to the heat absorbed at a given temperature, it is a *capacity* property, i.e. like mass, volume, energy and enthalpy, the amount of entropy in the system is proportional to the amount of material in the system. This contrasts with *intensive* properties such as pressure and temperature. We thus refer to the *entropy per mole* or *entropy per atom* or even to the *entropy density* in a system.

As an example, we expand and heat a mole of perfect gas from V_1, T_1 to V_2, T_2 along two different paths:

Path 1: Expand $V_1 \to V_2$ at constant T_1 and then heat $T_1 \to T_2$ at constant V_2.

Path 2: Heat $T_1 \to T_2$ at constant V_1 and then expand $V_1 \to V_2$ at constant T_2.

Consider the $T_1 \to T_2$ changes. Since the internal energy of a perfect gas is a function of temperature only and since no work $P \, dV$ is done during heating at constant volume, the entropy change due to $T_1 \to T_2$ is the same in both paths, i.e.

$$\Delta S_{T_1 \to T_2} = \int_{T_1}^{T_2} \frac{C_v \, dT}{T} = C_v \ln (T_2/T_1) \qquad \textbf{6.2}$$

where C_v is the specific heat at constant volume. Consider next the $V_1 \to V_2$ changes. The work done by the gas at temperature T is

$$\Delta W = \int_{V_1}^{V_2} P \, dV = RT \int_{V_1}^{V_2} \frac{dV}{V} = RT \ln (V_2/V_1) \qquad \textbf{6.3}$$

Since the internal energy of the gas is constant during this isothermal change, the heat ΔQ absorbed is also $RT \ln (V_2/V_1)$. The entropy change is then

$$\Delta S_{V_1 \to V_2} = \Delta Q/T = R \ln (V_2/V_1) \qquad \textbf{6.4}$$

and hence is the same whether $T = T_1$ or $T = T_2$. The entropy change is thus the same along both paths, even though more heat is absorbed along the second path than the first.

Let $S_1 = 0$ in eqn. 6.1, by definition. Then S_2 is the entropy S of the system in state 2. For a reason discussed later we choose state 1 to be the equilibrium state of the substance at 0 K (pure crystalline solid). The entropy of a system at temperature T is then

$$S = \int_0^T \frac{dQ}{T} \quad (reversible) \qquad\qquad \textbf{6.5}$$

where $dQ = C_p \, dT$ at constant P and $dQ = C_v \, dT$ at constant V. This formula enables entropies to be determined experimentally by calorimetry.

Because entropy is a function of state, a system in a given equilibrium state always has the same entropy by whatever path this state was reached. The path may have taken the system through the most violent and irreversible changes on the way. We might have raised the temperature of the system, for example, by stirring vigorously with paddles. The energy *supplied to the system as heat dQ* would then be zero, even though heat has appeared inside the system through the frictional conversion of work into heat. Since S is always the same, however, for the same equilibrium end state, we must have

$$\int_0^T \frac{dQ}{T} \, (reversible) = S > \int_0^T \frac{dQ}{T} \, (irreversible) \qquad\qquad \textbf{6.6}$$

If the system is *isolated*, then $dQ = 0$ and hence

$$dS \geqslant 0 \qquad\qquad \textbf{6.7}$$

i.e. the entropy of an isolated system can only increase (in an irreversible change) or stay constant (in equilibrium). This is the *second law* of thermodynamics.

As an example, consider a system of two parts A and B in thermal contact. Let heat dQ pass from A at temperature T_A to B at temperature T_B, where $T_A > T_B$. The entropy of A falls by dQ/T_A and that of B rises by dQ/T_B. The total entropy change is

$$dS = \frac{dQ}{T_B} - \frac{dQ}{T_A} \qquad\qquad \textbf{6.8}$$

and is *positive*, becoming zero only when $T_A = T_B$, i.e. when thermal equilibrium is reached.

6.3   Free energy

In practical problems systems are usually in thermal contact with their surroundings. It is obviously awkward to have to include all the surroundings in a criterion for equilibrium. We need a property belonging entirely to the system, not its surroundings, to define equilibrium. Suppose the system makes a small change whereby it absorbs heat dQ from its surroundings and does work dW on them. Its internal energy E then changes to $E + dE$ where $dE = dQ - dW$. Let the corresponding entropy changes in the system and its surroundings be dS and dS_x, respectively. Then $(dS + dS_x) \geqslant 0$. Let the surroundings be at the same temperature T as the system. Then $dS_x = -dQ/T$ and $[dS - (dQ/T)] \geqslant 0$. Substituting $dE + P\,dV$ for dQ and changing signs,

$$(dE + P\,dV - T\,dS) \leqslant 0 \qquad\qquad \textbf{6.9}$$

All the quantities in this belong to the system itself. The system is then in equilibrium when $dE + P\,dV - T\,dS = 0$.

Changes at *constant P and T* are of special interest. For these we can add $V\,dP - S\,dT$ to the left-hand side of the inequality in eqn. 6.9, since $dP = dT = 0$, so bringing it to the form $d(E + PV - TS) = 0$. We thus have a *new function of state*

$$G = E + PV - TS = H - TS \qquad\qquad \textbf{6.10}$$

called the *Gibbs free energy*. For spontaneous changes at constant P and T it always decreases, being releasable as work, and reaches its lowest value when the system comes to equilibrium. It is the most important thermodynamical property we shall use and it provides a guideline through innumerable metallurgical problems.

For changes at *constant V and T* the criterion becomes simpler. We have $P\,dV = 0$ and hence $(dE - T\,dS) \leqslant 0$. Adding $S\,dT$, since $dT = 0$, and developing as before, we are led to the *Helmholtz free energy*,

$$F = E - TS \qquad\qquad \textbf{6.11}$$

which plays the same role at constant V and T as the Gibbs free energy does at constant P and T. In solids and liquids at atmospheric pressure PV is often small compared with E and TS. It is then a fair approximation to neglect the PV term and so use F in place of G for changes at constant P.

Clearly, in reactions where $\Delta(TS) \ll \Delta H$ we have $\Delta G \simeq \Delta H$ and the heat of reaction then correctly points the direction of change (exothermic reactions). Since $TS \to 0$ as $T \to 0$ this condition is satisfied at low temperatures. Spontaneous endothermic reactions are those in which there is a large increase in the entropy of the system, sufficient to enable the $\Delta(TS)$ term make ΔG negative even though ΔH is positive.

To find the effect of temperature on the free energy of a system we consider an increase dT at constant P which increases the entropy and enthalpy by $dS = (C_p/T)\,dT$ and $dH = C_p\,dT = T\,dS$ respectively. Hence

$$dG = d(H - TS) = dH - T\,dS - S\,dT = -S\,dT \qquad\qquad \textbf{6.12}$$

so that the free energy falls as the temperature rises (at constant pressure), the rate of fall $-dG/dT$ being simply equal to the entropy. This result can be written in several other forms. Thus, substituting for S from eqn. 6.10, and replacing H and G by ΔH and ΔG, we obtain the *Gibbs-Helmholtz equation*

$$\frac{d(\Delta G)}{dT} = \frac{\Delta G - \Delta H}{T} \qquad \textbf{6.13}$$

for the effect of temperature on the free energy change at constant pressure.

6.4 Phase changes; vaporization

Suppose we are at a temperature and pressure such that two different *phases* (cf. § 5.1) coexist in equilibrium in the system, e.g. ice and water at 0°C and 1 atm pressure. Because they are at equilibrium the free energy of the system does not change if some of the system transforms from one phase into the other, isothermally, taking heat from or giving it to its surroundings in accordance with the latent heat of the change, ΔH per mole. It then follows from eqn. 6.10, since $\Delta G = 0$ and T is constant, that

$$\Delta S = \frac{\Delta H}{T} = \frac{\lambda}{T} \qquad \textbf{6.14}$$

i.e. the entropy change is the *latent heat* λ divided by the transition temperature. In evaluating the entropy of a system from eqn. 6.5 it is of course necessary to include any contributions ΔS due to phase changes which occur on heating the system from 0 to T.

When the two phases have different specific volumes their temperature of mutual equilibrium depends on the pressure, since the phase change can then cause work $P\Delta V$ to be done. Suppose that there is equilibrium at temperature T and pressure P. Then if a small amount of the system transforms isothermally from phase 1 to phase 2, the free energy change is zero, $\Delta G = 0$. Suppose now that this same change occurs at a slightly higher temperature $T + dT$. The free energy change is now, according to eqn. 6.13, $(-\Delta H/T)\,dT$, i.e. $(-\lambda/T)\,dT$. For equilibrium at this temperature the total free energy change has still to be zero, i.e. work $(\lambda/T)\,dT$ has to be done against a correspondingly increased pressure $P + dP$. The work done by the expansion of the system against the extra pressure dP is $(V_2 - V_1)\,dP$, where V_1 and V_2 are the volumes of the transformed material before and after the change. Equating these works,

$$\frac{dP}{dT} = \frac{\lambda}{(V_2 - V_1)T} \qquad \textbf{6.15}$$

This is the *Clausius-Clapeyron equation* for the relation between pressure and temperature at a phase change.

Let us apply it to vaporization of a solid or liquid. Usually then $V_1 \ll V_2$ and we can reasonably approximate $V_2 - V_1$ to V, the volume of the vapour. Both λ and V refer of course to the same standard amount of material, say

1 mole. Assume also that the vapour behaves as a perfect gas so that $V = RT/P$. Eqn. 6.15 then becomes

$$\frac{dP}{P} = \frac{\lambda \, dT}{RT^2} \qquad\qquad \textbf{6.16}$$

At least over small ranges of temperature λ is usually fairly constant. Assuming this constancy, we can then integrate this equation to the form

$$P = P_0 \exp\left(-\lambda/RT\right) \qquad\qquad \textbf{6.17}$$

which shows that the equilibrium vapour pressure rises, with exponential steepness, with increasing temperature. Fig. 6.1 shows this rapid rise, for the vapours of several common metals.

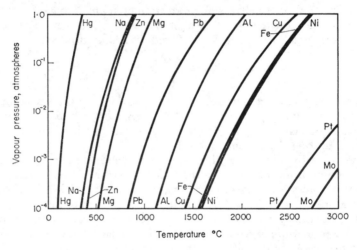

Fig. 6.1 Effect of temperature on the vapour pressures of metals

6.5 The physical nature of entropy

Thermodynamics gives a clear definition of entropy and shows us how to measure and make use of it, but does not tell us what it is. For this we have to turn to statistical mechanics. As a simple illustration suppose that we have to pack sixteen equal balls, eight black and eight white, into a box as in Fig. 6.2, while blindfolded.

Diagram (a) shows a typical distribution that might turn up in which the colours are *disordered*. Of course, the *ordered* distribution of diagram (b) might turn up but the odds against it are about 13,000 to 1. Notice that the *particular* distribution of diagram (a) is equally improbable. The difference is that this distribution is a representative of the whole large class of disordered distributions. The chance that *some* disordered distribution will turn up is very large, whereas the chance that *some* ordered distribution will turn up remains very small because there are so few of them.

We could regard this as a model of a very small alloy crystal, containing eight atoms each of two metals that are quite indifferent to one another as neighbours. When the argument is extended to sizeable crystals, with 10^{20} or more atoms, the number of disordered distributions is enormously increased but *not* that of the ordered ones, so that the crystal is then virtually *certain* to be disordered. The state of order could be determined experimentally, for example by x-ray diffraction (cf. § 17.7). Such observations would repeatedly show that the crystal was disordered. Moreover, if in some way the crystal were put into an ordered state and its atoms then allowed to exchange places randomly (which happens at high temperatures) we would see the ordered state change to the disordered one; but we would never see the reverse happen (or almost never!) since the chance that such random

(a)   (b)   (c)

Fig. 6.2

changes would extricate the crystal from the maze of disordered distributions is incredibly small. We would thus have seen a spontaneous, irreversible change from order to disorder and would have to conclude that the system had moved to an equilibrium state.

A purely statistical factor, the number of atomic distributions belonging to a thermodynamical state, thus plays a part in thermodynamics and favours disordered states. The principle applies to all kinds of disorder, e.g. the disorder of vibrational motion of atoms in a heated crystal and the disorder in atomic positions which appears when a crystal melts or a liquid boils. As a measure of disorder we could quote w, the number of distributions belonging to a thermodynamical state. However, w is often an enormously large number. Moreover, it is not a *capacity* property of a system (thus the chance of two 'ones' turning up when two dice are thrown is the *product* of the two separate probabilities, i.e. $\frac{1}{6} \times \frac{1}{6}$, not their *sum*). We use in fact $\ln w$, not w itself, to measure disorder. This is a capacity property since logarithms of products are additive. *Boltzmann* recognized that, since entropy is also a capacity property and since it is closely connected with heat, which is a disorderly form of mechanical energy, then $\ln w$ and S both refer to the same property of the system. This led to the *Boltzmann-Planck* relation

$$S = k \ln w \qquad\qquad\qquad \textbf{6.18}$$

where k ($= 1\cdot38 \times 10^{-23}$ J K^{-1}) is *Boltzmann's constant*, i.e. the gas constant R divided by Avogadro's number N_0.

As an illustration, let us use this relation to derive eqn. 6.4. We expand one mole of gas, N_0 molecules, from volume V_1 to V_2. Let the gas start in a small container 1, volume V_1, inside a large container 2, volume V_2. Remove the lid of the small container and let the gas spread into the large one. The chance that a given molecule will then afterwards be found in the small container is V_1/V_2. The chance that all will be found there is $(V_1/V_2)^{N_0}$. The relative probability of the two volume states V_2 and V_1 of the gas is then $w_2/w_1 = (V_2/V_1)^{N_0}$. Substituted into eqn. 6.18, with $N_0 k = R$, this gives

$$\Delta S = k \ln (V_2/V_1)^{N_0} = R \ln (V_2/V_1) \qquad \textbf{6.19}$$

We notice that w enters thermodynamics in the form *energy/temperature*. Why should a purely numerical factor be linked to energy and temperature? The answer lies in the following: (a) different distributions usually have different internal energies, (b) an exponential factor, like that of eqn. 6.17, reduces the chances of the high-energy distributions turning up. We discuss this factor in § 6.6.

Consider Fig. 6.2 again. If the black and white atoms belong to different elements they will show some preference, however slight, for either like or unlike neighbours; this preference appears in the interatomic bond energy and hence in the internal energy. Thus either diagram (b) or diagram (c) represents an ordered state with lower internal energy than the disordered state. The internal energy and entropy factors thus act in opposition. The first favours order, the second disorder. At sufficiently low temperatures TS is always small and the internal energy factor wins. The system is then highly ordered, in equilibrium. In the limit, at 0 K, only the most perfectly ordered state, i.e. the pure perfect crystal, is thermodynamically stable. No greater order than this is possible and so we take the entropy of this state to be zero. At higher temperatures the TS term becomes more important in the free energy and swings the equilibrium increasingly towards the more disordered, high-entropy states, even though these have higher internal energy and enthalpy.

6.6 The exponential energy distribution

The exponential energy distribution, of which eqn. 6.17 provided an example, is fundamental in all physico-chemical problems. Consider a collection of particles each of which can choose its energy from one of a set of energy levels which are, in increasing energy, $\epsilon_1, \epsilon_2, \epsilon_3 \ldots \epsilon_r \ldots$ etc. A typical distribution then has N_1 particles in ϵ_1, N_2 in $\epsilon_2 \ldots$, N_r in $\epsilon_r \ldots$ etc. where

$$\sum N_r = N \qquad \textbf{6.20}$$

summed over all r, is the total number of particles and

$$\sum \epsilon_r N_r = E \qquad 6.21$$

is the total energy.

We assume that the equilibrium distribution is that which can be formed in the greatest number of ways. Let us start with empty energy levels and put the particles in one by one. Begin with level ϵ_1. There are N choices for the first particle to go in it. Similarly, the second particle can be chosen in $N - 1$ ways. The number of ways of choosing the first two is only $N(N - 1)/2$, however, because it makes no practical difference whether particle 'a' goes in before 'b', or 'b' before 'a'. There are $N - 2$ candidates for the third choice but again we have to permute the particles in the level. Hence there are $N(N - 1)(N - 2)/3!$ ways of choosing the first three, where $3! = 3 \times 2 \times 1$. Continuing, we find that the number of ways of filling the first level is $N!/(N - N_1)!N_1!$. Similarly, the number for the second level is $(N - N_1)!/(N - N_1 - N_2)!N_2!$, the number for the third is $(N - N_1 - N_2)!/(N - N_1 - N_2 - N_3)!N_3!$, and so on up to the last occupied level, which can be filled in only one way. The total number w of ways of filling all the levels is the product of all these, i.e.

$$w = \frac{N!}{N_1!N_2!N_3!\ldots N_r!\ldots} \qquad 6.22$$

If $N_1, N_2 \ldots$ etc. are such that w is a maximum, *small* changes in their values must leave w unchanged. We can make only those changes that keep N and E constant. Consider any three equally spaced energy levels, $\epsilon_a, \epsilon_b, \epsilon_c$, where

$$\epsilon_c - \epsilon_b = \epsilon_b - \epsilon_a \qquad 6.23$$

To make such a change take two particles out of ϵ_b, put one in ϵ_a, the other in ϵ_c. This changes $N_a!N_b!N_c!$ into

$$(N_a + 1)!(N_b - 2)!(N_c + 1)!$$

If w is to remain unchanged, these must be equal, i.e.

$$(N_a + 1)(N_c + 1) = (N_b - 1)(N_b - 2)$$

If N_a, N_b, N_c, are large numbers, this becomes approximately $N_c/N_b = N_b/N_a$, i.e.

$$\ln N_c - \ln N_b = \ln N_b - \ln N_a \qquad 6.24$$

Eqns. 6.23 and 6.24 are simultaneously satisfied if $\ln N_a$ is proportional to $\epsilon_a \ldots$ etc. Thus, generalizing,

$$N_r = A \exp(-\mu\epsilon_r) \qquad 6.25$$

where A is obtained by summing over all energy levels, i.e.

$$\sum N_r = N = A \sum \exp(-\mu\epsilon_r) \qquad 6.26$$

and μ is a constant. The minus sign appears because for w to be a maximum the highest energy levels must be least populated.

We now find μ. First, we consider *two independent* systems which we denote by the superscripts $'$ and $''$ respectively. We repeat the above analysis and derive the above formulae for each. Now let the two systems be placed in thermal contact. We assume that they are in thermal equilibrium when $w'w''$ is a maximum. We move a particle in the first system from ϵ'_a to ϵ'_b and one in the second from ϵ''_x to ϵ''_y, where

$$\epsilon'_b - \epsilon'_a = \epsilon''_x - \epsilon''_y \qquad 6.27$$

This conserves N', N'', and the *total* energy $E(= E' + E'')$. We equate $w'w''$, before and after, and assume that all the N'_a and N''_x, etc. are large numbers to obtain in the same way as before

$$\ln N'_b - \ln N'_a = \ln N''_x - \ln N''_y \qquad 6.28$$

Using eqn. 6.25, this can be written in the form

$$-\mu'(\epsilon'_b - \epsilon'_a) = -\mu''(\epsilon''_x - \epsilon''_y) \qquad 6.29$$

and it then follows from eqn. 6.27 that $\mu' = \mu''$, i.e. μ *is the same for all systems in thermal equilibrium.*

It is sufficient then to find μ for one simple system. Consider a perfect gas at temperature T. The kinetic theory of gases shows that each particle on average has a kinetic energy $\frac{1}{2}kT$ for each of the three directions of space in which it moves. Consider one of these directions. The average energy is then, from eqns. 6.25 and 6.26,

$$\bar{\epsilon} = \frac{\sum \epsilon_r \exp(-\mu\epsilon_r)}{\sum \exp(-\mu\epsilon_r)} \qquad 6.30$$

We write $\epsilon_r = \frac{1}{2}mv^2$, where m is the mass and v the velocity of the particle, and replace the sums by integrals, to obtain

$$\bar{\epsilon} = \frac{\int_{-\infty}^{+\infty} \frac{1}{2}mv^2 \exp(-\frac{1}{2}\mu mv^2)\, dv}{\int_{-\infty}^{+\infty} \exp(-\frac{1}{2}\mu mv^2)\, dv} = \frac{1}{2\mu} \qquad 6.31$$

where we have used the standard integrals

$$\int_{-\infty}^{+\infty} x^2 \exp(-ax^2)\, dx = \frac{1}{2a}\left(\frac{\pi}{a}\right)^{1/2} \quad \text{and} \quad \int_{-\infty}^{+\infty} \exp(-ax^2)\, dx = \left(\frac{\pi}{a}\right)^{1/2} \qquad 6.32$$

Hence $\mu = 1/kT$ and the energy distribution law then becomes

$$\frac{N_r}{N} = \frac{\exp(-\epsilon_r/kT)}{\sum \exp(-\epsilon_r/kT)} \qquad 6.33$$

This is known as the *Maxwell-Boltzmann*, or *Gibbs*, or *classical*, energy distribution. It gives us a simple picture of *dynamical equilibrium* between the various parts of the system. Consider, for example, the vapour pressure formula, eqn. 6.17. Particles evaporate from liquid to vapour and condense from vapour to liquid. In equilibrium, the fluxes of particles in these two opposite directions are equal. A particle cannot leave the liquid unless it has acquired an extra energy, as represented by the latent heat λ per mole. Only those particles with energies $\epsilon_r \geqslant (\lambda/N_0)$ can leave. The flux of those leaving the liquid is thus proportional to $n_L \exp(-\lambda/RT)$, where n_L is the number of particles per unit volume of liquid. There is no exponential energy restriction on the passage of particles from the vapour to the liquid since these are moving into a state of lower energy. Hence, for equal fluxes, the number n_V of particles in unit volume of vapour is smaller than that in the liquid by the factor $n_V/n_L = \exp(-\lambda/RT)$ approximately. As a second example, consider the endothermic dissociation of molecules. Only those that have energies greater than the heat of dissociation can dissociate. Again the exponential energy factor correspondingly reduces the flux of dissociations. For equilibrium the number per unit volume of dissociated molecules has to be correspondingly reduced.

6.7 Activation energy

Systems often exist for long times in states *not* of lowest free energy. For example, mixed hydrogen and oxygen gases can stay uncombined at room temperature even though the stable state is water. Again, the structure of quench-hardened steel is unstable, yet steel swords hardened in ancient times keep their structures and properties to this day. Such systems are said to be *metastable*. Metallurgy itself depends on metastability since the equilibrium state of most metals in the atmosphere is the oxide.

It frequently happens that to rearrange a group of atoms into a new and more stable structure the atoms must first pass through some states of higher energy; for example, states in which the initial atomic bonds have been broken and the new ones not yet formed. As a simple illustration, let Fig. 6.3 represent the potential energy of an atom as a function of its position. The movement of the atom from the metastable position (a) to the stable one (c) is opposed by the energy barrier (b). In general, the *tunnel effect* (cf. Fig. 3.4) plays little part in such problems because the mass of the atom is large enough to make the tunnelling rate negligible for the types of energy barriers commonly encountered in physico-chemical problems. Thus, until the atom can temporarily acquire the necessary extra energy to carry it *over* the barrier, it must remain in the metastable position. The smallest energy ϵ which will allow it to go over is the *activation energy* of the reaction.

In most problems of practical interest, $\epsilon \gg kT$. The rate of escape over such a barrier is then very small and a particle held up by the barrier spends most of its time near the minimum at (a). We consider only those movements of the particle in the direction of the barrier and hence analyse Fig. 6.3 as a one-dimensional problem. For small displacements, x, about the minimum at (a), the potential energy $\epsilon(x)$ rises as x^2. This gives a *simple harmonic motion*, since the restoring force, $d\epsilon/dx$, then increases linearly with x. We thus approximate the actual motion in (a) by a simple harmonic motion with the same frequency ν at small amplitudes. The particle thus makes ν runs per second at the barrier. Such a run is successful when the particle has an energy ϵ or more. A property of simple harmonic motion is that the frequency ν is independent of amplitude. This leads to the result, in quantum mechanics, that the quantized energy levels of the motion are spaced at intervals $h\nu$, where h is Planck's constant. We wish to know the probability

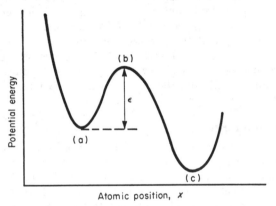

Fig. 6.3 A simple energy barrier

that the particle has an energy $\epsilon + nh\nu$, where $n = 0, 1, 2$ etc. Substituting into eqn. 6.33 we obtain for this probability

$$\frac{\sum \exp\left[-(\epsilon + nh\nu)/kT\right]}{\sum \exp\left(-nh\nu/kT\right)} = \frac{\exp\left(-\epsilon/kT\right)\sum \exp\left(-nh\nu/kT\right)}{\sum \exp\left(-nh\nu/kT\right)} \qquad \textbf{6.34}$$
$$= \exp\left(-\epsilon/kT\right)$$

Hence the frequency, f per second, of crossing the barrier is

$$f = \nu \exp\left(-\epsilon/kT\right) = \nu \exp\left(-E/RT\right) \qquad \textbf{6.35}$$

where E is the activation energy per mole of particles and R is the gas constant.

This formula is of great importance in many problems. To fit it to experimental observations we write it in the form,

$$\ln(\text{rate}) = \text{constant} - E/RT \qquad \textbf{6.36}$$

to show that the logarithm of the rate of change should vary linearly with the reciprocal of the absolute temperature. The slope of this line gives E/R and the intercept at $T^{-1} = 0$ gives the constant.

The great effect of temperature on reaction rates is due to the exponential energy factor. Suppose for example that $E = 166\,000$ J mol^{-1}, which is a fairly typical value in metallurgical processes. Taking $R = 8 \cdot 314$ J mol^{-1} K^{-1}, we obtain exp $(-E/RT) \simeq$ exp $(-166\,000/2500) \simeq 10^{-29}$ at 300 K and exp $(-166\,000/8314) \simeq 10^{-9}$ at 1000 K. A reaction taking 1 sec at 1000 K would then take about 3×10^{12} years at room temperature! This is the basis of *quenching* for producing metastable systems. When a system exists at high temperatures in a state different from that at low temperatures, and the activation energy for the change of state is large enough, it is possible by cooling at a sufficient speed from the high temperature to preserve the high-temperature state in a metastable form at room temperature. In the quench-hardening of carbon steel the red-hot metal has to be plunged into cold water or brine. In the cooling of hot glass to the vitreous state, even furnace cooling is usually fast enough to act as a quench because crystallization (i.e. *devitrification*) is too sluggish in this material.

Eqn. 6.35 applies strictly only when the number of energy states available to the particle is the same at (a) and (b). If there are w states at (b) for every one at (a), then the probability of the particle acquiring the activation energy E is correspondingly increased by the factor w. Using eqn. 6.18 we can interpret this as an *entropy of activation*. To take proper account of such effects, in fact, we must replace the concept of *potential energy* of activation E, by that of *free energy of activation* G, where G is the work per mole to transfer atoms from (a) to (b) (reversibly and isothermally). Then eqn. 6.35 becomes, using eqn. 6.10,

$$f = \nu \exp\left(-G/RT\right) = \nu \exp\left(S/R\right) \exp\left(-H/RT\right) \qquad \textbf{6.37}$$

where S is the *entropy of activation*.

The quantity measured from the plot of $\ln(rate)$ against *reciprocal temperature* is then

$$\frac{d \ln f}{d(1/T)} = -\frac{1}{R}\frac{d(G/T)}{d(1/T)} = \frac{T^2}{R}\frac{d(G/T)}{dT} \qquad \textbf{6.38}$$

This can be simplified. Since $G = H - TS$ and $dH/T = dS$ we have $d(G/T) = dH/T + Hd(1/T) - dS = -(H/T^2)dT$ and hence

$$\frac{d \ln f}{d(1/T)} = -\frac{H}{R} \qquad \textbf{6.39}$$

The experimental plot thus always gives the *enthalpy of activation* or *heat of activation* at the temperature concerned, even when H varies with temperature. The effect of a change $(dH/dT)\Delta T$ in H over a small temperature range ΔT is exactly compensated by the effect on the rate of the corresponding change $(dS/dT)\Delta T$ in S.

FURTHER READING

As for Chapter 5; also:

GURNEY, R. W. (1949). *Introduction to Statistical Mechanics*. McGraw-Hill, New York and Maidenhead.

HINSHELWOOD, C. N. (1951). *The Structure of Physical Chemistry*. Oxford University Press, London.

LANDAU, L. and LIFSHITZ, E. M. (1959). *Statistical Physics*. Pergamon Press, Oxford.

SLATER, J. C. (1939). *Introduction to Chemical Physics*. McGraw-Hill, New York and Maidenhead.

ZENER, C. (1950). Chapter in *Thermodynamics in Physical Metallurgy*. American Society of Metals.

Chapter 7

Free Energies of Metallic Compounds

7.1 The Ellingham diagram

Suppose that a moles of a reactant A, with free energy G_A per mole, etc., react as follows,

$$aA + bB \rightarrow cC + dD \qquad \textbf{7.1}$$

Then the *free energy of reaction*, ΔG per mole of A, is given by

$$\Delta G = \frac{c}{a} G_c + \frac{d}{a} G_D - G_A - \frac{b}{a} G_B \qquad \textbf{7.2}$$

This is, of course, negative for an irreversible, spontaneous reaction. Free energies of most simple substances have now been measured, for example by calorimetry, electrochemistry, vapour pressure methods and spectroscopy, and tabulated in several books: e.g. *Metallurgical Thermochemistry*, Kubaschewski and Evans; *Selected Values of Thermodynamical Properties of Metals and Alloys*, Hultgren *et al.* They are invaluable for studying the feasibility of metallurgical processes. Knowing the free energies of the reactants and products, the direction and free energy change of an unknown reaction can then be found from eqn. 7.2 or its like.

The most convenient form of presenting the free energy data is by means of *Ellingham diagrams*, as in Figs. 7.1 and 7.2. An advantage of these is that they give the free energy released by the combination of a *fixed amount* (1 mol) of the oxidizing agent. The relative affinities of the elements for this agent are thus shown directly. In other tabulations, e.g. based on unit amount of *product*, these affinities are not always immediately obvious. For example the values

$$\text{Mg} + \tfrac{1}{2}\text{O}_2 = \text{MgO}; \qquad \Delta G^{\ominus}_{298} = -600\ 000 \text{ J mol}^{-1} \text{ oxide}$$
$$2\text{Al} + \tfrac{3}{2}\text{O}_2 = \text{Al}_2\text{O}_3; \qquad \Delta G^{\ominus}_{298} = -1\ 600\ 000 \text{ J mol}^{-1} \text{ oxide} \qquad \textbf{7.3}$$

disguise the fact that magnesium can, in principle, reduce aluminium oxide at room temperature.

We see from the Ellingham diagram that metals mostly *release* free energy during oxidation. They thus mostly occur in nature in the oxidized state. The easily reducible, *noble* metals occur at the top of Fig. 7.1 and the highly reactive metals towards the bottom. This is not far from the historical order of discovery and use of the metals. The sulphide diagram is fairly similar,

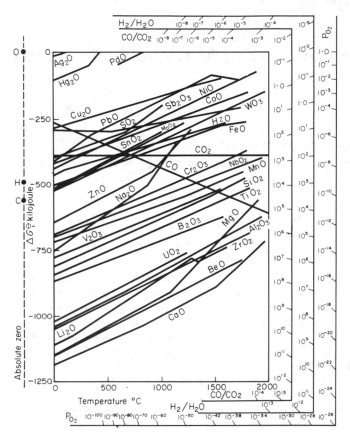

Fig. 7.1 Increase $\Delta G^{\ominus}$ in free energy when one mol of gaseous oxygen (O_2) at 1 atm pressure combines with a pure element to form oxide. (H. J. T. Ellingham, *J. Soc. Chem. Ind. Trans.* **63**, 125, 144; F. D. Richardson and J. H. E. Jeffes, 1948, *J. Iron Steel Inst.*, **160**, 261; L. S. Darken and R. W. Gurry, 1953, *Physical Chemistry of Metals*, McGraw-Hill; D. T. Livey, 1959, *J. less common Metals*, **1**, 145.)

with generally weaker chemical binding in accordance with the smaller electronegativity of sulphur compared with oxygen (cf. Table 4.1). Most metals form strongly ionic oxide crystals but their sulphides are not particularly ionic, apart from those of the very electropositive metals and a few transition metals (e.g. MnS).

The free energies of chlorides and a number of other metallic compounds have also been measured. The order of increasing thermodynamical stability of the chlorides at room temperature is:

Au, H, As, Hg, Cu, B, Bi, Ag, Sb, Sn, Fe, Si, Ni, Co, Pb, Tl, Cd, Zn, Cr, Ti, Al, Be, Mn, U, Zr, V, Mg, Th, Ce, Ca, Sr, Li, Na, Ba, K.

We notice that the monovalent alkali metals, which bond equi-atomically to the halides, form very stable chlorides, whereas their affinity for oxygen is not outstanding (apart from lithium).

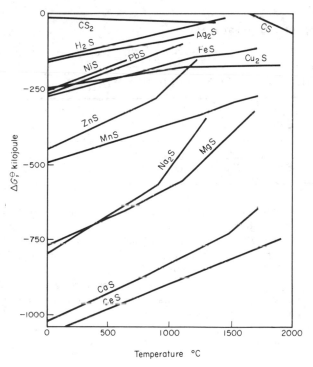

Fig. 7.2 Increase $\Delta G^\ominus$ in free energy when one mol of sulphur (S_2) combines with a pure element to form sulphide. (F. D. Richardson and J. H. E. Jeffes, *loc cit.*)

Several metals at both the top and bottom of the oxide diagram resist oxidation in air at room temperature. At the top this is simply because of their low affinity for oxygen. Those at the bottom (i.e. Al, Ti, Zr) owe their resistance to the impermeability of the thin, but very stable and coherent, oxide films that form on them. The affinity of these metals for oxygen is vividly demonstrated by their pyrophoricity when powdered and heated.

Reactive metals are used as oxygen 'getters' and deoxidizers. For example, titanium is sometimes used in vacuum systems to remove the last traces of oxygen; aluminium and silicon are likewise added to steel at the end of the steelmaking process. In purification by *selective oxidation*, oxygen is used to remove reactive impurities from a relatively noble metal (*fire refining*). The removal of antimony from molten lead is an example. The Ellingham diagram shows that the antimony impurity can capture the oxygen that enters the lead, when this is heated in air, and that the lead itself will remain unoxidized until the antimony has gone (floating off as oxide). Steelmaking is a sophisticated form of refinement by selective oxidation.

7.2   Effect of temperature

We see from Fig. 7.1 and Table 7.1 that the free energy approximates

TABLE 7.1. HEAT OF OXIDATION, $-\Delta H^{\ominus}_{298}$, kJ mol^{-1} O_2

Cu_2O	CO_2	PbO	FeO	SnO_2	ZnO
340	393.5	436	536	579	698

Na_2O	TiO_2	UO_2	Al_2O_3	MgO	CaO
850	930	1100	1113	1204	1271

to the heat of oxidation for several reactions at room temperature; this is thus a 'low' temperature for these reactions in the sense that TS plays only a minor role. At high temperatures things are very different. Most oxides are much *less* stable, some considerably so (e.g. MgO, ZnO), whereas CO_2 is unchanged and CO is much *more* stable.

These effects of temperature are due to the different entropies of the various participants in the reactions and their effects on the free energy, acting through eqn. 6.12. We recall that a pure perfect crystal at 0 K has no entropy so that near absolute zero the free energy is independent of temperature. As the temperature is raised the entropy also rises, as

$$S = \int_0^T \frac{C_p\, dT}{T} + \sum \frac{\text{latent heats}}{\text{transition temperatures}} \qquad \textbf{7.4}$$

at constant pressure. The free energy correspondingly falls, slightly at first and then more and more steeply, since $dG/dT = -S$. This fall is sharply accentuated at the melting and boiling points, in proportion to the latent heat divided by the transition temperature.

Table 7.2 gives some entropies at room temperature. We notice the small values for solids and the large ones for gases, as expected. The particularly low value for diamond is due to the strong atomic forces and small atomic mass in this crystal, which give a very low specific heat (cf. § 19.2).

TABLE 7.2. STANDARD ENTROPIES, $S_{298}^{\ominus}$, J mol^{-1} K^{-1}

C (diamond)	C (graphite)	Fe	Al	Cu	Al$_2$O$_3$	FeO	H$_2$O (liquid)
2·4	5·7	27·3	28·3	33·1	50·9	57·5	69·9

Hg (liquid)	H$_2$	Hg (vapour)	H$_2$O (vapour)	N$_2$	CO	O$_2$	CO$_2$
76·0	130·6	174·8	188·7	191·5	197·6	205·0	213·6

Clearly, if the free energy of reaction is to change strongly with temperature, *gas* must be either produced or consumed in the reaction. When

$$\text{metal} + \text{oxygen} \rightarrow \text{oxide} \qquad\qquad \textbf{7.5}$$
$$\text{\footnotesize(solid)}\quad\ \ \text{\footnotesize(gas)}\qquad\ \ \ \text{\footnotesize(solid)}$$

we have $\Delta G = -T(S_{oxide} - S_{metal} - S_{oxygen})$ and, since the entropy of oxygen gas is high, ΔG *rises* with rising temperature, i.e. the oxide becomes less stable. The entropy of solid oxide usually exceeds that of the solid metal by not more than about 20 J mol^{-1} K^{-1} so that $d(\Delta G)/dT$ is generally about 180 J mol^{-1} K^{-1}, i.e. almost equal to the entropy of the oxygen. Thus most of the free energy lines in Fig. 7.1 slope upwards at about this value. The sudden changes of slope are due to changes of state. Melting has a fairly small effect, and boiling and sublimation have large effects (cf. Mg, b.p. = 1103°C; Zn, b.p. —907°C; Li, b.p. = 1330°C; H$_2$O, b.p. = 100°C; PbO, b.p. = 1470°C; Li$_2$O, subl. p. —1300°C).

The slopes of the carbon oxidation lines are particularly important. Consider C + O$_2$ → CO$_2$. From Table 7.2 we have $\Delta S = 213\cdot6 - 5\cdot7 - 205\cdot0 = 2\cdot9$. This is very small, since one mole of gas is consumed and one mole produced, so that the CO$_2$ free energy line is nearly horizontal. However, in 2C + O$_2$ → 2CO, two moles of gas are produced and the reaction is favoured by high temperature. We have $\Delta S = (2 \times 197\cdot6) - (2 \times 5\cdot7) - 205\cdot0 = +179$ and the CO free energy line has a *downward* slope of this value.

This is of the utmost importance in extraction metallurgy. It means that carbon, a cheap abundant fuel, has a greater affinity for oxygen than most substances at high temperatures and so can be widely used as a *reducing agent* for extracting metals from oxide ores. By contrast, the H$_2$O free energy line slopes upwards since, even above 100°C, three moles of gas are required to produce two, i.e. 2H$_2$ + O$_2$ → 2H$_2$O. Thus hydrogen is a poor reducing agent in high-temperature metallurgy even though, as the low-temperature data show, its inherent affinity for oxygen is greater than that of carbon.

7.3   Dissociation temperature and pressure

At its *standard dissociation temperature* an oxide is in equilibrium (i.e. $\Delta G = 0$) with the pure element and oxygen at 1 atm pressure (i.e.

$\Delta G = \Delta G^{\ominus}$). It is thus the temperature at which $\Delta G^{\ominus} = 0$ in Fig. 7.1. Metallic oxides can in principle be reduced to the metal simply by heating in air at this temperature. Some values are: Au < 0°C, Ag 185°C, Hg 430°C, Pt group of metals 800–1200°C, most others 2000–5000°C. Apart from the noble metals there is thus no possibility of reducing the oxides by simple heating in ordinary furnaces. However, dissociation is possible in electric arcs, mirror furnaces and plasma jets.

Since one at least of the participants is a gas (O_2), the dissociation temperature generally depends on pressure. From eqn. 6.3, the work to compress 1 mol of perfect gas isothermally from pressure P_1 to P_2 is $RT \ln (P_2/P_1)$. Hence, the free energy $G(P)$ of the gas at pressure P (atm) is given by

$$G(P) = G^{\ominus} + RT \ln P \qquad\qquad 7.6$$

per mol. Consider reaction 7.5 again. At 1 atm O_2 we have

$$\Delta G^{\ominus} = G^{\ominus}_{oxide} - G^{\ominus}_{metal} - G^{\ominus}_{oxygen}$$

Repeat the reaction at P atm oxygen. This has a negligible effect on $G^{\ominus}_{oxide}$ and $G^{\ominus}_{metal}$. Hence, practically,

$$\Delta G = \Delta G^{\ominus} - RT \ln P \qquad\qquad 7.7$$

The system is in equilibrium when $\Delta G = 0$. Thus

$$P_{O_2} = \exp (\Delta G^{\ominus}/RT) \qquad\qquad 7.8$$

is the *equilibrium dissociation pressure* of the oxide at the temperature T. At higher oxygen pressures the oxide is stable; at lower, the metal.

Consider, for example, Ti + O_2 = TiO_2 at 1250°C. We have $\Delta G^{\ominus} \simeq 600\,000$ J mol^{-1} from Fig. 7.1. and $RT \simeq 12\,000$ J. Hence $P_{O_2} \simeq \exp(-600\,000/12\,000) \simeq 10^{-22}$ atm. This is far beyond the scope of present-day vacuum equipment so that titanium will always oxidize if heated at this temperature in the absence of a strong reducing agent.

It is often necessary to anneal metals at high temperatures without disfiguring their surfaces by oxidation, i.e. to *bright anneal* them. The common metallic oxides however generally have very low dissociation pressures at ordinary annealing temperatures. For instance, $\Delta G^{\ominus} \simeq -376\,000$ for 2Fe + O_2 = 2FeO at 900°C so that, with $RT \simeq 9830$, $P_{O_2} \simeq 10^{-17}$ atm. Simple heating in a conventional vacuum or inert gas of conventional purity is then unsatisfactory. It is necessary to include a reducing gas in the furnace atmosphere. Thus *controlled atmospheres* are generally necessary for bright annealing; the main ones are *cracked ammonia* (75 per cent H_2, 25 per cent N_2), *burnt ammonia* (20 per cent H_2, 80 per cent N_2) and *partly burnt fuel gas* (CO, CO_2 and H_2 in N_2).

7.4   The equilibrium constant

The standard free energy change can be used to calculate the *equilibrium constant* of a reaction. The method is a direct generalization of that leading to eqn. 7.8. Consider a reaction between various gases, which we assume to be perfect. For gas I in this mixture we have

$$P_I V = N_I RT \text{ and } G_I = G_I^\ominus + RT \ln P_I$$

where N_I is the number of moles of I present, P_I is the *partial pressure* of I and G_I the free energy per mol of I. The total pressure is $P = \Sigma\, P_I$, summed over all gases. Suppose that, at constant T and P, a moles of $A \ldots$ etc. are consumed and m moles of $M \ldots$ etc. produced in the reaction. The free energy change is

$$\Delta G = mG_M + nG_N + \cdots - aG_A - bG_B - \cdots \qquad \textbf{7.9}$$

i.e.

$$\Delta G = \Delta G^\ominus + \sum_i iRT \ln P_I \qquad \textbf{7.10}$$

where $I = A, B \ldots M, N \ldots$ etc. and $i = -a, -b \ldots +m, n \ldots$ etc. Rearranged, this is

$$\frac{P_M^m P_N^n \cdots}{P_A^a P_B^b \cdots} = \exp\left[(\Delta G - \Delta G^\ominus)/RT\right] \qquad \textbf{7.11}$$

If the gases are in equilibrium $\Delta G = 0$ and

$$\frac{P_M^m P_N^n \cdots}{P_A^a P_B^b \cdots} = \exp\left(-\Delta G^\ominus/RT\right) = K_p \qquad \textbf{7.12}$$

This is the *law of mass action* and K_p is the *equilibrium constant at constant pressure*.

The important relation

$$\Delta G^\ominus = -RT \ln K_p = -19 \cdot 14\, T \log_{10} K_p \qquad \textbf{7.13}$$

is called the *van't Hoff isotherm*.

Suppose that there is some solid or liquid M in the system, as well as the gases, in equilibrium with the vapour M. Because of this equilibrium, P_M is the equilibrium vapour pressure of solid or liquid M and its value is independent of the reacting gases since it is a property of the substance M alone. The gas equilibrium has to include this value of P_M, which is constant at constant temperature. We thus have $G_M \equiv G_M^\ominus$ for the vapour M. It follows that we must replace mG_M by $mG_M^\ominus$ in eqn. 7.9 and this removes the term $mRT \ln P_M$ from the sum in eqn. 7.10 and so removes P_M^m from eqn. 7.11 and 7.12. As an example, consider reaction 7.5 again. We can regard this as a gas reaction between metal and oxide vapours and oxygen. We have

$$P_{oxide}/P_{metal}P_{oxygen} = \exp\left(-\Delta G^\ominus/RT\right)$$

but when the metal and oxide are present also as pure solids or liquids this reduces, by the above argument, to eqn. 7.8.

7.5 Oxidizing—reducing gas mixtures.

Let us now heat a metal in a given CO/CO_2 mixture. Does it oxidize? The reaction is

$$\text{metal} + 2CO_2 = \text{oxide} + 2CO - \Delta G^\ominus \qquad \textbf{7.14}$$
$$\text{(solid)} \quad \text{(gas)} \qquad \text{(solid)} \quad \text{(gas)}$$

per mol of oxygen. We break this down into three reactions, all represented in Fig. 7.1,

$$\begin{aligned}
&(1)\ \text{Metal} + O_2 = \text{Oxide} - \Delta G_1^\ominus \\
&(2)\ 2C + O_2 = 2CO - \Delta G_2^\ominus \\
&(3)\ C + O_2 = CO_2 - \Delta G_3^\ominus
\end{aligned} \qquad \textbf{7.15}$$

thus,

$$\underbrace{\text{metal} + O_2}_{(1)} + \underbrace{2C + O_2}_{(2)} + \underbrace{2CO_2}_{(3)} = \underbrace{\text{oxide}}_{(1)} + \underbrace{2CO}_{(2)} + \underbrace{2C + 2O_2}_{(3)} \quad \textbf{7.16}$$

to obtain $\Delta G^\ominus = \Delta G_1^\ominus + \Delta G_2^\ominus - 2\Delta G_3^\ominus$. From eqn. 7.12 the system is in equilibrium at the CO/CO_2 ratio given by

$$\Delta G_1^\ominus = 2\Delta G_3^\ominus - \Delta G_2^\ominus - 2RT \ln (P_{CO}/P_{CO_2}) \qquad \textbf{7.17}$$

This is written in a form convenient for use with the Ellingham diagram. We can plot the right-hand side against temperature T in Fig. 7.1; where this straight line crosses the $\Delta G_1^\ominus$ line is the equilibrium point. Below this temperature the metal is oxidized; above, it is not. The CO/CO_2 scale round the sides of Fig. 7.1 is based on this plot. We draw a straight line from point C, through the $\Delta G_1^\ominus$ value for the oxide concerned at the temperature of interest, to the CO/CO_2 scale, where it gives the equilibrium CO/CO_2 ratio at this temperature. As an example, for $Zn + CO_2 = ZnO + CO$ at 640°C this ratio is 1000:1.

Fig. 7.1 also gives similar scales for the P_{O_2} equilibrium pressure, based on eqn. 7.8, and for H_2/H_2O mixtures. These scales are used in the same way as the CO/CO_2 scale, but with straight lines drawn from points O and H respectively. The H_2/H_2O scale refers to the

$$\text{metal} + 2H_2O = \text{oxide} + 2H_2 \qquad \textbf{7.18}$$

equilibrium.

7.6 Solutions

The Ellingham diagram does not deal with solid or liquid *solutions*, since these are not standard states. When a substance exists as a *solute* we can regard it as in a kind of 'gaseous' state, with the *solvent* providing the 'space' in which it exists. The vapour pressure of the solute, above the solution, is generally lower than that of the same substance as pure solid or liquid since there are fewer particles of the substance per unit volume in the solution to maintain dynamical equilibrium with the vapour (cf. § 6.6).

It is convenient then to define the *activity* a_I of the dissolved substance I as

$$a_I = \frac{\text{Vapour pressure of } I \text{ above solution}}{\text{Vapour pressure of } I \text{ in standard state}} \qquad \textbf{7.19}$$

the standard state being at the same temperature as the solution. If the activities are such that the vapours of several solutes are in mutual equilibrium, then of course the solutes are also in mutual equilibrium inside the solvent itself.

Solutions are usually *non-ideal*, i.e. they behave as if they contain either more or less solute than they actually do. In an *ideal* solution the mutual interactions between the constituent particles are such that a solute I has a vapour pressure P_I directly proportional to its *concentration* c_I, defined by

$$c_I = \frac{i}{\sum i} \qquad \textbf{7.20}$$

where the sum is taken over the number of moles, $i = a, b, c \ldots i \ldots$ of every component $A, B, C \ldots I \ldots$ in the solution. For an ideal solution, then,

$$P_I = P_I^{\ominus} c_I \qquad \textbf{7.21}$$

where $P_I^{\ominus}$ is the vapour pressure of I in the standard state at the same temperature. At high temperatures (e.g. in steelmaking at 1600°C) many alloy solutions, e.g. Fe-Ni, Fe-Mn, Fe-Co, and many ionic melts, e.g. FeO-MnO, FeS-FeO, Fe_2SiO_4-Mg_2SiO_4, behave almost ideally.

For a *non-ideal* solution we replace c_I by the activity a_I, so that

$$P_I = P_I^{\ominus} a_I \qquad \textbf{7.22}$$

We can thus think of a_I as an '*effective concentration*'. Consider, for example, copper dissolved in molten iron at 1600°C. One atomic per cent (i.e. $c_{Cu} = 0.01$) gives a measured copper vapour pressure over the solution of 8.8×10^{-5} atm. Pure liquid copper (the standard state) at this temperature has a vapour pressure of 1.1×10^{-3} atm. Hence $a_{Cu} = 8.8 \times 10^{-5}/1.1 \times 10^{-3} = 0.08$, i.e. the effective concentration of copper is eight times the real concentration.

Fig. 7.3 shows other examples. Manganese dissolves almost ideally in iron, i.e. it obeys *Raoult's law*

$$a = c \qquad \textbf{7.23}$$

Copper gives a strong *positive* deviation from ideality, showing that Fe and Cu atoms have little mutual affinity. Silicon gives a strong *negative* deviation, showing a tendency to form an FeSi compound.

As $c \to 1$ these solutions eventually become dilute solutions of Fe in Cu, Mn and Si. In this limit, the activities of Cu, Mn and Si, all become ideal, i.e. $a \to c$ as $c \to 1$, and Raoult's law is obeyed. At the other extreme, $c \to 0$, we have dilute solutions in iron. In this limit, a and c become proportional,

$$a_I = \gamma_I^\circ c_I \qquad\qquad 7.24$$

where γ_I° is independent of c_I. This is *Henry's law*. It greatly simplifies the theory of dilute solutions, such as molten steel.

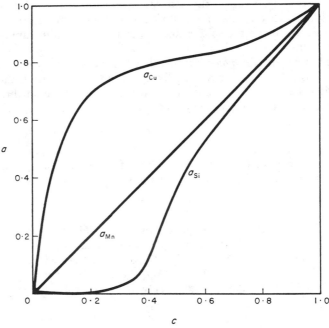

Fig. 7.3 Activities of copper, manganese and silicon in liquid iron at 1600°C.

To find the equilibrium among the components of a solution we substitute $P_M = P_M^\ominus a_M$ etc. in eqn. 7.12 and put $P_M^\ominus = 1$ since this refers to the standard state of M. The mass action law then becomes

$$\frac{a_M^m a_N^n \cdots}{a_A^a a_B^b \cdots} = \exp\left(-\Delta G^\ominus / RT\right) = K) \qquad 7.25$$

with a_I^i replaced by unity for any component present in equilibrium as pure solid or liquid I.

7.7 Refractories

Some of the ideas developed in this chapter have a bearing on the choice of *refractory* materials for crucibles and furnace linings. A refractory must obviously have a high melting point and be chemically inactive. Metallic oxides with large heats of formation are thus particularly suitable. Some melting points are given in Table 7.3.

TABLE 7.3. MELTING POINTS OF OXIDES (°C)

ThO_2	ZrO_2	MgO	BeO	CaO
3500	2800	2640	2520	2500
$ZrO_2 \cdot SiO_2$	Cr_2O_3	UO_2	Al_2O_3	$MgO \cdot Al_2O_3$
2430	2280	2230	2020	2000
TiO_2	SiO_2	Fe_3O_4	Fe_2O_3	FeO
1850	1713	1600	1460	1370

A refractory oxide should obviously have a free energy line, in the Ellingham diagram, below that of the metal it is to meet. It would be disastrous to melt titanium in a silica crucible, for example; the SiO_2 would be reduced and the Ti contaminated with O and Si.

This criterion alone is insufficient, however, since the mass action law always allows the reaction between a metal and its refractory container to occur to a certain extent. For example, suppose we melt aluminium in a magnesia crucible (MgO) at 1000°C. We consider the following possible reaction,

$$\underset{\text{(solid)}}{2MgO} + \underset{\text{(liquid)}}{\tfrac{4}{3}Al} = \underset{\text{(gas)}}{2Mg} + \underset{\text{(solid)}}{\tfrac{2}{3}Al_2O_3} \qquad \textbf{7.26}$$

There are, of course, other possibilities (e.g. the Mg may dissolve in the Al). From Fig. 7.1, the free energies of the reactions $2MgO \rightarrow 2Mg + O_2$ and $\tfrac{4}{3}Al + O_2 \rightarrow \tfrac{2}{3}Al_2O_3$ are $+941\,400$ and $-849\,400$ respectively. Hence $\Delta G^{\ominus} = +92\,000$ for reaction 7.26. We have $P_{Mg}^2 = \exp(-\Delta G^{\ominus}/RT)$. With $RT - 10\,450$ this gives $P_{Mg} \simeq 0.0123$ atm. This is a fair pressure so that we anticipate considerable attack of the crucible and loss of Mg vapour when melting in vacuum.

As another example, consider the effect on silica firebricks of manganese in a molten alloy steel (e.g. *Hadfield's steel*, 12 per cent Mn) at 1600°C. The Ellingham diagram suggests that the silica might not be attacked. However, the activity of silicon in iron is very low, whereas that of manganese is almost ideal (cf. Fig. 7.3). For instance, $c_{Si} \simeq 0.2$ gives $a_{Si} \simeq 0.000\,15$. This deviation from standard conditions raises the free energy of the $Si + O_2 \rightarrow SiO_2$ reaction by $-RT \ln(0.000\,15)$, i.e. by about $138\,000$ J mol^{-1}
· above that of the standard reaction at 1600°C. The free energy of the $2Mn + O_2 \rightarrow 2MnO$ reaction is, by comparison, not much altered from its standard value. Hence the SiO_2 free energy line is in effect lifted, due to the low activity of the Si, well above that of the MnO line at 1600°C so that the Mn in the steel will reduce the SiO_2 refractory. Because of this, high-manganese steels cannot be made in silica-lined furnaces.

Refractory oxides are classified as *acidic* or *basic*, according as they give, if soluble, acidic or basic solutions in water. The acidic refractories (silica, quartz, ganister) are rich in SiO_2. The basic ones contain mainly lime (CaO), magnesia (MgO) and *calcined dolomite* ($CaO.MgO$). Alumina (Al_2O_3) is fairly neutral. Where chemical reaction with the charge or furnace gases rules out both acidic and basic refractories, it may be necessary to use one of the more expensive *neutral* refractories (e.g. *chromite* $Cr_2O_3.FeO$, graphite, SiC, TiC and metals such as Mo).

Besides melting point and chemical stability, the other important properties of refractories are as follows:

(1) *Mechanical strength and dimensional stability at high temperatures.* Compressive loads usually of order 50 psi (pounds/sq in.), i.e. of order $350\,000$ N m^{-2} (newtons per square metre), have to be supported.

(2) *Thermal shock resistance (spalling resistance).* Temperature gradients produce thermal strains which can break brittle materials. Good thermal conductivity (e.g. SiC, graphite, metals) or a small coefficient of thermal expansion (e.g. SiC, Si_3N_4, pyrex glass) give resistance to thermal shock. Magnesia is susceptible because of a high expansion coefficient. Silica is susceptible at low temperatures because it undergoes changes of crystal structure but good at high temperatures because it has a low expansion coefficient.

(3) *Thermal conductivity.* For furnace linings a low conductivity (e.g. *porous* brick) is required. For recuperators, crucibles and muffles the conductivity should be high. Refractories that withstand the highest temperatures (as used in steel-making furnaces) are usually moderately good conductors. It is thus necessary to use a series of refractory linings, the most refractory ones on the inside, the most insulating ones on the outside.

The most important acidic refractories are silica bricks. These are cheap, high-temperature refractories which, when fairly pure (> 96 per cent SiO_2), remain strong up to about 1680°C. They are widely used for the roofs of steel-making furnaces. The *fireclays* (*china clay*; *kaolinite*, $Al_2O_3.2SiO_2.2H_2O$) are based on silica and alumina. Fireclay itself contains 20–45 per cent Al_2O_3, is neutral and when fairly pure can be used up to 1500°C. Low-grade fireclays contain FeO and other impurities which lower the melting point and are usable only up to 1200°C. *Sillimanite* (45–70 per cent Al_2O_3) and *mullite* (70–100 per cent Al_2O_3) are more expensive but can be used in the range 1700–1800°C and have good spalling resistance.

Lime is not used much, despite its refractoriness, because it hydrates. *Magnesite* (80–90 per cent MgO) can withstand very high temperatures but its high-temperature strength is low because the $MgO/SiO_2/FeO$ glass needed to bond its crystal grains together softens at high temperatures. Magnesite has good thermal conductivity but its spalling resistance is reduced by its high coefficient of thermal expansion. It has excellent chemical resistance to *basic slags*, as used in steel-making. Dolomite is a cheaper alternative to magnesite but has a tendency to hydrate. *Chrome-magnesite* (Cr_2O_3.MgO) is used for the roofs of basic open-hearth steel-making furnaces. It has better high-temperature strength and spalling resistance than MgO, although it tends to swell by absorbing FeO if used near the hearth.

The problem of refractory containers for molten *reactive* metals at high temperatures is difficult. The following refractory oxides, when pure, are generally favoured:

(1) *Thoria*, ThO_2; stable in oxidizing atmospheres up to its melting point ($\simeq 3500°C$) and involatile in vacuum up to 2000°C.

(2) *Magnesia*, MgO; useful in oxidizing atmospheres but easily reduced in reducing atmospheres at temperatures well above the boiling point of magnesium (1100°C).

(3) *Zirconia*, ZrO_2; a chemically stable high-temperature refractory but tends to change its crystal structure.

(4) *Alumina, Corundum*, Al_2O_3; stable in oxidizing atmospheres up to 1800°C, but slightly reactive in reducing atmospheres and volatile in vacuum.

(5) *Urania*, UO_2; stable in reducing atmospheres and in vacuum up to 2000°C, but forms higher oxides in oxidizing atmospheres.

It is often necessary, to avoid contaminating a reactive metal at high temperatures, to use the pure oxide of the metal itself as the refractory, wherever possible; or to use one of the methods described at the end of § 5.5.

FURTHER READING

As for Chapters 5 and 6; also:

BISWAS, A. K. and BASHFORTH, G. R. (1962). *The Physical Chemistry of Metallurgical Processes*. Chapman and Hall, London.

FARADAY SOCIETY (1961). Discussion on *Physical Chemistry of Process Metallurgy*. Butterworth, London.

HOPKINS, D. W. (1954). *Physical Chemistry of Metal Extraction*. J. Garnet Miller, London.

HULTGREN, R., ORR, R. L., ANDERSON, P. D. and KELLEY, K. K. (1963). *Selected Values of Thermodynamical Properties of Metals and Alloys*. John Wiley, New York and London.

KINGERY, W. D. (1960). *Introduction to Ceramics*. John Wiley, New York and London.

KUBASCHEWSKI, O. and EVANS, E. Ll. (1951). *Metallurgical Thermochemistry*. Butterworth-Springer, London and Berlin.

NORTON, F. H. (1942). *Refractories*. McGraw-Hill, New York and Maidenhead.

Chapter 8

Extraction of Metals

8.1 Metallic ores

We saw in § 2.6 that many of the familiar metals occur sparsely in the earth's crust. Fortunately, geological processes have partly concentrated them in workable *ore deposits*. The value of a mineral ore has, of course, to be judged against the general demand and availability of the metal and of alternative sources of supply as well as on characteristics of the ore itself, such as the chemical state of the metal, nature and content of impurities, physical state of the mineral and accessibility of the deposit. For example, a deposit containing less than 20 per cent Fe has little value as an iron ore since there are many with 30–50 per cent Fe and a few with 70 per cent, but one containing 4 per cent Cu is a good copper ore.

Table 8.1 gives the world production and prices of metals. These values are very approximate and subject to wide fluctuations, but they indicate the very large differences that exist. The choice of extraction processes has to be judged in the light of these economic factors. The elaborate chemical solvent treatments used, for example, in the extraction of beryllium would be

TABLE 8.1. APPROXIMATE ANNUAL WORLD OUTPUT OF METALS AND UNITED KINGDOM PRICES, 1972

	Steel	Mn	Al	Cu	Zn	Pb	Cr	Ni
10^9 kg:	630	23	11·5	8	5·5	4	2	0·6
£ per 10^3 kg:	50	270	230	430	150	105	930	1400

	Mg	Sn	Mo	W	Sb	Ti	U	Co
10^6 kg:	230	200	130	55	40	25	24	21
£ per kg:	0·35	1·5	1·5	3·0	0·4	3·0	7·5	4·5

	Cd	V	Hg	Ag	Se	Be	Au	Pt
10^6 kg:	16	14	11	9	5	3	1·2	0·1
£ per kg:	2·5	60	6	20	10	50	900	1800

prohibitively expensive for metals such as iron or lead, where cheap bulk methods have to be used.

We saw in § 2.6 that there is plenty of oxygen on earth to form oxides. However, at the very high temperatures at which the earth is presumed to have formed, most metals would have been chemically uncombined. The heavier metals, particularly much of the iron, would then have been pulled by gravity to the centre, so forming a metallic core about 4000 miles across. At the lower temperatures reached in the surrounding envelope of mainly light elements (O, Si, Al) a semi-molten *magma* of silica and silicates formed, about 2000 miles thick, and near the surface crystallized into the solid *crust*. This is about 30 miles deep, although much thinner underneath the oceans, and consists mainly of an upper layer of *granite* (Al and Ca silicates) and a lower layer of *basalt* (Mg and Fe silicates). The *mantle* beneath the crust consists mainly of *olivine* (Mg and Fe silicates) and other mixed silicates such as *garnet*.

As the surface layers of the magma cooled, the most refractory compounds solidified first. Heavy ones, such as chromium compounds, then sank down too deeply to be mined. Some of these primary deposits have since been brought to the surface by other processes. High pressures in the magma have occasionally forced the lower layers up through crevices to form *injection* deposits; examples are the very rich copper and nickel ores at Sudbury in Canada and iron ore at Kiruna in Sweden. The *folding* of the crust to form mountains and the subsequent erosion of these by *weathering* has also exposed some initially deep layers.

During the freezing of the crust most of the elements in dilute solution tended to concentrate in the last remaining traces of liquid. At about 600°C this liquid solidified to form the *pegmatite* rocks, a valuable source of many metals. Escaping gases often forced this remaining liquid up into fissures, so producing rich vertical *veins* of ore through the silicate rock. Many metals are found in these, e.g. Pb, Zn, Cu, Ag, Au, Co, As, Sb, Bi, Hg. Where such deposits encountered water descending from the surface, hot water solutions were formed and then crystallized out in cooler regions to give *hydrothermal* deposits; e.g. *pitchblende* (uranium ore).

The weathering effects of rain, wind, ice and temperature changes gradually disintegrated the early rocks over geological times and *alluvial placer* deposits were then formed from the gravel and mud washed down. Heavy minerals collected together in the slower regions of streams and were then often left behind as *placer beds* when these streams afterwards changed course. Many rich and easily mined deposits have formed in this way; e.g. Sn, Fe, Ti, Zr, Cr, U ores, also metallic Au and Pt. Weathering also led to *solution* and *deposition* and so produced the *sedimentary* rocks and ore deposits; e.g. some rich Fe, Mn and Al ores. Living organisms have also helped to form some deposits. For example vanadium, often found in coal and oil fields, is believed to have been collected and concentrated by bacteria and marine organisms.

8.2   Concentration of ore

The excavation of an ore varies from the simple dredging of sedimentary deposits (e.g. tin ores in South-east Asia) and *open-cast* mechanical digging of surface deposits (e.g. iron ores) to shaft and tunnel mining, often through hard quartz rock, for deep deposits (e.g. gold in South Africa).

Except for certain rich iron and aluminium ores it is generally necessary to *concentrate* the mined ore by *mineral dressing*, to remove some of the unwanted *gangue* material, before attempting to extract the metal. The processes of mineral separation have to work on a vast scale, to discard cheaply most of the material in the ore and so simple physical methods are used as far as possible. *Selective settling* is widely used, based on differences in the specific gravities of the constituents. The rock is crushed to powder and washed in a stream of water, the light particles then being swept along and the dense ones left behind. Many devices, for example *jigs* and inclined vibrating *tables* with roughened surfaces or corrugated with *riffles* to catch the separating particles, have been designed to make this process very efficient. In special cases, *magnetic separation* is used to separate magnetic compounds such as *magnetite* Fe_3O_4 and *ilmenite* $FeTiO_3$.

Flotation is widely used. It can be made extremely sensitive and selective and can be used to concentrate very low-grade ores. The principle is to use the forces of *surface tension* to float off from finely powdered ore the particles of the required mineral on to the surface of water in a tank, where they can be removed separately from the remaining mineral which falls to the bottom of the tank. To float on water, a dense particle has to be *hydrophobic*, i.e. not wetted, so that surface tension can support it at the air-water interface against its weight. Most hydrophobic substances, e.g. greases and waxes, are covalently (*non-polar*) bonded. Metallic minerals are, however, mainly ionically (*polar*) bonded and so electrostatically attract the dipolar water molecule. They are thus mainly *hydrophilic*, i.e. wetted.

It is therefore necessary to add a substance, i.e. a *collector*, to the water that will attach itself specifically to the required mineral particles and form a non-wetting layer on them. Such particles will then float off into a *froth* if the water is vigorously agitated to bring in small air bubbles to attach themselves to the particles and lift them up. The collectors are *polar/non-polar* molecules. One end of such a molecule is ionic and attaches itself to the surface of the mineral particle. The other end is a saturated covalent substance such as a paraffin and is hydrophobic. Two important types of substances used as collectors are the *xanthates*,

$$R—O—C \overset{\displaystyle S}{\underset{\displaystyle S—X}{{\Big\langle}}}$$

and the *aerofloats*

$$
\begin{array}{ccc}
\text{R—O} & & \text{S} \\
& \diagdown \ \diagup\!\!\diagup & \\
& \text{P} & \\
& \diagup \ \diagdown & \\
\text{R—O} & & \text{S—X}
\end{array}
$$

where X is usually a sodium or potassium ion and R a paraffin chain radical, e.g. C_2H_5 or C_5H_{11}. These dissolve in water and the alkali ion dissociates, leaving a negative sulphur ion at the polar end of the molecule to attach itself to the mineral. Xanthates are particularly effective for concentrating sulphides. For oxide minerals, a mixture of soap and oleic acid is used as the collecting agent.

It is generally necessary also to add a *frothing* agent, for example pine oil or cresylic acid, to the water. This concentrates at the air-water interface and lowers the surface tension, so enabling many small stable air bubbles to form and give a voluminous foam. *Activating* or *depressing* agents may also be added. These are adsorbed on the surface of the mineral particle, where they increase or reduce the attraction of the collecting agent. In this way it is possible to obtain a more complete and selective separation and also to separate more than one mineral. In mixed lead, zinc and iron sulphide ores, for example, sodium cyanide is first added to depress the action of the xanthate on the zinc and iron, so that only the lead sulphide is floated. When this has been separated and removed, copper sulphate is added to the flotation bath. This activates the zinc sulphide by forming a thin layer of copper sulphide on its surface and these particles can then be floated and separated from the iron sulphide.

8.3 Slags

The concentrated ore usually still contains some gangue. Except when dealing with highly reactive or refractory metals it is usual to remove this gangue during the reduction process itself. Sometimes the reduced metal is produced as vapour, e.g. zinc. Separation is then easy, by distillation. Most metals are not sufficiently volatile for this, however, and other methods of separating them from the gangue must be sought.

Molten ionic compounds are generally immiscible with molten metals and float on them. If then the gangue can be melted at the temperature of the reduction process it can be run off separately from the molten metal. The temperature of reduction should, of course, be as low as possible for thermal economy and also to minimize the reduction of substances in the gangue. However, the oxides and silicates of the gangue generally have high melting points (cf. Table 7.3). It is necessary then to add a *flux*, a substance that combines with the gangue to produce a fusible *slag*.

Most slags are impure silicate glasses. Silica, being acidic, combines with basic oxides, e.g. CaO, MgO and amphoteric oxides, e.g. Al_2O_3, Fe_2O_3, to form various silicates, mixtures of which have relatively low melting points. Limestone is the main flux for acidic gangue and silica sand likewise for basic gangue. Iron oxide gives a very fusible slag with silica, melting at 1180°C at compositions near Fe_2SiO_4, and is an important constituent of some slags in non-ferrous metallurgy but the easy reducibility of iron oxide limits its wider use. The lowest melting point of CaO-SiO_2 slag is 1430°C and that of Al_2O_3-SiO_2 is 1545°C, but lower melting points can be got by dissolving several oxides in one another.

8.4 Tin

As an example of a fairly easily reducible metal that occurs in nature as the oxide, we shall consider the metal *tin*. The main ore is *cassiterite*, SnO_2, found in South-East Asia, Bolivia, and to a small extent in Cornwall. Impurities are removed by washing the powdered ore, leaving the heavy tin oxide behind.

According to the Ellingham diagram, Fig. 7.1, the extraction of tin by *smelting* is straightforward in principle. The SnO_2 free energy line lies high in the diagram and carbon can reduce it at temperatures above about 600°C. The method used is to heat the washed and dried ore with powdered coal, together with lime to form a slag, in a reverberatory furnace at 1200°C. This temperature is needed to melt the slag and also to produce a reasonable rate of reaction. The mixture is stirred and when fully molten is run off into a pot where the slag is allowed to overflow.

In practice, tin smelting is not quite so straightforward. One reason is that tin oxide is amphoteric. As a result it tends to be absorbed into acidic slags as a silicate and into basic slags as a stannate. These slags have therefore themselves to be smelted afterwards for tin, sometimes twice.

The first stage in refining the impure smelted tin makes use of its low melting point. The tin is melted on a sloping hearth and runs away (*liquation*) leaving behind an unmelted alloy of tin and impurities (Cu, Fe, As) which is afterwards smelted for further tin extraction. The next stage, selective oxidation, makes use of the fairly low affinity of tin for oxygen. The metal is melted in an open crucible (a *kettle*) and stirred by injecting compressed air or steam (*poling*). The oxidized impurities such as zinc float to the surface and are skimmed off as a *dross*. This gives tin better than 99 per cent pure. If higher purity is required, e.g. 99·9 per cent, a final stage of *electrolytic refining* may be used; cf. § 9.6.

8.5 Sulphide ores

Several metals occur naturally as the sulphide; for example, *galena* (PbS) found in Australia, Canada and Mexico; *copper pyrites* and *chalcopyrite* (Cu_2S with FeS) in Africa, the U.S.A., the U.S.S.R., Canada and South America; *pentlandite* (NiS with Cu_2S and FeS) in Canada; and *zinc blende* and *sphalerite* (ZnS) commonly associated with lead ores.

Fig. 7.2 shows that carbon is not a good reducing agent for sulphides since CS is unstable and the free energy of formation of CS_2 is very low. Furthermore, the free energy curves of many metal sulphides are closely bunched, so that selective reduction would be difficult. It is necessary then to change the sulphides into oxides (or some other compound), in a preliminary process called *roasting*, before reduction. The powdered sulphide ore is roasted to oxide by heating at 600–1000°C, a temperature below the melting point of either sulphide or oxide, in good contact with air. A typical roasting reaction is

$$2PbS + 3O_2 = 2PbO + 2SO_2 \qquad \text{8.1}$$

Roasting reactions are often exothermic and the heat released then provides much or all of that needed to keep up the temperature during the roast. During roasting, the particles of powder may become stuck together, i.e. *sintered*, so forming a porous lumpy mass. This can be a valuable feature of roasting, particularly as a preliminary to blast furnace smelting where a free passage of the furnace gases through the charge is essential, but it is important not to allow the sintering to develop too quickly since otherwise the oxygen will fail to reach all particles and some sulphide will remain.

In the *Dwight-Lloyd* sintering process the charge is carried, in open grates on an endless belt, through a short hot igniter. The surface ignites and the combustion is then drawn downwards through the bed by suction from below. By adjusting the rate of travel, the process can be controlled to the required degrees of roasting and sintering. A more recent method is *fluidized bed roasting* in which the powdered charge is suspended in an upward stream of hot gas injected from below.

Fig. 7.1 shows that oxygen and sulphur have a fair affinity for each other. It is thus possible in some cases to use oxygen as a *reducing agent* (!) for the sulphide, e.g.

$$PbS + 2PbO = 3Pb + SO_2 \qquad \text{8.2}$$

The condition for such a *roast-reduction* to occur is that $-\Delta G^\ominus$ for $S + O_2 \to SO_2$ shall exceed the sum of $-\Delta G^\ominus$ for the reaction *metal* $+ O_2$ $\to$ *oxide* and $\frac{1}{2}(-\Delta G^\ominus)$ for the reaction *metal* $+ \frac{1}{2}S_2 \to$ *sulphide*. This is satisfied for copper above 800°C and for lead above 930°C.

8.6 Lead

The extraction of lead provides a simple example of sulphide metallurgy. *Galena* ore, PbS, is widely distributed, the main sources being the U.S.A., Australia, Spain, Mexico and Germany. The initial PbS content is about 1–10 per cent and flotation is used to concentrate it to 60–80 per cent and also to separate ZnS. Very rich concentrates are sometimes converted directly to lead by *roast-reduction* but the presence of FeS_2 impurities makes this process hard to control. Complete roasting to oxide, followed by reduction with carbon, is generally preferred. A small blast furnace is used, running at about 900°C, using coke as fuel and reducing agent and CaO and FeO to flux the SiO_2.

The lead so produced is first refined by heating in air at 400°C and stirring, when insoluble oxides and other compounds float up to the surface and are skimmed off. Sulphur is sometimes added to help eliminate copper as the sulphide. Afterwards, tin, antimony and arsenic are removed from the lead by selective oxidation in a reverberatory furnace. Alternatively, the molten lead is purified by contact with an oxidizing mixture of sodium nitrate, salt and caustic soda.

Lead ores generally contain useful amounts of *silver*. The next stage of refining is to extract this silver from the lead. This is done in the *Parkes process*, which makes use of the following facts:

(1) near the melting point of zinc (419°C), molten zinc and molten lead are almost insoluble in each other and the zinc floats on the lead;

(2) silver is more soluble in molten zinc than lead and distributes itself between these two solvents in a proportion of about 300 to 1;

(3) zinc boils readily (b.p. = 907°C) but silver does not.

The silver is separated from the lead in a tall vessel, fed from the top, the lower half of which is filled with lead, the upper half with zinc. The bottom of the vessel is held at the melting point of lead (327°C) and the top at 600°C. A siphon removes the desilvered lead from the bottom of the vessel. The silver which has passed into the zinc forms a solid crust of silver-zinc alloy contaminated with lead. This crust is removed, the zinc is distilled from it and recovered and the lead then separated by *cupellation*, a process of selective oxidation in which the lead oxide is absorbed into a porous refractory hearth. A final treatment with boiling sulphuric acid, in which silver is soluble, enables the silver to be parted from any gold which may also be present.

The desilverized lead may still contain bismuth as an impurity. This can be removed either by adding a calcium-magnesium alloy to the molten lead, which forms a solid intermetallic compound with bismuth, or by electrolytic refining (cf. § 9.6).

8.7 Copper

The main source of copper is the sulphide, generally with a large excess of iron sulphide. Flotation enables low-grade ores (less than 2 per cent Cu) to be worked successfully on a large scale.

Fig. 7.1 shows that copper is an easily reducible metal. Clearly we could make copper, like lead, simply by roasting to oxide and then reducing with carbon. This was once done but the metal produced was very impure and much was lost in the slag. To appreciate the modern method we consider the thermodynamic relationships between Cu, Fe, S and O. We see from Figs. 7.1 and 7.2 that at high temperatures sulphur has a slightly greater affinity for copper than iron and that oxygen has a greater affinity for iron than copper. The iron can thus be removed from the copper by a partial oxidation treatment followed by fluxing with silica to form iron silicate slag.

The first stage is to concentrate the copper sulphide by *matte-smelting*, a roasting process in which the prepared ore is melted in a reverberatory furnace with silica flux. Some iron is converted to oxide and then to silicate slag, which floats on top of the heavy, molten *sulphide matte*. The matte, which contains 40–45 per cent Cu, is run off separately and charged into a *converter*. This is a large cylindrical or pear-shaped vessel, open at the top, with tuyères for introducing an air blast in the side near the bottom. It is an adaptation of the steel-making *Bessemer converter* (see Fig. 11.3). Air is blown through the molten matte and the temperature maintained at about 1250°C by the heat of reaction. The iron is oxidized preferentially and the FeO then removed as silicate slag by the addition of silica-rich copper ore which acts as a flux. The slag is poured off by tilting the converter and more matte added until the converter is nearly full of molten copper sulphide. Air is then blown through to produce a reduction reaction of the type

$$Cu_2S + O_2 = 2Cu + SO_2 \qquad \qquad 8.3$$

thus forming crude *blister copper*.

The first stage of refining exploits the low affinity of copper for oxygen. The blister copper is run into a reverberatory furnace and refined by selective oxidation, being stirred by injected air. This oxidizes Fe, S, Zn, Pb and other impurities and forms a slag of mixed oxides. The addition of soda-ash (sodium carbonate), lime and sodium nitrate to the slag helps to remove As, Sb and Sn. The sulphur is not sufficiently removed until the oxygen content of the copper reaches about 0·9 per cent. The Cu_2O formed in solid copper from this oxygen makes the metal very weak and brittle (cf. § 21.12). It is necessary then, after removing the slag, to deoxidize the copper. This is done by *poling*, i.e. stirring the copper with green wood poles which liberate reducing hydrocarbon gases. This is a critical operation in which the aim is to bring the oxygen content down to 0·03 to 0·05 per cent. If not taken far enough, the copper is embrittled by its Cu_2O content. If taken too far, it becomes porous by the formation of *steam* pockets through the reaction of hydrogen and cuprous oxide during solidification. The optimum point is indicated when the surface of the metal remains level during solidification, the small expansion due to a slight evolution of steam balancing the normal shrinkage due to crystallization. This refined metal is cast into ingots of *tough-pitch* copper. High grade copper, particularly for electrical uses, is made by electrolytic refining (cf. § 9.6).

8.8 Nickel

The extraction metallurgy of nickel begins like that of copper. The sulphide ore is concentrated by flotation and smelted to a matte of nickel, copper and iron sulphides. If the iron content is high this matte is then treated in a converter to remove the iron as silicate slag. Unlike copper, however, the converter process cannot be continued beyond this stage. Fig. 7.1 shows that, at temperatures of interest (about 1200°C), the free energy of nickel oxide is over 80 000 J mol^{-1} O$_2$ lower than that of copper oxide. Quite apart from the problem of separating the copper from the nickel, this oxygen affinity of nickel means that large amounts of nickel oxide form before the sulphur is fully oxidized.

A different approach is thus needed. The problem has been solved by the *Mond process*, which exploits the fact that nickel forms a volatile *carbonyl* Ni(CO)$_4$ (cf. § 4.4). The nickel sulphide matte is crushed and heated to drive off sulphur. After treatment with sulphuric acid, to remove some of the copper as copper sulphate, and drying, the residue is exposed to water gas, i.e. H$_2$ + CO, in a *reducing tower* at 300°C. This reduces the NiO to Ni. The charge is then taken to a *volatilizing tower* where it is exposed to water gas at 50°C and the volatile nickel carbonyl is formed. This vapour is then passed to a *decomposing tower* at 180°C where the carbonyl decomposes to Ni and CO and the nickel is deposited on nickel shot. This gives metal of 99·8 per cent purity.

Monel metal, an alloy of 70 per cent Ni 30 per cent Cu, is sometimes made. The nickel and copper sulphide matte is fully roasted to oxide and then reduced with carbon in a blast furnace, the two metals being reduced simultaneously to the alloy.

An important *hydrometallurgical* process has recently been developed in Canada for mixed sulphide ores of nickel, copper, cobalt and iron (*Sherritt-Gordon Process*). The starting point was the discovery by *F. A. Forward* that these sulphides are soluble in aqueous ammonia in the presence of oxygen. The metals are selectively reduced from ammoniacal solution by treatment with hydrogen at about 170°C under a pressure of over 30 atm. Powdered nickel, copper and cobalt are now produced in this way.

8.9 Zinc

The extraction metallurgy of zinc is dominated by the low boiling point of the metal, 907°C. The sulphide ore is first concentrated by flotation and roasted to oxide. We see in Fig. 7.1 that the CO and ZnO lines intersect at 935°C, corresponding to the equilibrium

$$ZnO + C = Zn + CO \qquad\qquad 8.4$$

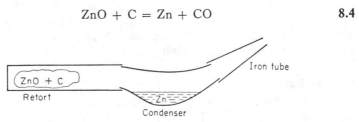

Fig. 8.1 A zinc retort

at *two* atmospheres total pressure (1 atm CO and 1 atm Zn vapour). At 1 atm total pressure the minimum reduction temperature is 897°C but this is still above the boiling point of zinc, 840°C at 0·5 atm partial pressure.

Reduction at atmospheric pressure is done in retorts and the zinc vapour condensed (Fig. 8.1). In this process fireclay retorts, about 1·2 m long and 20 cm diameter, charged with zinc oxide and powdered coal or coke, are inserted into a furnace at 1200–1400°C. The mouths of the retorts project from the furnace. Fitted on to them are clay *condensers*, in which the zinc collects as liquid at about 500°C, and iron tubes called *prolongs* to catch zinc which escapes the condensers. The reduction must be done at high temperatures, otherwise the CO_2 content in the retorts would become sufficiently high to oxidize the zinc appreciably. This makes the process thermally inefficient. There is much loss of heat by radiation and loss of fuel value by the necessarily high CO content of the retort gases. The heat from the furnace has to pass through the fireclay walls of the retorts and so is inefficiently used. The heat transfer difficulties also make it necessary to use small retorts, which results in an inefficient, small-scale, intermittent and manually arduous process. Some of these disadvantages have been overcome in the *vertical retort* process in which a 10 m tall narrow retort is used. The retort material is silicon carbide which is expensive but a good thermal conductor. Sintered bricks of ZnO and coal are fed in at the top and removed at the bottom, the zinc vapour being led from the top to a condenser. The process is continuous.

Zinc produced by the retort process is fairly pure but needs to be refined to 99·99 per cent purity for use both in die-casting alloys and in high quality brass. Fractional distillation and electrolytic refining are used.

8.10   The zinc blast furnace

Many people have tried to smelt zinc in a blast furnace. In the early attempts the furnace was run at high pressure, up to 5 atm., in the hope of producing liquid zinc but these proved costly failures. A thermodynamic analysis explains why (*C. G. Maier*, 1930, *U.S. Bureau of Mines Bulletin No. 324*).

The reaction 8.4 occurs in two steps,

$$ZnO + CO = Zn + CO_2 \qquad\qquad \textbf{8.5}$$

$$C + CO_2 = 2CO \qquad\qquad \textbf{8.6}$$

i.e. the zinc is actually reduced by CO, not C directly, and the supply of the CO is controlled by reaction 8.6. To obtain $\Delta G^{\ominus}$ for reaction 8.5 we regard this reaction as

$$ZnO + CO + (C + O_2) = (Zn + \tfrac{1}{2}O_2) + (C + \tfrac{1}{2}O_2) + CO_2 \quad \textbf{8.7}$$

and hence find

$$\Delta G^{\ominus} = \Delta G_2^{\ominus} - \tfrac{1}{2}\Delta G_1^{\ominus} - \tfrac{1}{2}\Delta G_3^{\ominus} \qquad\qquad \textbf{8.8}$$

in terms of the following standard free energies,

$$2Zn + O_2 = 2ZnO \qquad (\Delta G_1^\ominus)$$
$$C + O_2 = CO_2 \qquad (\Delta G_2^\ominus) \qquad \qquad \textbf{8.9}$$
$$2C + O_2 = 2CO \qquad (\Delta G_3^\ominus)$$

as given in the Ellingham diagram. Eqn. 7.13 then gives us the equilibrium constant, K_1, for reaction 8.5.

Next, we find the equilibrium vapour pressures, p_{CO}, p_{CO_2} and p_{Zn}. We have three equations for these,

$$p_{CO} + p_{CO_2} + p_{Zn} = p = total\ pressure\ (assuming\ no\ N_2)$$
$$p_{CO_2} = p_{Zn} \qquad\qquad from\ eqn.\ 8.5 \qquad\qquad \textbf{8.10}$$
$$p_{CO_2}p_{Zn}/p_{CO} = K_1 \qquad (assuming\ pure\ solid\ ZnO)$$

Solving these, we obtain

$$p_{Zn} = p_{CO_2} = -K_1 + \sqrt{(K_1^2 + pK_1)}$$
$$p_{CO} = p - 2p_{Zn} \qquad\qquad\qquad\qquad\qquad \textbf{8.11}$$

For reaction 8.6 we have $\Delta G^\ominus = \Delta G_3^\ominus - \Delta G_2^\ominus$. From this we obtain an equilibrium constant K_2,

$$p_{CO}^2/p_{CO_2} = K_2 \qquad\qquad\qquad\qquad \textbf{8.12}$$

It happens that reaction 8.5 is fast and reaction 8.6 slow, so that the actual gas concentrations in the mixture must be near those given by reaction 8.5.

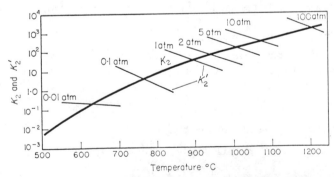

Fig. 8.2 Minimum temperatures for continuous reduction of zinc oxide by carbon

Let K_2' be the p_{CO}^2/p_{CO_2} ratio given by reaction 8.5. Then if $K_2' > K_2$ reaction 8.6. consumes CO and produces CO_2, thus halting the reduction of ZnO. Conversely, if $K_2' < K_2$, reaction 8.6 consumes CO_2 and the reduction of ZnO continues. The *minimum temperature of continuous reduction* is that at which $K_2' = K_2$. In Fig. 8.2 we plot K_2' and K_2 for various total pressures. The points of intersection give the minimum reduction temperatures for these pressures.

Let us now try to make liquid zinc by CO reduction. We put ZnO in a gas-tight retort into which we pump CO; and for each temperature studied we find the equilibrium p_{Zn}. When this exceeds the vapour pressure of condensed zinc at the temperature concerned, zinc is produced in a condensed form. Fig. 8.3 compares these pressures at various temperatures and total pressures. At 1 atm total pressure it is impossible to obtain liquid zinc. The lowest temperature at which reaction 8.5 goes at a reasonable rate is 700°C and the total pressure required in this case is 50 atm ($\simeq 5 \times 10^6$ N m^{-2}). Assuming that this pressure could be achieved it would nevertheless still be necessary to remove gas, rich in Zn vapour, to keep the pressure constant so that only part of the zinc would be condensed.

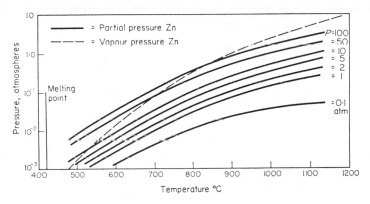

Fig. 8.3 Partial pressure of zinc in equilibrium with CO; and vapour pressure of condensed zinc

Let us now try to make liquid zinc by carbon reduction. The reaction 8.6 now helps to reduce the CO_2 content, which permits a higher Zn concentration in the equilibrium gas mixture. The total pressure now need not be quite so high. Fig. 8.4 shows the results. Curve A is the minimum temperature for continuous reduction, as obtained from Fig. 8.2. Under the actual conditions of a continuous blast furnace this temperature is displaced a little, to curve B. Curve C gives the temperature below which p_{Zn}, as produced by reactions 8.5 and 8.6, exceeds the vapour pressure of condensed zinc. The shaded region is that in which some liquid zinc could be produced.

We conclude that it is technically difficult to produce liquid zinc directly in a blast furnace. Futhermore, a blast furnace method for producing zinc vapour faces the difficulty that, in the normal way, much of this vapour would be converted back to ZnO as it cooled in the presence of the CO/CO_2 mixture.

This last problem has now been solved in the *Imperial Smelting Process.* Hot Zn and CO vapour from a blast furnace are passed through a chamber drenched by a spray of molten lead at 550°C. At this temperature zinc dissolves appreciably in lead. The zinc vapour is then collected by the lead droplets and swept down with them, too quickly to oxidize, into a molten lead bath below. This lead is circulated through a separate cooler chamber, at 440°C, and the reduced solubility at this temperature causes the zinc to separate from the lead. The molten zinc is run off and the lead returned to the spray chamber. This successful process now enables zinc to be produced

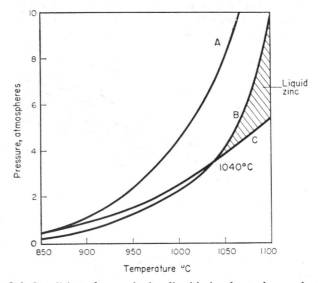

Fig. 8.4 Conditions for producing liquid zinc by carbon reduction

continuously on a large scale in a blast furnace. The process is applied to mixed zinc and lead ores, both metals being reduced simultaneously, the liquid lead being run off from the bottom of the blast furnace and the zinc vapour being blown out at the top into the spray chamber.

FURTHER READING

As for Chapter 7; also:

BAILEY, A. R. (1960). *A Textbook of Metallurgy.* Macmillan, London.
IVES, D. J. G. (1960). *Principles of the Extraction of Metals.* Royal Institute of Chemistry.
HAYWARD, C. R. (1952). *An Outline of Metallurgical Practice.* Van Nostrand, New York and London.
NEWTON, J. (1959). *Extractive Metallurgy.* John Wiley, New York and London.
ROBERTS, E. R. (1950). *The Extraction of Non-Ferrous Metals.* Temple Press, London.

Chapter 9

Electrochemical Extraction and Refining Processes

9.1 Electrolysis

Electrochemical methods are widely used for both extraction and refining of metals, particularly in places where electricity is cheap and abundant, near large hydroelectric power stations. In the simplest processes the metal is dissolved as positive ions in a liquid *electrolyte*, an ionic conductor, and is subjected to an electric field applied between two electrodes immersed in the electrolyte. The positive ions are drawn by electrostatic attraction to the negative cathode where they take up electrons, pumped into this electrode by a battery or generator, and then deposit on it as neutral metal atoms. The electrons removed from the electrolytic cell through the positive anode are supplied to the anode either from the valency electrons of the anode substance itself, as the atoms of this dissolve in the electrolyte ionically and leave their valency electrons behind (*soluble anode*), or from the discharge of negative ions in solution which are attracted electrostatically to the anode.

The amount of electricity required is given by *Faraday's law*. One mole of singly-charged ions is equivalent to one Faraday F of electricity, where F = 96 500 coulombs (i.e. ampere-seconds). Multiply-charged ions are equivalent to the corresponding multiples of this. For example, the reaction $MO_2 \rightarrow M + O_2$ needs $4F$ per mole O_2 since each of the two oxygen ions involved is doubly-charged. In practice, of course, rather more electricity is needed because some current leaks through the cell by other processes. The weight W of ions transferred by a current i amperes flowing for t seconds, assuming no losses, is given by

$$W = Ait/nF \qquad \textbf{9.1}$$

where A is the atomic weight and n the multiple-charge number of the substance concerned. From this we deduce the *electrochemical equivalents* of various elements, as in Table 9.1.

TABLE 9.1.   ELECTROCHEMICAL EQUIVALENTS

Kg deposited per 1000 ampere-hour:

Al	Cd	Cr	Cu (ous)	Cu (ic)	Au (ic)	H	Fe (ous)	Pb	Ni	Pt	Ag	Sn (ous)	Zn
0·34	2·1	0·32	2·37	1·19	2·45	0·038	1·04	3·86	1·09	1·82	4·02	2·21	1·18

In extraction metallurgy the electrolytic method is invaluable for reducing reactive metals whose free energy curves lie too far down, in Fig. 7.1, for reduction by carbon at ordinary smelting temperatures. Suppose that a chemical reaction is blocked by an unfavourable free energy change ΔG. If we can arrange to do sufficient electrical work on the system, suitably directed onto the reaction, then we can make this reaction occur despite its unfavourable free energy. The electrical work done is the product of the electrical charge transported, nF coulombs per mole of substance concerned, and the potential difference V volts through which it moves. The minimum voltage required is thus given by

$$\Delta G = -nFV \qquad\qquad 9.2$$

Here the energy unit is the joule (cf. *electron-volt*, §2.2). The free energy changes in Fig. 7.1 refer to 1 mole O_2, i.e. to $n = 4$ since each of the two oxygen ions is doubly-charged. Hence the thermodynamically ideal voltage needed to make a reaction go forward, against a free energy change $\Delta G^\ominus$ J mol^{-1} O_2, is $\Delta G^\ominus/386\,000$ V. Quite small voltages are thus equivalent to large free energy changes. In practice, of course, a rather higher voltage is needed to overcome resistance losses and polarization.

Eqn. 9.2 forms the basis of an important method for measuring the free energies of reactions from the potentials developed between electrodes in electrolytic cells working under thermodynamical reversible conditions.

Reactive metals cannot be extracted from aqueous solutions since they displace hydrogen from solution. *Ionic melts* are therefore used as electrolytes, particularly *chlorides* and *fluorides*, molten at temperatures below about 800°C. Metals of low melting point (e.g. Al, Mg, Ca, alkalies) are most suitable for electrolytic extraction since they can be collected in a liquid metal bath at the cathode and run off from the cell at intervals. More refractory metals (e.g. Ti, Zr, U, Th) are less suitable. They deposit in the form of powdery crystals entangled with electrolyte and then oxidize rapidly when removed.

9.2   Aluminium

The outstanding example of a metal produced by fused-salt electrolysis is aluminium, the annual world tonnage of which is now second only to that of iron and ferro-alloys. It is the most abundant metal in the earth's crust but cheap methods of extracting it from alumino-silicate clays have not yet been found. The workable ore is *bauxite*, hydrated Al_2O_3, found mainly in Guiana, Europe, Jamaica, North America and the U.S.S.R. Before reduction the alumina is purified chemically by the *Bayer process*; the oxide is dissolved in hot caustic soda, under pressure, as sodium aluminate, filtered off from the insoluble iron oxide, silica and other impurities and then precipitated as pure hydroxide by *seeding* the cooled solution with freshly prepared aluminium hydroxide. The precipitate is washed and dried at about 1100°C to give 99·4 per cent pure Al_2O_3.

The large-scale production of aluminium has grown entirely out of the discovery in 1886 by *Hall and Héroult* that molten *cryolite*, Na_3AlF_6, is a good solvent for alumina and that the solution can be electrolysed to produce aluminium without appreciably decomposing the solvent. Fig. 9.1 shows a

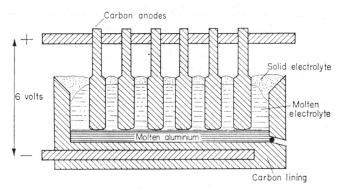

Fig. 9.1 Electrolytic cell for aluminium

Hall-Héroult reduction cell. It operates at about 6 V, 7000 amp per sq m of cathode, and 960°C. A typical cell produces about 300 kg of aluminium daily and consumes 26 kWh of electricity per kg produced.

The carbon anodes burn away by anodic oxidation and are renewed continuously by feeding coke and pitch into an aluminium mould which slowly descends into the electrolyte. The oxidation of the anodes plays an important part, both in heating the cell and in the thermodynamics of reduction. The primary oxidation product at the anodes is CO_2, not CO, despite the temperature; thus the effective reaction is

$$\tfrac{2}{3}Al_2O_3 + C = \tfrac{4}{3}Al + CO_2 \qquad\qquad 9.3$$

At 960°C, $\Delta G^\ominus$ for the Al/Al_2O_3 reaction is $-853\ 530$ J mol^{-1} O_2 and that for the C/CO_2 reaction is $-397\ 480$ J mol^{-1} O_2. Hence the carbon oxidation supplies almost half of the total free energy required for the alumina reduction. The remainder, $456\ 050$ J mol^{-1} O_2, is supplied electrically. The thermodynamically ideal voltage is thus $456\ 050/386\ 000 = 1\cdot18$ V.

The aluminium produced is about 99·6 per cent pure and is often further refined to about 99·99 per cent, in a three-layer electrolytic cell (*Hoopes cell*). The bottom layer (the anode) is the impure aluminium, alloyed with copper and silicon to form a heavy liquid of low melting point, the middle layer is the electrolyte, made heavier than pure aluminium by adding barium fluoride to the aluminium and sodium fluoride mixture, and the top layer is the refined aluminium which is periodically ladled out for casting.

9.3   Magnesium, calcium and alkali metals

Electrolytic reduction is widely used for the production of other reactive metals of low melting point. Magnesium is extracted from deposits of magnesite ($MgCO_3$), dolomite ($MgCO_3, CaCO_3$), brucite ($MgO . H_2O$), carnallite ($MgCl_2 . KCl . 6H_2O$) and from sea water. The chloride is used for electrolysis and oxide ores are first converted to chloride by heating with carbon and chlorine or by treatment with hydrochloric acid. The magnesium chloride is dissolved in an electrolyte of mixed calcium and sodium chlorides in a cell with carbon anodes and steel cathodes. The reduced magnesium metal floats above the electrolyte and is periodically ladled off. Chlorine is evolved at the anodes.

Sodium is made by electrolysing the molten hydroxide at 315°C, using iron electrodes (*Kastner cell*). The sodium metal (melting point 97·9°C) floats up into a small iron chamber from which it is ladled off. Alternatively, sodium is made by electrolysing a mixture of NaCl and $CaCl_2$ at 600°C, using a carbon anode and copper cathode (*Downs cell*). The $CaCl_2$ is necessary to decrease the melting point, since that of pure NaCl is 808°C. The higher temperature is a disadvantage compared with the hydroxide process but this is offset by the cheapness of the raw NaCl charge and by the value of the chlorine produced.

The other alkali metals are made by similar processes. The electrolytic reduction of calcium is slightly more difficult because of the higher melting point of the electrolyte, $CaCl_2$ (MP 851°C).

9.4   Electrochemistry of aqueous solutions

Because of its dipolar structure, water has a high dielectric constant, $\kappa = 80$ (cf. §4.5). It is thus a good solvent for both positive and negative ions, because the large electrostatic energy which is normally associated with a free ion in empty space can be almost totally eliminated by polarization of the water molecules round the ion, their negative oxygen atoms pointing towards a positive ion and their positive hydrogen atoms towards a negative ion.

The weakening of electrostatic forces between charged particles in water is sufficient to overcome the cohesion of many ionic solids such as the alkali halides, so that water is a good solvent of these. The water molecule itself dissociates to some extent, forming a dilute solution of hydrogen H^+ and hydroxyl OH⁻ ions in water. At room temperature a small fraction of the molecules are so dissociated. This is usually expressed by saying that the concentrations of hydrogen and hydroxyl ions, denoted as [H⁺] and [OH⁻] respectively, in pure water at 25°C are both equal to 10^{-7} moles litre⁻¹, or that the pH value of pure water, defined as $-\log_{10}$ [H⁺], is equal to 7. In dilute solutions the product [H⁺] [OH⁻] remains constant at 10^{-14} although the hydrogen and hydroxyl ion concentrations themselves vary according to the acidity or alkalinity of the water. The pH value ranges from about 1 in strong acids to about 14 in strong alkalies.

Water is not an electronic conductor—its molecules are chemically saturated and a free electron in water would be too mobile to give time to allow the molecules to polarize themselves around it—but it is an electrolytic conductor. The conductivity of pure water is low, about 4×10^{-6} ohm⁻¹ m⁻¹ at room temperature, but the addition of strongly ionized solutes produces a large increase, in proportion to the increased concentration of ions present.

When a piece of metal is dipped into an aqueous solution containing its own ions, some of its atoms detach themselves from its surface and go into solution as ions, leaving their valency electrons behind. Some of the ions in solution also attach themselves to the metal piece and become part of it. The rate of dissolution depends on the binding forces holding an atom in the surface and the rate of deposition depends on the concentration of ions in solution. Initially these two rates are generally different. If the first predominates, the metal is corroded. In the reverse case, chemical plating occurs. The rates quickly change towards each other, however, bringing the exchange into dynamical equilibrium, because of the build-up of electrical charge in the metal piece; positive if deposition predominates, due to the positive ions acquired from solution; negative, if dissolution predominates, due to the valency electrons left behind. This charge acts electrostatically on the ions in the nearby solution so as to oppose the predominating process. The electrical potential of the metal, relative to the solution, when this state of dynamical equilibrium is reached is the *electrode potential* of the metal or the *half-cell potential*.

The half-cell potentials given in Table 9.2 are *standard electrode potentials* measured on the *hydrogen scale*, in which the concentration of metal ions in solutions is such as to give *unit activity* (approximately one mole of ions per litre) and the potential is measured at 25°C across a voltaic cell with a vanishingly small current, the other electrode being a *standard hydrogen electrode*, i.e. a platinum electrode coated with finely divided 'black' platinum immersed in hydrogen gas at 1 atm in contact with HCl solution of 1·2 moles litre⁻¹. These values allow us to determine the voltages developed in galvanic couples. For example, in the *Daniell cell*, i.e. a copper electrode in copper sulphate solution in electrolytic contact with a zinc electrode in zinc sulphate solution, zinc goes into solution and leaves a negative charge of -0.76 V on its electrode and copper deposits from solution,

TABLE 9.2.   STANDARD ELECTRODE POTENTIALS, HYDROGEN SCALE, 25°C

Metal		Ion	Volts	Metal		Ion	Volts
Cs		Cs^+	$-3\cdot02$	Ni		Ni^{2+}	$-0\cdot25$
Li		Li^+	$-3\cdot02$	Sn		Sn^{2+}	$-0\cdot136$
Rb		Rb^+	$-2\cdot99$	Pb		Pb^{2+}	$-0\cdot126$
K		K^+	$-2\cdot922$				
Na		Na^+	$-2\cdot712$	H		H^+	O
Ca		Ca^{2+}	$-2\cdot5$				
Mg		Mg^{2+}	$-2\cdot34$	Sb		Sb^{3+}	$+0\cdot11$
Al		Al^{3+}	$-1\cdot67$	Cu		Cu^{2+}	$+0\cdot34$
Ti	Reactive metals	Ti^{2+}	$-1\cdot63$	Hg	Noble metals	Hg_2^{2+}	$+0\cdot7986$
Zn		Zn^{2+}	$-0\cdot762$	Ag		Ag^+	$+0\cdot7995$
Cr		Cr^{2+}	$-0\cdot6$	Pd		Pd^{2+}	$+0\cdot82$
Cr		Cr^{3+}	$-0\cdot5$	Hg		Hg^{2+}	$+0\cdot86$
Fe		Fe^{2+}	$-0\cdot44$	Pt		Pt^{2+}	$+1\cdot20$
Cd		Cd^{2+}	$-0\cdot4$	Au		Au^{3+}	$+1\cdot50$
Co		Co^{2+}	$-0\cdot29$				

giving a positive charge of $+0\cdot3$ V on its electrode. Under standard condition a potential difference of $1\cdot1$ V is developed between the electrodes.

We see in Table 9.2 that the reactive metals, with easily ionized atoms, have large negative potentials—i.e. will dissolve even in concentrated solutions of their own ions—whereas the noble metals have positive potentials and so are deposited from dilute solutions. These differences depend mainly on the average distances of the valency electrons from the positive cores in the atoms. Far-ranging electrons, as in sodium, are weakly held and easily ionized. The atomic diameter of the gold atom, by contrast, is not much bigger than the ionic diameter, so that its valency electrons are more firmly held and ionization is difficult.

Arranged in order of their electrode potentials, the elements form the *electrochemical series*. Any metal will displace from aqueous solution the ions of a metal more noble than itself in the series; e.g. when iron is put into copper sulphate solution it becomes coated with metallic copper. Hydrogen is more noble than many of the metals in the series. In such cases, when positive ions arrive from solutions at a cathode and collect electrons there hydrogen gas is formed rather than metal deposited. This limits aqueous extraction and refining processes to the more noble metals.

Care is needed when interpreting Table 9.2 in the above way because the potentials there refer only to the particular conditions stated. Several factors can change the electrode potentials developed under other conditions. Large changes usually occur when the electrolyte contains substances which form *complex ions* (cf. § 4.4) with the metal ions. For example, we might expect cadmium ($-0\cdot4$ V) to displace copper ($+0\cdot34$ V) from solution, according to Table 9.2. So it does in simple inorganic solutions, e.g. sulphate, but in a solution rich in cyanide CN^- ions, which form complex ions with both copper and cadmium, both metals develop electrode potentials of about $-1\cdot 0$ v and there is little or no displacement.

The *concentration* c(moles litre^{-1}) or *activity* a of ions in solution affects the electrode potential developed. A high concentration produces a more positive potential since a greater proportion of positive ions is deposited; vice versa for a low concentration. It follows from the mass action, cf. eqn. 7.25, and from eqn. 9.2 that, if $E^\ominus$ is the standard electrode potential, corresponding to unit activity, and E is the equilibrium potential for an activity a of metallic ions in solution, then

$$E = E^\ominus + \frac{RT}{nF} \ln a \qquad\qquad 9.4$$

In dilute solutions we can usually replace a by c. We then have, at 25°C, the following simplified form of this equation,

$$E = E^\ominus + \frac{0\cdot 059}{n} \log_{10} c \qquad\qquad 9.5$$

Thus, a tenfold increase in ionic concentration shifts the potential by $0\cdot 059/n$ volts in the positive direction, e.g. by $0\cdot 0295$ V in the case of the Cu^{2+} ion.

Applied to hydrogen ions, eqn. 9.5 reduces to $E = -0\cdot 059$ pH. Thus, in neutral water, hydrogen should deposit on the cathode at $-0\cdot 413$ V at 25°C; in highly acidic solutions (pH $\simeq 1$) this falls to about $-0\cdot 06$ V. We might thus expect not to be able to deposit a metal such as zinc (standard potential $= -0\cdot 76$ V) from aqueous solution, hydrogen being deposited instead. In practice, however, the evolution of hydrogen bubbles on a cathode is delayed until the (negative) voltage is increased a certain amount above the values estimated above. This effect, known as *overvoltage*, makes possible the electrodeposition of metals such as zinc from aqueous solution.

Overvoltage appears only when a finite current is passing through a cell and is the numerical difference between the actual potential developed for deposition at a certain rate and the ideal value as calculated above for thermodynamically reversible conditions. It exists in varying degrees for all substances deposited. For example, the standard potential of nickel is $-0\cdot 25$ V but nickel plating does not begin until a potential of about $-0\cdot 45$ V is applied. The nickel overvoltage is thus about $0\cdot 2$ V. Overvoltages decrease as the temperature is increased; that of nickel, for example, approaches zero at about 95°C.

The cause of overvoltage lies in the detailed atomic processes by which an ion becomes reduced and deposited when it reaches the cathode. These cathodic reactions are opposed by an activation energy barrier (cf. Fig. 6.3) and, except at high temperatures where thermal activation is strong, a large potential is necessary to pull the ions across the barrier at a reasonable rate. The *hydrogen overvoltage* for the evolution of hydrogen on a cathode is particularly important and its value varies considerably from one cathode metal to another. Some typical values are given in Table 9.3. To show the use of these, suppose that we try to plate zinc on to a zinc cathode from a solution containing 1 mole litre^{-1} of zinc sulphate and 0·1 mole litre^{-1} sulphuric acid. The function of the acid is to suppress precipitation of zinc hydroxide. The overvoltage for zinc deposition is small so that this metal requires a potential of about $-0·76$ V. The reversible potential for hydrogen evolution at a pH $\simeq$ 1, corresponding to this acid solution, is about $-0·06$ V. The overvoltage is about $-0·7$ V so that the potential for hydrogen evolution is about $-0·76$ V. Zinc will thus deposit at the same voltage as hydrogen is evolved under these conditions. If the current density through the cell is increased, the hydrogen overvoltage is raised higher and a greater fraction of the current then deposits zinc.

TABLE 9.3.   TYPICAL OVERVOLTAGES FOR EVOLUTION OF HYDROGEN BUBBLES ON METAL CATHODES AT ROOM TEMPERATURE

Ag	Co	Fe	Ni	Cu	Cd	Sn	Pb	Zn	Hg
0·15	0·16	0·18	0·21	0·23	0·48	0·53	0·64	0·70	0·78

We notice that the hydrogen overvoltages on some metals are small. It would thus be very difficult to deposit zinc on a copper or nickel cathode for example. It would be necessary then, as a preliminary step, to put a thin coating of zinc on the cathode, known as a *strike* deposit, using a special electrolyte. Once this layer was formed, zinc could then be deposited from a simple solution in the manner described above.

9.5   Electrolytic extraction from aqueous solutions

It is clearly feasible to extract the less reactive metals by electrolysing aqueous solutions of their salts. Processes of this type have been developed and used. Some zinc ores are treated in this way. The concentrated sulphide is roasted to oxide and *leached*, i.e. dissolved, in dilute sulphuric acid. The solution is then purified by the addition of precipitation agents, e.g. zinc dust is added to displace copper, cadmium etc. from solution, and finally electrolysed in a cell with aluminium cathodes and lead anodes. The zinc is stripped off the cathodes daily.

Except where local circumstances are particularly favourable, e.g. suitable ore, often low-grade, and cheap electrical power, electrolytic extraction of metals from aqueous solutions is usually uneconomic. It is difficult to develop solutions that are efficient both for leaching and for electrolysis. The removal of various impurities taken into solution is often both necessary and troublesome. Furthermore, it is generally not possible economically to extract more than about one-half of the required metal from the electrolyte.

9.6   Electrolytic refining in aqueous solutions

In principle, electrolytic refining is simple. An impure anode of the metal is dissolved electrolytically and the metal deposited on a pure cathode. Impurities less noble than this metal stay in solution, in accordance with the electrochemical series, and those more noble are precipitated near the anode as *anode mud* (usually taking some of the metal with them). The anode mud frequently contains precious metals (Ag, Au, Pt, etc.) and is an important by-product of the process. Some of the electrolyte is removed and purified at intervals to reduce the concentration of soluble impurities. The refined

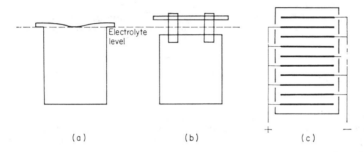

Fig. 9.2 Electrolytic refining of copper: (a) anode, (b) cathode, (c) electrical connections in cell

metal on the cathode is usually 99·9 to 99·99 per cent pure. Impurities with deposition potentials similar to it tend to be co-deposited with it, e.g. tin impurity in lead refining; they must be removed during the preliminary fire refining treatments, before electrolysis.

Since the anode dissolves as the cathode builds up, the overall chemical work done in the bath is virtually zero. The driving force for solution provides the potential for deposition, so that the applied voltage is required solely to overcome resistance losses and the effects of polarization and overvoltage. The main polarization effect is caused by concentration differences which build up near the electrodes. Due to slowness in the migration of ions across the cell, the ions dissolved at the anode build up to a locally high concentration round the anode. Similarly, the electrolyte is locally depleted of metal ions near the cathode. The changes of potential resulting from this (cf. eqn. 9.5) oppose the applied potential. It is important to reduce this *concentration polarization* by stirring and warming the electrolyte. Sometimes insoluble films build up on the anode, e.g. silver chloride formed from silver in gold refining. They are removed by superposing an alternating current on the electrolysing current.

Electrolytic refining is mainly used to make pure copper for the electrical industry. Fig. 9.2 shows some features of the cell. Fire-refined blister copper is cast into 250 kg anodes, about 1 m square, and these are arranged alternately with cathodes of pure sheet copper in a lead-lined tank of copper sulphate solution in dilute sulphuric acid. The cell is operated at 50°C, with good circulation of the electrolyte, and at about 0·3 V and a current density of 180 amp m^{-2} of cathode. Copper of about 99·98 per cent purity is deposited, with a current efficiency (i.e. actual deposit/ideal Faraday deposit) of better than 90 per cent.

FURTHER READING

DARKEN, L. S. and GURRY, R. W. (1953). *Physical Chemistry of Metals*. McGraw-Hill, New York and Maidenhead.

GILCHRIST, J. D. (1967). *Extraction Metallurgy*. Pergamon, Oxford.

IVES, D. J. G. (1972). *Principles of the Extraction of Metals*. Chemical Society Monographs, London.

MANTELL, C. L. (1950). *Electrochemistry*. McGraw-Hill, New York and Maidenhead.

METALS BULLETIN HANDBOOK (1973).

POTTER, E. C. (1956). *Electrochemistry: Principles and Applications*. Cleaver-Hume Press, London.

POURBAIX, M. J. N. (1949). *Thermodynamics of Dilute Aqueous Solutions with Applications to Electrochemistry and Corrosion*. Edward Arnold, London.

WEST, J. M. (1965). *Electrodeposition and Corrosion Processes*. Van Nostrand, New York and London.

Chapter 10

Extraction of Reactive and Refractory Metals

10.1 Introduction

We shall consider now the special problems which are set by the high reactivity or high melting point, or both, of the metal to be extracted and how these are overcome. One of the successes of modern metallurgy has been the discovery of satisfactory ways of producing and handling these metals on an industrial scale. As a result, many 'new metals', e.g. Ti, Mo, Zr, Nb, U, Be, have now come into much wider use.

How are we to reduce metals whose free energy lines lie near the bottom of Fig. 7.1, other than by the electrolytic method? In principle, carbon can reduce them all at sufficiently high temperatures. Magnesium, for example, is occasionally produced by carbon reduction in an electric arc (the *carbothermic* process). But straightforward carbon reduction generally has too many disadvantages when applied to the reactive metals; there is the expense of high-temperature heating and refractory problems; the difficulty of preventing the reaction going into reverse on cooling; and the reactive transition metals have a strong tendency to combine with carbon to form *carbides*. Broadly, the two main lines of approach for reducing these metals by methods other than electrolytic are (1) the use of stronger reducing agents than carbon, (2) the conversion of the charge to a *halide*, before reduction.

One fairly common feature of the extraction metallurgy of the reactive and refractory metals is the partial or complete disappearance of the final refining stage, apart from a few special treatments. Simple fire-refining is impossible for the reactive metals; and the problems of chemical reaction with furnace linings rule out the conventional methods of Chapter 8. It is therefore generally necessary to convert the ore, *before reduction*, to a fairly pure metallic compound by treatment with chemical reagents. Similarly, because the highly refractory metals are generally reduced direct to the solid state and since the solid cannot be refined by conventional methods, nor melted in conventional crucibles and furnaces, it is necessary in this case also to reduce purified ore to the pure metal.

10.2 The Pidgeon process for magnesium

Even though much magnesium is extracted electrolytically, an ingenious thermal method has also been developed. Although magnesium has a very high affinity for oxygen, the oxide is rather easier to reduce at high temperatures than those of most other reactive metals because of the effect of magnesium vaporization on the free energy line, cf. Fig. 7.1. However, we want a manageable reduction temperature, e.g. 1200°C. In the *Pidgeon process* the reducing agent is *silicon*, used in the form of *ferro-silicon*, an iron–silicon alloy (cf. § 11.11). We see from Fig. 7.1 that at 1200°C the reaction

$$2MgO + Si \rightarrow 2Mg + SiO_2 \qquad\qquad 10.1$$

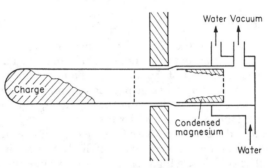

Fig. 10.1 Retort for magnesium reduction

has an *unfavourable* free energy of about 250 000 J mol^{-1} O$_2$ under standard conditions. The reaction can still be made to go forward, however, if the activities of the reaction products can be sufficiently reduced. In the Pidgeon process the pressure of magnesium vapour is kept low (2·5 × 10^{-2} mm Hg) by vacuum pumping and the activity of the silica is reduced by adding CaO to convert SiO$_2$ to CaSiO$_3$.

Fig. 10.1 shows the general arrangement. Dolomite (CaO . MgO) is made up into bricks with ferro-silicon and charged into horizontal steel retorts which project through the wall into the furnace. The reduced magnesium vapour, better than 99·95 per cent pure, is condensed on cool sleeves in the ends of the retorts.

10.3 Use of halides

In the extraction metallurgy of the reactive metals it is often useful to convert the starting material to a *halide*. The elimination of oxide from the reacting system enables the oxygen content of the reduced metal to be kept small; the halides of the multivalent metals are usually much more reducible than the oxides (cf. § 7.1); and several of these halides, being more covalent than ionic (cf. § 4.4), are highly volatile.

A simple example of the use of halides is provided by the *van Arkel* or *iodide process* for the purification of titanium (and zirconium). The impure titanium is held against the inside face of a silica vessel by a molybdenum cage. The vessel, which contains a small amount of iodine, is evacuated and heated to 170°C. The iodine vapour attacks the titanium and forms TiI_4 which is volatile at this temperature. Down the centre of the vessel is a titanium filament, electrically heated to incandescence. The molecules of TiI_4 vapour are decomposed by heat when they touch this filament. Crystals of pure titanium metal grow on the filament by deposition from this vapour and the iodine vapour returns to the vessel and continues the process.

10.4   Titanium

The van Arkel process is too expensive for the primary production of titanium. It is necessary to replace iodine by the much cheaper *chlorine* but $TiCl_4$ is far less decomposable than TiI_4. It is therefore necessary to use one of the cheaply produced reactive metals as a reducing agent.

In the *Kroll process*, magnesium is used. Chlorine gas is passed over a mixture of *rutile* (TiO_2) and coke in a steel vessel at 800°C to produce the chloride, thus

$$TiO_2 + 2Cl_2 + 2C = TiCl_4 + 2CO \qquad 10.2$$

The $TiCl_4$ is purified by fractional distillation and the vapour then passed over molten magnesium in a steel vessel. The magnesium reacts to the chloride and a spongy mass of titanium is produced. This is cleaned, compressed into the shape of an electrode, and then melted and cast to an ingot under high vacuum, usually in a cooled-crucible consumable-electrode arc furnace (cf. Fig. 5.5).

The *I.C.I. process* is similar but uses sodium in place of magnesium. The $TiCl_4$ is reduced by reaction with liquid sodium on solid NaCl. The vessel contains NaCl powder under an inert gas. Molten sodium is fed in from the top and the $TiCl_4$ vapour carried upward through the vessel by the inert gas stream which levitates the NaCl powder. To start the process the reaction chamber is heated to 600°C. The temperature is then maintained by the heat released from the reaction. Some of the powder is removed at intervals and heated, without exposure to air, at 800°C to consolidate the titanium metal. After cooling, the NaCl is then dissolved out in water.

10.5   Uranium

Uranium has a high free energy of oxidation and combines with carbon to form carbide, so that reduction by carbon at high temperatures is impracticable. Its melting point is rather too high (1132°C) for fused salt electrolysis. The iodide is fairly involatile and tends to attack silica. It is necessary then to reduce a halide with a reactive metal. The fluoride UF_4 is easiest to prepare and handle. It also has the most favourable energy of reaction and can be reduced by both calcium and magnesium.

In the *autothermic calcium process* the UF_4 powder, chemically purified by hydrometallurgical processes, is mixed with calcium chips in a calcium fluoride crucible. The mixture is ignited, either electrically or from a magnesium ribbon, and the heat of the reaction

$$UF_4 + 2Ca = U + 2CaF_2 \qquad\qquad 10.3$$

raises the temperature to about 1800°C. A billet of molten uranium covered with a protective CaF_2 layer is formed in the crucible. No special atmospheric control is needed because the uranium is protected from oxidation by a blanket of calcium vapour.

The *magnesium process* is very similar. Magnesium is a cheaper reducing agent than calcium but its heat of reaction with UF_4 is lower and so some external heating must be provided. Also, because of the low boiling point of magnesium, the reduction has to be done in a closed vessel.

10.6 Refractory metals

The melting points, in deg C, of some of the main refractory metals are as follows: W 3410; Re 3170; Os 3000; Ta 2980; Mo 2610; Nb 2470. Some of these, e.g. W and Mo, are easily reduced from their oxides; cf. Fig. 7.1. Others, e.g. Nb and Ta, are more reactive and it is usual to reduce their fluorides with sodium. Conventional crucible methods are not possible at the high temperatures needed to produce these metals as liquids. Inevitably, then, the method is to reduce the metal direct to the solid. This leads to two main features of the processes: first, the need to purify the metal compound by treatment with chemical reagents, usually in aqueous solution, before reduction; second, the need to consolidate and densify the reduced metallic powder by *sintering* in the solid state.

Advantage can sometimes be taken of the very high melting points to purify thin wires or sheets of these metals by heating them near their melting points in high vacuum. At such temperatures impurities can migrate through the solid (cf. § 19.8) and when they reach the surface most of them volatilize. This process is sometimes aided by deliberately adding traces of substances; e.g. rhenium, which can deoxidize other refractory metals by forming a volatile oxide (Re_2O_7).

10.7 Tungsten and molybdenum

Tungsten is obtained mainly from *scheelite*, $CaWO_4$, and molybdenum from *molybdenite*, MoS_2. In both cases the first stage is to prepare pure oxide powder by chemical treatments. This is then reduced to the metal by heating in a stream of hydrogen gas. Hydrogen is an effective reducing agent for these metals, particularly when used in flowing form so that the reaction product (H_2O) is swept away and kept at low activity. It gives a clean metal and there is no problem of carbide formation.

Tungsten is made by heating the purified WO_3 powder at 800°C in nickel boats under flowing hydrogen. The reduced metal is then cooled to room temperature under hydrogen, before opening to the air, to avoid oxidation. Most of it is used for electric lamp filaments. For this, it is first put under high pressure at room temperature in a steel mould to *compact* it into the form of a porous but coherent bar. This is slowly heated to 1300°C in hydrogen which allows the first stage of *sintering* to begin by the joining together of the grains of powder through the migratory movements of their atoms, mainly in the surface layers of the grains. An electric current is then passed through the bar, in hydrogen, to bring the temperature slowly up to 3000°C. This enables sintering to continue until the bar reaches almost the full density of the metal, very little porosity then remaining. The bar is then mechanically worked to shape, first to a rod at about 900°C, then to a wire by drawing through holes in diamond dies at room temperature. To prevent mechanical weakening and rapid failure of the filaments in service, a small amount of *thoria* (ThO_2) is usually added to the reduced powder, to become finely disseminated in the finished wire (cf. § 21.13).

FURTHER READING

FINNISTON, H. M., Editor (1954 *et seq.*). Series *Metallurgy of the Rarer Metals.* Butterworth, London.

INSTITUTION OF MINING AND METALLURGY (1951). *Extraction and Refining of the Rarer Metals.*

MENT, J. DE, DAKE, H. C. and ROBERTS, E. R. (1949). *Rarer Metals.* Temple Press, London.

ROLSTEN, R. F. (1961). *Iodide Metals and Metal Iodides.* John Wiley, New York and London.

Chapter 11

Iron and Steel Making

11.1 Introduction

The various irons and steels are all forms of the metal iron that differ in their content of carbon and other alloying elements and in the ways in which they are made. Carbon is by far the most important alloying element, being mainly responsible for the immense range of strengths and other useful properties that can be developed in steels.

The main ores are *haematite* (Fe_2O_3), *magnetite* (Fe_3O_4), *limonite* or *bog iron* (hydrated haematite) and *siderite* ($FeCO_3$). They are widely deposited, often in large surface deposits sometimes very rich in iron (e.g. 60 per cent Fe). The better deposits require little or no treatment before smelting. We see from Fig. 7.1 that the oxide is easily reducible, the temperature at which C can reduce FeO being about 750°C (the *additional* oxygen of the higher oxides is more easily removed). It is thus not surprising—in view of the abundance, richness and reducibility of iron ores and of the great strength of the metal and the variety of useful properties developable by alloying —that iron far outstrips all other metals in cheapness, production and general engineering use.

In addition to the abundance, richness and reducibility of the ore, one other fact plays a crucial part in the technology of iron and steel making: the high melting point of the metal (1535°C). In the presence of carbon the oxide is reduced to the *solid* metal, in the loose granular form called *sponge iron*, when its temperature reaches about 800°C. Since this is so far below the melting point, an attractive possibility is to reduce pure oxide direct to the solid by a stream of pure CO in a simple heating chamber such as a kiln. There is considerable interest at present in such a method. However, the technology of iron-making developed historically along quite a different direction. The earliest iron-making processes were, of course, necessarily limited to the solid state. Once the great technical barrier of reaching the high temperatures needed to melt the metal was overcome, however, the technology was transformed. The advantages of producing the metal in liquid form for large-scale continuous production, for separating it from the gangue so that a pure FeO starting material was not required, for casting and for controlling the composition, proved overwhelming. Despite the difficulties of the higher temperatures, the liquid state processes have now become so efficient that the alternative solid state processes have little chance of competing with them except in a few special circumstances.

11.2　The iron-making blast furnace

The decisive step came in the 14th century. For the preceding 3000 years sponge iron had been made in small hearths and shaft furnaces in which ore and charcoal were fed in at the top and an air blast provided by small manually driven bellows. The grains of metallic iron collected at the bottom, mixed up with slag in loose porous aggregates called *blooms*. The metal itself was fairly pure and because of this it was possible to *forge* these white hot and pasty blooms, by hammering, into compact dense bars with much of the slag squeezed out, so forming a type of *wrought iron*.

This decisive step was the introduction of more powerful bellows, driven by water mills, to give a much stronger air blast. This was done originally to increase production. It allowed much larger shaft furnaces to be built, up to 10 m high and 3 m internal diameter. But it also raised the temperature in the hearth sufficiently to produce a complete change in the process. As the sponge iron, reduced by CO, moved down the shaft into the white hot combustion zone it became hot enough to absorb *carbon*, up to 3 or 4 per cent, and this brought its melting point down to about 1150°C, well below the temperature of the hearth. A pool of *molten* iron-carbon alloy, *pig iron*, thus collected at the bottom of the furnace, separately from the slag which floated above it. This molten iron could then be periodically run out of the furnace and cast. Because of its high carbon content this pig iron could not be shaped by forging; it lacked the malleability of the carbon-free iron. But its low melting point enabled it to be shaped by casting into moulds, from which the *cast iron* industry has since developed. Its low melting point also enabled it to be melted in hearth furnaces, open to the air, so that the carbon could be removed by oxidation and carbon-free iron then produced as a new source of wrought iron.

Iron-making developed rapidly from this point onwards. By 1600 there were about a hundred iron blast furnaces at work in Britain, mainly near large forests (e.g. Forest of Dean and the Sussex Weald) from which charcoal could be obtained for the fuel and reducing agent. The British ores, being not very pure, gave a low-quality iron mainly used for making cannon balls. Some European countries, e.g. Sweden, possessed very pure ores from which they were able to make irons that were more malleable and also suitable for primitive steel-making processes.

The modern iron-making blast furnace, shown in Fig. 11.1, has evolved directly from these early shaft furnaces. Some of the main features developed very early, particularly the tall *stack*, the *bosh*, the ring of *tuyères* for the air blast and the *hearth* with its *slag* and *iron tapholes*. Another early feature was the conical shape of the stack and bosh, the furnace reaching its greatest diameter where they meet. The stack widens towards its base to allow the charge to fall and expand freely. This expansion is due to soot deposited on the charge by the reaction $2CO \rightarrow C + CO_2$ in the furnace gas rising upwards into the cooler regions of the stack. In the bosh, where the sponge iron and slag melt and drip down channels through the fuel to the hearth below, the shaft narrows to compress the charge and support it against the weight of material above.

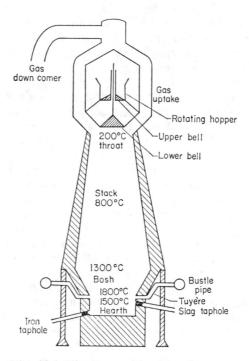

Fig. 11.1 The iron-making blast furnace

The charcoal used in the early furnaces was an expensive material, produced by consuming wood badly needed for other purposes, e.g. shipbuilding. Also, being mechanically weak, it could only be used in small low-shaft furnaces less than about 10 m high. Taller furnaces, when filled with charge, became choked up and impenetrable by the air blast because the charcoal in the bosh was crushed to a dense pad by the weight of material above. It is extremely important in a blast furnace to have a uniform network of clear passageways through the charge for the furnace gases to pass up the shaft. Air is by far the main constituent of a blast furnace. For example, 1 kg of iron can be made from 2 to 3 kg of ore and 1 kg of carbon; the oxidation of this carbon,

$$2C + O_2 = 2CO \qquad\qquad \textbf{11.1}$$

in the combustion zone at the tuyeres requires 1·3 kg of oxygen, i.e. nearly 6 kg of air. In a modern furnace air is supplied at over 3000 m³ per minute (at about 1·3 atm pressure).

Coal was tried as an alternative to charcoal but without great success. It was heavy, lacked porosity and strength, and contaminated the iron with sulphur which made it brittle. The decisive step was the use of *coke*, by *Abraham Darby* at Coalbrookdale (Shropshire) in 1709. This was mechanically strong, light and porous, could be made much more cheaply than charcoal and contained much less sulphur than coal. Coke smelting rapidly replaced the previous methods and the iron industry then moved from forested regions to the coalfields.

The furnaces grew larger and taller since the coke could withstand much greater weights than charcoal, a trend which has continued up to the present-day shafts, 30 m high and 8 m internal diameter. Hearth temperatures also increased and this led to some reduction of silica (cf. Fig. 11.2) to *silicon*, which dissolved in the molten iron. As a result, the *casting* qualities of the metal were greatly improved; partly because the silicon lowered the melting point so that the iron could run further along the channels of a mould; and partly because the silicon caused the dissolved carbon to precipitate out as *graphite* flakes in the iron as it solidified and the expansion due to this partly compensated for the normal freezing contraction of the metal, so that the casting was sounder and would take a good impression of the mould. Intricate iron castings of high quality could thus be made.

The coke-fired blast furnace has since been improved in the following ways:

(1) The air blast is heated to 600–800°C to reduce coke consumption (cf. § 5.4).

(2) The firebrick refractory lining round the hot bosh zone is water-cooled to prolong its life and so allow the furnace to be run continuously for about ten years. The hearth is often lined with carbon blocks.

(3) The exit gas from the furnace, which contains about 30 per cent CO, is collected, led through a *downcomer* pipe to a cleaning chamber and electrostatic dust precipitator and then used as fuel gas to heat the Cowper stoves and drive the blowing engines, etc.

(4) The charge is fed, from *skip* cars on an inclined ramp, into the top of the furnace through a gas-tight valve formed by two *bells* which are lowered and raised independently and consecutively. The upper bell and its hopper are rotated to spread the charge uniformly round the lower bell.

These improvements, together with the general increase of size of the furnace and other improvements discussed in § 11.4, have enabled modern furnaces to make over 2 million kg of iron per day. The molten iron is tapped off about four times a day by driving through a clay plug in the taphole with the aid of hot FeO flux in a burning oxygen jet. The slag is tapped more frequently. The iron is either run into a machine to be cast into *pigs*, i.e. bars about 2 m long and about 0·1 m deep and wide, or is run into a *mixer*, a large storage furnace in which it is kept molten, ready for steel-making furnaces. About 85 per cent of the pig iron is made into steel.

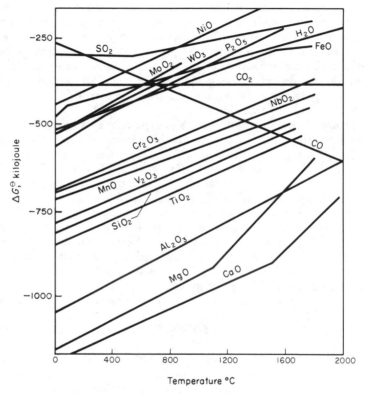

Fig. 11.2 Ellingham diagram for elements in iron and steel-making

11.3   Chemistry of iron-making

Iron-making is a *counter-current* process. The hot gas, rich in CO produced in the combustion zone, rises up the shaft and heats and reacts with the charge descending towards the hearth. Part of the iron oxide is reduced in the upper half of the shaft, at temperatures below 800°C, by *indirect reduction reactions* such as

$$FeO + CO = Fe + CO_2 \hspace{3cm} \textbf{11.2}$$

and the remainder is reduced further down the shaft by *direct reduction* reactions such as

$$FeO + C = Fe + CO \qquad \textbf{11.3}$$

the higher oxides in both cases having been first reduced through the sequence $Fe_2O_3 \rightarrow Fe_3O_4 \rightarrow FeO$, by CO or by C. The gas passes up through the furnace quickly (about five seconds) and so does not come into equilibrium with the cooler parts of the charge high in the stack. It is because of this that there is a large CO content in the exit gas even though the temperature in the *throat* of the furnace is only about 200°C.

Let us consider reaction 11.2 at 700°C (973 K), which is about the lowest temperature at which the Fe, FeO, CO, CO_2 system reaches equilibrium. From eqns. 7.13 and from the data in Fig. 11.2, we obtain for this reaction

$$\Delta G^{\ominus} = -19\,450 + 22 \cdot 6T$$
$$\log_{10} K = (1018/T) - 1 \cdot 181 \qquad \textbf{11.4}$$

and hence $K = 0 \cdot 73$ when $T = 973$. Since $K = p_{CO_2} p_{Fe} / p_{CO} p_{FeO}$ and assuming that we have pure solid Fe and FeO, then

$$p_{CO}/p_{CO_2} = 1 \cdot 37 \qquad \textbf{11.5}$$

Reduction occurs when this ratio is exceeded. For the *Boudouard reaction*

$$C + CO_2 = 2CO \qquad \textbf{11.6}$$

we similarly obtain

$$\Delta G^{\ominus} = +\,170\,700 - 174 \cdot 5T$$
$$\log_{10} K = -(8910/T) + 9 \cdot 11 \qquad \textbf{11.7}$$

and hence

$$p_{CO}^2/p_{CO_2} = 0 \cdot 887 \qquad \textbf{11.8}$$

at 973 K. Consider a blast furnace with a shaft pressure of $1 \cdot 3$ atm. Calculation based on the chemistry of the reactions shows that about 40 per cent of this is $CO + CO_2$, i.e.

$$p_{CO} + p_{CO_2} = 0 \cdot 52 \qquad \textbf{11.9}$$

Solving eqns. 11.8 and 11.9, we obtain $p_{CO}/p_{CO_2} = 2 \cdot 47$. This is higher than the value in eqn. 11.5 and so the FeO can be reduced at this temperature and pressure.

Limestone is added to the furnace *burden*, i.e. the charge, to flux the gangue into a fluid slag and also to remove sulphur through the reaction

$$FeS + CaO + C = Fe + CaS + CO \qquad \textbf{11.10}$$

and through a similar reaction with MnS. The lime ($CaCO_3$, which *calcines* to CaO in the stack) combines with silica and other unreduced oxides (Al_2O_3) to form a molten slag in the bosh. This drips down into the hearth and so passes through the combustion zone too quickly to be reduced appreciably. Unfortunately, the P_2O_5 is reduced and the phosphorous absorbed into the iron before the slag-forming temperature is reached.

Because it produces CO, reaction 11.10 is endothermic and so is favoured by high temperature. The temperature of the hearth is determined partly by the proportion of coke charged with the burden, and the air blast, and partly by the *free-running* temperature of the slag, i.e. the temperature at which the slag becomes fluid. If the rate of combustion at the tuyères is increased, this does not greatly raise the temperature there because the height in the bosh at which the slag drips away is then increased and this slag still falls into the hearth at its same free-running temperature. Since the aim is to remove sulphur, using lime, the free-running temperature is increased by deliberately adding excess lime, i.e. a *basic* slag with a *basicity ratio* $(CaO + MgO)/SiO_2 \simeq 1.5$ is made. The high temperature also favours the reduction of MnO and SiO_2, cf. Fig. 11.2, and the reduced manganese and silicon dissolve in the iron. A highly basic slag thus tends to produce an *acid iron*, i.e. rich in Si, despite the attraction of the CaO for SiO_2 in the slag.

Two main types of pig iron are made: *acid* (or *haematite*) iron, high in Si and low in P; and *basic* iron, low in Si and high in P. Their compositions, in weight per cent, are as follows:

	C	Si	P	S	Mn
Acid	3–4	2–3	<0.04	<0.04	1–2
Basic	2.5–3.5	<1.2	<2.5	<0.08	0.5–1.5

Since all phosphorus in the charge enters the iron, acid irons must be made from low-P ore. They are used for high quality cast iron and for acid steel-making. Most iron made in the United Kingdom is basic and is used for basic steel-making. A slag low in lime is used. This has a low free-running temperature, which saves coke, but little sulphur is removed unless the ore is high in manganese (to form MnS).

11.4 Recent developments in the blast furnace method

To improve efficiency, particularly in dealing with lower grade materials, increasing effort is being put into *preparing* the charge before it enters the furnace. Low-grade ores (20–30 per cent Fe) are concentrated by drying and calcining, a mild heating which drives off water and carbon dioxide, and are sometimes given more extensive *beneficiation* treatments, involving roasting to magnetite, followed by magnetic concentration, to improve their iron content. Low-grade United Kingdom ores are also mixed with rich imported ores (60 per cent Fe). The powdered mixture is then agglomerated into hard, strong, permeable lumps. In the *sintering* process this is done by mixing the prepared ore with powdered coke and lime and treating in a Dwight-Lloyd or similar sintering machine (cf. § 8.5). Alternatively, the powder may be pressed into *pellets* or small *bricks* and then baked hard.

Most of the other modern developments are aimed at reducing the consumption of strong coke and increasing the furnace output. The injection of fuel oil, gas and powdered coal with the air blast through the tuyères is now being practised. Production can be increased by 'driving the furnace harder', i.e. by increasing the blast pressure. Unfortunately, the tendency then is for the strong air blast to blow large channels for itself upwards through the charge, so giving very non-uniform conditions in the furnace. If the *top pressure* (i.e. pressure high in the stack) is also increased, however, e.g. to 2 atm instead of its normal value of 1·3 atm, the increased overall pressure allows the mass of air through the furnace to be increased without increasing the air velocity. Taking the $CO + CO_2$ content as 40 per cent of the total in the stack, eqn. 11.9 is replaced by

$$p_{CO} + p_{CO_2} = 0.8 \qquad\qquad \textbf{11.11}$$

for a 2 atm total top pressure. Solving eqns. 11.8 and 11.11 gives $p_{CO}/p_{CO_2} = 1.75$. Increasing the furnace pressure thus reduces the CO content of the exit gas and so reduces the consumption of coke. Production rates have been increased 20 per cent and coke consumption reduced 13 per cent by operating at increased top pressure.

Coke consumption can also be reduced by adding steam to the air blast. In the combustion zone this produces hydrogen through the reaction

$$H_2O + C = H_2 + CO \qquad\qquad \textbf{11.12}$$

and the preheating temperature of the blast has to be increased to compensate for the endothermic nature of this reaction. The hydrogen so produced reduces some of the iron oxide in the stack and, in so doing, releases heat there. An additional advantage of the steam addition is that it enables the oxygen content of the air blast to be increased. Normally, 0·8 of the air going into the furnace is nitrogen. This takes no part in the chemistry of the process and carries away much heat from the combustion zone, although some of this is given back at lower temperature to the charge in the stack. Reduction of the nitrogen content by enriching the blast with oxygen enables higher temperatures to be reached in the combustion zone, with a saving of coke. This can cause difficulties, however, by altering the heat balance in various parts of the furnace. If steam is also added, with the oxygen, the coke can still be saved without raising the temperature in the combustion zone. A further advantage is that adjustments of the oxygen-steam composition of the blast enable the furnace temperature to be changed quickly, so that close control of the operating conditions is then possible.

11.5   Wrought iron

The sponge iron produced by the earliest iron-making processes had the advantage of being *malleable*. This allowed it to be shaped and welded by forging at white heat with a hammer. The forged metal was also ductile, strong and tough when put to use at ordinary temperatures.

These properties were lost in pig iron, because of the embrittling effect of the high sulphur, phosphorus and carbon contents. Sulphur forms a network of sulphide films through the metal. At forging temperatures these films melt and allow the iron grains separated by them to fall apart (*hot shortness*). Phosphorus forms a hard Fe_3P compound which embrittles the iron at ordinary temperatures (*cold shortness*). Carbon similarly forms an embrittling Fe_3C compound. However, if the silicon content of the iron is high and the cooling rate of the casting is low, the Fe_3C is converted to *graphite* in the iron. The metal is then fairly soft and much less brittle (*grey cast iron*).

The charcoal-smelted pig iron produced from non-phosphoric ore was fairly free of sulphur and phosphorus. The 3 per cent or so carbon which it contained could be removed by melting and stirring in a crucible exposed to air, i.e. by selective oxidation. The iron of course solidified as it purified, through the rise $1150 \rightarrow 1535°C$ in its melting point. The solid was removed from the furnace and hammered to consolidate it into a dense form, so making a wrought iron. The *puddling process* for making wrought iron from coke-smelted phosphoric iron was perfected by *Henry Cort* in 1784. A low-sulphur pig iron is melted in a reverberatory furnace (cf. Fig. 5.3) at about 1300°C and oxidized, partly from air passing through the furnace and partly from iron oxide added to the charge. The C, Si, P and Mn contents are then reduced to low values ($\simeq 0·05$ per cent) by selective oxidation; the SiO_2, P_2O_5 and MnO so formed join with the remaining Fe_2O_3 to make a fluid slag. The iron solidifies during this refining process to form a pasty mass which is worked up into a large ball by manipulation with a *rabble* through the furnace doors. This is then taken from the furnace, hammered to squeeze out some slag and rolled to bars. The slag remaining in the metal is elongated into longitudinal threads and distributed more uniformly by putting rolled bars alongside one another and rolling them together while hot, to weld into a single bar.

The wrought iron produced in this way was the main metal used in structural engineering until the discovery of cheap steel-making processes. It has excellent ductility and toughness, fair mechanical strength and excellent resistance to atmospheric corrosion. Although not much used now, the puddling process was the starting point of the modern *open-hearth* steel-making process.

11.6   Early steel-making processes

Wrought iron is soft because it contains almost no carbon; and pig iron is brittle because it contains too much. Early experience showed that irons with intermediate carbon contents, i.e. *carbon steels*, could be superior to both in their combination of hardness, strength and toughness. An old method of achieving this was to heat forged sponge iron bar in charcoal for several days in red-hot air-tight boxes (*cementation* or *case-hardening*). Carbon atoms diffused into the surface of the solid iron and produced a hard surface layer on a soft malleable interior, an excellent structure for a material used for cutting tools and swords. This *blister steel* could then be improved into *shear steel* by hot forging and welding several bars together under a hammer, so giving a finely interlaced structure of hard and soft lamellae.

The interest in steel as a material for tools was greatly enhanced by the discovery, some 3000 years ago, that it could be hardened by *heat-treatment*, in particular by heating to a bright red heat (900–1000°C) and *quenching*, i.e. cooling quickly, by plunging into water; and that the quench-hardened steel could then be toughened by mild heating (*tempering*) or softened by heating to redness again (*annealing*). Early steel was prized mainly for this property and was produced as a rare, expensive metal almost entirely for high-quality tools and weapons.

In 1746 *Benjamin Hunstman*, making clocks in Sheffield, came to the conclusion that his steel clocksprings broke because the carbon in the metal was not uniformly distributed. To overcome this problem he melted the blister steel to improve homogeneity and so invented the *crucible steel* process, still used occasionally today. He was able to do this because crucible fireclays and coke-fired crucible furnaces had, by that time, developed to the level at which temperatures of 1600°C could be reached. Crucible steel-making in essence consists simply of adjusting the carbon content closely to the required value, e.g. 0·9 per cent C for cutlery steel, by mixing and melting together appropriate amounts of pure high-C and low-C irons, e.g. acid pig iron and wrought iron.

11.7   The Bessemer process

The whole concept of steel changed dramatically in the nineteenth century. An entirely new constructional material was discovered: *low-carbon steel*, strong, ductile, tough, not quench-hardenable but capable of being produced and shaped by mechanical working cheaply on an enormous scale. The modern age of steel was begun. The starting point was an experiment made by *Henry Bessemer*. Trying to improve the puddling process by applying a hot air blast to molten pig iron he made two discoveries. First, the air quickly removed the carbon and silicon from the iron. Second, and more striking, the heat released by this oxidation of the impurities in the iron was sufficient to keep the metal molten and *raise its temperature into the steel-making range* (1600–1650°C). He immediately saw the significance of this and, working with great speed and boldness, developed his *converter* process which he announced to the British Association in 1856.

The process is spectacular in both conception and operation. (The converter is an egg-shaped refractory-lined vessel capable of holding several tons of molten iron; cf. Fig. 11.3. It has an open mouth at the top, tuyères for introducing an air blast at the bottom and can be rotated into the horizontal or vertical position. The method is to set it horizontally, pour in a charge of molten pig iron through the mouth, turn on the air blast and then rotate into the vertical position so that the molten iron is supported by the stream of air coming into it from below. After 15–20 minutes, the converter is returned to the horizontal position, the air blast shut off and the steel is then ready for removal.

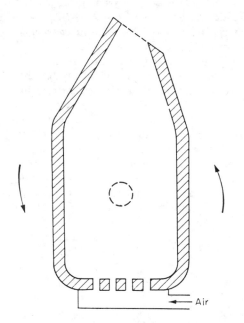

Fig. 11.3 Bessemer converter

This still remains the cheapest steel-making process. (The vessel is simple. No fuel is required, other than that provided by the impurities in the metal. Very large outputs are achieved, e.g. 25 tons of steel made in 20 minutes.

Bessemer's original product could not be cast satisfactorily into ingot moulds; it contained too much oxygen. This combined with carbon in the metal on cooling and the CO bubbles so formed caused the metal to froth up in its mould. The problem was solved in 1857 by *Robert Mushet*, who *deoxidized* the steel by adding a little manganese to it, in the form of *ferromanganese* or *spiegeleisen* (cf. § 11.11), before casting. The final stage of removing oxygen, called *killing*, is important in all steel-making processes. The carbon content of the deoxidized metal could then be adjusted to the required level by small additions and the cast metal then solidified quietly in its ingot mould. Manganese, which remains in the steel after deoxidation, has another important function. It combines with sulphur and converts the harmful iron sulphide films into harmless globules of manganese sulphide, MnS.

Bessemer was fortunate in using a non-phosphoric pig iron. When his process was applied to the more common phosphoric irons it failed completely because the phosphorus stayed in the steel and made it brittle. Compared with the puddling process, the vital difference here is the much higher temperature. We see from Fig. 11.2 that at high temperatures phosphorus is not oxidized in the presence of iron.

Bessemer could not solve this problem and had to confine his process to non-phosphoric irons because he used a fireclay refractory lining in the converter and was operating an *acid* process. Phosphorus cannot be removed from iron under these conditions because its own oxide is acidic and so cannot be slagged with acidic fluxes. The problem was not solved until

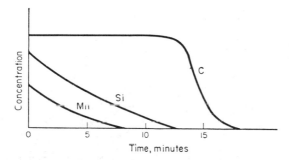

Fig. 11.4 Removal of elements in the acid converter

1878, when *Thomas and Gilchrist* used a *basic* lining of calcined dolomite (MgO . CaO) in order to be able to make a lime-rich slag and so remove the phosphorus as calcium phosphate. Steel-making by the *Thomas* or *basic Bessemer* process then became very popular in Europe although it was not taken up to any great extent in Britain.

In the acid Bessemer process most of the heat required is supplied by the oxidation of silicon, the content of which should be at least 2 per cent. An acid pig iron is thus used. Fig. 11.4 shows the sequence of removal of elements during the process. This can be understood from Fig. 11.2. At first, at a temperature of about 1200°C, the silicon and manganese are selectively oxidized and produce a short flame, with sparks, from the mouth of the converter. As the concentration of these elements falls and the temperature rises towards about 1650°C, the carbon in the metal begins to oxidize. The flame grows long and bright due to the burning of CO. When the carbon is exhausted the flame sharply *drops*, i.e. becomes short, and this is the signal for swinging the converter horizontally and shutting off the air blast. The manganese deoxidizer is now added. The carbon in this material helps bring the carbon content of the steel towards its final value (e.g. 0·5 per cent for railway track steel), but anthracite may also be added where necessary. A slag, mainly of SiO_2 with some FeO and MnO, which forms during the converter blow, is skimmed off and the metal then poured into a ladle.

In the basic converter lime is charged together with a basic phosphoric pig iron and some of the heat required is supplied by the oxidation of phosphorus, the initial content of which should be at least 1·7 per cent. Fig. 11.5 shows the sequence of changes. The first stage is completed earlier because there is less silicon in basic iron. After the CO flame drops the process is continued for 3–4 more minutes (the *afterblow*) to remove the phosphorus. Dense brown fumes of iron oxide smoke are emitted during this stage, since it is not possible to oxidize phosphorus without oxidizing iron as well (cf. Fig. 11.2). The phosphorus cannot be removed earlier

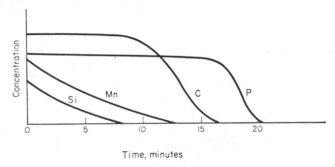

Fig. 11.5 Removal of elements in the basic converter

because the lime cannot assimilate the P_2O_5 until it has become a fluid slag and FeO is required to make such a slag with CaO. The FeO content of the slag cannot rise until the carbon content is low.

Following the afterblow stage the converter is brought horizontal and the air shut off. The slag is then removed (before deoxidizing, otherwise the phosphorus would be reduced and returned to the iron) and the metal deoxidized with manganese. Because of the almost complete removal of carbon which is necessary before the phosphorus is removed, the basic Bessemer process is used mainly for making low-carbon steels (e.g. 0·1 per cent C).

Little sulphur is removed in the Bessemer process. A method for de-sulphurizing pig iron, before making into steel, has been developed for dealing with irons not sufficiently low in sulphur. The molten metal from the blast furnace is poured on to *soda-ash* (Na_2CO_3) at the bottom of the ladle. The heat decomposes the Na_2CO_3 and sulphur is extracted as sodium sulphide in a slag of Na_2O and SiO_2 which is formed. In good quality steel the sulphur should be below 0·04 per cent. The acid Bessemer process cannot remove phosphorus and so, for good steel made by this process, the pig iron should contain less than 0·04 per cent P. Converter steels tend to contain about 0·015 per cent N, absorbed from the air blown through them. This nitrogen helps to harden the metal, which is useful in railway tracks, but otherwise it is an undesirable impurity. It is bad for sheet steel used for pressings, e.g. motor car bodies, and makes large steel plates and girders liable to brittle fracture. To reduce the nitrogen content, the air blast is sometimes partly or wholly replaced by a mixture of oxygen and steam.

11.8 The open hearth process

At present, much of the steel in Britain is made by the open hearth process and almost all of this by the basic open hearth process. This mainly produces *plain carbon steels* with compositions in the range (wt per cent): C 0·1 to 1·7; Si 0·02; Mn 0·4; S 0·05; P 0·05. The process was invented by *Siemens and Martin* in 1865–7. This was half-way in time between Bessemer's first discovery of the converter process and Thomas and Gilchrist's solution of the phosphorus problem. It thus appeared at a time when people were critical of the Bessemer process and became widely adopted as the basis of the then rapidly expanding steel industry in Britain. It also had the advantage, compared with the converter process, that it could use large amounts of steel scrap (e.g. 50 per cent scrap and 50 per cent pig iron) which by 1870 was already becoming widely available. By the time that the Thomas converter was invented the open hearth process had already become well established in Britain. Europe was less developed industrially at that stage and when its own steel industry formed it was able to base it on the Thomas process.

The open-hearth process was developed by applying the regenerator principle of pre-heating (cf. § 5.4) to the wrought iron reverberatory furnace. The higher temperatures which could then be reached enabled even low-carbon steel to be maintained in the molten state. In the gas-fired open-hearth process, producer gas and air are separately pre-heated in two chambers containing a checkerwork of firebricks and the outgoing hot gases similarly pass through and heat the other two chambers. Fig. 11.6 shows the general arrangement. In some furnaces oil fuel is used instead of gas.

The largest O.H. furnaces make some 500 000 kg of steel per charge and can usually be tilted to discharge. The process is slow, about 5 hours being spent in charging the furnace, 4 in melting and forming the slag and 3 in refining, finishing and discharging. Because of this and the greater cost of the plant, it is a more expensive process than the Bessemer. On the other hand, its slowness allows time to make chemical analyses of metal samples taken from the bath and generally enables better, more closely controlled steel to be made.

The process is one of controlled oxidation by the addition of iron oxide to the bath. Slag plays an essential part and the aim is to control the activity of oxygen in the slag by adding iron ore (Fe_2O_3) or mill-scale (Fe_3O_4), so that the oxygen in the iron, in approximate equilibrium with the slag, brings the composition of the metal to the required value. The shallow bath, with its large surface area, gives the necessary close contact of metal and slag.

In the *acid* process a silica-lined furnace is used, with an iron silicate slag. A non-phosphoric pig iron is charged, followed by steel scrap, and the charge then melted. This is a fairly expensive process and is used for making high-quality and alloy steels. The *basic* process, which produces cheaper structural steel, uses a magnesite, dolomite, or chrome-magnesite, hearth with a silica or chromite roof. The slag consists of limestone and iron oxide with some fluorspar, CaF_2, to increase its fluidity. Limestone, ore and steel scrap is charged first, heated to about 1200°C and molten pig iron is then charged, usually from a mixer. The high temperatures reached (1650°C)

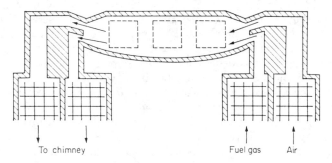

Fig. 11.6 Gas-fired open-hearth process

cause the furnace linings, particularly in the roof, to be severely attacked. The sequence of charging and melting is carefully designed to minimize this. It is necessary to repair the furnace lining between heats by *fettling*, i.e. throwing silica sand (acid) or dolomite (basic) into the damaged regions.

The refining sequence is broadly similar to that in the corresponding Bessemer process. First, the silicon and manganese are oxidized and removed. The bath is quiescent at this stage. Then comes the *boil*, in which the carbon is oxidized and the eruption of CO bubbles at the surface gives the impression of boiling. The bath becomes quiescent again when most of the carbon has gone. The final adjustments are then made by deoxidizing with ferromanganese and ferrosilicon (cf. § 11.11) and by adding coke or pig iron to bring the carbon to the required level.

The basic O.H. process has an important advantage over the basic Bessemer in that the phosphorus can be removed before all the carbon. Because the slag is made by charging iron oxide with the lime, this slag can form while there is still carbon in the metal. Its formation no longer depends on making iron oxide within the metal, which in turn depends on the elimination of the carbon. Moreover, since the furnace flame plays directly on to the slag, floating on the metal, fully molten slag can be produced at an earlier stage of the process. This early removal of the phosphorus increases the flexibility of the O.H. process and enables steels of a wider range of carbon contents to be made (e.g. 0·06 to 0·8 per cent C). Much of the phosphorus (and silicon) can be eliminated from the furnace during the boil by removing the slag and then making a new slag. This phosphorus-rich slag sometimes has value as a fertilizer. Some deoxidation and finishing additions are made while the metal is in the furnace, although the final additions are usually made afterwards when it is in the ladle.

Some sulphur can be removed in the basic O.H. process as MnS if the manganese content is high and also by using a large volume of lime-rich slag. Molten iron from a mixer is often desulphurized with soda ash before being charged into the O.H. furnace.

11.9   Killed and rimming steels

Deoxidation with manganese is often reinforced by deoxidation with silicon and aluminium. These are more powerful deoxidizers than manganese and the term *killing* usually refers to their use. *Silicon-killed* steel is made by adding ferrosilicon (75 per cent Si 25 per cent Fe) together with some silico-manganese which helps to float the silica away as manganese silicate. Aluminium-killed steel is made by adding a small amount of aluminium to the steel in the ladle. Al-killed steel is tougher than Si-killed steel at low temperatures (e.g. 0°C) but Si-killed steel is stronger at high temperatures (e.g. 400°C).

Most steel is cast into *ingot moulds* and then rolled into bars and sheet. Ingot moulds are vertical, approximately cylindrical, shells of grey cast iron, open at the top, square or polygonal in cross-section and capable of holding a few 1000 kg of steel. The steel is *teemed*, i.e. run out from a nozzle in the bottom of the ladle, either into the top of the ingot mould below it, or down a *trumpet* and along a *runner* into the bottom of the mould. To permit easy separation of the ingot, the moulds are slightly tapered from top to bottom. Depending on whether the ingot or mould is to be lifted off, the mould is set with either its wide or narrow end uppermost. Ingots are mainly cast with their wide ends up, to reduce the depth of *pipe*.

When a killed steel solidifies, the contraction which the metal undergoes produces a large contraction cavity or pipe, as shown in Fig. 11.7 (a). This unsound part of the ingot has to be cut off and returned for re-melting, so that the productivity is reduced. The effect can be minimized by delaying the solidification at the top of the mould, so that a reservoir of liquid is maintained there to feed into the ingot below as this freezes and contracts. This is done by fitting a thermally insulating refractory or even exothermi-cally-lined *hot top* or *feeder head*, as shown in Fig. 11.7 (b). Although this gives a considerable improvement, the metal discarded from the hot top and the cost of the hot top itself still add to the expense of the ingot.

The alternative is to prevent the metal from shrinking by allowing a small amount of CO to form as bubbles, when it freezes. This is done in *rimming steels*. These are low-carbon (< 0·15 per cent C) steels usually used for sheet steel pressings. Only low-carbon steels contain enough oxygen to rim satisfactorily. Silicon and aluminium are not used but some manganese is added to the ladle. With careful control, the CO effervescence is delayed until about 0·025 m of iron has solidified against the wall of the mould. The bubbles then form and compensate for the freezing contraction, so that no pipe is formed and none of the ingot has to be discarded. The CO blow-holes weld up when the ingot is afterwards forged. Another advantage is that the

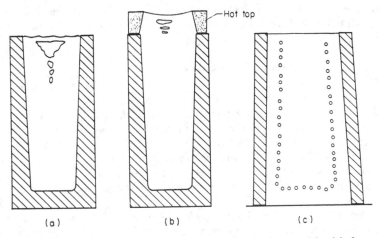

Fig. 11.7 Steel ingots (a) killed steel, (b) killed steel in mould with hot top, (c) rimming steel

oxygen in the metal captures its carbon from the first layers to solidify, so that these become very ductile carbon-free iron. This soft skin gives the metal a good smooth surface when it is afterwards rolled into sheet and pressed.

Cheap steel for girders, plates, etc. is usually made by a *balanced* or *semi-killed* method. It is slightly deoxidized with silicon or aluminium but not enough to require a hot top.

11.10 Electric steel-making

Electric arc furnaces of the type shown in Fig. 5.4 are now widely used for making high grade steels, especially alloy steels, from steel scrap. The main elements added to steel are Ni, Cr, Mn, V, Mo, W, Nb, Ti, mostly as ferro-alloys. We see from Fig. 11.2 that many of them are more oxidizable than iron. It is thus impossible to make most alloy steels under the oxidizing conditions that exist in a Bessemer converter or an O.H. furnace. For example, vanadium added to an O.H. bath would be oxidized away, $FeO + V \rightarrow Fe + VO$.

To make alloy steels, then, we have to reduce the oxygen activity in the furnace and also to work at higher temperatures so that the alloying elements become more reducible relative to carbon (cf. Fig. 11.2). The electric arc furnace enables both of these to be done. The hot carbon arc provides a slightly reducing atmosphere. High metal temperatures (e.g. 1800°C) can be reached in the hearth without the furnace roof being unduly damaged because, compared with the O.H. furnace, there is little flame inside the arc furnace.

The first stage in electric steel-making is to charge steel scrap and melt this down with the arcs. Some silicon, phosphorus and carbon may then be removed from the melt by directing a stream of oxygen on to the metal through a *lance*. This has the advantages of refining the metal very rapidly, which is important because the electric process must have high productivity to offset its intrinsic expensiveness, and of raising the metal to about 1800°C, the temperature at which CO is more stable than the alloy oxides, so that the alloy content of special scrap need not be lost. Slag formed during this refining stage is poured away and a new *reducing* slag is then made, mainly of calcium carbide CaC_2, by throwing in lime and anthracite. In this reducing environment the necessary alloys, ferro-manganese, ferro-chromium, ferro-vanadium, etc. can be fed into the bath without loss due to oxidation.

Large electric arc furnaces, e.g. 150,000 kg capacity, have recently come into wide use for making high grade plain carbon steels from steel scrap. They are now economically competitive with the acid O.H. method.

11.11   Ferro-alloys

Substances such as manganese, silicon, chromium, etc. are produced in large amounts, not as the elements themselves, but alloyed with iron, i.e. as *ferro-alloys*, for steel-making. The main alloys produced are *ferro-silicon*, (15 to 90 per cent Si), *ferro-manganese* (80 per cent Mn, 5 per cent C, 1 per cent Si), *spiegeleisen* (about 18 per cent Mn, 5 per cent C), *silico-manganese* (70 per cent Mn, 20 per cent Si, 1·5 per cent C), *ferro-chromium* (70 per cent Cr, 5 per cent C), *ferro-tungsten, ferro-molybdenum, ferro-vanadium* and *ferro-titanium*.

Some ferro-alloys are made in blast furnaces but the high reduction temperatures needed and the high carbon content of the product are unsatisfactory features. The other method now used is the *electrothermic* process. The charge is contained in a crucible and heated by a submerged arc. The charge consists of the mineral to be reduced, e.g. quartz for ferro-silicon, pyrolusite (MnO_2) for ferro-manganese, chromite ($FeO.Cr_2O_3$) for ferro-chromium, ilmenite ($FeO.TiO_2$) for ferro-titanium, together with coke, steel scrap or iron ore as required, and limestone flux where necessary. Carbon is an effective reducing agent for most of the ferro-alloys at the high temperatures achieved in the electric furnace, although aluminium or ferro-silicon is used to reduce ilmenite, in the presence of iron oxide, to ferro-titanium.

11.12   Oxygen steel-making

The availability in recent years of cheap 'tonnage' oxygen has revolutionized steel-making. Several new processes have been developed, with great success, for making steel by directing a stream of oxygen downwards on to molten iron. Fig. 11.8 shows the three main types; the *Kaldo* process, developed in Sweden; the *Rotor* process, developed in Germany; and the L.D. process, developed at Linz and Donawitz in Austria. The capacity of the vessels used in these ranges from 20 000 to 200 000 kg.

The Kaldo process is based on a slag-metal refining action, like the open-hearth, but the oxidizing slag is made by blowing oxygen on to the surface

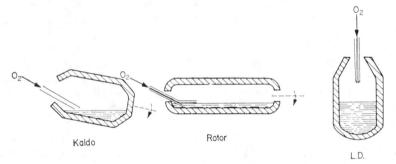

Fig. 11.8 Oxygen steel-making processes

instead of throwing in iron oxide. The sequence follows that of the open-hearth process but takes only about 35 minutes. Heat losses are small and no external heating is required, enough heat being generated by the reaction theoretically to allow 40 per cent steel scrap to be included in the charge. The burning of CO released into the vessel from the iron contributes substantially to the heating of the vessel. The vessel is rotated to stir the bath for heat transfer and to prevent the refractories from being locally over-heated by the CO flame. The process has advantages over both the O.H. and Bessemer processes. The slowness and expensive heating of the O.H. are avoided. The nitrogen problem of the Bessemer process is also avoided and a large proportion of steel scrap can be included in the charge.

The Rotor process is somewhat similar, but some oxygen is injected directly into the metal to give rapid oxidation and turbulence and some is blown into the vessel above to release heat by burning the CO. The long vessel provides a large slag-metal interfacial area to assist the removal of sulphur and phosphorus and it is slowly rotated to give a good heat transfer. Even simpler and more successful is the L.D. process. A simple crucible-like vessel is used and a high-speed jet of oxygen is injected into the molten iron from a water-cooled nozzle above. This avoids the problem of over-heated tuyères in the oxygen-enriched Bessemer process and enables nitrogen-free steel to be produced at the same rate as the Bessemer. The oxygen jet penetrates deeply into the bath and produces rapid oxidation throughout. The early formation of iron oxide under the jet enables the FeO-CaO slag to form rapidly, so that phosphorus is removed from the iron at an early stage. Because a blast of pure oxygen is used no heat is carried away by nitrogen and this allows a higher proportion of steel scrap (25 per cent) to be included, compared with the Bessemer process.

The L.D. process was originally developed in Austria to deal with low-phosphorus irons and it there proved an immediate success. It was not at first successful, however, when applied to the high-phosphorus irons of Britain and Western Europe. To deal with these a large amount of lime was needed, which increased the time required to form a fluid slag and remove the phosphorus.

The less enterprising steel-makers were at first discouraged by this, but the problem was solved in Luxembourg and France by injecting powdered lime with the oxygen jet from a *dispenser*, so producing the *L.D.A.C.* process. The lime goes in with the oxygen to the centre of the oxidation reactions where the temperature is extremely high and where the FeO concentration is high. The lime is thus continuously fluxed, as it enters, to produce a strongly dephosphorizing slag in the reaction zone, right from the start of the blow. In this way more phosphorus can be removed before the required carbon content is reached. The penetration of the oxygen jet is also less obstructed by slag during the early stages of the blow.

The L.D.A.C. process, because of its simplicity, economy, and ability to deal with a wide range of irons, has become extremely successful. The tendency now in new steelworks is to install L.D. vessels for converting iron into steel and electric furnaces for melting excess scrap and for making alloy steels. Existing Bessemer and open-hearth plants are still being widely used, of course, but are now gradually becoming obsolescent.

Very recently a new steel-making process, *spray-refining*, has been developed by the British Iron and Steel Research Association. A stream of molten iron from a blast furnace falls through a ring of jets, which breaks it up into droplets, about 1 mm diameter, and a second ring of jets which inject powdered lime and other fluxes. Silicon, manganese and phosphorus are removed and the carbon content is reduced to the required level, controlled by regulating the flow of oxygen and lime. The process is extremely rapid and promises to be very cheap.

FURTHER READING

BAILEY, A. R. (1960). *A Textbook of Metallurgy*. Macmillan, London.
BASHFORTH, R. G. (1956). *The Manufacture of Iron and Steel*. Chapman and Hall, London.

BODSWORTH, C. and BELL, H. B. (1972). *Physical Chemistry of Iron and Steel Manufacture.* Longmans Green, London.

BRAY, J. L. (1954). *Ferrous Process Metallurgy.* Chapman and Hall, London.

CHARLES, J. A., CHATER, W. J. B. and HARRISON, J. L. (1956). *Oxygen in Iron and Steelmaking.* Butterworth, London.

HAYWARD, C. R. (1952). *An Outline of Metallurgical Practice.* Van Nostrand, New York and London.

HOPKINS, D. W. (1954). *Physical Chemistry and Metal Extraction.* J. Garnet Miller, London.

JACKSON, A. (1964). *Oxygen Steelmaking for Steelmakers.* George Newnes, London.

PHYSICAL CHEMISTRY OF STEELMAKING COMMITTEE (1951). *Basic Open Hearth Steelmaking.* American Institute of Mining and Metallurgical Engineers, New York.

UNITED STATES STEEL CORPORATION (1957). *Making, Shaping and Treating of Steel.* Pittsburgh.

WARD, R. G. (1962). *An Introduction to the Physical Chemistry of Iron and Steel Making.* Edward Arnold, London.

Chapter 12

Kinetics of Metallurgical Reactions

12.1 Introduction

Thermodynamics tells us which way a reaction will go but not how fast it will go. A proposed scheme of metallurgical reactions might appear attractive thermodynamically yet be quite useless in practice because it involved a very slow reaction. We must therefore study the *kinetics* of metallurgical reactions as well as the thermodynamics. Kinetic effects can be either advantageous or disadvantageous. We have seen that advantage is taken of the rapid separation of zinc in lead droplets, in the zinc blast furnace process (cf. § 8.10), to suppress the thermodynamically-predicted oxidation of zinc vapour on cooling in CO and CO_2. Similarly, in basic O.H. and L.D. steel-making the removal of phosphorus from the metal is speeded up by bringing the lime and iron oxide together in the hottest part of the system. As examples of disadvantages we note the high temperatures required in tin smelting to produce a fair rate of reaction and the high CO content of the outgoing gas from the iron-making blast furnace, which increases the consumption of high-quality coke.

12.2 Theory of reaction rates

The classical theory of reaction rates deals with reactions between molecules in a homogeneous solution or gas at a constant temperature. Although this is an idealized situation and although the rate-determining factors in metallurgical reactions usually involve quite different things such as diffusion, heat transfer and nucleation, we shall nevertheless start with it to lay the ground-work for more realistic theories.

In a reaction in which, for example, atom B breaks away from atom A to join atom C, the reactants pass through an intermediate configuration, corresponding to the top of the activation energy hill in Fig. 6.3, in which the AB bond is partly broken and the BC bond partly formed, thus

$$AB + C \rightarrow ABC \rightarrow A + BC \qquad \textbf{12.1}$$

This intermediate configuration ABC is called an *activated complex*. It is like an ordinary molecule except for its instability in respect of the reaction 12.1. In the *transition state* theory of reaction rates it is assumed that, at least approximately, a chemical equilibrium exists between the reactants and the activated complex. This assumption allows the law of mass action (cf. § 7.4 and § 7.6) to be applied to the problem. To express it more generally, suppose that a particles of reactant A, and b of B, etc. unite into one 'molecule' of activated complex X. Then, from eqn. 7.25, the concentration c_X of activated complexes in the system is given by

$$c_X = (c_A^a c_B^b \ldots) W \exp{(-\Delta G/RT)} \qquad \textbf{12.2}$$

where c_A is the concentration of A, etc., ΔG is the *free energy of activation* per mole (cf. § 6.7) and W, often assumed to be unity, is an *activity factor* given by $W = (\gamma_A^a \gamma_B^b \ldots)/\gamma_X$ where $c_A\gamma_A$ is the activity of A, etc. The *rate of reaction* is then c_X/τ where τ is the transit time of the activated complex from (b) to (c) in Fig. 6.3. Clearly, τ^{-1} is of the order of magnitude of an atomic vibrational frequency, about 10^{13} sec^{-1}. Some theories assume $\tau^{-1} = h/kT$ where h is Planck's constant.

The sum $a + b + \cdots$ over all reactants is called the *order of the reaction*. For example, reaction 12.1 is second order. Simple radioactive decay (cf. eqn. 2.10) is first order. In chemical reactions in gases and solutions the order of reaction is generally the number of different molecules which must come together to react. Three-molecule collisions in dilute systems occur very rarely so that an apparent third or higher order reaction is usually a sequence of first and second order reactions. When one reaction in the sequence is very much slower than any of the others, it becomes the *rate-determining step* governing the speed of the overall reaction. The apparent order of reaction also depends on the relative abundances of the reactants. If one reactant is present in very small amount compared with the others, the rate of reaction is then determined effectively by the concentration of this reactant only, not the others, so that the reaction may appear to be first order.

To illustrate these effects, consider the reaction

$$\text{Si} + 2\text{FeO} \rightarrow 2\text{Fe} + \text{SiO}_2 \qquad \textbf{12.3}$$

in the steel bath. This appears to be third-order since two FeO molecules combine with one Si atom, dissolved in the molten iron. Since three-molecule collisions are rare, we must consider this reaction as occurring in stages. One possible sequence is:

$$
\begin{aligned}
&\text{(a)} \qquad \text{Fe} + \text{Si} \rightarrow \text{FeSi} \\
&\text{(b)} \qquad \text{FeSi} + \text{FeO} \rightarrow 2\text{Fe} + \text{SiO} \qquad \textbf{12.4} \\
&\text{(c)} \qquad \text{FeO} + \text{SiO} \rightarrow \text{Fe} + \text{SiO}_2
\end{aligned}
$$

If (a) is the slow step, the rate of oxidation of Si is proportional to c_{Si} only and independent of c_{FeO}. The overall reaction then appears to be first order since c_{Fe} = constant, Fe being the solvent. If (b) is the slow step, reaction (a) comes to an equilibrium in which $c_{FeSi} \propto c_{Si}$ according to the mass action law. The overall reaction then goes at a rate proportional to $c_{FeSi}c_{FeO}$, i.e. to $c_{Si}c_{FeO}$, and so appears to be second order. If (c) is the slow step, reaction (b) also comes to equilibrium, with $c_{SiO} \propto c_{FeO}c_{FeSi} \propto c_{FeO}c_{Si}$. The overall reaction then goes as $c_{FeO}c_{SiO}$, i.e. as $c_{FeO}^2 c_{Si}$, and so appears to be third order.

When a reacting system is near equilibrium we must take the rate of the *reverse* reaction into account. An example of this is provided by the oxidation of carbon in an open-hearth steel bath. This reaction takes place in two steps,

(a)	Oxygen in slag $\rightarrow$ oxygen in metal	
(b)	C + O $\rightarrow$ CO	**12.5**

both of which are usually near equilibrium in practice. The first reaction takes place at the slag-metal interface, the second probably at the surface of gas bubbles in the metal. The observed oxygen content of such a steel bath lies between the respective values for equilibrium with the slag and with the carbon, so that neither (a) nor (b) goes so fast, relative to the other, that its contribution to the overall rate can be ignored.

A reaction near to equilibrium goes at a rate proportional to the deviation Δc of the actual concentration of reactant from its equilibrium value. For reactions 12.5 let c be the actual concentration of O in the Fe, let c_s be the concentration of O in Fe for equilibrium with the slag and c_c likewise for equilibrium with the C in the Fe. Then reaction (a) goes at the rate $k_a A_s(c_s - c)$ and (b) likewise at $k_b A_g(c - c_c)$, where k_a and k_b are rate constants and A_s and A_g are the respective areas of the slag-metal and gas-metal interfaces. The amount of oxygen in the metal at any instant is generally very small compared with that required to remove the carbon, so that there is opportunity during the boil to develop a steady rate of overall reaction in which the two steps go at more or less the same speed, i.e.

$$\frac{c - c_c}{c_s - c} = \frac{k_a A_s}{k_b A_g} \qquad \textbf{12.6}$$

This analysis, based on the assumption of near-equality of the two reactions (a) and (b), is consistent with observed steel-making practice. Thus the rate of carbon removal can be speeded up either by adding more iron oxide or ore to the slag (together with more heat because reaction (b) is endothermic), so speeding up reaction (a); or by throwing lime into the bath, which forms gas bubbles in the metal through the reaction

$$CaCO_3 \rightarrow CaO + CO_2 \qquad \textbf{12.7}$$

and so, by increasing A_g, speeds up reaction (b), giving the so-called *lime boil*.

12.3   Homogeneous and heterogeneous reactions

At the high temperatures typical of metallurgical processes most chemical reactions go rapidly once the reactants are brought together. The activation energy for reaction is usually less than 200 000 J mol^{-1} and even with this energy the frequency of successful attempts at the energy barrier is extremely high at a temperature of, say, 1600°; thus

$$\nu \exp \left(-\Delta G/RT\right) \simeq 10^{13} \exp \left(-200\ 000/15\ 000\right) \simeq 10^7 \text{ sec}^{-1}$$

This high rate is confirmed by practical observations. For example, when Si-rich iron is thrown into a steel-bath to consume oxygen through the reaction 12.3 and so temporarily slow up the oxidation of carbon ('*blocking the heat*') to allow time for taking chemical samples and for adding alloys, the carbon boil ceases as soon as the Si-iron melts and dissolves in the bath.

Because of this speed the important rate-controlling factor in practice is the *transport* of reactants, products and heat, to and from the site of reaction. Many important reactions occur *heterogeneously*, i.e. at the interfaces between different phases in the system; examples are the slag-metal reactions in refining processes and the reactions in the Bessemer process between the metal and the air bubbles passing up through it. Another example is the *calcination* of lime thrown into the steel bath. The equilibrium pressure of CO_2 generated by reaction 12.7 far exceeds 1000 atm at steelmaking temperatures. The reaction is first order and we might expect it to produce a CO_2 explosion. In fact, however, the reaction goes slowly because it is endothermic. The layer of CaO formed on the surface of a limestone in the bath acts as a thermal insulator and restricts the flow of heat needed to allow the $CaCO_3$ inside to decompose.

12.4   Diffusion and heat conduction

Consider a *still* gas, liquid or solid. Although there is no overall motion of the medium, individual atoms or molecules are nevertheless moving about in it. The *diffusion* of a substance through the medium, from high concentrations to low, is brought about by the haphazard individual wanderings of the atoms and molecules. The path of an individual particle is an unpredictable zigzag. The length of an individual step in this zigzag is determined in a gas by the *mean free path* between successive collisions with other molecules, collisions which randomly change the direction of motion of the particle; and in a solid or liquid it is the *jump length*, i.e. the distance the particle is able to move when it acquires enough activation energy to make one jump. The jump length is usually of order of one atomic spacing $(\simeq 3 \times 10^{-10} \text{ m})$. Although an individual particle moves haphazardly, when large numbers move in this way they produce a systematic flow down a concentration gradient. Consider a plane perpendicular to the gradient. Because there are more of the particles on one side of this plane than the other, more will wander across this plane from the high concentration side than the other, and a statistical drift down the gradient will result.

Consider a diffusing substance with a volumetric concentration c (number of particles per unit volume) that varies along the x coordinate axis through the medium. Suppose that a particle moves in randomly directed steps of average length b at an average step frequency f. Let the concentration gradient dc/dx be small enough for the change of concentration over the distance b to be regarded as infinitesimal; this assumption is almost always valid in practical problems of diffusion.

In Fig. 12.1 we show a sequence of planes of unit area and of spacing b, normal to the concentration gradient. If c_1 and c_2 (where $c_1 < c_2$) are the concentrations of diffusing particles in planes 1 and 2 respectively, the corresponding numbers of such particles in the planes are $n_1 = c_1 b$ and $n_2 = c_2 b$, respectively. Then $n_1 f \, \delta t$ particles jump off plane 1 in a small time δt, where $f \, \delta t \ll 1$. On average, one-half of these jumps are in each direction, so that the number of particles which make the transition $1 \rightarrow 2$

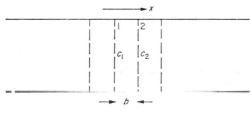

Fig. 12.1

is $\frac{1}{2} n_1 f \, \delta t$. Similarly the number of $2 \rightarrow 1$ transitions is $\frac{1}{2} n_2 f \, \delta t$. Hence the *flux J* of particles, i.e. the net number which pass through a plane of unit area in unit time, is given by

$$J = \tfrac{1}{2} n_1 f - \tfrac{1}{2} n_2 f = \tfrac{1}{2} b (c_1 - c_2) f \qquad \textbf{12.8}$$

Setting $c_1 - c_2 = -b(dc/dx)$, this becomes

$$J = -D \frac{dc}{dx} \qquad \textbf{12.9}$$

where

$$D = \tfrac{1}{2} b^2 f \qquad \textbf{12.10}$$

Eqn. 12.9 is *Fick's first law of diffusion* and D is the *diffusion coefficient*, usually measured in $m^2 \, s^{-1}$. The rate of diffusion is proportional to the gradient and is directed down the gradient, both of which follow directly from the statistical nature of the process.

We now develop the diffusion equation in a more general form, to deal with *non-stationary* flows in which the concentration at a point changes with time. Consider a volume element between two planes, as in Fig. 12.1, now spaced some larger distance l. Although $l > b$ we nevertheless keep l small enough for the difference in concentration across the volume element to be infinitesimal. If the concentration is c at plane 1, then that at plane 2 is $c + l(\partial c/\partial x)$. The flux across 1 is $J = -D(\partial c/\partial x)$ and that across 2 is

$$J + l\frac{\partial J}{\partial x} = -D\frac{\partial c}{\partial x} - l\frac{\partial}{\partial x}\left(D\frac{\partial c}{\partial x}\right) \qquad \textbf{12.11}$$

We use partial differentials now because we have to distinguish between the change of c with x at a given time t and the change of c with t at a given place x. The difference in the two fluxes at 1 and 2 gives the rate of accumulation of particles in the volume element. This is equal to $l(\partial c/\partial t)$. We thus obtain *Fick's second law*

$$\frac{\partial c}{\partial t} = \frac{\partial}{\partial x}\left(D\frac{\partial c}{\partial x}\right) \qquad \textbf{12.12}$$

When D is constant this reduces to

$$\frac{\partial c}{\partial t} = D\frac{\partial^2 c}{\partial x^2} \qquad \textbf{12.13}$$

The *conduction of heat* in a still medium follows the same mathematical laws as those of diffusion. In fact, Fick's laws were modelled on *Fourier's law*,

$$Q = -\kappa\frac{dT}{dx} \qquad \textbf{12.14}$$

for the heat Q conducted in unit time through a unit surface normal to a temperature gradient dT/dx in a medium of *thermal conductivity* κ (watt per metre kelvin, i.e. $\text{W m}^{-1}\,\text{K}^{-1}$). If we replace D by the *thermal diffusion coefficient* D_h, given by

$$D_h = \kappa/C\rho \qquad \textbf{12.15}$$

where $C\rho$ is the *specific heat per unit volume*, and replace c by T, then eqns. 12.9 to 12.13 apply also to thermal conduction.

The generalization of these equations to two or three dimensions is straightforward. In three dimensions eqn. 12.12 becomes

$$\frac{\partial c}{\partial t} = \frac{\partial}{\partial x}\left(D_x\frac{\partial c}{\partial x}\right) + \frac{\partial}{\partial y}\left(D_y\frac{\partial c}{\partial y}\right) + \frac{\partial}{\partial z}\left(D_z\frac{\partial c}{\partial z}\right) \qquad \textbf{12.16}$$

where D_x, D_y and D_z are the diffusion coefficients along the x, y and z axes. In an *isotropic* medium

$$D_x = D_y = D_z = D \qquad \textbf{12.17}$$

The equations can also be transformed to suit particular geometries. If the concentration (or temperature) gradients have *spherical symmetry* about a point, i.e. c varies only with the radial distance r from this point, eqn. 12.13 for an isotropic material with constant D becomes

$$\frac{\partial c}{\partial t} = D \left(\frac{\partial^2 c}{\partial r^2} + \frac{2}{r} \frac{\partial c}{\partial r} \right) \qquad \textbf{12.18}$$

Similarly, when the diffusion field has *radial symmetry* about a cylindrical axis, the equation becomes

$$\frac{\partial c}{\partial t} = D \left(\frac{\partial^2 c}{\partial r^2} + \frac{1}{r} \frac{\partial c}{\partial r} \right) \qquad \textbf{12.19}$$

where r is the distance from the axis.

For a *gas* we have, from the kinetic theory of gases,

$$D \simeq D_h \simeq \nu \simeq \tfrac{1}{3} \bar{v} \bar{l} \qquad \textbf{12.20}$$

Here ν is the *kinematic viscosity* η/ρ, where ρ is the *density* and η is the *dynamical viscosity*, i.e. *applied shear stress* divided by the *rate of shear* produced by it in a laminar flow of the medium; $\bar{v}$ is the *root mean square velocity* of the gas molecules, given by $\sqrt{(3kT/m)}$ where m is the molecular mass; and $\bar{l}$ is the mean free path of a molecule between successive collisions, given approximately by $kT/\pi\sigma^2 P$ in a gas of pressure P, where σ is the molecular diameter in collisions. For $T \simeq 1300$ K, $P \simeq 10^5$ N m^{-2} (1 atm), $m \simeq 5 \times 10^{-26}$ kg, and $\sigma^2 \simeq 2 \times 10^{-19}$ m^2, eqn. 12.20 gives $D \simeq D_h \simeq \nu \simeq 10^{-4}$ m^2 s^{-1}.

The fact that $D \simeq D_h$ for gases has some interesting consequences. Consider the passage of air bubbles through a molten metal, as in a Bessemer converter. If these bubbles are large, there is time only for the oxygen near their surfaces to reach the metal and react. It might be thought that the unreacted air inside would absorb heat and cool the metal. However, because $D \simeq D_h$, the heat from the metal penetrates by conduction only those same surface layers from which the oxygen is removed. The unreacted regions inside remain practically unheated.

For solids and liquids eqns. 12.20 do not apply even approximately. There is an *inverse* relation between diffusion and viscosity in liquids, given by the *Stokes-Einstein* formula

$$D = \frac{kT}{6\pi\eta r_a} \qquad \textbf{12.21}$$

where r_a is the radius of the diffusing particle and η is the dynamical viscosity. For liquid metals $D \simeq 10^{-9}$ to 10^{-8} m^2 s^{-1}. Slags, because of their higher viscosities, have lower diffusion coefficients. The diffusion coefficient in a solid is much smaller than that in the corresponding liquid, because the atoms are held in place by the crystal structure (cf. § 19.8). For a metal near its melting point $D \simeq 10^{-13}$ m^2 s^{-1}. Diffusion in a solid is a thermally activated process and follows the rate equation

$$D = D_0 \exp \left(-E_D/RT \right) \qquad \textbf{12.22}$$

so that D becomes extremely small at low temperatures. For metals D_0 is usually from 10^{-5} to 10^{-3} m^2 s^{-1} and $E_D \simeq 20RT_m$ where T_m is the melting point (K).

For solids and liquids $D_h \gg D$ because heat can migrate by processes not dependent on the migrations of molecules. The high thermal conductivities of metals are due to the movements of free electrons (cf. § 4.6). In non-metals the heat is conducted along the network of atomic bonds by oscillations of the atoms. Some orders of magnitude of the thermal conductivity κ, in W m^{-1} K^{-1}, are: *solid metals*, 10 to 400; *liquid metals*, 10 to 200; *slags*, 1·0 to 50; *firebricks*, 1·0 to 50; *insulating materials*, 0·05 to 1·0; *still gases*, 0·05 to 2.

Stationary or *steady-state* solutions of the diffusion equations are important for analysing the conduction of heat through furnace walls, recuperator tubes, gas ducts, etc. For one-dimensional diffusion through a flat *plate*, the steady-state condition $\partial c/\partial t = 0$ reduces eqn. 12.13 to $d^2c/dx^2 = 0$, where x is distance into the plate. This has the solution $c = Ax + B$. To find A and B we introduce the *boundary conditions* $c = c_0$ and $c = c_1$ respectively at the two surfaces $x = 0$ and $x = l$ of the plate. Substituting these values of c and x and solving for A and B, we obtain the solution

$$\frac{c - c_0}{c_1 - c_0} = \frac{x}{l} \qquad\qquad \textbf{12.23}$$

The flux J through the plate is then, from eqn. 12.9,

$$J = -D\frac{c_1 - c_0}{l} \qquad\qquad \textbf{12.24}$$

For radially-symmetrical diffusion through the wall of a hollow cylinder, the steady-state form of eqn. 12.19 is

$$\frac{d^2c}{dr^2} + \frac{1}{r}\frac{dc}{dr} = 0 \qquad\qquad \textbf{12.25}$$

which has the solution $c = A \ln r + B$. Introducing boundary conditions $c = c_0$ and $c = c_1$ respectively, at the two surfaces $r = r_0$ and $r = r_1$ of the cylinder wall, and solving for A and B, the solution becomes

$$c = \frac{c_0 \ln (r_1/r) + c_1 \ln (r/r_0)}{\ln (r_1/r_0)} \qquad\qquad \textbf{12.26}$$

The flux through any shell of radius r is $-2\pi r D(dc/dr)$, i.e.

$$J = -2\pi D\frac{(c_1 - c_0)}{\ln (r_1/r_0)} \qquad\qquad \textbf{12.27}$$

Since J decreases only logarithmically with increase in r_1, building up an excessively thick insulating lagging (e.g. $r_1 \gg 3r_0$) is a rather ineffective way of conserving heat in a pipe; it is better to use a thinner coat of a more highly insulating material. By contrast, a plate can be effectively insulated by a thick layer because J varies directly as l^{-1} in eqn. 12.24.

Non-stationary solutions of the diffusion equations are important in innumerable problems of metallurgical reactions; e.g. heterogeneous reactions at interfaces between phases, dissolution, precipitation, etc. The simplest problem is one-dimensional diffusion across a plane interface between columns of solution and solvent, as shown in Fig. 12.2. If the columns are sufficiently long for the concentrations at their outer ends to remain constant during the diffusion period, a solution of eqn. 12.13 can then be found, which is

$$c = \frac{c_0}{2}\left[1 - \frac{2}{\sqrt{\pi}}\int_0^{\frac{x}{2\sqrt{Dt}}} e^{-y^2}\, dy\right] \qquad \textbf{12.28}$$

where c is the concentration at time t at a distance x from the interface and c_0 is the initial concentration in the solution. The second term in the brackets is the *probability integral* or *Gauss error function* and its value depends

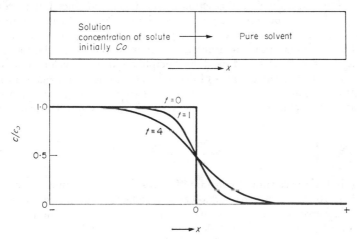

Fig. 12.2 Diffusion across a boundary from a solution into a solvent

entirely on the value of the upper limit $x/2\sqrt{Dt}$. Examples of the type of concentration curve obtained from this formula are shown in Fig. 12.2. As follows directly from eqn. 12.13, the concentration rises with time in those regions where the curvature $\partial^2 c/\partial x^2$ is positive, falls in those regions where the curvature is negative, and stays constant where the curvature is zero.

Many other non-stationary solutions have been derived for various geometrical and physical conditions. Naturally, they become increasingly complex as the conditions themselves become more complicated. In metallurgical reactions, where irregularly shaped lumps or drops of many phases may be reacting together under variable conditions, exact solutions of the diffusion equations are neither possible nor necessary. Fortunately, order of magnitude values can be obtained by a simple general argument. Diffusion is the overall result of *random walks* made by the individual migrating particles. The theory of statistics enables the average behaviour in a random walk to be analysed and shows that in a time t such a particle migrates on average a distance $\sqrt{Dt}$ from its starting point. This then is the distance over which the concentration (or temperature, with D_h replacing D) can change substantially in a time t. In the problem of Fig. 12.2, for example, $\sqrt{Dt}$ is the distance from the interface at which the concentration is changed by $0{\cdot}26c_0$.

12.5   Effects of mixing

The D values of liquids and solids are such that diffusion can occur through only short distances in most metallurgical processes. For example, if $D = 10^{-9}$ m^2 s^{-1}, then $\sqrt{Dt}$ is about 10^{-3} m in 15 minutes and 10^{-2} m in 1 day. These rates are impractically slow for slag-metal reactions between bulk phases. In practice, the rate of reaction is enormously enhanced by increasing the interfacial area between the reacting phases and by turbulent mixing. Very high rates can be attained if one reacting phase (e.g. molten metal) is allowed to fall as a spray of droplets through another (e.g. slag) and new metallurgical processes based on this principle are being developed. There is a limit, of course, to the fineness of droplet that can be used since very small ones settle too slowly. From *Stokes law*, a sphere of radius r_a and density ρ_1 falls under gravity through a fluid of dynamical viscosity η and of density ρ_2 at a speed

$$v = \frac{2}{9} \frac{(\rho_1 - \rho_2)}{\eta} g r_a^2 \qquad\qquad 12.29$$

where g is the gravitational constant ($9{\cdot}8$ m s^{-2}). For example, a sphere of radius 5×10^{-4} m and density 8000 kg m^{-3} falls through a slag, of viscosity 1 poise (i.e. viscosity at which a shear stress 1 newton per square metre produces a shear rate 1 s^{-1}) and density 3000 kg m^{-3}, at about $0{\cdot}025$ m s^{-1}.

Turbulent convection in fluids greatly speeds up reactions. Fresh, un-reacted material is continually swept into the reaction zones and sharp concentration gradients are repeatedly re-established there. For example, the agitation produced by the carbon 'boil' during steel-making, although not violent enough to stir the slag into the metal, does mix up the slag and the metal within themselves and in this way greatly speeds up the diffusion-controlled slag-metal reactions. Turbulence does not completely disperse the molecules of fluids amongst one another. It is a more macroscopic type of process than diffusion and individual masses of the fluid to some extent preserve their identity as they move about. The mixing occurs by individual regions becoming stretched out into thin irregular sheets and filaments. Diffusion has still to occur between neighbouring regions but the greatly increased interfacial areas and concentration gradients enable this to occur much more rapidly. In a slag-metal reaction in a well-stirred bath, the composition of the liquid is kept fairly constant by the turbulent mixing and the rate-controlling process is then the diffusion through a thin interfacial zone between the slag and metal. This zone appears to be typically about 10^{-5} to 10^{-3} m thick.

L. S. Darken (Basic Open Hearth Steel Making, p. 592) has applied these ideas to estimate the rate of removal of carbon from an open-hearth steel bath. He assumes that oxygen, from the slag floating on the metal, diffuses down through an interfacial zone of thickness l into the metal and is then immediately mixed in the metal bath below by turbulent convection. From eqn. 12.24, the flux of oxygen atoms through the zone is $D(c_s - c)/l$, where c_s is concentration of oxygen in iron for equilibrium with the slag and c is the actual concentration in the metal. Assuming that one oxygen atom removes one carbon atom, this also gives the rate of removal of carbon atoms from the metal, per unit area of slag-metal interface. If the concentrations are measured by weight, then multiplication by 12/16 gives the rate of removal of carbon by weight. The carbon removed per unit area of interface is removed in effect from the column of metal, of unit area and bath depth L, beneath that surface. The rate of carbon removal from the metal, in weight per unit volume, is then

$$Rate = \frac{12}{16} \frac{D(c_s - c)}{Ll} \qquad \qquad 12.30$$

Darken uses the values $D = 3 \times 10^{-8}$ m s^{-1}, $c_s = 0.23$ per cent, $c = 0.04$ per cent, $l = 3 \times 10^{-5}$ m, and $L = 0.76$ m, to obtain a rate of carbon removal of 0.67 per cent per hour, which agrees remarkably well with the observed rate during the period when the oxygen content of the slag is fully maintained by additions of iron oxide. During the normal carbon boil stage, the value $c_s - c = 0.04$ per cent gives carbon removal at the observed rate of 0.14 per cent per hour.

12.6 Nucleation

When a reaction leads to the formation of a new phase in a system, this new phase often appears first as small *nuclei* in the old phase, which then grow by the addition of more material from the old phase. The formation of raindrops in a cloud is a familiar example. Many metallurgical processes occur by nucleation and growth, e.g. the formation of CO bubbles in a steel-making bath during the carbon boil.

The theory of nucleation is difficult, but the following simplified argument brings out the essential features. Consider the formation of one nucleus by a thermodynamically reversible process at some particular place in the system. We shall see below that in general some work W has to be done to make this nucleus. The probability of the nucleus being formed is therefore reduced by the usual factor $\exp(-W/kT)$. Suppose now, however, that there are N similar places in the system where the nucleus could form. If N is sufficiently large, then we can expect to find nuclei at some of them; i.e. the number n of nuclei in the system is given, at least approximately, by

$$n = N \exp(-W/kT).\qquad\qquad 12.31$$

If the nucleus is very small and can form equally well at every place in the system, we can expect N to be of the order of magnitude of the number of atoms or molecules in the system. When the probability of forming a nucleus is the same everywhere, the system is in a state suitable for *homogeneous nucleation*. Unless specially arranged, however, this state of affairs hardly ever exists in practice. In large practical systems, e.g. in baths of molten metals and slags, there are usually some places where the work of nucleation is much smaller than average, e.g. on rough surfaces on the wall of the container or specks of foreign solid bodies in the reacting medium. Because W is small at these places, nucleation can occur preferentially on them and the system is then in a state suitable for *heterogeneous nucleation*.

The work of nucleation, W, is of course a free energy, ΔG, and we write this as follows,

$$\Delta G = a^3 \Delta G_V + \alpha a^2 \gamma.\qquad\qquad 12.32$$

Here, a is a linear dimension of the nucleus, so that a^3 is the *volume* and αa^2 is the *surface area* where α ($\simeq 6$, typically, for homogeneous nucleation) is a factor related to the *shape* of the nucleus. The *bulk* or *volume* free energy change ΔG_V is the free energy per unit volume required to form one large piece of the new phase. In Chapters 5, 6 and 7, it was implicitly assumed that all phases were present in large pieces, so that we could neglect the small proportion of atoms in the surfaces of the phases and hence could neglect their *surface energies*. Since a nucleus begins at an extremely small size we can now no longer neglect surface energy. As is shown by the positive *surface tensions* of liquids and solids, atoms and molecules in the surface layer of a phase generally have higher free energies than those in the interior. We shall see the reason for this in § 19.4. In eqn. 12.32, $\alpha a^2 \gamma$ is the surface energy term and γ is the surface free energy per unit area. We assume that ΔG_V and γ are independent of a and α.

The surface energy dominates the nucleation process. As eqn. 12.32 shows, it varies as the square of the nuclear dimensions whereas the bulk free energy varies as the cube. This brings in a *size effect*; ΔG depends on the absolute size of the nucleus. Since ΔG_V has to be negative to drive the nucleation reaction forward, ΔG must vary with a as shown in Fig. 12.3. The positive γ term dominates when a is small and the negative ΔG_V term dominates when a is large, so that ΔG reaches a maximum, $\Delta G = A$, at a critical size $a = a_0$. This is the smallest size for a self-sustaining nucleus, since the change of ΔG with a favours the disappearance rather than the growth of nuclei smaller than this. This critical size corresponds to an *unstable* equilibrium between the nucleus and its parent phase; ΔG makes larger ones grow and smaller ones shrink.

To find the work of nucleation A we have at $a = a_0$,

$$\frac{d}{da}\Delta G = 3a^2\Delta G_V + 2\alpha a\gamma = 0 \qquad\qquad \textbf{12.33}$$

so that

$$a_0 = -\frac{2\alpha\gamma}{3\Delta G_V} \qquad\qquad \textbf{12.34}$$

and

$$A = \frac{4\alpha^3\gamma^3}{27(\Delta G_V)^2} \qquad\qquad \textbf{12.35}$$

The *rate of nucleation*, J, is then given by

$$J = BN\exp(-A/kT) \qquad\qquad \textbf{12.36}$$

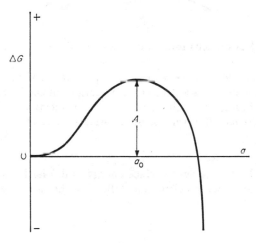

Fig. 12.3 The free energy of nucleation in relation to the size of the nucleus

where B is the frequency of encounters of a nucleus with atoms or molecules of the nuclear substance in the parent phase.

At the equilibrium point as defined by bulk free energy we have $\Delta G_V = 0$, $A = \infty$, and $J = 0$. Before nuclei can form the system has to *overshoot* this equilibrium point, i.e. pass into the thermodynamically unstable region where ΔG_V is negative. There is then a finite critical size at which a nucleus, if so formed, can grow. Such nuclei can be provided artificially by *seeding* the system with externally supplied fragments of the new phase, or they can be formed spontaneously within the system by thermal fluctuations at the rate given by eqn. 12.36. This is the basis of the familiar *undercooling*, *superheating* and *supersaturation* effects widely observed in nucleation processes. The greater the overshoot, the smaller is A and the faster is the nucleation (unless, of course, by extreme supercooling the molecules become so sluggish in their movements that B is severely reduced).

The overshoot needed is often large. For example, CO bubbles appear not to form homogeneously in a steel bath until a supersaturation equivalent to about 10 atm gas pressure is reached. This reluctance of CO bubbles to form within the bath (or at the slag-metal interface) is of great importance in open-hearth steel-making. The bubbles usually form *heterogeneously* on the rough solid surface at the bottom of the bath, like CO_2 bubbles in a glass of soda water. As they rise up through the metal and slag these CO bubbles produce the deep stirring necessary to promote a uniform and rapid rate of reaction throughout the bath and to get the heat from the top surface down into the metal.

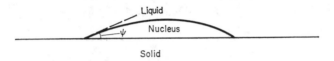

Fig. 12.4 A nucleus formed heterogeneously on a solid surface

A simple example of heterogeneous nucleation is shown in Fig. 12.4. Let γ_{SL} and γ_{SN} be the respective surface energies of the interfaces between the solid and liquid, and solid and nucleus, respectively. If the nucleus forms on a flat surface the equilibrium relation between the surface energies is

$$\gamma_{SL} = \gamma_{SN} + \gamma \cos \psi \qquad\qquad 12.37$$

where γ is the liquid–nucleus surface energy and ψ is the *contact angle*. A calculation of the critical nucleus size, following the same method as above, then gives

$$A = \frac{\beta \gamma^3 (2 + \cos \psi)(1 - \cos \psi)^2}{(\Delta G_V)^2} \qquad\qquad 12.38$$

where $\beta \simeq 4$. This is similar to eqn. 12.35 except for the factor $(2 + \cos \psi)$ $\times (1 - \cos \psi)^2$. When $\psi = 0$ nucleation can occur without overshoot since the original solid–liquid interface can be replaced by solid–nucleus and nucleus–liquid interfaces without increasing the surface energy. When $\psi > 0$ there is an activation energy and some overshooting. In this case, however, nucleation is further favoured if it occurs in a rentrant corner or *crevice* in the surface of the solid. In a steel bath the metal probably does not fully penetrate into sharp crevices and cracks in the refractory furnace lining. The CO can collect in these spaces, expand and then periodically break away as bubbles.

FURTHER READING

CARSLAW, H. S. and JAEGER, J. C. (1959). *Conduction of Heat in Solids*. Oxford University Press, London.

CRANK, J. (1956). *Mathematics of Diffusion*. Oxford University Press, London.

DARKEN, L. S. (1951). *Basic Open Hearth Steel Making*. American Institute of Mining and Metallurgical Engineers, New York.

— (1958). Chapter in *The Physical Chemistry of Steelmaking*. Chapman and Hall, London.

DARKEN, L. S. and GURRY, R. W. (1953). *Physical Chemistry of Metals*. McGraw-Hill, New York and Maidenhead.

FRANK-KAMENETSKII, D. A. (1955). *Diffusion and Heat Exchange in Chemical Kinetics*. Princeton University Press, New Jersey.

GLASSTONE, S., LAIDLER, K. J. and EYRING, H. (1941). *The Theory of Rate Processes*. McGraw-Hill, New York and Maidenhead.

HINSHELWOOD, C. N. (1951). *The Structure of Physical Chemistry*. Oxford University Press, London.

SZEKELY, J. and THEMELIS, N. J. (1971). *Rate Phenomena in Process Metallurgy*. Wiley-Interscience, New York.

WAGNER, C. (1958). Chapter in *The Physical Chemistry of Steelmaking*. Chapman and Hall, London.

WARD, R. G. (1962). *An Introduction to the Physical Chemistry of Iron and Steel Making*. Edward Arnold, London.

Chapter 13

Solids, Liquids and Solidification

13.1 Metal crystals

A liquid metal normally solidifies by *crystallization*. Metals can be produced in the *glassy* or *amorphous* solid state—hard but non-crystalline—but only with the utmost difficulty (cf. § 13.5) and in practice 'solid' is almost synonymous with 'crystalline' in metals. Crystallization ideally is a sharp change of state from the disordered atomic arrangement of the liquid to the ordered arrangement of the crystal, at a single temperature (*ideal freezing point = ideal melting point*).

We saw in § 4.6 that the free electron bond allows atoms in metals to pack together like close-packed spheres. Although not all metals behave in exactly this way it is nevertheless useful to examine the close packing of equal spheres. Fig. 13.1(a) shows a fully close-packed plane. The spheres, or atoms, lie in three sets of close-packed lines, physically equivalent and symmetrically orientated to one another. There are *two* fully close-packed crystal structures, known as *face-centred cubic* (F.C.C.) and *close-packed hexagonal* (C.P. Hex.), both formed by stacking a number of close-packed planes on one another in a certain *stacking sequence*.

Fig. 13.1(b) shows an element of the stacking sequence, a layer B laid on a similar layer A in the most closely packed position. The layers have the same orientation and each B atom rests symmetrically in a hollow provided by three adjoining A atoms. We notice that only one-half of the hollows are used by B atoms and that an equally close-packed arrangement would result if we had used the other hollows, i.e. the C positions in diagram (c), instead. Any stacking sequence chosen from the A, B and C positions is fully close-packed provided positions of the same type are not used by neighbouring planes. For example, ABACBABAC is close packed but ABBAAACCBBC is not.

When a certain stacking sequence repeats itself regularly in successive planes the structure is a crystal. The two simplest repetitions are ABABAB... etc. (or BCBC... or CACA...), which is the C.P. Hex. structure, and ABCABCABC... etc., which is the F.C.C. These sequences appear very consistently in many metals and *stacking faults*, such as for example ABABACABABA, are fairly uncommon. This shows that the cohesion of a metal cannot be entirely due to non-directional interactions between nearest-neighbour atoms. If it were, all close-packed sequences would then be equally favoured. On the argument of § 6.5 the regular sequences would then hardly ever appear.

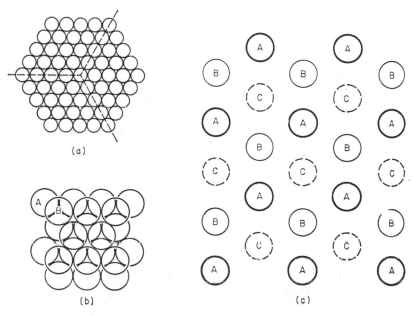

(a)

(b)

(c)

Fig. 13.1 Arrangement of atoms in (a) a close-packed plane, and (b) and (c) in successive planes

The pattern of atoms in a C.P. Hex. metal is best shown by a *structure cell* as in Fig. 13.2. The points in this diagram indicate *atomic sites*, i.e. the centres of positions occupied by atoms. The pattern of the whole crystal is made by continuing the pattern of the structure cell, in all directions, through a large number of atomic sites. This pattern is a *layer* structure of *basal planes*. The three atomic sites inside the structure cell belong to one basal plane layer and those on the top and bottom hexagonal faces of the cell belong to the two basal plane layers immediately above and below this one. Any apparent differences, which may be suggested by the fact that some sites are inside the cell and others are on its surfaces, are lost in the crystal, which consists entirely of the alternating ABAB ... sequence of basal planes.

In C.P. Hex. crystals the *axial ratio* c/a gives the ratio of the height c of the structure cell to the distance a between neighbouring sites in the basal plane. For ideal close-packing of equal spheres, $c/a = 1\cdot633$. Each atom then has twelve equidistant nearest neighbours, i.e. the *coordination number* is twelve. In practice a distorted form of the C.P. Hex. structure, in which $c/a \neq 1\cdot633$, is usually observed. Examples of C.P. Hex. metals are given in Table 13.1.

The F.C.C. structure, also with coordination number twelve, is shown in Fig. 13.2. The atoms lie at corners and face-centres of a cube. In diagram (c)

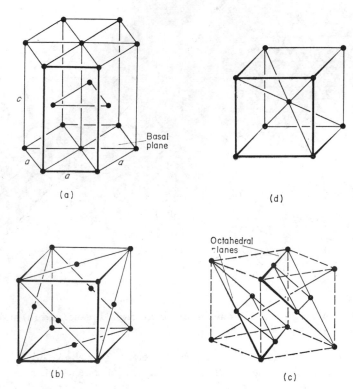

Fig. 13.2 Crystal cells, (a) close-packed hexagonal, (b) and (c) face-centred cubic, (d) body-centred cubic

one set of close-packed octahedral planes is emphasized. There are four distinct sets of such planes, with different orientations. Each of the other three can be generated from diagram (c) by rotating the cube 90° about one of the three cube edges. Table 13.1 gives some F.C.C. metals.

The third main crystal structure found in metals, the *body-centred cubic* (B.C.C.), is not built from close-packed planes. Because of this it is, for the same sphere size, less closely packed than the F.C.C. and ideal C.P. Hex. structures. Its structure cell, shown in Fig. 13.2(d), has atoms at the corners

TABLE 13.1. COMMON CRYSTAL STRUCTURES OF METALS

Close-packed hexagonal:

	Be	Cd	Co	Hf	Mg	Os	Re	Tl	Ti	Zn	Zr
b:	2·225	2·979	2·506	3·15	3·196	2·675	2·74	3·407	2·89	2·664	3·17
c/a:	1·57	1·88	1·62	1·58	1·62	1·58	1·62	1·60	1·59	1·86	1·59

Face-centred cubic:

	Ag	Al	Au	Ca	Ce	Co	Cu	Fe[1]	Ir
b:	2·888	2·862	2·884	3·94	3·64	2·506	2·556	2·585	2·714
	Li[2]	Ni	Pb	Pd	Pt	Rh	Sc	Sr	Th
b:	3·11	2·491	3·499	2·750	2·775	2·689	3·211	4·31	3·60

Body-centred cubic:

	Ba	Cr	Fe	K	Li	Mo	Na
b:	4·35	2·498	2·481	4·627	3·039	2·725	3·715
	Nb	Rb[3]	Ta	Ti[4]	U[5]	V	W	Zr[6]
b:	2·859	4·88	2·86	2·89	3·02	2·63	2·739	3·13

(b = distance of closest approach, in Ångströms; c/a = axial ratio; (1) at 950°C; (2) at −195°C; (3) at −173°C; (4) at 900°C; (5) at 800°C; (6) at 867°C; all others at room temperature)

and body-centre of a cube. The coordination number is only eight but this is compensated by six next nearest neighbours at a distance only $2/\sqrt{3} = 1.16$ larger than that of the nearest neighbours. Table 13.1 gives some B.C.C. metals.

Some metals crystallize in more than one structure and each of the exhibited structures exists (in equilibrium) exclusively over a certain temperature range. This *polymorphism* is fairly common, particularly amongst the transition metals. For example, crystalline iron is stable as B.C.C. below 910°C and above 1400°C; and as F.C.C. between 910°C and 1400°C.

The simple crystal structures of Fig. 13.2 only appear widely in the 'true' metals, i.e. with one, two or occasionally three valency electrons. Some metals form much more complicated structures. One form of manganese, for example, has a structure rather like B.C.C. but with twenty-nine atoms at every cube corner and cube centre. The simple B.C.C. form of uranium is stable only at temperatures above 775°C. At lower temperatures more complicated structures are formed.

The less metallic elements, e.g. Bi, As, Se, generally form more complicated structures. This is because the metallic (free electron) bond is partly replaced by the covalent (electron pair) bond. In some of these structures atoms are bound to some neighbours by covalent bonds and to others by metallic bonds. The atoms are then bonded covalently into chains or sheets and these are then held together by metallic and van der Waals bonds. In As, Sb and Bi, *layer* structures are formed and in each layer an atom has three close neighbours, in agreement with the chemical valency of these elements. Some elements of high valency do behave as true metals, however. An obvious example is F.C.C. lead with four valency electrons per atom. This is usually explained on the basis of incomplete ionization, i.e. some electrons are not given up to the free electron gas or used for bonding but are held back within the parent atom.

13.2   Liquid metals

When a crystal melts, or a liquid crystallizes, practically all the properties of the substance change sharply by finite amounts. The most spectacular is of course the change in resistance to shear stress; the *rigidity* of the crystal changes to the *fluidity* of the liquid and vice versa. In terms of viscosity this is a change by a factor of order 10^{20}. Other properties change less dramatically. For example, most metals increase in volume some 2 to 4 per cent on melting (a few, with crystal structures of low coordination number, such as bismuth and gallium, contract on melting). We can think of the latent heat of melting as the energy required to pull the atoms apart to the more openly packed structure of the liquid. The latent heat of melting varies from about 2000 J mol^{-1} for metals with weak cohesive forces, such as the alkalies, to about 20 000 J mol^{-1} for those with strong forces, such as tungsten and similar transition metals. These values are only about 3 to 4 per cent of the corresponding latent heats of vaporization, in which the atoms are pulled apart completely. The melting points of metals, given in Table 13.2 also increase with the cohesive forces. It follows that the entropy of melting (i.e. latent heat divided by absolute melting point) is fairly constant for all metals. For most metals its value ranges from 5 to 20 J mol^{-1} K^{-1}. The free energy of the liquid is higher than that of the crystal at low temperatures, where the internal energy largely determines the free energy, but it falls more steeply than that of the crystal as the temperature is increased, due to the extra entropy of the liquid, and crosses that of the crystal at the melting point, as in Fig. 13.3.

The smallness of the volume change shows that the atoms in a liquid metal must be fairly closely packed, at least at temperatures near the melting point. Any fairly close packed structure *cannot be completely random*. If the atomic spheres were randomly packed at this high density they would overlap one another extensively. The interatomic forces in fact fix a characteristic radial distance from the centre of each atom at which the nearest neighbours to this atom are most likely to be found. This is not a random arrangement and in this respect the structure of a liquid (or a very dense gas) differs from that of a typical gas. The liquid, in so far as it has this *short-range order*, i.e. correlation in the positions of neighbouring atoms, bears a slight resemblance to a crystal. This regularity has nothing to do with a *crystalline pattern*, however, i.e. with the *long-range order* or correlation in the positions of distant atoms found in the crystal. It is a simple consequence of the dense packing and the finite sizes of the atoms.

This conclusion is borne out by determinations of the structure. The arrangement of atoms in both crystals and liquids is determined from the scattering of x-rays (or electrons, or neutrons). Each atom scatters an in-

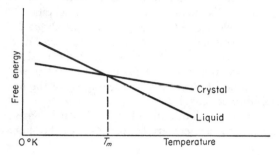

Fig. 13.3 Intersection of free energy/temperature curves of liquid and solid at the melting point T_m

coming beam radially outwards in all directions. The scattered waves from different atoms produce interference effects where they overlap and so develop into well-defined beams radiating outwards in certain directions from the scattering substance. By analysing the directions, amplitudes and

TABLE 13.2.   MELTING POINTS (°C)

Ag	Al	Au	B	Ba	Be	Bi	C	Ca	Cd
960·8	660	1063	(2100)	714	1277	271·3	3700[S]	838	320·9
Co	Cr	Cs	Cu	Fe	Ga	Ge	Hf	Hg	In
1495	1875	28·7	1083	1536	29·8	937	2250	−38·36	156·2
Ir	K	La	Li	Mg	Mn	Mo	Na	Nb	Ni
2455	63·7	920	180·5	650	1245	2610	97·8	2470	1453
Os	Pb	Pd	Po	Pt	Pu	Ra	Rb	Re	Rh
(3000)	327	1552	250	1769	640	700	38·9	3170	1965
Ru	Sb	Sc	Se	Si	Sn	Sr	Ta	Te	Th
(2500)	630·5	1540	217	1410	231·9	768	2980	499·5	1750
Ti	Tl	U	V	W	Y	Zn	Zr		
1670	303	1132	1860	3410	1510	419·5	1852	(S) = *sublimes*	

sharpness of the scattered beams the structure of the substance can be deduced. The long-range order of the crystal leads to the sharply defined diffracted beams discussed in § 17.7. Diffuse haloes are obtained from liquids and these enable the average distances between atoms to be obtained.

Fig. 13.4 shows some typical results, on liquid gold. The curve gives the *radial density function* $\rho(r)$, i.e. the numbers of atoms per unit volume at various radial distances from any atom, relative to the average number per unit volume in the liquid. The vertical lines indicate the number $n(r)$ of neighbours at various distances in the F.C.C. gold crystal. The strong peak (representing about eleven nearest neighbours) at about 3 Å corresponds closely with the nearest-neighbour distance in the crystal. Successive, less well-defined, peaks occur at the distances expected for other shells of atoms round the nearest-neighbour shell. They do not correlate with the positions of atoms in the crystal, showing that the regularity of the liquid is different from the crystallinity of the solid.

We prefer in fact to think of a liquid as similar to a dense gas rather than a 'blurred' crystal. It is a necessary consequence of the short-range charac-

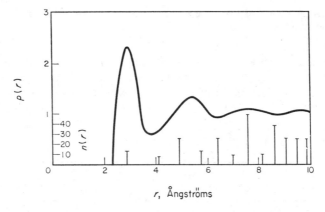

Fig. 13.4 The relative density $\rho(r)$ of atoms in liquid gold at various radial distances r from any atom (after *H. Hendus*, 1947, *Zeit. Naturforsch.*, **2A**, 505)

ter of interatomic forces that there should exist a condensed state of matter which merges *continuously* into the gaseous state at a certain temperature and pressure. This is seen most simply in the *van der Waals* theory of the critical point in imperfect gases. A crystal is totally divorced, by its long-range order, from the type of structures—isotropic and random—possessed by gases and there can be no continuous transition from one to the other. A simple liquid is isotropic in structure, with no long-range correlations between the positions of its atoms. Its short-range regularity can diminish continuously as it expands and gains more free volume between its atoms. We should therefore picture a liquid as a dense gas in which the atoms are held together not by external pressure but by the attractive forces between the atoms.

According to this view, the structure of the liquid is fundamentally different from that of the crystal. The discontinuous change in properties (except free energy) which occurs on melting or freezing is then not surprising. Nor is it surprising that no critical point has ever been discovered at which the liquid and crystalline states would merge together. These states are completely separated from each other, as are the gaseous and crystalline states, by the crystalline symmetry of the crystal and the isotropic symmetry of the liquid.

Evidence has sometimes been quoted, e.g. from changes of viscosity, to suggest that 'pre-crystallization' and 'pre-melting' take place, i.e. that the old phase begins to change its structure towards that of the new phase *before* the thermodynamic transition temperature is reached. This would contradict our conclusion about the fundamental dissimilarity of the liquid and crystalline states and we must regard such evidence as doubtful. Some of the observed effects may be due to impurities and disappear when the purity is increased.

13.3 Nucleation of crystals from a melt

A theory of the nucleation of crystals from a melt has been given by *D. Turnbull and J. C. Fisher*, (1949, *J. chem. Phys.*, **17**, 71). The following is a simplified argument. Consider *homogeneous nucleation*, i.e. in which groups of atoms in the liquid spontaneously rearrange themselves into the crystal structure of the solid. At least one nucleus must be formed. Hence, with $n = 1$ in eqn. 12.31 the work of nucleation could not greatly exceed about $kT \ln N$, i.e. $W \simeq 50kT$ if $N \simeq 10^{23}$. From eqn. 6.12 we have

$$d(\Delta G_V)/dT = \Delta S,$$

where ΔS is the extra entropy of the liquid relative to the crystal, per unit volume. We neglect the (small) change of ΔS with temperature and use the value at the true melting point T_m, i.e. $\Delta S = \Delta Q/T_m$, where ΔQ is the latent heat of melting per unit volume. We write $\Delta Q = L_m/M$, where L_m is the latent heat per mole and M the molar volume; also let θ be the *degree of undercooling*, i.e. $\theta = T_m - T$ where T is the nucleation temperature. Then

$$\Delta G_V = \theta \frac{d(\Delta G_V)}{dT} = \frac{\theta L_m}{M T_m} \qquad\qquad \textbf{13.1}$$

and hence, from eqn. 12.35,

$$A = \beta \gamma^3 \left(\frac{M T_m}{\theta L_m} \right)^2 \qquad\qquad \textbf{13.2}$$

Freezing by homogeneous nucleation should then begin approximately at the temperature T given by

$$\beta \gamma^3 \left(\frac{M T_m}{\theta L_m} \right)^2 = 50kT \qquad\qquad \textbf{13.3}$$

If we knew the interfacial energy γ between the crystal and its melt, we could now predict the undercooling. In practice we have to see what value of γ is implied by observed undercoolings. The experiments mentioned below have shown that $\theta \simeq 0.2T_m$. For typical metals, where $T_m \simeq 1000$ to 2000 K, we take $L_m/M \simeq 1.5 \times 10^9$ J m^{-3} and $kT_m \simeq 2.5 \times 10^{-20}$ J. This gives $\gamma \simeq 0.1$ J m^{-2}. We shall see below that this value is reasonable. The critical nucleus, corresponding to these values, contains about 200 atoms.

Turnbull's measured values of θ, and more precisely deduced values of γ, are given in Table 13.3. We see from these and from $\theta \simeq 0.2T_m$ that γ (J m^{-2}) $\simeq 10^{-4}\ T_m$ (K). Thus, if we take 10^{-19} m^2 as the area of an atom on the interface, then γ (J atom^{-1}) $\simeq 10^{-23}\ T_m$ (K). The average entropy of melting in metals is about 10 J mol^{-1} K^{-1}. Hence L_m (J mol^{-1}) $\simeq 10\ T_m$ (K). Converting to joule per atom this becomes L_m (J atom^{-1}) $\simeq 2 \times 10^{-23}\ T_m$ (K). The value of the interfacial energy between the crystal and liquid is thus equivalent to 'half-melting' each atom of the interface, which seems reasonable since such atoms are in a state intermediate between that of the liquid and that of the solid.

The method used to measure the degrees of undercooling in Table 13.3

TABLE 13.3. DEGREES OF UNDERCOOLING OF PURE METALS AND CALCULATED CRYSTAL–LIQUID INTERFACIAL ENERGIES

	Hg	Ga	Sn	Bi	Pb	Sb	Ge	Ag
θ, K:	58	76	105	90	80	135	227	227
γ, 10^{-3} J m^{-2}:	24.4	56	54.5	54	33	101	181	126

	Au	Cu	Mn	Ni	Co	Fe	Pd	Pt
θ:	230	236	308	319	330	295	332	370
γ:	132	177	206	255	234	204	209	240

(D. Turnbull, 1950. J. chem. Phys. **18**, 769)

is based on a technique of B. Vonnegut (1948, J. Colloid Sci., **3**, 563). The problem is to avoid heterogeneous nucleation on foreign inclusions suspended in the liquid. This is solved by breaking the liquid up into a large number of small droplets, 1 to 5×10^{-5} m in diameter, which are separated from one another by being suspended in some other liquid such as oil or slag, or by being spread out on a supporting plate of some non-nucleating material. The idea is to break up the liquid into more drops than there are foreign particles. Then, although some drops contain particles which cause freezing at small undercoolings, others are free from them and do not freeze until the temperature range of homogeneous nucleation is reached.

The freezing points of particular droplets were well-defined and repeatable in repeated cycles of melting and freezing. Each droplet possessed its own characteristic undercooling temperature so that a 'spectrum' of undercooling temperatures could be constructed as in Fig. 13.5. This regularity suggests that the foreign particles nucleate at definite degrees of undercooling, presumably according to the efficacy of their surfaces for heterogeneous nucleation.

Once a highly undercooled droplet is nucleated, freezing is then very rapid. This differs from the casting of metals under practical conditions in a foundry where heterogeneous nucleation occurs at temperatures within 10° of the ideal freezing point. The rate of growth of the solid on the nucleus in this case is much smaller, being controlled by the rate of removal of the latent heat released at the solid–liquid interface.

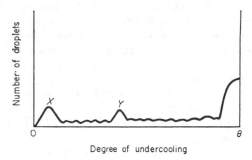

Fig. 13.5 The number of droplets which freeze at a given undercooling; specific catalysing particles cause freezing at X and Y of those droplets which contain them; homogeneous nucleation of remaining liquid droplets occurs when the degree of undercooling is θ

13.4 Nucleation of melting

It might be thought that when a crystal is heated melting should be delayed until a superheating has occurred comparable with the undercoolings discussed above. This is not so. Melting can begin at the surface of the crystal without significant superheating. To understand this we consider a situation, analogous to that of Fig. 12.4, in which a nucleus of liquid forms on the surface between the solid and its vapour. In general, the solid–vapour interfacial energy is greater than the sum of the solid-liquid and liquid–vapour interfacial energies, so that the formation of a film of liquid on the surface of the solid is *not* opposed by surface energy. In gold, for example, the three energies are, in J m^{-2}: solid-vapour 1·4, solid–liquid 0·132, liquid–vapour 1·128. In fact the gain in surface energy, e.g. 0·140 J m^{-2} for gold, implies that a thin film of liquid should form on the surface of the crystal *before* the melting point is reached.

Superheating can be produced, however, by heating the interior of a crystal, e.g. ice heated internally by focused infra-red rays, so that the surface is kept cold. It is also possible in certain circumstances for solid in fine cracks in the wall of the crucible or container, or in crevices in foreign inclusions in the material, to become superheated without melting (cf. § 13.6). This is of some practical importance because these unmelted fragments act as solidification nuclei when the melt is afterwards cooled down again and they cause the liquid to solidify into crystals of the *same orientation* as those which existed before melting. To change the crystal orientation it is therefore usually necessary to overheat the liquid to well above the true melting point, so as to melt the remaining fragments of solid in cracks and crevices.

13.5   Glasses

The high viscosity of a silicate, as in a slag for example, is due to the network of strong chemical bonds which link the molecules together. It is difficult for a molecule to break away from its neighbours. As a result, all processes of molecular movement such as diffusion and crystallization are slowed down, and it is easy to undercool the liquid to such a degree that, while still non-crystalline, it becomes a *solid* in its mechanical behaviour (i.e. its viscosity exceeds 10^{14} N s m^{-2}). This same high viscosity prevents the material from crystallizing and keeps it in this metastable *glassy state*. Broadly, the easy glass-forming liquids are those with activation energies for viscous flow greater than about 25 RT_m per mole. This includes silica, sulphur, selenium and glycerol.

Liquid metals, with their low viscosity (activation energy $\simeq$ 3 RT_m), are about the least suitable substances for forming glasses. Metal glasses can in fact be made by allowing atoms of a vaporized metal to condense on a very cold surface at the temperature of liquid helium (4 K). The thermal energy at this temperature is too low to permit any significant readjustment to take place in the positions of the atoms, which must therefore remain largely in the random positions in which they landed on the cold surface. These amorphous metals prove to be very unstable, however, and usually crystallize when warmed up, even to as little as 20 K. It appears that the liquid structure in a metal is exceptional in the smallness of its viscosity ($\simeq 10^{-3}$ N s m^{-2}) at most temperatures. In fact viscosity is one of the least variable factors in the technology of metal casting. The ability of a liquid metal to flow into a mould and penetrate all the corners is generally limited not by viscosity but by factors such as heat transfer and rate of crystallization (see § 13.10).

There is thus little possibility of making metal glasses from pure metals by rapid cooling; or of making them stable at room temperature. Extremely high rates of cooling, of order 10^6 deg sec^{-1}, can be achieved by shooting small drops of liquid at speed against a cold rotating metal cylinder but experiments of this type have always failed to produce metal glasses. The position is quite different for certain alloys, however, and some alloy glasses can be made and a few are even stable at room temperature.

13.6   Growth of crystals from a melt

We now consider the crystallization of pure metals. At the interface between a nucleated crystal and its melt there is, because of the small activation energies involved, considerable activity amongst the atoms. Some atoms break away from the crystal, at the surface, to join the liquid; atoms in the liquid crystallize on the surface and become part of the crystal. If the rates of these two opposing processes are equal, then the surface is at the crystal–liquid equilibrium temperature. This is not necessarily the ideal thermodynamical melting point because the surface may be curved. If the centre of curvature of the surface lies inside the crystal, then atoms of the crystal at the surface are on average less surrounded by neighbouring atoms of the crystal than otherwise and their escaping tendency into the liquid is increased. The equilibrium temperature is then *lower* than the ideal value; and conversely if the centre of curvature lies outside the crystal

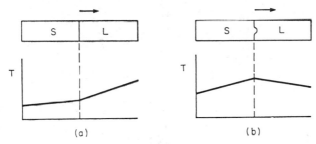

Fig. 13.6 Crystallization by the advance of a solid (S)—liquid (L) interface; (a) normal temperature gradient, (b) inverted temperature gradient in the liquid

(cf. superheated crystal in a crevice). These effects are small unless the radius of curvature r is very small, the deviation ΔT from the ideal temperature T_m (i.e. the temperature for a plane interface) being given by

$$\Delta T = \frac{2\gamma}{r}\frac{MT_m}{L_m}\qquad\qquad 13.4$$

in the notation of § 13.3.

For crystal growth to occur, the actual temperature or the interface must lie below the equilibrium temperature. Experiments have shown that the *rate of growth*, i.e. the speed at which the solid–liquid interface advances into the liquid, increases approximately as the square of the degree of undercooling. In pure metals very fast rates as possible; e.g. $0 \cdot 1$ m s^{-1} in tin and lead undercooled 8°C; 40 m s^{-1} in nickel undercooled 175°C. The atomic rearrangements required for crystallization at the interface thus occur easily, with little activation energy. It follows that under most practical conditions of solidification, e.g. in foundries, the temperature of the interface is always near that of the ideal melting point. The rate of growth is then mainly determined by the rate of removal of the latent heat of solidification from the interface.

Consider in Fig. 13.6 crystallization by the advance of a plane interface from the solid into the liquid. Diagram (a) shows the normal case in which the heat is all extracted through the solid. The temperature gradient is somewhat steeper in the liquid because the thermal conductivity of this is usually lower than that of the solid. Apart from small-scale crystallographic steps, the advancing interface then remains plane and perpendicular to the temperature gradient. If one part of the interface were to move ahead of the rest it would then enter hotter liquid and would slow down. The planar form is stable in this case.

In diagram (b) we show the liquid with an *inverted* temperature gradient. This could be arranged in a special experiment by undercooling the liquid (without nucleation) ahead of the interface; or it might occur in a casting if the heat is extracted *slowly* through the surface of the metal into the mould;

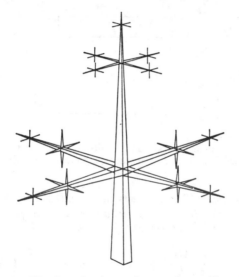

Fig. 13.7 A dendrite

crystallization begins, after undercooling, at the surface and, since the region of the solid–liquid interface warms up due to the release of the latent heat, this interface then advances into colder liquid ahead. Under these conditions if a small part of the interface runs on ahead of the rest, bulging out as in diagram (b), this region moves into colder liquid and so runs on ahead still faster. The planar form is now *unstable* and spikes will grow out from it into the liquid. Similarly, other spikes may branch out sideways from the primary spikes. Such a set of branching spikes, as shown in Fig. 13.7, is called a *dendrite*.

These dendrites grow in preferred crystallographic directions, indicating that crystal growth occurs preferentially on certain planes. In F.C.C. and B.C.C. crystals the dendrites grow in the directions of the cube edges of the crystal cell. Individual dendrites grow rapidly until obstructed by running into other dendrites or into the walls of the container (or less undercooled regions of the liquid) and dendritic growth as a whole is brought to a halt, usually at a fairly early stage in the solidification process, by the release of the latent heat which warms up the liquid and reduces the undercooling. The final stage of freezing then occurs more slowly by the crystallization of the remaining liquid on to the arms of the dendrites.

Dendritic growth occurs very commonly in cast impure metals, mainly because soluble impurities and alloy elements in the metal very strongly promote dendritic crystallization even when the temperature gradient is *not* in verted. We shall discuss the mechanism for this in § 13.8.

13.7 The grain structure of metals

Fig. 13.8 shows in a simple way how the structure of a solid metal develops by solidification of the liquid. First, dendrites grow outwards from each crystal nucleus until they meet other dendrites from neighbouring nuclei. Then the remaining liquid crystallizes on the dendrites until all is solid.

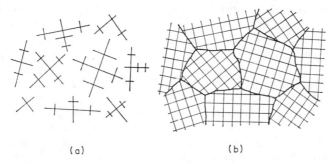

(a)   (b)

Fig. 13.8 Formation of dendrites (a) leading to a polycrystalline structure (b)

This produces a *polycrystalline* structure, with one crystal from each nucleus. It is possible with care to grow *single* metal crystals (e.g. by slow directional solidification along a thin column of a pure liquid metal) but under ordinary conditions of metal casting many nuclei are formed and grow into crystals.

Each crystal of the polycrystal is called a *grain* and its interface with neighbouring grains is a *grain boundary*. Each grain inside the metal is joined to its neighbours at *all* points on its boundary and the position of the boundary is determined by where the separately growing crystals happen to meet. The grains thus generally have irregular and uncrystallographic outlines. Flat crystal faces cannot form when each grain takes up a shape dictated by the chance arrangement of neighbouring dendrites. This lack of well-defined crystal shapes, together with the facts that a cast metal as a whole takes up the shape of its mould and that many metals are malleable and can be altered in shape by hammering, led many people in the nineteenth century to think wrongly that solid metals were non-crystalline.

The sizes and shapes of grains vary enormously. The 'diameter' of a grain in a worked and heat-treated metal is usually about 10^{-5} to 10^{-4} m. As-cast metals generally have coarser grains, 10^{-4} to 10^{-2} m, sometimes even larger. The grains can occasionally be seen on the surfaces of everyday objects, e.g. on the zinc coating of galvanized steel sheet and on well-worn cast brass door handles. Usually, however, they are invisible, being hidden by surface films or by the mirror glitter of a polished surface. The grains of a brittle metal can sometimes be seen by breaking the metal at a sharp notch and looking at the fracture. Bright flat facets, variously orientated, are produced when the fracture follows crystallographic *cleavage planes* through the grains. An *intercrystalline* fracture, i.e. along the grain boundaries, shows sharp ridges and valleys.

The grain structures of metals are studied mainly by *metallography*. A small specimen, e.g. $10^{-2} \times 10^{-2} \times 10^{-2}$ m, is cut from the metal with a saw or water-cooled slitting wheel and a flat smooth surface is then prepared on it by filing and repeated grinding with successively finer grades of abrasive (e.g. emergy or carborundum). The specimen may be embedded in a thermosetting plastic mount, or electroplated, if its edges are to be examined. The grinding is done on abrasive papers, by hand or on wheels, or on lapping wheels coated with lead impregnated with abrasive. After all grinding scratches down to a fine grade (000 paper or 400 mesh grit) have been removed, the specimen is cleaned and then polished by rubbing it on a soft moist velvet cloth with a polishing paste, e.g. fine alumina, magnesia or diamond dust, to a mirror scratch-free finish. This produces a smooth surface but a smeared surface layer has then to be removed by chemical attack with an *etching reagent* (e.g. *nital*, 2 per cent nitric acid in alcohol for steels; *picral*, 4 per cent picric acid in alcohol for steels; hydrofluoric acid for aluminium and its alloys; ammonia and ferric chloride solutions for copper and its alloys; chromic acid for brasses; aqua regia for gold and platinum; hydrochloric acid for tin; sodium hydroxide for zinc). The specimen is immersed in or swabbed with the reagent until the polished surface becomes very slightly 'frosted' or discoloured. The reagent is then thoroughly washed off, first with water, then with alcohol, and the surface is dried in warm air. The prepared surface is now viewed in a microscope with reflected light, as shown in Fig. 13.9. With some reagents and very light etching only the grain boundaries are attacked, grooves being formed along them, and they can be seen as a network of dark lines on a light featureless surface, as in diagram (b) and Plate 1A. More generally the surfaces of the grains themselves are also attacked, forming crystallographic terraces as in diagram (c) and Plate 1B. These reflect the light in various directions, according to their orientation, and so some appear dark and others bright in the image. A linear magnification of 100 to $500\times$ is generally most useful for a metallurgical microscope, although coarse cast structures are best seen at lower magnifications, e.g. $30\times$, and very fine-grained structures may require magnifications of up to about $2000\times$. The grain size of a metal can be reported in several ways: as the number of grains counted in unit area of cross-section; as the mean 'diameter' of the grains; or as the *ASTM grain size index* N, where 2^{N-1} is the number of grains per square inch at a magnification of $\times 100$.

The orientations of the crystal axes in the different grains are often distributed randomly so that most of the grain boundaries are *large-angle*

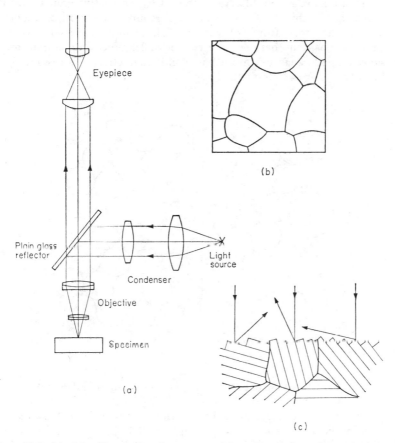

Fig. 13.9 (a) Metallurgical microscope (b) polygonal grain structure (c) reflection from differently oriented grains

boundaries between grains some 20° or more apart in orientation. In some cases, however, the grain orientations are grouped closely about a common mean value; this is called a *preferred orientation* or *texture*. In a pure metal the atomic structure of the crystal continues unchanged right up to the interface itself and the boundary is a narrow *transition region*, not more than about two atoms thick, across which the atoms change over from the crystal sites of the one grain to those of the other, somewhat as shown in Fig. 13.10.

For many years the evidence for this type of grain boundary structure was only circumstantial. Arguments of the following type had to be used:

(1) The same forces which cause crystallization also act on atoms at the grain boundaries. If a thick amorphous layer were to exist between the grains the atoms in it should crystallize on their respective grains until the layer became so thin that its remaining atoms came within range of the interatomic forces from *both* crystals. These atoms necessarily take up compromise positions between the two crystals. Since interatomic forces are short-ranged, weak beyond a few atomic diameters, this residual layer can be only one or two atoms thick.

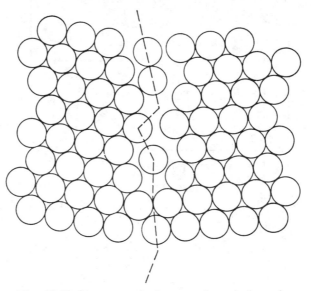

Fig. 13.10 Structure of a large-angle grain boundary

(2) The energies of large-angle boundaries have now been measured (see § 19.5). The value for copper, for example, is $0.55 \, \text{J m}^{-2}$. The latent heat of melting is of order $10^{-20} \, \text{J atom}^{-1}$ and, if the disorder in a boundary is like that in the liquid, the energy of an atom in the boundary should exceed that of one in the crystal by the same amount. Let the thickness of the boundary be d m. There are about $10^{29}d$ atoms in 1m^2 of boundary. The energy of the boundary is then about $10^{-20} \times 10^{29}d = 10^9d$. Equating this to 0.55, we obtain $d \simeq 5$ Å.

It has recently become possible to view the atomic structures of grain boundaries directly, using a *field-ion microscope* (cf. § 17.10). We shall discuss grain boundaries further in § 18.9.

13.8 Solidification of solutions and impure metals

Small amounts of substances in solution, e.g. alloys and impurities, profoundly alter the freezing behaviour of liquids. Usually they tend to remain in the liquid, rather than join with the solvent atoms in freezing on the crystal. The result then is that the freezing point of the liquid is lowered and the crystals are purer than the liquid from which they grow. We shall discuss the general theory of this in Chapter 15 and merely note here that a simple cause of this behaviour, in some systems, is a *difference in size* between the atoms (or molecules) of solute and solvent. There is little difficulty in mixing such atoms together in the liquid because the irregular structure of this allows them to fit together easily. But in the crystal, when some of the atomic sites are occupied by solute and the others by solvent, an oversized (or undersized) solute atom is a source of localized distortion in the crystal round it and the energy associated with this distortion adds to the free energy of the crystal. This has two effects. The concentration of solute in the crystal is reduced, relative to that of the liquid, to keep this extra free energy as small as possible; and the liquid must be cooled further before its own free energy matches that of the crystal, so that the freezing point is thereby lowered.

An important consequence of this is *constitutional undercooling* (*J. W. Rutter and B. Chalmers*, 1953, *Can. J. Phys.* **31**, 15). Consider again the problem of Fig. 13.6(a), taking account now of the effect of solute. Since relatively pure solid is formed, the liquid ahead of the freezing interface becomes enriched with solute rejected by the solid. The composition of the liquid must then vary with distance from the interface in the manner shown in Fig. 13.11. It further follows, since the freezing point is lowered by the solute, that the temperature at which the liquid can freeze must also vary with this distance in the manner shown. As a result, there is a layer of liquid out to the point P in which the *actual* temperature T is below the *freezing* temperature T_L, i.e. this liquid is *undercooled*.

This constitutional undercooling makes a plane solid–liquid interface unstable even when the temperature gradient is *positive*. Suppose a small bulge forms. The solid there projects out into the more highly undercooled liquid and the rapid freezing of this enables it to grow still more. One general effect of this is the formation of a *cellular structure* in the growing crystal, as shown in Fig. 13.12.

A growing bulge (diagram (a)) rejects solute at its sides, so producing a high concentration in the liquid near its base. This liquid remains unfrozen as the rest of the interface moves forward and convexities are thereby formed at P and Q on the interface. These are starting points for other bulges and in this way the whole interface develops the form of diagram (c). The last traces of solute-rich liquid in the channels between the bulges eventually freeze, at a much lower temperature, and a crystal with a periodic columnar distribution of solute or impurity is produced. If the liquid is poured away,

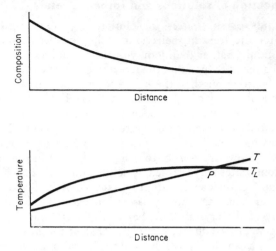

Fig. 13.11 The variation in composition, actual temperature (T) and freezing temperature (T_L), with distance into the liquid from the solid–liquid interface

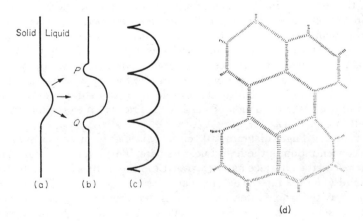

Fig. 13.12 Cellular structure of a crystal grown from an impure liquid; (a), (b), (c) formation (d) appearance of the crystal at the solid–liquid interface

during freezing, and the solid interface examined, this periodic structure has the honeycomb or cellular appearance of diagram (d).

If the degree of constitutional undercooling is increased, e.g. by reducing the gradient of the actual temperature T in Fig. 13.11, the above cell structure develops into *cellular dendrites*, as shown in Fig. 13.13. The branches or webs of the dendrites are interconnected and are an extreme development of the bulges of the cell structure in crystallographic directions of rapid growth.

These interconnected cellular dendrites are different from the 'free' dendrites of § 13.6. They probably form in a small *positive* temperature gradient and occur commonly in cast alloys. Free dendrites form when

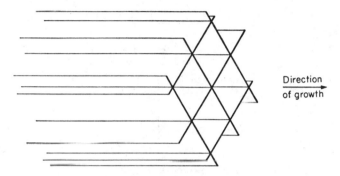

Direction of growth

Fig. 13.13 Cellular dendrites

there is some conduction of latent heat into the liquid and they grow in much the same way as in pure metals, although their growth is slowed down by the accumulation of solute in the liquid near them.

13.9 Structure of cast metals

There is enormous variety in the structures of cast metals. Some of the main features can be discussed from Fig. 13.14, which shows a simple idealized structure of an ingot. When the liquid is poured into a cool mould (particularly a metal mould) the layer next to the mould wall is cooled very rapidly. Crystals nucleate (heterogeneously) and grow rapidly in this *chill zone* but the undercooling which results from the high cooling rate speeds up nucleation even more than crystal growth, so that many fine-grained crystals are formed. These grains in the chill zone have similar dimensions along all axes, i.e. are *equi-axed*.

Crystallization continues in the outer layers where the temperature is low. As the shell of solid metal thickens, the temperature gradient from the liquid to the mould becomes less steep and the liquid next to this shell cools more slowly and undercools less, with the result that crystal *growth* rather than nucleation is then favoured. The crystals on the inside of the chill zone begin to grow inwards, so forming long *columnar* crystals against the direction of heat flow. Crystals with orientations favourable for rapid growth in this direction grow faster than their neighbours and widen out at their expense until these latter have disappeared, so that the finally established columnar grains are much wider, as well as longer, than those of the chill zone. This *selective growth* produces a preferred orientation in the columnar grains, e.g. crystal cube edges along the columnar axis in metals of cubic crystal structure. In most alloys and impure metals the columnar crystals grow as cellular dendrites.

The columnar crystals grow inwards until they meet the dendritic skeletons of crystals growing in the centre of the ingot. These central

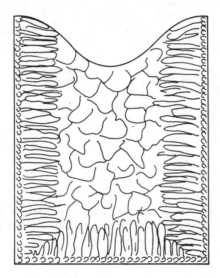

Fig. 13.14 Section of an ingot

dendrites develop into coarse, randomly oriented, equi-axed crystals, as shown in Fig. 13.14. Their nuclei may come from the chill zone, small crystals formed there initially having been swept into the centre of the ingot. In at least some cases however, *constitutional undercooling* seems to be an essential requirement for their formation. It is probable then that the nuclei form (heterogeneously) in the constitutionally undercooled regions of the liquid ahead of the growing columnar crystals. In Fig. 13.15 we see that as the gradient of the actual temperature is lowered, i.e. $1 \rightarrow 2 \rightarrow 3$, the width of the constitutionally undercooled zone and the degree of undercooling are both increased. Lowering the gradient thus increases the possibility of nuclei forming in the liquid ahead of the advancing solid–liquid interface. It is just this condition which is expected in the later stages of solidification, when the temperature gradients have flattened out and all of the remaining liquid is practically at the same temperature as the solid–liquid interface.

The actual grain structure of an ingot or casting may differ considerably from that of Fig. 13.14. If the liquid is *superheated*, i.e. is poured at a temperature well above its melting point, the initially formed chill zone may be remelted and disappear. When freezing begins again, at the surface, the cooling is much slower and only a few nuclei are formed, so that a coarse columnar structure grows inwards from the surface. If, additionally, the actual temperature gradient is prevented from becoming too flat, e.g. by water cooling the mould, then constitutional undercooling is reduced and the columnar structure may continue right to the centre.

Another extreme structure is developed in *chill castings*. If a thin section of metal is cast in a cold metal mould, the rate of cooling may remain so high throughout that the entire section freezes as fine-grained, equi-axed chill crystals.

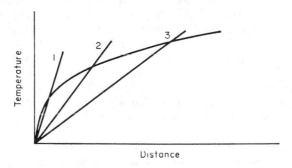

Fig. 13.15 Increase of constitutional undercooling with decrease of actual temperature gradient from 1 to 3

An important practical problem is to produce uniform, equi-axed, fairly fine crystals in large castings, for this structure gives metal with the best strength and other mechanical properties. The aim is to promote as much nucleation as possible in the liquid immediately ahead of the solidification front. This is done by adding some solute to the metal that produces intense constitutional undercooling and also a *nucleating agent*, the grains of which are good centres for heterogeneous nucleation. Sometimes high-frequency mechanical vibrations (ultrasonic) are also applied to help promote nucleation. The combined effect of the solute and nucleation additions enables many small crystals to be nucleated in the constitutionally under-cooled zone immediately ahead of the solidification front. For example, the grain structure of cast aluminium can be greatly refined by adding copper, to provide constitutional undercooling, and titanium, to provide titanium carbide nucleation centres (*A. Cibula*, 1949–50, *J. Inst. Metals*, **76**, 321). Various nucleating agents for different metals and alloys have now been discovered but the principles governing their effectiveness are not yet fully clear. The most important example of grain size control in practice is the production of *fine-grained steel* by the addition of about 0·5 kg of aluminium per 1000 kg of steel in the ladle. The main effect here however is the control of *grain growth* in the *solid* metal, during working and annealing, by finely dispersed particles of alumina (cf. § 19.5).

Apart from grain structure, several other factors determine the quality of a cast metal. The main ones are *shrinkage, segregation* and *gas evolution*.

The shrinkage which occurs on solidification is a source of many practical difficulties in the casting of metals. We have already discussed *pipes* in ingots and their elimination by the use of hot tops and by gas evolution in rimming steel in § 11.9. Shrinkage is equally important on a microscopic scale, between dendrites and grains, and can lead to an *unsound* casting, full of fine-scale porosity, unless the interdendritic liquid channels can all be continuously fed with liquid from a central supply to make up the volume difference when they freeze. For a sound casting the metal should freeze progressively from surface to centre with no trapping of liquid in pockets totally enclosed by solid; and the zone in which freezing takes place, between the fully solid and fully liquid regions, should be *thin* so that the feeding liquid is not required to flow along *long* narrow interdendritic channels. This last condition is best obtained by the addition of solute, for constitutional undercooling, and of nucleating agents, as described above. The casting must of course be provided with a *feeder head*, like a hot top, to maintain a pool of liquid to feed into the metal below as it freezes and shrinks.

Another practical problem of shrinkage is *hot tearing*. If, for example, a spoked wheel is cast and the rim solidifies first, then, as the spokes freeze, they contract in length while constrained at their ends. The tensile stresses thereby set up may then break them by tearing their partly solid grains apart. Careful design of the casting and its mould are necessary to overcome this problem. In a sand mould metal *chills* may be inset in certain places to ensure that the various parts of the casting freeze in the required order.

The name *segregation* refers to all non-uniformities of composition in an ingot or casting. There are many types of segregation. *Normal segregation* is produced by the large-scale migration of solute through the liquid along the direction of solidification, so that the solute is segregated on a macroscopic scale to the centre of the ingot. *Gravity segregation* occurs when the solid has different *density*, as well as composition, from the liquid and so sinks to the bottom of the mould or floats to the top (e.g. solidified Sb in liquid Pb-Sb). More surprising is *inverse segregation* in which beads of solute-rich liquid appear on the *outer* surface of the casting (e.g. *tin sweat* on cast bronze). This seems to be due to the combined effects of shrinkage and segregation. Consider a casting in which, at a certain stage of solidification, narrow interdendritic channels full of solute-rich liquid of very low freezing point extend right through to the surface of the mould wall. As metal freezes and shrinks on the sides of these channels, and especially when the shrinking casting begins to pull away from the mould wall, the solute-rich liquid is sucked along the channels and in extreme cases right out on to the surface of the casting (*E. Scheil*, 1947, *Metallforschung*, **2**, 69).

There are also several types of *microscopic segregation*; to grain boundaries; to cell walls (cf. Fig. 13.12); and between dendrites. Fig. 13.16 shows the typical appearance of dendritic segregation. This type occurs very commonly in cast alloys and is usually referred to as *coring*. When dendritic segregation occurs strongly, particularly between cellular dendrites, normal segregation is reduced because the interdendritic liquid, including the solute, then freezes locally on to its nearby dendrites.

Macroscopic segregation is controlled by promoting equi-axed solidification, rather than columnar, through the use of nucleating agents and constitutional undercooling. Microscopic segregation is almost unavoidable but it can be reduced or eliminated afterwards by annealing.

Liquid metals usually contain some dissolved gases. The gas solubility generally falls steeply with fall in temperature and drops sharply at the freezing point, cf. Fig. 13.17, so that there is an evolution of gas during solidification. Near the surface this gas may escape, but inside the metal (except in those cases where it forms a compound with the metal, e.g. Cu_2O in Cu) it precipitates out as gas bubbles (*blowholes*, *pinholes*, etc.), particularly in regions of high segregation. Hydrogen in steel is especially harmful because at low temperatures it produces small thin cracks in the iron crystals (*hairline cracks* or *flakes*) which make the steel mechanically weak and brittle. In thin steel sections hydrogen is not a great problem because it can be removed from the solid metal by diffusion to the surface. Hairline cracking is mainly a problem of large steel forgings, e.g. for steam turbine rotors.

An important recent development in steel-making is *vacuum degassing*, a process for removing hydrogen, oxygen and nitrogen from the metal. There are several methods. In *stream degassing* the nozzle in the bottom of the ladle is opened into a large vacuum chamber below. The molten steel streams through, losing gas as it falls into a second ladle or an ingot mould in the chamber. In *ladle degassing* the ladle of molten steel is put into a chamber which is then evacuated while the metal is stirred (e.g. electromagnetically). In *incremental degassing* a small vacuum chamber is placed over the ladle. Two tubes extend down into the molten steel from this chamber. The steel is forced up one of them by injecting argon into it, becomes degassed in the chamber and then returns to the ladle down the other tube. A 10^5 kg heat of steel can be treated in about 15 minutes in this way.

The hydrogen content of steel can be reduced to about 1 part per million by vacuum degassing. This eliminates the hairline cracking problem in large forgings and considerably improves the ductility and toughness of the metal. The oxygen content is also greatly reduced, both by direct removal as oxygen gas and also by removal as CO (i.e. by *carbon deoxidation*). The need

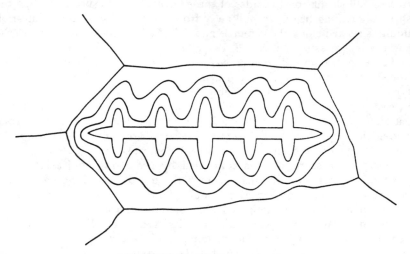

Fig. 13.16 Dendritic segregation

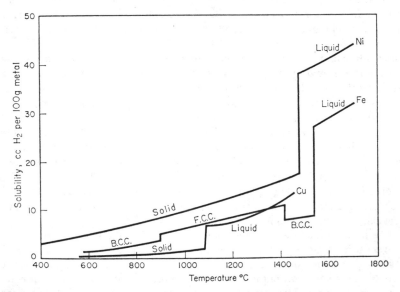

Fig. 13.17 Solubility of hydrogen in metals, in equilibrium with 1 atm H_2

for Mn, Si and Al deoxidizers is then greatly reduced. The advantage of this is that the steel is then much *cleaner*, i.e. freer from small particles of deoxidation products, e.g. silicates and alumina. The mechanical properties of the metal, e.g. ductility and fatigue strength, are greatly improved.

13.10   Casting and related processes

There are many different kinds of casting processes. *Ingots* and *billets* are castings of simple shape, usually suitable for mechanical working, e.g. forging, rolling, extrusion and drawing. We discussed steel ingots in § 11.9. Non-ferrous metals such as zinc, lead and copper are often cast in horizontal open metal trays. Freezing occurs from the bottom upwards and there is little porosity since gas can escape from the top surface; and no pipe

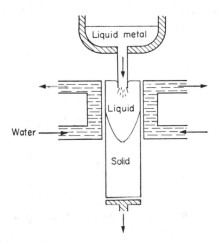

Fig. 13.18 Continuous casting

since the shrinkage is accommodated over the whole surface. Water-cooled metal moulds are also used extensively. The *Junkers mould* is lined with high-conductivity copper and cooled by water circulated in an outer shell of cast iron. Pipes, tubes and other axially symmetric hollow castings are made by *centrifugal casting*. The cylindrical mould, usually horizontal, is rotated about the cylindrical axis and the molten metal poured into it is thrown against the wall by centrifugal force, so producing a tubular casting.

Continuous casting is now widely used. Fig. 13.18 shows the general arrangement. The liquid metal is poured into a vertical mould, open at top and bottom, and withdrawn as solid metal from the bottom. The advantages of this method are its greater productivity, the elimination of hot top devices, directionality of solidification and reduction of segregation. The mould is usually made of copper, water-cooled, and may be lined with chromium plate. The cast bar is withdrawn at speeds of up to about 10 m s^{-1} and cooled by a water spray beneath the mould. Continuous casting was invented by Bessemer but taken up seriously only recently. It is widely used for non-ferrous metals and is the basis of continuous processes for making sheet and wire. Steel is also now being cast continuously, in bars up to 0·25 m square in section.

Castings of more complex shape are made in many ways. In *sand casting* a wooden pattern of the required shape, slightly enlarged to allow for shrinkage of the casting, is firmly packed in sand in a *moulding box*. Fig. 13.19 shows a typical arrangement. The box is usually split to facilitate the removal of the pattern from the sand. To make castings with re-entrant hollow regions, sand *cores* are fitted in the mould. A *green sand casting* is made in sand bonded with clay. Alternatively, the sand may be bonded with molasses or other substances and then baked hard. In *shell moulding* a thin strong envelope of sand and thermosetting resin is made on a removable metal pattern and then supported in a vessel of metal shot to receive the metal casting poured into it.

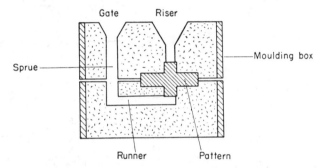

Fig. 13.19 A simple sand mould before removal of the pattern

The metal is poured into a sand mould through a *gate*, down a *sprue* and along a *runner*. A *riser* has also to be provided, leading upwards from the top of the casting, to provide a pool of molten metal to feed the casting as it freezes and shrinks. Risers are sometimes lined with *exothermic compounds*, based generally on the oxidation of aluminium powder, which give off heat and keep the metal molten. Metal *chills* also may be fitted at suitable places in the mould to promote rapid cooling locally and produce the required freezing sequence in the casting.

An advantage of sand casting is the complexity of shape that can be cast, since the sand can afterwards be removed piecemeal from re-entrant parts of the casting. Disadvantages are the low productivity, since a separate mould has to be made for each casting, and the roughness of the surface produced on the casting. The surface can be improved by lining the mould with fine *facing sand* and a thin layer of smoothed graphite powder.

Where a given object is to be made in large numbers by casting a *permanent mould* of metal is used. The rapid cooling obtained in such a mould gives a fine-grained and mechanically strong casting. A good surface is also usually produced. In *pressure die casting* the molten metal is forced under pressure into a permanent mould. This process, which is applied particularly to zinc, aluminium, tin, lead and magnesium alloys, enables castings to be made rapidly by an automatic method. Good surfaces and precise dimensions are obtained and the high pressures (e.g. 700 atm.) force the metal into thin sections and sharp corners. In a *slush casting*, as used for making toy soldiers, etc. the metal mould is turned over immediately after casting to pour back all remaining liquid metal, so leaving the casting itself as a thin shell.

In *investment casting* (or *lost wax* or *precision casting*) a pattern of wax or other fusible material is die cast and a plaster of Paris mould is then made round it. When the plaster of Paris is 'fired' to harden it the wax evaporates and burns away, leaving a clean cavity for the metal casting. The process is expensive but produces castings with precise dimensions and good surfaces. It is used particularly for casting refractory alloys for gas turbine blades and also for jewellery, ornamental and dental castings.

In all casting processes care must be taken to avoid splashing the metal on the sides of the mould; otherwise splashed solid films form and then become coated with oxide which prevents them from afterwards uniting with the rest of the casting (*cold shuts*). Oxide films on the liquid surface may also get folded into the metal unless the pouring is done with care. Special casting devices are sometimes used to transfer the metal to the mould smoothly for this purpose (*Durville casting*). The oxide film on aluminium is particularly troublesome, partly for this reason and partly because, being a continuous and mechanically strong film, it prevents the metal from filling sharp corners and narrow channels in the mould. The ability of a cast metal to flow along a narrow channel and take a sharp impression is roughly indicated by the *casting fluidity*, which is measured by the distance the metal will flow along a standard spiral channel. Casting fluidity is not the inverse of viscosity. It is determined mainly by the rate at which the metal freezes inwards from the walls of the channel while the liquid is still flowing through. This depends particularly on the *degree of superheat*, i.e. the amount by which the temperature of the cast liquid exceeds the freezing point.

Many of the processes used for *joining* one piece of metal to another are based on casting. In *fusion welding* the metal pieces themselves are melted locally at the joint. In *brazing* and *soldering* only the *filler metal* is melted, at a temperature below the melting point of the metals being joined. *Copper-zinc* brazing alloys are used at temperatures in the range 800–1100°C, *silver solders* (i.e. silver-copper-zinc alloys) from 600–800°C and *soft solder* (tin-lead alloy) at about 200°C. It is essential to make true metal-metal contact between the braze or solder and the metals being joined, so that oxide films and dirt must first be removed with a flux (e.g. borax). The filler metal will then 'wet' the cleaned surface and be sucked into the joint by capillary forces.

There are many welding processes. In *gas welding* the filler metal and edges of the joint are melted by a flame, usually oxy-acetylene. Heavy steel sections are sometimes joined by *thermit* welding in which the heat is supplied by the exothermic oxidation of aluminium powder in iron oxide. *Resistance* welding depends on the resistance heating effect of a large electrical current passed through the joint. In *spot* and *seam* welding, for example, sheets of the metal to be joined are pressed together between thick copper electrodes and fused together locally by the current passed between the electrodes. For certain special purposes, e.g. the welding of hard alloys on the leading edges of steam turbine blades, an *electron beam* is used to make the weld. The most widely used welding process is *shielded arc* welding in which an electrical arc is struck between an electrode of filler metal and the pieces to be joined, the heat of which melts the tip of the rod and the edges of the joint, so forming a small pool of liquid metal there. The arc is slowly moved along the line of the joint to form a continuous deposit of weld metal. The electrode consists of a rod of metal, usually similar to that of the joint, coated with a substance that acts as a flux and forms a protective cover of molten slag over the weld deposit. *Submerged-arc* or *electro-slag* welding is a variant of the process, generally used for welding thick steel sections automatically, in which the arc is formed beneath a pool of molten slag.

A weld is a miniature casting into a (relatively) large metal mould. Cooling thus occurs very rapidly and a fine grained columnar structure is usually produced in the weld. In some steels the rapid cooling produces a rather brittle micro-structure (cf. § 20.4) and there is then a danger of *cracking* in the weld. This is aggravated by the large stresses set up by the thermal contraction of the joint under the severe mechanical constraint imposed by the general rigidity of the welded pieces. Some additional heating is often necessary to reduce the rate of cooling and to relieve the shrinkage stresses.

Closely related to fusion welding are the various *fusion cutting* processes. Carbon and metal arcs are used to cut metals by localized melting. In *flame cutting* on the other hand, e.g. with an oxy-acetylene torch, the main cutting action is by the *burning* of the hot metal in an oxygen-rich flame.

13.11   Growth of single crystals

Single crystals of metals have been grown for research purposes for many years; and, with the development of semi-conductors and other electrical materials, the growth of single crystals of substances such as germanium and silicon has become an important industrial technique. Many such crystals are grown by controlled solidification of a melt.

In the *Chalmers* method the melt is supported in a horizontal 'boat' (e.g. of graphite), as shown in Fig. 13.20, and is made to freeze progressively from one end by slowly moving an electrical furnace along the boat. The leading end of the boat is tapered to reduce the number of crystal nuclei formed there. Sometimes only one nucleus is formed and the entire melt freezes on to it, forming a single crystal. If several nuclei are formed, then usually one will generally grow faster than the others, due to a favourable orientation for growth, and will gradually take over the entire growth front. It can then be used as a *seed crystal* for growing another crystal of a required orientation by setting it at a suitable angle to the melt in another boat and then melting the interface, to join it to the melt, before starting the progressive solidification, as shown in Fig. 13.20.

There are several other methods for growing crystals from a melt. The *Andrade* method, similar to the above, is suitable for growing single crystal *wires*. The metal, about 10^{-3} m dia, is supported in a horizontal tube

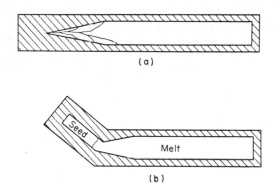

Fig. 13.20 Growth of single crystals from a melt; (a) boat with nucleation and competitive growth from pointed end; (b) growth of crystal of controlled orientation from an inclined seed crystal

of slightly larger diameter. Surface tension holds the molten wire in shape. The advantage is that there is very little mechanical constraint on the wire from the mould, so that a crystal very free from strain can be grown. In the *Bridgman* method a vertical mould is lowered slowly through a tubular furnace. The *Czochralski* method, also known as *crystal pulling*, is widely used for growing crystals of semi-conducting materials. The melt is held in a crucible and a seed crystal is dipped into it from above and then slowly withdrawn, so that the crystal is 'pulled' out of the melt. The crystal is rotated about its vertical axis as it is withdrawn to give it a symmetrical cylindrical shape. A *floating zone* method, in which a zone of liquid is suspended by surface tension between two halves of a solid bar and slowly moved along the bar, is also sometimes used. It is also possible to grow a crystal as a '*ribbon dendrite*' by pulling from a melt.

FURTHER READING

CHARLES BARRETT and MASSALSKI, T. B. (1966). *Structure of Metals*. McGraw-Hill, New York and Maidenhead.

CHALMERS, B. (1964). *Principles of Solidification.* John Wiley, New York and London.

CHALMERS, B. and QUARRELL, A. G. (1960). *The Physical Examination of Metals.* Edward Arnold, London.

COTTRELL, A. H. (1964). *The Mechanical Properties of Matter.* John Wiley, New York and London.

DAVIES, G. J. (1973). *Solidification and Casting.* Applied Science Publishers, London.

FLINN, R. A. (1963). *Fundamentals of Metal Casting.* Addison-Wesley, London.

HEINE, R. W. and ROSENTHAL, P. C. (1955). *Principles of Metal Casting.* McGraw-Hill, New York and Maidenhead.

KEHL, G. L. (1949). *Metallographic Laboratory Practice.* McGraw-Hill, New York and Maidenhead.

MODIN, H. and MODIN, S. (1973). *Metallurgical Microscopy.* Butterworths, London.

RUDDLE, R. W. (1957). *The Solidification of Castings.* Institute of Metals.

WILLIAMS, R. S. and HOMERBERG, V. O. (1948). *Principles of Metallography.* McGraw-Hill, New York and Maidenhead.

WINEGARD, W. C. (1964). *An Introduction to the Solidification of Metals.* Institute of Metals.

Chapter 14

Alloys

14.1 Types of alloys

An alloy is a metallic solid or liquid formed from an intimate combination of two or more elements. Any chemical element may be used for alloying, but the only ones used in high concentrations are metals.

The intimate combination is usually brought about by dissolving the alloy elements in one another in the liquid state. The *parent metal* or *solvent*, in largest concentration, is melted first in a crucible and solid pieces of the alloy additions in weighed amounts are then dropped in, dissolved and stirred. Minor amounts of an element are more accurately added in the form of *master alloys*, i.e. more concentrated alloys of the element with the parent metal. Where possible, a master alloy is conveniently made up as a *brittle* intermetallic compound (see below) so that it can readily be broken up into pieces for weighing out. Special difficulties appear with volatile and reactive alloy additions. When zinc is added to molten copper to make brass, for example, it is put in as large pieces which are pushed and held under the surface to minimize its loss by volatilization. Even so, it is generally necessary to add up to about 2 per cent more zinc than the alloy requires, to meet such losses.

We now consider solid alloys. Suppose that, by freezing a liquid alloy or by some other means, we introduce into a pure solid metal some atoms of another element. The atoms are then in some way allowed to move about, rearrange themselves and change the crystal structure until they come to thermodynamical equilibrium amongst themselves. What is the structure of this alloy? To study this question it is convenient to think of the two types of atoms as (a) indifferent to one another, (b) attracting one another, or (c) repelling one another. In other words, the internal energy (a) stays unchanged, (b) decreases or (c) increases, when the atoms rearrange themselves to increase the number of unlike nearest neighbours.

If the atoms are indifferent to one another they behave as if belonging to the same species and they become mixed together so thoroughly that the alloy is homogeneous right down to the atomic scale. This structure, already anticipated in Fig. 6.2(a), is a *random solid solution*. Many alloys approximate to it.

If unlike atoms are attracted together, the nature of the resulting alloy varies widely according to the factors determining the attraction. When formed from true metals the structure is often an *ordered solid solution* or *superlattice*, in which the two species are arranged in some regular alternating pattern. If the components differ electrochemically, the bond between their atoms becomes partly *ionic* and the structure is usually termed an *intermetallic compound*. In the limit, when one component is strongly electronegative (e.g. S, O, Cl), a true chemical compound is formed and the substance is no longer metallic.

If unlike atoms are attracted less than like, the two types tend to separate into distinct and different crystals joined at mutual grain boundaries. Some grains are rich in atoms of one type, others are rich in those of the other. These microscopically heterogeneous mixtures are termed *phase mixtures*. A phase mixture is, however, not always an indication that unlike atoms are attracted less than like. We shall discuss the thermodynamical principles of its formation in § 14.7.

14.2 Solid solutions

A solid solution can exist over a *range of composition*. Anywhere within this range the material is fully homogeneous and its structure and properties differ only infinitesimally from those of neighbouring compositions. A few solid solutions extend continuously all the way from one pure metal to the other, e.g. Cu-Ni, Ag-Au, Au-Pt, K-Cs, Ti-Zr, As-Sb, and the components are then said to be *completely miscible*. A necessary condition for this is, of course, that the components must have the same crystal structure; but there are other conditions to be satisfied as well. More commonly the range of homogeneous composition is limited. When this limited range includes one of the pure components the solution is a *primary* one, with that component as the *solvent* and the other as *solute*. Solid solutions in alloys are also formed in intermediate ranges of composition which do not include the pure components. These are *secondary* solid solutions. They usually have different crystal structures from those of their components. Secondary solutions and intermetallic compounds are both referred to as *intermediate phases*.

The prevalence of solid solutions in alloys is due to the fact that atoms in metals are held together mainly by their attraction (as positive ions) to the free electrons moving between them (cf. § 4.6). This type of bond is largely indifferent both to the precise proportions of the component atoms and also to their precise distribution in the crystalline array of atomic sites. Random solutions over wide ranges of composition are thus possible.

Solid solutions can be either *substitutional* or *interstitial*. As Fig. 6.2(a) shows, in a substitutional solution the atoms share a single common array of atomic sites. In interstitial solutions the atoms of one component are small enough to fit into the interstitial spaces between those of the other, which are themselves arranged in a complete crystalline array (Fig. 14.1).

The equilibrium distribution of atoms in a substitutional solution generally depends on temperature. Various distributions are possible, ranging from the fully random state, which is most nearly obtained at high temperatures, to *clustered* (like neighbours preferred) and *ordered* (unlike neighbours preferred) states at lower temperatures. *Short-range order*, in which there is some tendency for unlike atoms to be neighbours but there is no long-range correlation in the distribution of atoms amongst the array of atomic sites, occurs fairly commonly. In addition, some alloys undergo a transformation below a critical *ordering temperature* to an ordered state in

Fig. 14.1 An interstitial solid solution

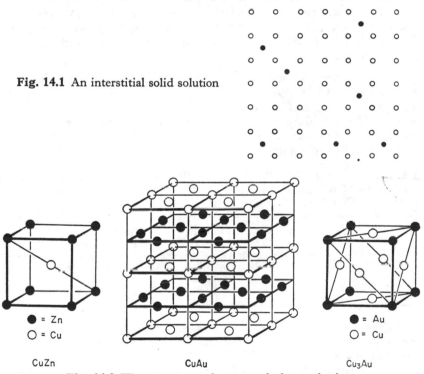

● = Zn
○ = Cu

CuZn

CuAu

● = Au
○ = Cu

Cu₃Au

Fig. 14.2 The structures of some typical superlattices

which a *superlattice* is formed by the regular alternation of unlike atoms through the entire crystal, or at least through large regions of it. A perfect superlattice is of course possible only at a critical and simple proportion of atoms, e.g. 1 to 1, or 3 to 1. Systems which form superlattices at such ratios also show imperfect superlattices or long-range partial order at neighbouring compositions. Superlattices are found in both primary and secondary solutions. Some typical ones are shown in Fig. 14.2.

14.3 Primary substitutional solid solutions

What determines the solid solubility of one metal in another? The pioneering work of *Hume-Rothery* and his research school has led to the following general answers:

(1) *The atomic size factor.* The more the solute atom differs in size from that of the solvent, the smaller is the range of primary solution. If the atomic diameters differ by more than 14 per cent of that of the solvent the solubility is small; the size factor is then *unfavourable.*

(2) *The electrochemical factor.* The more electropositive is one component and the more electronegative is the other, the greater is the tendency to form compounds rather than solid solutions and the smaller is the solubility.

(3) *The relative valency factor.* Other things being equal, a metal of lower valency is more likely to dissolve one of higher valency than vice versa. This rule is valid mainly for alloys of copper, silver or gold with metals of higher valency.

The size factor rule was deduced from the study of the solubilities of various elements in copper but later work has shown that it applies fairly well also to other metals. A favourable size factor does not, of course, ensure a high solubility; if the atoms differ electrochemically a compound may be formed instead. The rule enables us to make sense of many otherwise perplexing solubility effects. For example, Table 14.1 shows that,

TABLE 14.1. SOLUBILITIES (ATOMIC PER CENT) AND ATOMIC SIZE FACTORS (PER CENT) OF BERYLLIUM, CADMIUM AND ZINC, IN COPPER AND SILVER

Solute	Maximum solubility		Atomic size factor	
	Cu	Ag	Cu	Ag
Be	16·6	3·5	−12·9	−22·9
Cd	1·7	42·5	+16·5	+ 3·1
Zn	38·4	40·2	+ 4·2	− 8·0

taking the divalent solutes beryllium, cadmium and zinc, the first dissolves well in copper but not in silver, the second does the opposite, and the third dissolves well in both.

The size factor effect derives from the *strain* which exists in the crystal round a misfitting solute atom. The atoms of the solvent are pushed out or pulled in, according as the solute atom is larger or smaller than its crystal site, and this alteration of the interatomic spacings from their ideal values increases the energy of the crystal (cf. Fig. 4.1). This extra energy then limits the solubility through the usual Maxwell-Boltzmann factor, as discussed at the end of § 6.5. As mentioned in § 13.8, solubility is usually higher in the liquid than in the solid state because the liquid with its irregular structure can more easily accommodate atoms of different sizes.

To apply the size factor rule we need to know the atomic sizes. Experience has shown that the most practical and useful measure is the closest distance of approach of atoms in the crystal of the pure element. The metallic atomic radii in Table 3.2 and the atomic size factors in Table 14.1 were derived this way. There are difficulties, however, especially when the solute atom changes its electronic state on alloying, as for example when changes occur in the degree of ionization, electron concentration or crystal structure. There are also some anomalies. For example, an atom of antimony distorts (expands) copper more than an atom of cadmium does, if the distortion is measured in terms of atomic radii (cf. Table 3.2), but copper nevertheless can dissolve more antimony (5·9 atomic per cent) than cadmium (1·7 atomic per cent). The explanation seems to lie in the influence of the *ionic radius* (Table 3.2). The ionic core of the antimony atom, inside the valency electron cloud, is smaller than that of cadmium; and in copper the ions are packed close together, which favours the substitutional solution of atoms with smaller ions.

Considering now the electrochemical factor we see from Table 4.1 that, with the exception of the electropositive alkali and alkaline earth metals, and of the electronegative semi-metallic elements, most of the metals are fairly similar to one another electrochemically. The elements S, Se and Te show little solubility in most metals and prefer to form sulphides, selenides and tellurides; P, As, Sb and Bi generally form compounds with metals such as Mg and Li, but with the less strongly electropositive metals such as Cu and Ag there is some solubility when the size factor is favourable.

The relative valency effect is mainly a reflection of the tolerance of the metallic bond to changes of *electron concentration* (i.e. the ratio of the number of free electrons to atoms), compared with the strict valency laws obeyed in covalently bonded solids. Copper, for example, can accept substitutional silicon atoms—they merely increase the number of free electrons moving through the structure—but when silicon atoms are replaced by copper in a silicon crystal the single valency electron of copper cannot form four covalent bonds and the crystal structure is weakened.

Hume-Rothery and his associates have observed that, in certain alloy systems, the limit of primary solubility is set by electron concentration. If we take copper or silver as solvents and add to them elements of higher valency, choosing only those with favourable size and electrochemical factors, we find that a definite relation exists between the limit of primary solubility and the valency of the solute. If the composition is measured as electron concentration the solubility limit is at about 1·4 electrons per atom in all cases. For example, divalent zinc dissolves in copper up to about 40 atomic per cent, trivalent aluminium and gallium up to about 20 per cent, and tetravalent silicon and germanium up to about 13 per cent.

The Hume-Rothery rules show the broad pattern of alloy behaviour. From the free electron standpoint, we expect the alkali metals and the copper group to be good solvents for other metals. However, alkali metal atoms are large and electropositive, so that only rarely is there good solubility in an alkali metal (e.g. complete miscibility in K-Rb, K-Cs, Rb-Cs systems; extensive solubility of Mg in Li).

The copper group are excellent solvents for many metals. Copper itself, for example, lies well in the middle of the metals as regards both atomic size (1·27 Å radius) and electrochemical factor. It forms extensive primary solid solutions, dissolving at least 5 per cent and often very much more of many metals (e.g. Al, As, Au, Ga, Ge, Mn, Ni, Pd, Pt, Sn, Zn). These solutions form the basis of many commercially important alloys, such as the brasses, bronzes, and cupro-nickels. Silver also dissolves many metals (e.g. Al, Au, Cd, Hg, Mg, Pd, Pt, Sn, Zn), particularly those of slightly larger atomic size than the copper solutes.

Extensive solid solutions are also found amongst the transition metals, particularly between those which are neighbours in the same row of the periodic table and so have similar atomic sizes (Table 3.2). The size factor of iron is favourable for many metals (e.g. Al, Be, Co, Cr, Cu, Mn, Mo, Ni, Pd, Pt, V, W) and these dissolve extensively in F.C.C. or B.C.C. iron.

In multi-valent non-transition metals the broadly increasing electrochemical and size factors, and the trend in some towards a less ideally metallic bond, generally restrict the opportunities for forming extensive solid solutions. Beryllium is a poor solvent for most metals and this has restricted the development of useful alloys based on this metal. To a lesser degree this is also true of zinc and magnesium as solvents. Extensive solutions nevertheless still occur where the size and electrochemical factors are favourable, e.g. Mg-Cd, Cd-Hg, Ca-Sr, As-Sb. Aluminium dissolves a few metals extensively (Zn, Mg) and others (e.g. Cu) in smaller though useful amounts. Tin dissolves some metals (e.g. Bi, Sb). The lead atom is too large for much solubility, although a few metals are dissolved in useful amounts (Sn, Bi).

Primary solid solutions based on metals of cubic crystal structure, particularly F.C.C., are usually suitable for practical development as *wrought alloys*, which are shaped and improved in mechanical properties by plastic working (cf. Chapter 22). Alloys with sparingly soluble (e.g. below 5 per cent) or nearly insoluble (e.g. below 1 per cent) added elements are also often of considerable practical value. A small solubility, which decreases at lower temperatures (cf. § 14.11), is the basis of many *heat-treatable alloys* in which great mechanical strength and hardness is developed by quenching and ageing treatments (cf. § 20.1). Insoluble mixtures in many cases form *eutectics* (cf. § 15.3) which, because of their low melting points, are often good *casting alloys*.

14.4 Intermediate phases

Intermediate phases in alloys can be classified broadly into *electrochemical compounds*, *size factor compounds* and *electron compounds*. There is no absolute distinction between these classes, however, and some phases belong partly to one and partly to another. The word 'compound' is widely used, even though only the electrochemical compounds obey the ordinary chemical valency laws and even though many of the electron compounds are in fact secondary solid solutions of variable composition.

Electrochemical compounds are formed between several electropositive and electronegative elements; e.g. Mg_2Si, Mg_3Sb_2, ZnS. Their compositions satisfy the valency laws, the range of solubility in them is generally small and many have simple structures of the types found in ionic crystals. They often have high melting points. Electrochemical compounds apart, it is best to regard intermediate phases in alloys as structures similar to primary solid solutions, with their atoms held together mainly by free electrons. In some intermediate phases the distribution of atoms among the crystal sites is disordered and there is an appreciable range of solubility; these are secondary solid solutions, not really different from primary ones except that their range of composition does not include one of the pure components. In others the atoms are ordered; they may be regarded either as ordered secondary solid solutions or as intermetallic compounds. When the electrochemical factor is small, the formation of these intermediate phases depends on the size factor and electron concentration such that, if a crystal structure is possible in which the atoms pack well together (size factor compounds) or has a low free electron energy (electron compounds), this structure is likely to have a lower free energy than any other structure of similar composition and is therefore likely to appear as a stable phase.

In size factor compounds the composition and crystal structure are chosen in such a way as to allow the component atoms to pack together well. An important group of these are the *Laves phases*, with compositions based on the chemical formula AB_2; e.g. $MgCu_2$, KNa_2, $AgBe_2$, $MgZn_2$, $CaMg_2$, $TiFe_2$, $MgNi_2$. They exist mainly because, when the component atoms differ in size by about 22·5 per cent, they can pack together neatly in crystal structures of *higher coordination* than the maximum (12) possible in crystals of close-packed equal spheres. In a Laves phase AB_2, each A atom has 16 neighbours (4A and 12B) and each B atom has 12 neighbours, so that the average coordination number is 13·33. These high coordination numbers are of course favoured by the free electron bond (cf. § 4.6).

There are several other types of size factor compounds. Interstitial compounds, in which one atom is much smaller than the other, are discussed below. In *Zintl compounds*, e.g. NaTl, LiCd, the atoms are fairly equal in size and form a crystal structure rather like B.C.C., in which each atom is surrounded by four like and four unlike neighbours in a superlattice pattern. An important group are the *sigma phases*, which tend to form in certain heat-resisting alloys and high-alloy steels, with harmful effect because of their brittleness. Examples are FeCr, CoCr, Mn_3Cr, FeV, FeW, Mn_3V and Co_2Mo_3. Their structure is based on polyhedral figures, such as the icosahedron, which provide at their vertices various patterns for closely packing 12, 14, 15 and 16 spheres round a central one. Such polyhedra cannot, however, be stacked together without distortion to fill space in a crystalline array. In sigma phases the necessary distortion is provided by the slight difference in size of the regularly arranged alloy atoms; and so extended crystal structures, complex but very densely packed, are built up. Some of the complex crystal structures observed in certain metals (e.g. β-uranium, α-manganese) are similar to sigma phase and Laves phase structures, which suggests that the metal atom adopts different radii so as to be able to form an 'intermetallic size factor compound' *with itself* in one of these densely packed structures.

Turning now to *electron compounds*, *Hume-Rothery* and *Westgren* first observed that in many alloy systems phases of similar crystal structures are formed at the same ratios of valency electrons to atoms. The electron concentrations concerned are 3/2, 21/13 and 7/4. Some examples are:

(1) A. *Body-centred cubic (β-brass) structure*; *electron–atom ratio* = 3/2.
         (Cu, Ag or Au) Zn; CuBe; AgMg; Cu_3Al; Cu_5Sn; (Co, Ni or Fe) Al.
(1) B. *Complex-cubic (β-manganese) structure*; *electron–atom ratio* = 3/2.
         (Ag or Au)$_3$ Al; Cu_5Si; $CoZn_3$.
(1) C. *Hexagonal close-packed structure*; *electron–atom ratio* = 3/2.
         AgCd; Cu_5Ge; Ag_7Sb.
(2)   *Complex-cubic (γ-brass) structure*; *electron–atom* = 21/13.
         (Cu, Ag or Au)$_5$ (Zn or Cd)$_8$; Cu_9Al_4; $Cu_{31}Sn_8$; (Fe, Co, Ni, Pd or Pt)$_5$ Zn_{21}.
(3)   *Hexagonal close-packed (ϵ-brass) structure*; *electron–atom ratio* = 7/4.
         (Cu, Ag or Au) (Zn or Cd)$_3$; Cu_3Sn; $CuBe_3$; Ag_5Al_3.

These electron concentrations are based on the following numbers of valency electrons assumed to be contributed by each atom: Cu, Ag, Au, 1; Mg, Zn, Cd, Be, 2; Al, 3; Sn, Si, Ge, 4; Sb, 5; Fe, Co, Ni, Pt, Pd, 0. The transition metals are given *zero* valency, which is justified on the assumption that, in the alloys, the partly-empty d-states in the electron structure (see § 3.6) absorb valency electrons when the electron concentration is increased. There are also indications that transition metals absorb electrons in aluminium alloys (*Raynor, Progress in Metal Physics I*).

Electron compounds are also formed in certain *ternary* (i.e. three component) alloy systems at the appropriate compositions and they usually exist over a small range of composition about the exact electron–atom ratio. Both ordered and disordered structures are observed in electron compounds.

The existence of electron compounds is a consequence of the nature of the metallic bond. The stability of such a phase depends mainly on the electron concentration and on the pattern of atomic sites in the crystal structure, while the actual distribution of the different atoms in these sites is less important. This agrees with the conclusions of the electron theory of metals (§ 19.1).

14.5   Interstitial phases

An important group of interstitial phases consists of the hydrides, nitrides, carbides and borides of the transition metals. The non-metal atoms in these, being small, go into interstitial sites in the structure, between the metal atoms which themselves form a complete and often fully close-packed crystal structure. These interstitial phases are true alloys with metallic properties. By contrast, more highly electropositive metals tend to form non-metallic compounds, e.g. calcium carbide.

These phases have simple crystal structures when the interstitial atom has a radius less than 0·59 of that of the metal atom. The phases usually form at compositions approximating to the formulae M_4X, M_2X, MX, MX_2 (M = metal, X = non-metal). They are mostly F.C.C. or C.P. Hex, occasionally B.C.C., with the metal atoms filling the normal atomic sites of the crystal structure and the non-metal atoms in sites between them. When the ratio of radii exceeds 0·59 the non-metal atom is too big for its

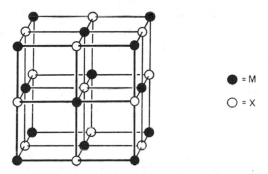

● = M

○ = X

Fig. 14.3 Sodium chloride structure of MX carbides

interstitial site and a more complex crystal structure is usually then formed to accommodate the atom. In carbon steel the ratio is 0·63 and the carbon atom is too big to fit easily into the B.C.C. crystal structure. The intermediate phase formed in this case is *cementite*, Fe_3C. This has metallic properties but a complex crystal structure. Various *carbo-nitride* phases can also be formed, with compositions in the range Fe_2X to Fe_3X, in which X stands for freely interchangeable carbon and nitrogen atoms.

The transition metal carbides are of great practical importance. Some of them are amongst the hardest and most refractory of all known substances; some melting points, for example, in deg C are: NbC, 3500; TaC, 3800; TiC, 3150; VC, 2800; WC, 2750; ZrC, 3500. They are the basis of most high-speed cutting tools (WC and TaC) and are also used as finely dispersed second phases (see § 25.3) in heat-resisting steels and other alloys where mechanical strength is required at high temperatures. Some transition metals (Ti, V, Zr, Nb, Hf, Ta) form extremely stable carbides of the type MX with the sodium chloride structure (Fig. 14.3). The carbides

MoC, Mo_2C, WC, W_2C, and Ta_2C, have a hexagonal structure. The affinity of chromium for carbon is important in alloy steels. Chromium is soluble in cementite and it also forms three carbides of the type $M_{23}C_6$, M_7C_3 and M_3C_2, which have considerable solubility for iron and other transition metals. For example the $M_{23}C_6$ phase, which has a complex cubic structure, can be formed from mixed carbides of the type $(Cr, Fe, W, Mo)_{23}C_6$.

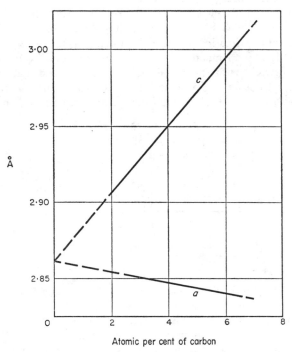

Fig. 14.4 Variation of c and a axes with carbon content in martensite (after Kurdjumov)

As well as forming intermediate phases, H, N, C, and B dissolve appreciably in transition metals, forming primary interstitial solid solutions. The size factor appears to play a large part in determining solubility. In iron there is more room for carbon atoms in the interstices of the F.C.C. crystal structure than in those of the B.C.C., so that the solubility is greater in F.C.C. (maximum of 1·7 weight per cent) than B.C.C. (0·02 weight per cent). This difference of solubility is extremely important in the heat-treatment of carbon steels (cf. § 20.4).

When carbon or nitrogen is dissolved interstitially in B.C.C. iron (*ferrite*) the crystal structure is distorted, either locally or generally, into the *body-*

centred tetragonal form in which one of the cube axes of the B.C.C. cell becomes slightly longer and the other two (equal) cube axes become slightly shorter. The effect is best seen in *martensite*, a supersaturated solution of carbon in ferrite which is prepared by rapid quenching from a temperature of about 900°C. The length c of the long axis (the *tetragonal* axis) and the length a of the short axis, as determined by x-ray analysis, are given in Fig. 14.4.

The local distortion round an atom of carbon or nitrogen in dilute solution in ferrite can also be detected by *internal friction* (*J. L. Snoek*, 1941, *Physica*, **8**, 711). If one of the cube axes of B.C.C. iron is stretched slightly by a mechanical force applied to the crystal, carbon and nitrogen atoms will, by jumping from one site to another within the crystal, gradually gather preferentially in those particular interstitial sites that make the stretched axis the tetragonal axis, so that the crystal can stretch slightly further in the direction of the force. If the force is oscillated, so that first one crystal axis and then another is stretched, and at such a rate as will

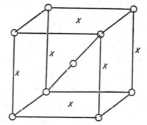

Fig. 14.5 Interstitial positions for carbon and nitrogen in ferrite and martensite

just allow the interstitial atoms to transfer themselves to the most favoured sites during each cycle, mechanical work will then be done repeatedly as the crystal yields to the applied force through these changes of the tetragonal axis. This work dissipates the vibrational energy and so *damps* the mechanical oscillations. By studying the amplitude and frequency of the damped oscillations and their dependence on interstitial content, crystal orientation and temperature, the local distortion can be determined as well as the movements and solubility of the interstitial elements.

The tetragonal distortion occurs because the interstitial atoms occupy positions of type x in Fig. 14.5. Only a few of these positions are occupied, of course, because the amount in solution is always small. The positions in the top and bottom faces are structurally identical with those in the vertical cell edges. In any of these positions the interstitial atom pushes apart the two very close iron atoms directly above and below it and so expands the cell along the vertical axis. The positions shown in the figures represent one set of sites in the B.C.C. cell; the other two sets, which correspond to tetragonal axes along the two horizontal cube edges, are based on the positions in the horizontal edges and in the side faces of the cell.

An examination of the B.C.C. crystal structure shows surprisingly that these interstitial positions are *not* the largest interstitial holes in the cell. The elastic properties of the B.C.C. structure suggest, however, that the two iron atoms above and below each x site in Fig. 14.5, which are mainly responsible for the smallness of these sites, can in fact be pushed apart rather easily (*Zener, Imperfections in Nearly Perfect Crystals*, Chapter XI).

14.6 The free energy of solid solutions

If we had a complete theory of alloys it would be possible to deduce the structure of an alloy from first principles, given only the necessary information about the component atoms. We are at present far from being able to do this. Our theory is able to deal only with broad general effects, such as are represented by the Hume-Rothery rules, and these only in a fairly qualitative way. The individual peculiarities of particular alloy systems are at present largely beyond its reach. Moreover, it is doubtful whether *precise* theoretical predictions could ever be made with confidence because different alloy phases formed from the same component metals tend, from the nature of the free electron bond, to have rather similar free energies, so that the theoretical calculations would have to be extremely accurate before they could predict reliably which alloy structure would be stable.

The results of experimental studies of alloy systems are usually assembled into *phase diagrams* which give the ranges of composition and temperature in which various phases are stable. Such diagrams show that most alloy systems have many features in common. The diagrams are often qualitatively very similar. Other alloy systems have more individualistic diagrams but these are always made up from certain standard features, common to many diagrams, arranged in various ways. These standard features are largely independent of the particular properties of the atoms concerned and are due instead to the general thermodynamical laws that govern the equilibrium of alloys. They can be explained from very simple assumptions about the behaviour of the atoms. These assumptions, which we shall now make, are certainly not valid for every alloy but they nevertheless provide a framework for discussing the common features of all alloy systems and a useful starting point for exploring the peculiarities of individual systems.

We shall neglect pressure as a thermodynamical variable because of the small effect of atmospheric pressure on the free energies of condensed phases. We thus use the Helmholtz free energy F as an approximation to G for equilibrium at constant pressure.

First, we consider the *entropy of mixing* or the *configurational entropy* of a random solution. We discussed this problem qualitatively in § 6.5. This entropy comes from the large number of atomic distributions which belong to the disordered state. Consider then a framework of N crystal sites into which we shall put n atoms of kind A and $N - n$ of kind B randomly among the sites. The first A atom can enter in N different ways since there are N sites to choose from at this stage. The second A atom has $N - 1$

places open to it. The number of distinguishable ways of putting in the first two A atoms is only $N(N - 1)/2$, however, since the arrangement in which the first atom went into site p and the second into q is physically indistinguishable from that in which the first went into q and the second into p. Continuing, the number of ways of placing the first three A atoms in is $N(N - 1)(N - 2)/3!$, where $3!$ is the number of ways of permuting the three atoms amongst any three given sites. The number of ways of placing all the A atoms in is then given by

$$\frac{N(N - 1) \ldots (N - n + 2)(N - n + 1)}{n!} = \frac{N!}{n!(N - n)!} \qquad \textbf{14.1}$$

For each way that we put the A atoms in, there is only one way to put the $N - n$ atoms of B in the remaining $N - n$ empty sites. Hence the formula 14.1 gives the number of distributions of the random solution. From eqn. 6.18 the entropy of mixing is then given by

$$S = k \ln [N!/n!(N - n)!] = k[\ln N! - \ln n! - \ln (N - n)!] \qquad \textbf{14.2}$$

To simplify this result we use *Stirling's approximation*

$$\ln x! = x \ln x - x \qquad \textbf{14.3}$$

which is very accurate when $x \gg 10$. It gives

$$S = k[N \ln N - n \ln n - (N - n) \ln (N - n)] \qquad \textbf{14.4}$$

If we now substitute $c = n/N$ for the atomic concentration of A in the solution and $(1 - c) = (N - n)/N$ for that of B, this expression reduces to

$$S = -Nk[c \ln c + (1 - c) \ln (1 - c)] \qquad \textbf{14.5}$$

It should be noticed that we would have obtained this same formula for the entropy of mixing had we assumed, wrongly, that we could distinguish between one A atom (or B atom) and another. The number of ways of building the crystal would then have been $N!$ but, to get the entropy of mixing, we would then have to subtract the configurational entropies of the pure A and pure B crystals, which on this basis can be built in $n!$ and $(N - n)!$ ways, respectively. This is also why the *entropy of isotope mixing*, which is a real entropy, does not enter our problem. If we make the alloy from mixed isotopes of A, for example, there is then the same isotopic contribution to the entropy of both the alloy and the pure A from which it is made. This cancels out when the entropy of mixing, which is the *extra* configurational entropy of the solution relative to the pure A and B components, is calculated.

We notice from eqn. 14.5 that the entropy of mixing is positive since the fractional quantities c and $(1 - c)$ give negative logarithms. If the crystal has 1 mole of atomic sites, i.e. $N = N_0 =$ Avogadro's number, we have $Nk = R = 8 \cdot 314 \text{ J K}^{-1}$. The entropy of mixing in this case is as given in Fig.

14.6. The curve is symmetrical about the mid-point, $c = 0.5$, where the entropy reaches its greatest value 5.77 J K^{-1}. The slope is very steep near $c = 0$ and $c = 1$. This explains why it is difficult to produce really pure materials since the free energy change $dF(=dE - T \, dS)$ caused by contamination is almost certainly negative, even if dE is large and positive, because of the large value of $T \, dS$ due to the rapid change of S with c near $c = 0$.

Let the internal energy of the solution be E_0 at 0 K and let the specific heat be C_p. The free energy at temperature T is then given by

$$F = E_0 + K(c,T) + NkT[c \ln c + (1 - c) \ln (1 - c)] \qquad \textbf{14.6}$$

where

$$K(c,T) = \int_0^T C_p \, dT - T \int_0^T \frac{C_p}{T} \, dT \qquad \textbf{14.7}$$

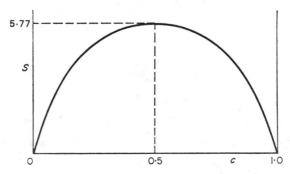

Fig. 14.6 The entropy of mixing (J K^{-1}) as a function of concentration

The first of these integrals is the contribution of C_p to the total internal energy at the temperature T; the second is its contribution to the total entropy.

The next problem is to find an expression for E_0. This is some function $f(c)$ of the concentration. If we expand $f(c)$ in powers of c and $(1 - c)$, the simplest form the function could take is

$$E_0 = \alpha c + \beta(1 - c) \qquad \textbf{14.8}$$

where α and β are constants. This describes an *ideal solution* in which E_0 is independent of the distribution of atoms. Expanding $f(c)$ to the second power, we have

$$E_0 = \alpha c + \beta(1 - c) + \gamma c(1 - c) \qquad \textbf{14.9}$$

Strictly, the appearance of a non-zero *heat of mixing* term, $\gamma c(1 - c)$, in the internal energy means that the equilibrium state of the solution is non-random and has more like or unlike neighbours according as γ is positive or negative. However, when the heat of mixing is small the solution may be so nearly random that its entropy of mixing can be fairly represented by eqn. 14.5. It is then a *regular solution*.

Eqn. 14.9 also follows if we assume that the atoms interact through bonds of constant energy with their nearest neighbours. Let the coordination number of the crystal structure be z. Then on average in the random solution there are zc atoms A and $z(1 - c)$ atoms B, next to any given atom. Since there are Nc atoms A and $N(1 - c)$ atoms B in the alloy, the numbers N_{AA} and N_{BB} and N_{AB} of AA, BB and AB nearest-neighbour bonds are given by

$$N_{AA} = \tfrac{1}{2}Nc.zc = \tfrac{1}{2}Nzc^2$$
$$N_{BB} = \tfrac{1}{2}N(1 - c).z(1 - c) = \tfrac{1}{2}Nz(1 - c)^2 \qquad \textbf{14.10}$$
$$N_{AB} = Nc.z(1 - c) = Nzc(1 - c)$$

The factor $\tfrac{1}{2}$ in N_{AA} allows for the fact that, in going round all the A atoms in turn and counting the number of A atom neighbours of each one, we count each AA bond twice. Similarly for the BB count.

Let the bond energies be V_{AA}, V_{BB} and V_{AB} respectively. Then

$$\begin{aligned} E_0 &= N_{AA}V_{AA} + N_{BB}V_{BB} + N_{AB}V_{AB} \\ &= \tfrac{1}{2}Nz[c^2V_{AA} + (1 - c)^2V_{BB} + 2c(1 - c)V_{AB}] \qquad \textbf{14.11} \\ &= \tfrac{1}{2}Nz[cV_{AA} + (1 - c)V_{BB} + c(1 - c)(2V_{AB} - V_{AA} - V_{BB})] \end{aligned}$$

which is eqn. 14.9 again, with α, β and γ expressed in terms of bond energies. The first two terms give the energies of the pure crystals before mixing. The term $2V_{AB} - V_{AA} - V_{BB}$ then determines the heat of mixing at 0 K. If $2V_{AB}$ is higher than $V_{AA} + V_{BB}$, the replacement of AA and BB bonds by AB bonds raises the internal energy, so that at low temperatures where $F \simeq E_0$ the solution is thermodynamically unstable relative to a *phase mixture* (cf. § 14.7) in which like atoms are grouped together. Conversely, when $2V_{AB}$ is lower than $V_{AA} + V_{BB}$ an ordered solution or a compound is favoured.

Fig. 14.7 shows various possibilities based on eqns. 14.6 and 14.11. In these we have not entirely excluded the variation of the specific heat term, $K(c,T)$, with composition. To a first approximation it varies linearly with c and this can be absorbed, along with the effect of $V_{AA} \neq V_{BB}$, in the slope from A to B.

Before we can decide on the stable state, i.e. the *equilibrium constitution*, of the alloy we must first find the conditions for homogeneous or heterogeneous equilibrium. We shall find that the curvature of F as a function of c is very important. In diagrams (a) and (b) above, the curvature d^2F/dc^2 is positive everywhere, but in (c) there are two minima, p and s, and a region of negative curvature between points of inflexion q and r ($d^2F/dc^2 = 0$).

We have assumed randomness in the above calculations but non-ideal, non-dilute solutions are not random. They deviate from randomness in that direction which reduces the internal energy. We shall discuss effects of these deviations later. Fortunately the error in our free energy expression, when applied to homogeneous and disordered but non-random solutions, is not large because we have overestimated both the heat and entropy of mixing and these make opposite contributions to the free energy. Furthermore, *strong* deviations from randomness lead either to ordered solutions or to phase mixtures and we can develop separate theories for these.

The assumption of nearest-neighbour bonds is unrealistic since the atoms in alloys are mainly held together by the communal free electron gas. Eqn. 14.9, to which the assumption leads, does have a wide significance however as a simple expression for the internal energy of a non-ideal solution. As regards the specific heat term, $K(c,T)$, in many alloys it proves

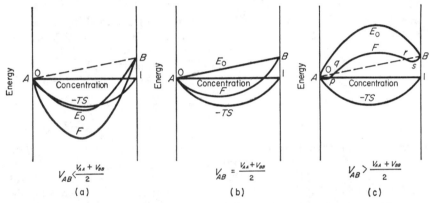

Fig. 14.7 The free energy of a solution as a function of concentration

a fair approximation to take the specific heat as the weighted average of those of the components (*Neumann-Kopp rule*). There is sometimes a deviation from this rule, however, due to misfitting solute atoms; this affects the thermal vibrations and hence the specific heat of these atoms.

14.7 Phase mixtures

Alloys often exist in *heterogeneous equilibrium* (cf. §5.1), i.e. as *phase mixtures* in which homogeneous and distinct crystal grains of two or more different phases are mixed together and joined to one another along well-defined, narrow interfaces. Before discussing the reason for this we shall set up the *lever rule*, which gives the amounts of the phases present. Consider a binary alloy of components A and B in concentrations c and $(1 - c)$ respectively. Let the alloy consist of two different phases, 1 and 2, in which the concentrations of A are c_1 and c_2, respectively. Let the proportion of phase 1 in the alloy be x; that of phase 2 is then $(1 - x)$. If there are N atoms in all, we then have for the A atoms, $Nc = Nxc_1 + N(1 - x)c_2$, i.e.

$$x = (c - c_2)/(c_1 - c_2) = m/l$$
$$(1 - x) = (c_1 - c)/(c_1 - c_2) = n/l \qquad \textbf{14.12}$$
$$x/(1 - x) = (c - c_2)/(c_1 - c) = m/n$$

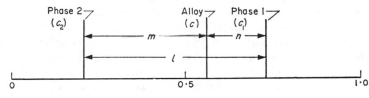

Fig. 14.8

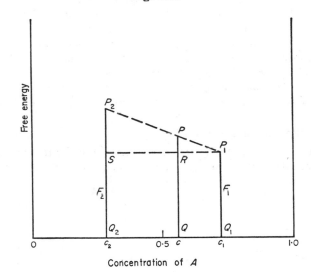

Fig. 14.9

where l, m and n are the lengths defined in Fig. 14.8. These relations are the lever rule and are necessary for determining the constitution of phase mixtures.

To find the average value of a *capacity* property (cf. § 6.2) it is sufficient, when the individual grains of each phase are large enough to allow surface effects to be neglected, to take a *weighted average* of the values belonging to the phases. For example, in Fig. 14.9 the free energy F of unit amount of alloy of average composition c is given by $F = F_1 + (F_2 - F_1)(QQ_1/Q_1Q_2)$, where $F_1 \,(=P_1Q_1)$ and $F_2 \,(=P_2Q_2)$ are the free energies of the constituent phases. We have $F_2 - F_1 = P_2S$ and $QQ_1/Q_1Q_2 = PR/P_2S$, so that $F = PQ$.

The individual grains of the various constituents in a phase mixture can usually be seen quite easily when a metallographically prepared section is examined under an optical microscope. They are often of order 10^{-5} to 10^{-4} m in size and generally stain differently when etched. We shall discuss some of their characteristic forms later. It is frequently necessary to determine the proportions of phases present in the material from a metallographic section. There are several methods. In the *area method* we take the *proportions by area* of the phases, as viewed and measured on a *random* section through the material. That these are identical with the *proportions by volume* can be seen by thinking of the section as a wafer slice of material, cut randomly. The composition of this slice is then identical with that of the whole alloy (provided that the area is so large that many grains of each phase are intercepted, to reduce statistical sampling errors). The second method is by *line intercepts*. Straight lines are scribed at random across a random metallographic section and the *proportions by length* of the phases intercepted by them are measured. Again, since the lines may be thought of as narrow tubes which randomly sample the material, the proportions by volume are directly given. The third method is by *point counting*. A grid is laid randomly on a random section and the proportions by volume measured from the proportions of grid points which fall on the various phases. Provided that the grid is not too fine compared with the micro-structure, this is usually the quickest method for a given accuracy.

14.8   The stable state of an alloy

We now deduce the stable state of an alloy from the free energies of its phases. Consider first systems with free energy curves as in Fig. 14.10.

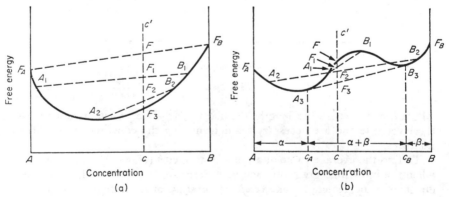

Fig. 14.10 Free energy curves which lead to (a) homogeneous, (b) hetero-geneous, equilibrium

In diagram (a) we have to decide whether the alloy c should exist as a homogeneous solution or as a phase mixture. As a mixture of pure components A and B, it has the free energy F, by the argument of § 14.7. As the components begin to dissolve in each other, so that $A \to A_1$ and $B \to B_1$, the free energy of the alloy falls, i.e. $F \to F_1$. The closer A_1 and B_1 approach. the lower is F_1, and in the limit when the alloy becomes homogeneous single phase the free energy reaches its lowest value F_3. This is the stable state. This conclusion remains valid even if one phase has to *increase* its own free energy, e.g. $A_2 \to F_3$, to reach the homogeneous state. The homogeneous state is stable because the free energy curve has a simple U shape, i.e. d^2F/dc^2 is positive everywhere, so that any straight line joining two points on the curve lies above the curve between the points.

In diagram (b) the alloy c has a free energy F when homogeneous. If it splits into two phases, e.g. A_1 and B_1, its free energy falls, $F \to F_1$. As A_1

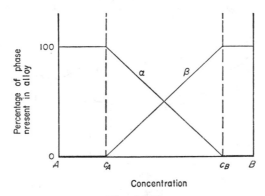

Fig. 14.11

and B_1 diverge in composition the free energy continues to fall at first, but eventually rises again as $A_1 \to F_A$ and $B_1 \to F_B$. For example, the mixture $A_2 + B_2$ has higher free energy than $A_3 + B_3$. The stable state is now the phase mixture whose representative points on the free energy curve are joined by the lowest straight line. This line is the *common tangent* which touches the free energy curve twice (at A_3 and B_3). Between the compositions c_A and c_B, which correspond to A_3 and B_3, the alloy consists of grains of A-rich phase (which we shall call α) of composition c_A intermingled with grains of B-rich phase (β) of composition c_B. From the lever rule we have,

$$\text{proportion of } \alpha = \frac{c_B - c}{c_B - c_A}; \quad \text{proportion of } \beta = \frac{c - c_A}{c_B - c_A} \quad 14.13$$

The proportions of α and β thus vary linearly with the alloy composition c, as shown in Fig. 14.11. From A to c_A and from c_B to B the homogeneous single phase, α or β, is stable. In each of these regions the primary metal, A or B, acts as a solvent for the other and dissolves it homogeneously. At the compositions c_A and c_B the *solubility limits* or *phase boundaries* are reached; each solvent is then *saturated* with solute. In alloys of composition c between c_A and c_B some of the A and B atoms group together to form grains of α, of composition c_A; the remainder similarly form β, of composition c_B. As c increases from c_A to c_B the proportion of β increases linearly at the expense of α, through the *two-phase* $\alpha + \beta$ region, as in Fig. 14.11.

The common tangent construction is basic to the analysis of all free energy curves. In Fig. 14.12 we show the curves of a system which forms intermediate phases β and γ, as well as primary phases α and δ. We construct the common tangents PQ, RS and TU, as shown. In the region $A'P$

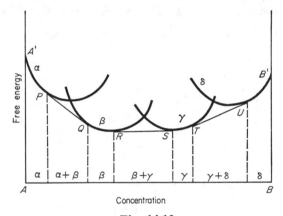

Fig. 14.12

the homogeneous single phase α is stable. Then a two-phase region, $\alpha + \beta$, follows from PQ, the compositions of the phases being given by the points P and Q respectively. This in turn is succeeded by a single-phase region, β, from Q to R, then a two-phase region, $\beta + \gamma$, from R to S, and so on. We thus have the following rules:

(1) At overall compositions where the lowest free energy is that given by a point on the free energy curve of a single phase, a homogeneous system is stable, consisting entirely of this phase.

(2) At overall compositions where the lowest free energy is given by a point on the common tangent to the free energy curves of two or more phases, the system is stable as a mixture of these phases.

(3) In a homogeneous system the overall composition is that of the single constituent phase. In a phase mixture the compositions of the phases are fixed by the points of contact of the common tangent with the free energy curves; the proportions of the phases vary with the overall composition in accordance with the lever rule.

The common tangent construction also explains why the boundaries between grains of different phases are extremely sharply defined. For reasons of continuity we expect that material in such a boundary has a structure and composition intermediate between those of its adjoining phases. Its free energy (per unit amount) must then be high since the free energy curves rise above the common tangent in this region. To minimize this excess interfacial free energy, the boundary becomes as thin as possible. As with grain boundaries, we expect interphase boundaries to be only one or two atoms thick.

14.9   Equations of phase equilibrium; the phase rule

We now set up the common tangent construction mathematically. Referring to Fig. 14.10, we obtain two conditions for this. The slope of the free energy curve must be the same at c_A and c_B; and the tangent there must be *common* to both parts of the curve. Thus, if F_A and F_B denote the free energies at c_A and c_B, we have *two* equations

$$\frac{dF_A}{dc_A} = \frac{dF_B}{dc_B} = \frac{F_B - F_A}{c_B - c_A} \qquad\qquad 14.14$$

In systems which contain many components, the number of equations of equilibrium is correspondingly larger. It is important to know how many there are. Suppose that we have r phases coexisting in equilibrium in a system of n components. In each phase there are $n - 1$ independent concentration variables since the concentration of one of the n components is always fixed, by difference, when the others are given. The free energy curve now becomes a surface in which the free energy is plotted against $n - 1$ composition variables. The slope of this surface at any point is defined by giving the values of $n - 1$ terms of the type $\partial F/\partial c_i$, where c_i is the concentration of component i. The common tangent has to touch at r such points on the free energy surface. The slopes must be equal at all of them. This requires $(n - 1)(r - 1)$ equations of the type $\partial F_A/\partial c_A = \partial F_B/\partial c_B$. For the tangent to be a common one, the heights of $r - 1$ of these points have to be adjusted relative to the remaining one so that they all lie in the same plane. This requires further $r - 1$ equations. Altogether, then, $n(r - 1)$ equations are required.

The famous *phase rule* of *Willard Gibbs* follows from this. How many variables (e.g. temperature, pressure, composition) have to be specified to fix the thermodynamic state of a system exactly? Of course this number can be very large if we admit the influence on the system of electrical, magnetic, gravitational and elastic forces, etc. and also include various surface effects. But in most problems of phase equilibrium we are not concerned with these additional variables. Let us consider then only temperature, pressure and composition. As we have seen, in each phase the concentration of $n - 1$ components can be varied independently. Hence, in all, there are $r(n - 1)$

independent concentration variables. Adding temperature and pressure, this makes $r(n - 1) + 2$ altogether. However, these cannot all be varied independently of one another, because they are linked by the $n(r - 1)$ thermodynamic equations of equilibrium. The number of *independent* variables, or *degrees of freedom*, v, as they are called, is thus given by

$$v = r(n - 1) + 2 - n(r - 1) = n - r + 2 \qquad \textbf{14.15}$$

This is the phase rule. For solids and liquids where we can neglect pressure as a variable it reduces to

$$v = n - r + 1 \qquad \textbf{14.16}$$

For a pure metal we have $n = 1$, so that $v = 2 - r$. If the metal is in a single-phase state, either solid or liquid, then $r = 1$ and $v = 1$. There is one degree of freedom left to the system, its temperature. This means that the same single-phase state can exist over a *range* of temperature. If the metal is in a two-phase state, i.e. solid and liquid coexist in equilibrium, then $r = 2$ and $v = 0$. There are no degrees of freedom, i.e. the temperature cannot vary. This state can therefore exist only at a single temperature, the melting point.

In binary alloys we have $n = 2$ and $v = 3 - r$. For a single-phase alloy we have $v = 2$ which means that this state can exist over a range of both temperature and composition. If two phases coexist, we have $v = 1$. The temperature and composition can still change, but *not independently* of each other. If we change the temperature, for example, and wish to maintain this two-phase state, then we must change the compositions and proportions of the phases in some particular and fixed way. Applied to the solid-liquid equilibrium, this degree of freedom means that melting can occur over a range of temperatures. If three phases coexist in the binary alloy, then $v = 0$. This can happen only at a particular temperature and particular composition of each phase. Such a singular point of phase equilibrium is known as a *eutectic* or *peritectic*. We shall discuss these in Chapter 15.

14.10   The free energy of intermediate phases

The simple U shape assumed for the free energy curves of the intermediate phases in Fig. 14.12 is reasonable since it gives a low energy over a certain range of composition, where the phase is likely to be stable, and then rises sharply at compositions unsuitable for the phase. If the phase is a solid solution we expect a broad shallow U curve and a wide range of composition, e.g. phase α in Fig. 14.13. If it is a compound, with a precise composition, the U curve is very narrow and rises sharply, away from the ideal stoichiometric value, e.g. the β phase $(A_x B_y)$ in Fig. 14.13. The phase cannot then appear over a wide range of composition.

However, if a phase appears over a narrow range of composition, it is not necessarily a compound. For example, in Fig. 14.14(a) the β solid solution is confined to a narrow range of composition despite its broad U curve, because the common tangents pq and rs are nearly parallel. Usually in such cases a small change of temperature reveals this situation; the slopes of the tangents change slightly with temperature and produce large changes in the solubility limits c_1 and c_2.

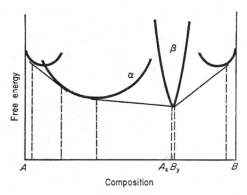

Fig. 14.13

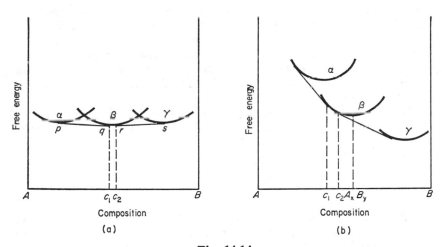

Fig. 14.14

It is frequently found that an intermediate phase, obviously based on a simple structural formula A_xB_y, exists only at compositions on one side of the ideal stoichiometric value. For example, in Cu-Al alloys the compound $CuAl_2$ exists only on the Al-rich side of the exact $CuAl_2$ composition. This effect can be explained by free energy diagrams as in Fig. 14.14(b). Here the α and β curves are such that the range of existence, c_1 to c_2, for the β phase lies to one side of the minimum point of the curve. It is clear from Fig. 14.12 that in systems which form several intermediate phases this must happen quite generally.

The Hume-Rothery electrochemical rule can be explained from free energy curves. For example, in Fig. 14.13 the greater the attraction of the A and B atoms in the compound $A_x B_y$ the lower is the position of the free energy curve of this phase. The common tangent joining this phase to the B-rich solution is then increased in slope, so that the solubility limit in B is shifted towards pure B. Thus, the greater the tendency to form compounds, the smaller is the extent of primary solubility.

14.11 Variation of solubility with temperature

Solubility usually increases with increasing temperature. This is because a disordered solution generally has higher entropy, other things being equal, than a phase mixture and so is favoured at high temperatures through the influence of the TS term in the free energy $F = E - TS$. On the other hand, if the free energy curves go as the α and β curves in Fig. 14.12, and the β phase has higher entropy than the α phase (e.g. because the concentration is higher), then, since $dF/dT = -S$ (cf. eqn. 6.12), the β curve moves downwards more rapidly than the α curve as the temperature is raised and the limit of primary solubility then *decreases*. This happens most commonly when the β phase is liquid solution and the α is solid solution, as discussed in the next chapter. It also sometimes happens in the solid state, however. In brass, for example, the primary solution of zinc in copper (α phase) decreases as the temperature rises above 450°C because the β phase in equilibrium with it is also a disordered solid solution above this temperature (see § 14.12).

We now examine further the system of Fig. 14.7(c). For simplicity we deal only with the symmetrical case where $V_{AA} = V_{BB} = V_0$. We write $2V_{AB} - V_{AA} - V_{BB} = 2V$, so that eqn. 14.11 becomes $E_0 = \frac{1}{2}NzV_0 + Nzc(1 - c)V$. The free energy, eqn. 14.6, then becomes

$$F = \tfrac{1}{2}NzV_0 + K(c,T) + Nzc(1 - c)V \\ + NkT\left[c \ln c + (1 - c) \ln (1 - c)\right] \qquad \textbf{14.17}$$

We also neglect the variation of $K(c,T)$ with c. Then, since the absolute level of free energy is not needed for determining the solubility limit, we can drop the terms $\frac{1}{2}NzV_0 + K(c,T)$. It is convenient to work with the free energy per atom, $f = F/N$. Hence

$$f = zVc(1 - c) + kT\left[c \ln c + (1 - c) \ln (1 - c)\right] \qquad \textbf{14.18}$$

We note that c and z are pure numbers and that zV/k has the dimensions of temperature. It is therefore convenient to measure temperature as the dimensionless quantity kT/zV. Fig. 14.15 shows the variation of f with c at temperatures defined by $kT/zV = 0$, 0·2, 0·3, 0·4 and 0·5. At low temperatures the free energy is high in the middle of the diagram because of the large number of high-energy unlike-atom bonds there. At higher temperatures this is counteracted by the entropy of mixing, giving two minima (e.g. c_1 and c_2) which converge as the temperature is raised. At the critical point $kT/zV = 0·5$ the region of negative curvature disappears and the homogeneous disordered solution becomes stable at all compositions (assuming that melting does not occur).

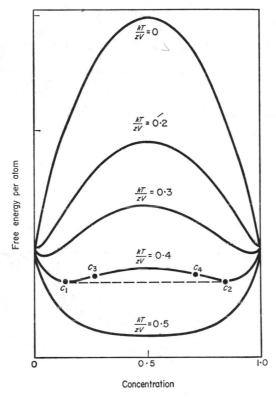

Fig. 14.15 Free energy curves at various temperatures, according to eqn. 14.18

Because of the simple symmetrical form of the curves in Fig. 14.15 the general conditions for equilibrium, eqns. 14.14, reduce *in this special case* to the simple condition $df/dc = 0$. We can then find the solubility limit simply from the positions of the minima in this case, as represented by the full line in Fig. 14.16.

Applying the condition $df/dc = 0$ to eqn. 14.18 we obtain

$$zV(1 - 2c) + kT\,[\ln c - \ln(1 - c)] = 0 \qquad\qquad \textbf{14.19}$$

8 + I.T.M.

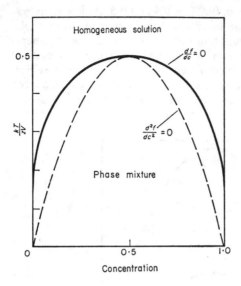

Fig. 14.16 Variation with temperature of the compositions at which $df/dc = 0$ and $d^2f/dc^2 = 0$ for the alloy system of eqn. 14.18. The full curve is the solubility limit

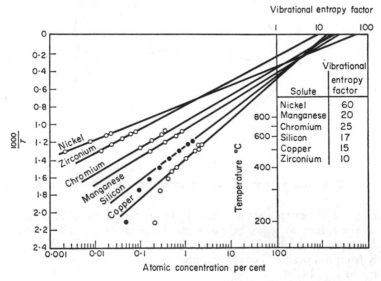

Fig. 14.17 Experimental data of W. L. Fink and H. R. Freche for the solubility of various metals in aluminium (as presented by C. Zener, 1950, Chapter II, Thermodynamics in Physical Metallurgy, *Amer. Soc. Metals*).

as the equation to the solubility curve. A few alloy systems, e.g. Au-Pt, have curves qualitatively similar to this but for most alloys $kT/zV = 0.5$ is well above the melting point. When the solubility is small, so that $(1 - c)$ and $(1 - 2c)$ can be approximated by unity, eqn. 14.19 can be simplified to

$$c = \exp\left(-zV/kT\right) \qquad\qquad \textbf{14.20}$$

giving a simple exponential dependence of solubility on temperature. In fact the solubility limits of many dilute solutions prove to be proportional to $\exp\left(-\Delta H/RT\right)$ where ΔH is the *heat of solution*, i.e. the heat absorbed when 1 mole of solute goes into dilute solution. Fig. 14.17 shows experimental data for solid solutions of several metals in aluminium. In the range 400–600°C the logarithm of solubility varies as the reciprocal of absolute temperature, in accordance with the theory. The deviations at lower temperatures are almost certainly due to the difficulty of establishing true equilibrium at temperatures where atomic diffusion is very slow.

We notice in Fig. 14.17 that the solubility lines overshoot the point $c = 100$ per cent at $1000/T = 0$. Zener (*loc. cit.*) interprets this as evidence for an additional *entropy factor* ΔS and writes the solubility as

$$c = \exp\left(\Delta S/R\right)\exp\left(-\Delta H/RT\right) \qquad\qquad \textbf{14.21}$$

His explanation is that the crystal distortion round a misfitting solute atom results in a lower frequency of atomic vibrations there. It follows from the theory of specific heats (cf. § 19.2) that this gives an additional *vibrational entropy*, which is ΔS. Since a large heat of solution is a fair indication of a large atomic misfit, we expect a large ΔS to be associated with a large ΔH. This is to some extent confirmed by Fig. 14.17; e.g. nickel has the largest ΔH and ΔS, copper has the smallest ΔH and one of the smallest ΔS values.

Returning to Fig. 14.15, we notice that the curves which rise in the middle of the diagram have two *points of inflexion* (e.g. c_3 and c_4). We find their positions by applying the condition $d^2f/dc^2 = 0$ to eqn. 14.8, which gives

$$c(1 - c) = kT/2zV \qquad\qquad \textbf{14.22}$$

This is plotted as the broken curve in Fig. 14.16. Curves for which $d^2f/dc^2 = 0$ are called *spinodal lines*.

14.12   Long-range order in solid solutions

We now consider the opposite case to that of the previous section; i.e. a solid solution in which $2V_{AB} < (V_{AA} + V_{BB})$ so that the internal energy favours the *ordering* of the atoms. X-ray analysis and other methods have shown that two kinds of order can exist. In *long-range* order the crystal can be regarded as two or more interpenetrating crystals, or *sub-lattices* as they are called. Fig. 14.18 shows for example how the B.C.C. crystal structure can be regarded as two interpenetrating simple cubic crystals, α and β. Most or all of the A atoms then lie on one sub-lattice; and the B atoms on the other There is thus a *long-range correlation* in the distribution of the atoms, i.e. a coherent scheme of order extending over a large volume of the crystal. Even though there may be disordered regions, it is possible to find long continuous paths through the ordered regions of the crystal.

Above a certain *critical temperature* long-range order is destroyed. The ordering force is not strong enough to preserve the coherent scheme in the face of the disrupting effect of intense thermal agitation. However, the ordering force is still present and is able to influence the *local* positions of the atoms and prevent them from becoming completely random. Thus a *short-range* order exists in which atoms have more unlike neighbours than expected for a truly random solution. These states of short-range and long-range order are reminiscent of the structures of substances just above and

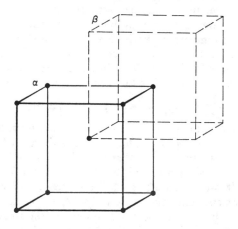

Fig. 14.18

below their melting points. In both, a coherent ordered scheme of atomic positions correlated over large distances is stable below a critical temperature, whereas above this temperature only small, dispersed ordered groups are formed.

For a simple theory we consider a crystal which is already in a state of long-range order and ask how much the two kinds of atoms separate out into their respective sub-lattices (*W. L. Bragg and E. J. Williams*, 1934, *Proc. Roy. Soc.*, **A145**, 699). We consider an alloy such as CuZn (β-brass) in which the pattern of atomic sites is B.C.C. Let there be N sites, $N/2$ copper atoms and $N/2$ zinc atoms. Referring to Fig. 14.18 we define the *degree of long-range order W* by saying that the number of A atoms in α sites is $(1 + W)N/4$. All states of long-range order then follow as W ranges from 0 to ± 1. When $W = 0$ the atoms are randomly distributed. When $W = \pm 1$ the copper atoms are all on either the α or β sub-lattice, giving perfect order in both cases.

To find the equilibrium W at any temperature we write down the free energy F as a function of W and then find the value of W which minimizes F, i.e. gives $dF/dW = 0$. Since $F = E - TS$ we have to find dE/dW and dS/dW.

For the internal energy term we have to find N_{AA}, etc. as in eqn. 14.11. Since each atomic site is surrounded by eight neighbouring sites belonging to the other sub-lattice, the number n_{AA} of AA bonds formed by one A atom in an α-site is eight times the fraction of nearest neighbours (in β-sites) which are A atoms. We *assume* that this fraction is equal to the overall atomic concentration of A atoms in β-sites, which is $(1 - W)/2$. Hence $n_{AA} = 4(1 - W)$. This assumption greatly simplifies the mathematics but is physically unconvincing. It implies that, once the atoms have sorted themselves out on the sub-lattices, their distribution within each sub-lattice is then quite random; i.e. notwithstanding the ordering force, every copper atom in an α-site attracts round itself a strictly average proportion of the copper and zinc atoms in the β sub-lattice. Short-range order is thus excluded by the assumption and this theory can deal with long-range order only.

The total number of AA pairs is given by

$$N_{AA} = n_{AA} \times \text{no. of } A \text{ atoms in } \alpha\text{-sites} = N(1 - W^2) \qquad \textbf{14.23}$$

Similarly

$$N_{BB} = N(1 - W^2) \quad \text{and} \quad N_{AB} = 2N(1 + W^2) \qquad \textbf{14.24}$$

The internal energy, in so far as it depends on W, is thus given by

$$
\begin{aligned}
E &= N(1 - W^2)(V_{AA} + V_{BB}) + 2N(1 + W^2)V_{AB} \\
&= N(V_{AA} + V_{BB} + 2V_{AB}) + 2NW^2V \qquad \textbf{14.25}
\end{aligned}
$$

where $2V = 2V_{AB} - V_{AA} - V_{BB}$. Hence

$$dE/dW = 4NWV \qquad \textbf{14.26}$$

We now consider the entropy, in so far as it depends on W. Each sub-lattice is a random solution of $N/2$ sites with concentrations $c = (1 + W)/2$ and $(1 - c) = (1 - W)/2$. The total entropy of mixing is twice that for each sub-lattice. Applying eqn. 14.5 we have

$$S = -Nk\left[\left(\frac{1 + W}{2}\right)\ln\left(\frac{1 + W}{2}\right) + \left(\frac{1 - W}{2}\right)\ln\left(\frac{1 - W}{2}\right)\right] \qquad \textbf{14.27}$$

and hence

$$\frac{dS}{dW} = -\frac{Nk}{2}\ln\left(\frac{1 + W}{1 - W}\right) \qquad \textbf{14.28}$$

The condition of equilibrium is then

$$\frac{dF}{dW} = 4NWV + \frac{NkT}{2}\ln\left(\frac{1 + W}{1 - W}\right) = 0 \qquad \textbf{14.29}$$

As the alloy is heated towards its critical temperature, i.e. $T \rightarrow T_c$, we have $W \rightarrow 0$ and $\ln [(1 + W)/(1 - W)] \rightarrow 2W$ since $\ln (1 + x) \simeq x$ when $x \ll 1$. Hence, as T rises towards T_c, eqn. 14.29 reduces to $4NWV + NWkT = 0$, giving

$$T_c = -4V/k \qquad\qquad\qquad 14.30$$

where V is, of course, negative. The equation of equilibrium can then be written in the form

$$2W \frac{T_c}{T} = \ln \left(\frac{1 + W}{1 - W} \right) \qquad\qquad 14.31$$

To show the dependence of W on T we choose values of W, substitute them in this equation, and find the corresponding values of T/T_c, as in Fig. 14.19. The order is nearly perfect at low temperatures and decreases with increasing temperature, slowly at first and then ever more steeply,

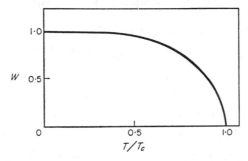

Fig. 14.19 Degree of long-range order as a function of temperature

until all long-range order is lost at the critical temperature. These are the typical features of a *co-operative* equilibrium. In a perfectly ordered alloy it is difficult for the first few atoms to go into 'wrong' positions because this is resisted by the ordering forces from all the neighbours, which are correctly placed. However, once there are some atoms in wrong positions the ordering forces from these atoms attract other neighbouring atoms into wrong positions. The more wrong atoms there are, the easier it is for others to go wrong—a typical social effect.

The total change of configurational entropy, from the fully ordered to the fully disordered state, is simply the entropy of mixing of an equi-atomic alloy, i.e. $5 \cdot 77$ J mol^{-1} K^{-1} (cf. §14.6). An experimental value has been obtained for β-brass from specific heat measurements and is $4 \cdot 23$ J mol^{-1} K^{-1} (*C. Sykes and H. Wilkinson*, 1937, *J. Inst. Metals*, **61**, 223). This is a little smaller and the difference is due to the fact that short-range order above the critical temperature reduces the configurational entropy change.

From eqns. 14.26 and 14.30 we obtain, for the total configurational energy change ΔE,

$$\Delta E = -NkT_c \int_{W=1}^{W=0} W \, dW = \tfrac{1}{2} NkT_c \qquad\qquad 14.32$$

When N = Avogadro's number this becomes $\tfrac{1}{2} RT_c$, i.e. 3100 J for β-brass (T_c = 743 K). By determining the *excess* specific heat due to disordering, Sykes and Wilkinson (*op. cit.*) obtained the experimental value ΔE = 2640 J. Again this is a little smaller; again due to short-range order above the critical temperature.

The excess specific heat due to disordering can be easily recognized experimentally as a 'peak' on top of a standard specific heat/temperature curve (cf. Fig. 19.4). If the entire order-disorder change occurred at a single temperature this peak would be sharpened up into a thin spike, i.e.

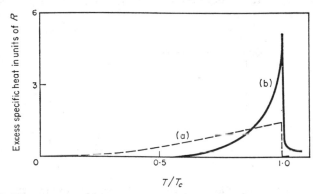

Fig. 14.20 The excess specific heat of beta brass (a) theoretical (Bragg-Williams), (b) experimental (Sykes-Wilkinson)

the excess specific heat would become a latent heat ΔE at T_c. In β-brass, however, in agreement with the predictions of the above theory, ΔE is absorbed gradually over a range of temperature. The excess specific heat is given theoretically by

$$\frac{dE}{dT} = \frac{dE}{dW}\frac{dW}{dT} = -NkW\frac{dW}{d(T/T_c)} \qquad\qquad 14.33$$

Taking $Nk = R$, this gives curve (a) in Fig. 14.20. The experimental curve, due to Sykes and Wilkinson, is similar but much sharper. In practice the change is confined to a narrower temperature range than the theory predicts. The shape of the specific heat curve in fact is extremely sensitive to small changes in the shape of the ordering curve (Fig. 14.19) and so this third comparison with experiment is really a searching test of the finer details of the theory. The effect of short-range order can again be seen, in Fig. 14.20, from the fact that the extra specific heat does not fall quite to zero at T_c.

Changes of the type shown in Fig. 14.19, in which the energy of transition is absorbed as an excess specific heat, not as a latent heat, are said to be *second-order*; they contrast with *first-order* changes such as melting and polymorphic changes of crystal structure. Order-disorder systems that show second-order transitions are those, like β-brass, based on the B.C.C. structure. In systems based on the F.C.C. structure (e.g. Cu₃Au, CuPd, AuMn) the transition takes on a first-order character, with a latent heat at the critical temperature; cf. Fig. 14.21.

It seems that this difference is due to the intimacy of atomic interconnections in the crystal structures (*L. Guttman*, 1956, *Solid State Physics*, **3**, 145). In the F.C.C. structure, but not the B.C.C., atoms which are nearest neighbours to any given central atom can also be nearest neighbours to one another. If one of these neighbours is a 'wrong' atom it is not only coupled to the central atom *directly* by its ordering force; it is also coupled

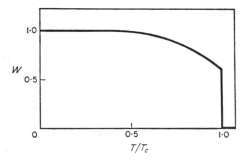

Fig. 14.21 Dependence of long-range order on temperature in a system such as Cu₃Au

to it *indirectly* by its coupling to its other neighbours, which are themselves nearest neighbours of the central atom. The chain of nearest-neighbour links, for this indirect coupling, is short and so we expect this indirect coupling to be strong. The cooperative forces are thus expected to be particularly strong in structures of the F.C.C. type. These structures should then persist in good long-range order on heating until the temperatures become high enough to disrupt the order locally; then the ordered state should disintegrate totally. In B.C.C. systems, however, the indirect couplings are weaker because they are transmitted through chains of at least three nearest-neighbour links. Disorder should then set in more readily, at low temperatures, but the scheme of long-range order should not be so vulnerable to local disruptions at these temperatures.

In some alloys, e.g. CoPt, the structure splits into a *phase mixture*, on ordering, of separate regions of ordered and disordered material. Such splitting is favoured when an alloy changes its crystal structure on ordering (e.g. CoPt changes from F.C.C. to F.C. tetragonal) or when the composition of the alloy is not the ideal one for the superlattice. In this second case, the ordered phase separates out with a composition closer to the ideal one.

14.13   Short-range order and anti-phase domains

The basic condition for ordering is that unlike atoms should prefer to be nearest neighbours. This can be satisfied to a high degree without having long-range order, merely groups or *domains* of local order. An example of two such domains is shown in Fig. 14.22. Each domain is a region of good order, so that there are many *AB* bonds, but the two are out-of-step or *anti-phase* in that the sub-lattice used by *A* atoms in one is used by *B* atoms in the other. Since the domains have equal sizes the degree of long-range order is *zero*, even though the only *AA* and *BB* bonds are those across the *domain boundary*.

On a criterion of *short-range* order, this is a highly ordered state. In the theory of short-range order (*H. A. Bethe*, 1935, *Proc. Roy. Soc.*, **A150**,

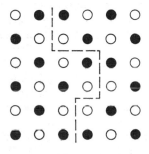

Fig. 14.22 Anti-phase domains

552), a *degree of short-range order* δ is defined such that on average an *A* atom has $z(1 + \delta)/2$ nearest neighbour *A* atoms and $z(1 - \delta)/2$ nearest neighbour *B* atoms, where z is the coordination number. There is no critical temperature above which δ becomes zero, because unlike atoms always tend to be nearest neighbours in spite of thermal agitation, so that although δ decreases at high temperatures it never reaches zero.

Since short-range order can, as in Fig. 14.22, so nearly satisfy the need to form *AB* pairs, why is long-range order ever formed? The answer lies in the fact that the domain boundaries are a source of excess free energy at low temperatures. The *AA* and *BB* bonds across them give them high internal energy. There is also a configurational entropy associated with them, due to the number of ways in which they can be distributed in the crystal, but at low temperatures the (negative) contribution of this to the free energy is outweighed by the (positive) contribution of the internal energy. This excess free energy provides a driving force for the *growth* of domains. Like bubbles in a soap froth, one may grow at the expense of another by rearrangement of atoms on one side of the domain boundary. The domains thus become coarser by the migration of their boundaries and the total area of domain boundary thereby decreases.

We tł ıs picture the sequence of changes, on cooling from a high temperature, as follows. At first small ordered groups continually form and break up again. This short-range order becomes more intense as the temperature falls until, at the critical temperature, the groups develop into well-defined domains which then grow by absorbing one another and remanent disordered regions until coherent schemes of order are established over large regions of the crystal. From a practical point of view, by the time that the domains exceed some 10^4 atoms in width, the crystal can be said to have long-range order. Experimental tests for long-range order (e.g. x-ray analysis) give positive answers at this stage. Our previous definition of long-range order is really too exacting. It is sufficient for the correlation on given sub-lattices to extend over distances large enough to make the numbers of atoms in domain boundaries negligible compared with those in the domains.

Below the critical temperature the domain boundaries have positive tensions and we may thus expect them to try to arrange themselves like films in a soap froth. From elementary experiences we know that in mechanically stable soap froths two bubbles (domains) meet across a surface, i.e. a soap film; three meet along a line, where three films join; and four meet at a corner, where four films meet at a point. This arrangement is impossible in an alloy such as CuZn, however, because there are only *two* different schemes of order, i.e. Cu on the α sub-lattice and Cu on the β sub-lattice, respectively. It is impossible for three different schemes of order to meet along a line, or four to meet at a point, because there are only two available. Hence a mechanically stable foam structure cannot be developed by the domain boundaries in CuZn. Domain growth can thus occur without hindrance once the temperature falls below T_c. Long-range order develops easily in this alloy.

In Cu_3Au, however, there are four different sub-lattices and hence four different schemes of order can develop. Conditions are now favourable for the formation of a stable foam structure and this provides a strong mechanical hindrance to the growth of domains, even though the thermodynamical driving force is directed towards the elimination of all domain boundaries (*W. L. Bragg*, 1940, *Proc. Phys. Soc.*, **52**, 105). In fact, long-range order develops only with some difficulty in Cu_3Au and prolonged heating at temperatures just below T_c (390°C) is necessary to produce a coarse domain structure (*C. Sykes and F. W. Jones*, 1936, *Proc. Roy. Soc.*, **A157**, 213; 1938, **A166**, 376).

14.14 Ordering of carbon atoms in martensite

The confinement of the carbon atoms in body-centred tetragonal martensite to one set of interstitial sites, associated with one tetragonal axis, may be thought of as a special type of order. The driving force for this structure is provided by the tetragonal distortion itself. The elastic strain energy is lowest when the tetragonal unit cells, one round each carbon atom, all have their tetragonal axes aligned in parallel since the stretching of the crystal in this direction, due to the presence of any one carbon atom, makes it slightly easier to fit another carbon atom into a site with this same tetragonal axis. Opposing this, of course, is the entropy of mixing, which is reduced when the carbon atoms are confined to only one-third of the total number of possible interstitial sites. As a result, there is a critical temperature T_c below which martensite is tetragonal and above which it becomes cubic, due to the disordering of the carbon atoms amongst all the interstitial sites (*C. Zener*, 1946, *Transactions A.I.M.E.*, **167**, 550). Calculation shows that

$$T_c = 0 \cdot 243 c E \lambda^2 / k \qquad\qquad 14.34$$

where c is the carbon concentration, E is Young's modulus of elasticity of iron along the cubic crystal axis, λ is the strain produced by the transfer of one carbon atom per unit volume from one tetragonal site to a perpendicular one, and k is Boltzmann's constant. This critical temperature coincides with that at which martensite forms (the M_s *temperature*; see § 20.4) at about 2·8 atomic per cent carbon.

FURTHER READING

ANDREWS, K. W. (1973). *Physical Metallurgy*. George Allen and Unwin, London.

AMERICAN SOCIETY FOR METALS (1950). *Thermodynamics in Physical Metallurgy*.
— (1956). *Theory of Alloy Phases*.

CHARLES BARRETT and MASSALSKI, T. B. (1966). *Structure of Metals*. McGraw-Hill, New York and Maidenhead.

CAMPBELL, A. N. and SMITH, N. O. (1951). *The Phase Rule and its Applications*. Dover Publications, London.

DARKEN, L. S. and GURRY, R. W. (1953). *Physical Chemistry of Metals*. McGraw-Hill, New York and Maidenhead.

HUME-ROTHERY, W. and RAYNOR, G. V. (1954). *The Structure of Metals and Alloys*. Institute of Metals.

LUMSDEN, J. (1952). *Thermodynamics of Alloys*. Institute of Metals.

RAYNOR, G. V. (1949). *Progress in Metal Physics I*. Butterworth, London.

REED-HILL, R. E. (1973). *Physical Metallurgy Principles*. Van Nostrand, New York.

SLATER, J. C. (1939). *Introduction to Chemical Physics*. McGraw-Hill, New York and Maidenhead.

SWALIN, R. A. (1962). *Thermodynamics of Solids*. John Wiley, New York and London.

WAGNER, C. (1952). *Thermodynamics of Alloys*. Addison-Wesley, London.

ZENER, C. (1952). *Imperfections in Nearly Perfect Crystals*. John Wiley, New York and London.

Chapter 15

The Phase Diagram

15.1 Introduction

Free energy diagrams, such as that of Fig. 14.12, give the compositions at which various phases or phase mixtures are stable at a given temperature. From several such diagrams at different temperatures we can construct the *phase diagram* (or *equilibrium diagram* or *constitutional diagram*) which marks out the composition limits of the phases in a given alloy system as functions of temperature. The full curve in Fig. 14.16 is a simple example of part of such a diagram. The phase diagram is of utmost importance in metallurgy. It gives a 'blueprint' of alloy systems from which we can anticipate at what compositions alloys are likely to have useful properties, what treatments we must give to alloys to develop their properties to best effect, and what treatments are likely to be harmful and must be avoided.

In this chapter we shall see how the various standard types of phase diagrams follow from a few basic arrangements of free energy curves. In practice, phase diagrams are usually determined directly, by finding the positions of the phase boundaries experimentally, for this is generally much easier than measuring free energies. The value of the free energy approach lies in the understanding it gives and in the way it coordinates the various types of phase diagrams, which might otherwise seem capricious and arbitrary, into a logical and coherent pattern.

We shall deal mainly with liquid–solid equilibria, although the method is quite general and applies equally well to other equilibria, e.g. those in the solid state. Although it is not at all essential, we shall restrict ourselves to systems in which there is complete miscibility in the liquid state. To ensure this we shall use a free energy curve for the liquid of the simple U shape shown in Fig. 15.1.

15.2 Complete miscibility in the solid state

The simplest systems are those in which the components are so similar that in the solid state they form homogeneous disordered solution at all compositions. The heat of solution term in the free energy of mixing is then very small and the free energy curve follows the curve of $-TS$ against composition, where S is the ideal entropy of mixing. The free energy curve of the solid solution is thus also like that of Fig. 15.1. To obtain the phase diagram we have to study the positions of two such curves, one for the liquid and one for the solid, at various temperatures.

This is done in Fig. 15.2. At high temperatures (e.g. T_1) all alloys are liquid and the free energy curve of the liquid phase lies wholly below that of the solid; vice versa at low temperatures (e.g. T_5). On cooling from T_1 to T_5 the free energy curve of the liquid moves upwards, across that of the solid. In the diagram we suppose that the melting point (T_2) of pure A is higher than that (T_4) of pure B. The curves are thus tilted slightly, to make ΔF_1 smaller than ΔF_2 in diagram (a). On cooling from T_1, the curves first meet,

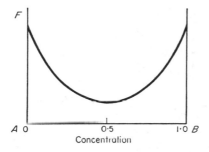

Fig. 15.1

at pure A, at the temperature T_2. All other compositions are still liquid, but pure A can solidify. As the temperature falls to T_3 the curves cross. The common tangent construction then shows that alloys of composition from A to c_1 exist in equilibrium as homogeneous solid solution, those from c_2 to B remain as homogeneous liquid and those between c_1 and c_2 exist as mixtures of solid (c_1) and liquid (c_2). As the temperature falls further, the crossing point moves towards the B-rich end of the diagram. Eventually, at T_4, c_1 and c_2 converge at pure B and all compositions except pure B are now necessarily solid. Collecting these results together, we obtain the phase diagram shown in diagram (f), which is typical of this type of system. The diagram is divided into three separate areas or *phase fields* by the two phase boundary lines known as the *liquidus*, above which all is liquid, and *solidus*, below which all is solid. We notice that the alloys in such a system must freeze over a range of temperature and that in the two-phase region the compositions of the co-existing phases must be different. Examples of alloy systems with this type of diagram are silver-gold (see Fig. 15.3), copper-nickel and gold-platinum.

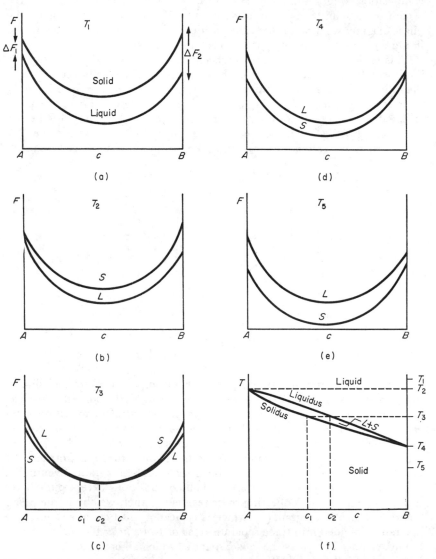

Fig. 15.2 (a) to (e) Free energy curves for the solid and liquid phases of an alloy system at temperatures T_1 to T_5, where $T_1 > T_5$, and (f) the phase diagram

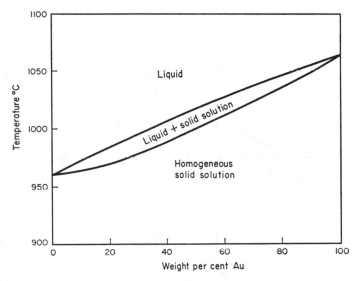

Fig. 15.3 The phase diagram of the silver-gold system

15.3   Partial miscibility in the solid state

Suppose now that the components are only partially soluble, in the solid state, and that no intermediate phases are formed. We use a free energy diagram like that of Fig. 15.4(a) when the crystal structures of the component metals are the same, and like that of Fig. 15.4(b) when they are different. The same general form of phase diagram is produced by both.

Fig. 15.5 shows the free energy curves of solid and liquid at various temperatures. On cooling from T_1, the liquid curve first rises above the solid at the A-rich end (T_2). A similar intersection then occurs at the B-rich end (T_3). At such temperatures (diagram c) there are two common tangents

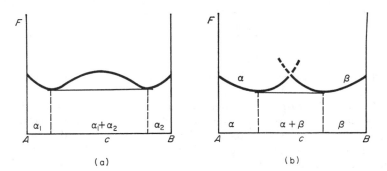

Fig. 15.4

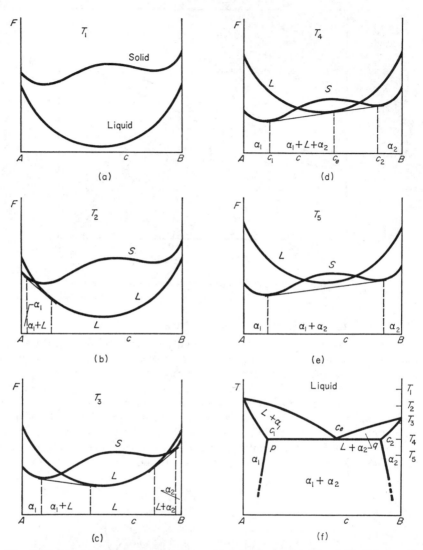

Fig. 15.5 (a) to (e) Free energy curves which lead to a eutectic phase diagram (f)

which give *two* two-phase regions between solid solution α_1, liquid L and solid solution α_2. Further cooling brings these two tangents together and narrows the composition range in which homogeneous liquid is still stable. Eventually the two tangents merge into one, at T_4, which simultaneously touches *three* points on the free energy curves. This is a *eutectic* and the temperature T_4—the lowest at which any liquid can exist—is the *eutectic temperature.* The composition c_e at which an alloy can still be completely liquid at this temperature is the *eutectic composition.*

The line pq in the eutectic phase diagram (f) is important. It is part of the solidus and marks the temperature below which no liquid can exist in equilibrium. It is necessarily an *isothermal* line because of the arrangement of the free energy curves near T_4; just above this temperature all alloys from c_1 to c_2 contain some liquid; just below, they are all solid. An alloy of composition c_e at the temperature T_4 can exist wholly as liquid, or as solid solutions c_1 and c_2 (in the proportions given by the lever rule), or as a

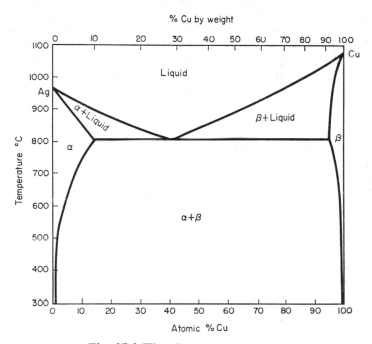

Fig. 15.6 The silver-copper system

three-phase mixture of liquid c_e and the solids c_1 and c_2. Since the weighted average of the free energies of the solids coincides with the free energy of the liquid, at c_e, the total free energy of the alloy is indifferent to the proportions of liquid and solid at this temperature. This is the singular point referred to in § 14.9. Below the eutectic temperature the solubility limits of the two solid solutions decrease with falling temperature in the manner discussed in § 14.11. Examples of simple eutectic systems are lead-tin, lead-antimony, aluminium-silicon, chromium-nickel and silver-copper (Fig. 15.6).

If the eutectic temperature is relatively low there is a chance that the effect of temperature on the shape of the solid solution free energy curve, which was ignored in Fig. 15.5 but in general is like that of Fig. 14.15, may cause the solid solubility in equilibrium with the *liquid* (e.g. α_1 in diagram (c) in Fig. 15.5) to *decrease* with falling temperature. This gives a phase diagram with a *retrograde* solidus curve, as shown in Fig. 15.7, with the striking property that alloys of certain compositions, e.g. c_1, can become completely solid and then partially melt again on cooling. Examples of systems with retrograde solidus curves are silver-lead and zinc-cadmium.

Some systems with free energy curves fairly similar to those of Fig. 15.5 develop a quite different form of phase diagram. This happens mainly when the melting points of the two components differ widely. As shown in Fig. 15.8, it is then possible for the β-phase to make its first appearance, on cooling, in the composition range previously occupied by the phase mixture $\alpha + liquid$. Geometrically, this is the opposite of the behaviour in Fig. 15.5, for now a free energy curve (β) breaks through a common tangent (between α and liquid) on *cooling*. This part of the phase diagram is therefore like a eutectic turned upside down. It is called a *peritectic*. The composition at r, in Fig. 15.8(d), is the *peritectic composition* and the isothermal line pq marks the *peritectic temperature*. Just above this temperature there is no β-phase in any alloy; just below, all alloys from p to q contain some β. The silver-platinum system has a simple peritectic diagram.

The comparative lowness of the temperatures at which eutectics generally freeze is of considerable practical importance. The resulting ease of melting and high fluidity make eutectic alloys particularly attractive for the casting of intricate objects. Thus *cast iron* is based on a eutectic in the iron-carbon system at 4·3 wt per cent C at 1130°C (cf. Fig. 15.13) and the die-casting aluminium-silicon alloys are based on a eutectic at 12 wt per cent Si at 577°C. *Tinman's solder*, used for solders because of its low melting point, is based on the eutectic of 40 wt per cent lead in tin at 180°C. *Plumber's solder* (33 wt per cent tin in lead) freezes over a range of temperature from the liquidus to the eutectic solidus (180°C) and while in the pasty semi-solid form can be conveniently 'wiped' on lead pipe joints. The fusible slags of § 8.3 are mainly based on eutectics between the component oxides. Many eutectics are of more complex types and contain three or more components (cf. § 16.2). *Linotype metal*, used for printing, contains 12 wt per cent antimony and 4 wt per cent tin in lead and melts at 240°C. *Fusible alloys*, for automatic sprinkler systems, for mould patterns in precision casting and for many other uses where really low melting points are required, are generally based on eutectics between bismuth, lead, tin and cadmium (e.g. *Wood's metal*, 50 per cent Bi, 27 per cent Pb, 13 per cent Sn, 10 per cent Cd; M.P. = 66°C).

The low melting points of eutectics can also cause difficulties, particularly when they lead to local melting due to impurities in alloys. The melting of a eutectic based on iron sulphide in grain boundaries causes some steels to break apart along such boundaries during rolling or forging at high temperature (*hot shortness*).

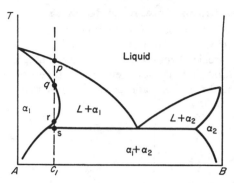

Fig. 15.7 A eutectic system with a retrograde solidus curve. In equilibrium, the alloy c_1 starts to freeze at p on cooling, is completely solid at q, then partially melts at *r*, and becomes completely solid again at s.

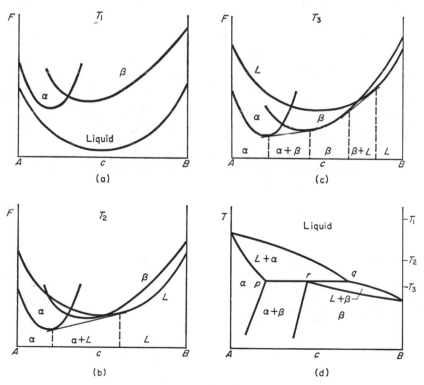

Fig. 15.8 (a)–(c) Free energy curves leading to a peritectic phase diagram (d)

15.4 Systems containing intermediate phases

A typical form of phase diagram obtained when an intermediate phase freezes directly out of the liquid is shown in Fig. 15.9. The free energy curve of this phase breaks through that of the liquid at a fairly high temperature (T_2). The common tangents then merge together to form eutectics between the intermediate phase and each of the primary solid solutions.

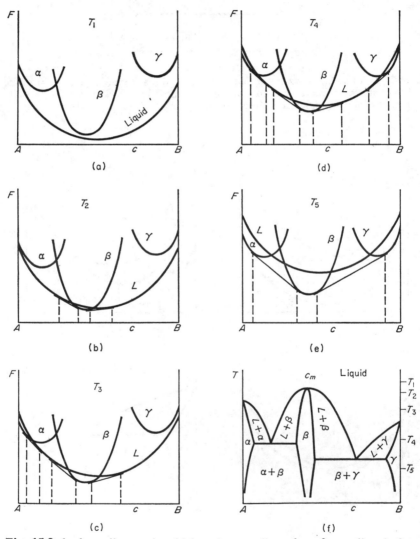

Fig. 15.9 A phase diagram in which an intermediate phase forms directly from the liquid

The eutectics are not inevitable in such systems. Peritectics may be formed instead. For example, if the γ-phase free energy curve were to break through the common tangent between the liquid and β phase, a diagram such as Fig. 15.10 would then result.

The local maximum at c_m in the liquidus and solidus curves is characteristic of this type of phase diagram. Examples can be seen in the bismuth-tellurium, magnesium-tin, magnesium-lead and magnesium-cadmium systems. The composition c_m usually does not coincide exactly with that of the minimum in the free energy curve of the intermediate phase because the free energy curve of the liquid is generally not horizontal there (cf. § 14.10).

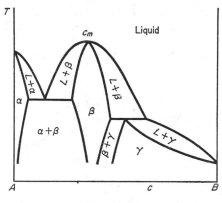

Fig. 15.10

An intermediate phase may also be formed through a reaction between solid and liquid in a partly solidified alloy, as shown in Fig. 15.11. Here the free energy curve of β breaks through the common tangent between the liquid and α. A peritectic is thus formed. Intermediate phases are formed by peritectic reaction in many alloys; simple examples are found in the lead-bismuth and sodium-potassium systems. Several variations of the phase diagram are possible. Fig. 15.12 shows an example in which the γ primary solution also forms peritectically.

15.5   More complicated phase diagrams

We have now examined some elementary forms of binary phase diagrams. Many alloy systems are complicated by the appearance of several intermediate phases and by phase changes in the solid state. Their phase diagrams all conform to the same principles, however, and no new features are involved. They can be broken down into sub-diagrams between the various intermediate phases, each of which is geometrically similar to one of the diagrams we have deduced above. Phase changes in the solid state also lead to the same arrangements of phase boundary lines. For example, *eutectoid*

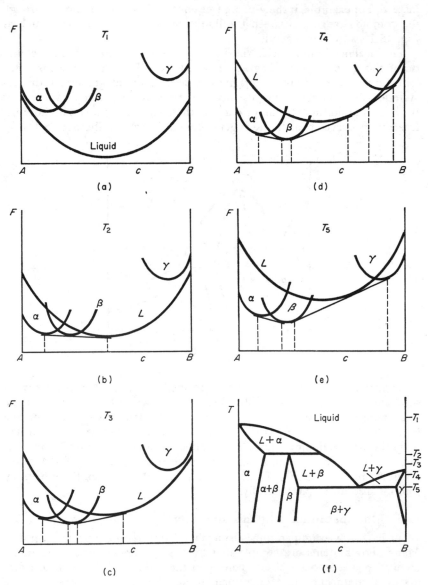

Fig. 15.11 A phase diagram in which an intermediate phase forms by a peritectic reaction

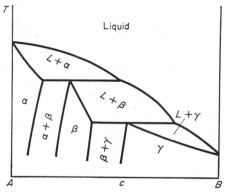

Fig. 15.12

and *peritectoid* changes occur in the solid state which are the exact analogues of the eutectic and peritectic forms of the liquid-solid change.

The most important example of a eutectoid change is that in the iron-carbon system, the phase diagram of which is shown in Fig. 15.13. It originates from the polymorphic change which occurs in pure iron at 910°C. Above this temperature the F.C.C. crystal structure is stable; and below, the

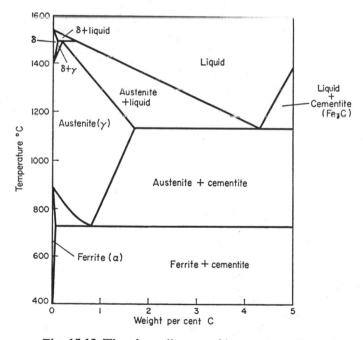

Fig. 15.13 The phase diagram of iron-carbon alloys

B.C.C. Solutions of carbon in these two forms of iron are known as γ-iron or *austenite* (F.C.C.) and α-iron or *ferrite* (B.C.C.), respectively; and the iron carbide is *cementite* (Fe_3C). From 700 to 1100°C and 0 to 2 per cent C the phase diagram is similar to the simple eutectic diagram of Fig. 15.5. As the temperature falls, the two common tangents, connecting the austenite free energy curve to those of ferrite and cementite, merge to give the eutectoid at 723°C. At higher temperatures, pure iron reverts to the B.C.C. form (δ-iron) before melting and this produces a peritectic between γ, δ and liquid iron. The remaining feature of interest is the eutectic between liquid, γ, and Fe_3C, at 4·3 per cent C.

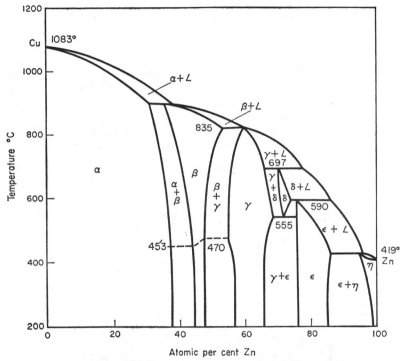

Fig. 15.14 The copper-zinc phase diagram

Fig. 15.13 is actually a diagram of *metastable* equilibrium because Fe_3C is normally unstable relative to graphite. However, graphite does not form readily in steels and Fig. 15.13 is usually the appropriate diagram for discussing the actual behaviour of carbon steels. The iron-graphite diagram, which we shall discuss in § 25.4, is nevertheless essential for the study of cast iron and certain aspects of steel.

A good example of a diagram with several intermediate phases is provided by the copper-zinc system, Fig. 15.14. A sequence of electron compounds is formed, through a cascade of peritectic reactions. The δ-phase, which forms peritectically at about 70 per cent zinc from a mixture of liquid and γ-phase, becomes unstable on cooling and decomposes in a eutectoid to γ + ε. The broken line crossing the β-phase marks the order-disorder change discussed in § 14.12.

15.6   Phase changes in alloys

To discuss how phase changes occur when an alloy is cooled or heated we draw a vertical line on the phase diagram at the composition of the alloy and consider the movement of a point along this line in accordance with the change of temperature. This representative point may then:

(1) pass through a single-phase field, with no changes of phase;
(2) pass through a two-phase field, in which the compositions and proportions of the phases both change;
(3) cross an isothermal phase boundary, as in eutectic and peritectic changes.

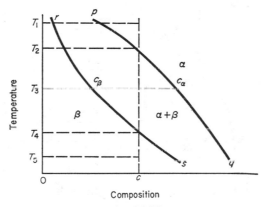

Fig. 15.15

Fig. 15.15 shows a two-phase field, $\alpha + \beta$. At the temperature T_1 the alloy c consists entirely of homogeneous α phase. As the temperature changes towards T_5 the two-phase field is entered at T_2. Nuclei of β then form in the α and grow and adjust their composition in accordance with the phase boundaries. The composition of the α phase moves along the line pq and that of the β along rs as the temperature changes from T_2 to T_4. For example, at T_3 the compositions of α and β are c_α and c_β respectively, and their proportions are given by the lever rule as $\alpha/\beta = (c - c_\beta)/(c_\alpha - c)$. Together, the phase diagram and the lever rule thus allow us to state the *equilibrium constitution* of the alloy, i.e. to state the phases present and give their compositions and amounts.

In this description we have assumed that the alloy remains in equilibrium throughout but this is possible only if the temperature changes sufficiently slowly to allow enough time for nucleation to occur and for the component atoms to redistribute themselves between the phases by *diffusion*. In practice, particularly when one or both of the phases is solid, only a partial redistribution usually occurs and the particles of β which are formed vary in composition about the mean value c. This is the origin of the *coring* effect discussed in § 13.9. The composition in the centre of each particle of β is less than c, since this is the first part to form, and the compositions of the outer layers are correspondingly greater than c (cf. Fig. 13.16 and Plate 2A).

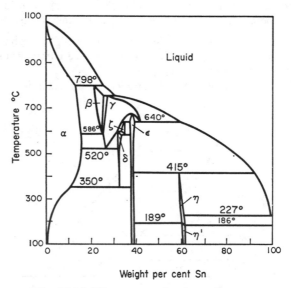

Fig. 15.16 The copper-tin phase diagram.

A simplified approach to the coring process is to assume that the liquid always remains completely mixed, by rapid diffusion and stirring, that no significant diffusion occurs in the solid and that the liquid remains in thermodynamic equilibrium with the surface of the cored solid grains. This means, however, that in a phase diagram like Fig. 15.15 the liquid can *never* completely freeze, for the surface layers of the solid *always* reject some of the solute into the remanent liquid, the freezing point of which then falls still lower (cf. § 13.8). The temperature T_4 is of no particular significance under these conditions. The composition moves down the line pq to much lower temperatures, the composition of each successive surface layer of solid likewise moves down rs, and this continues until some point is reached where the configuration of the phase diagram changes. In principle, in a diagram like that of Fig. 15.2(f) we expect freezing to continue down to the stage where the last traces of liquid are virtually pure B.

In other phase diagrams, such as those of Fig. 15.5 and 15.8, the remanent liquid may eventually reach a eutectic or peritectic line and a second phase may then form and freeze even when it is not predicted by the equilibrium relations. This happens, for example, in *cast tin bronzes*. Fig. 15.16 shows the copper-tin diagram. According to this, all bronzes containing less than 13·5 wt per cent Sn freeze in equilibrium as homogeneous primary solid solutions (α). In practice, coring develops extensively and bronzes containing about 10 wt per cent Sn, i.e. *gun-metals*, freeze into phase mixtures. The tin-rich liquid between the cored dendrites of α freezes peritectically to an intermetallic β phase which in turn decomposes in solid state reactions, eventually giving a eutectoid mixture of $\alpha + \delta$, the latter being a brittle intermetallic compound. Difficulties of casting due to excessive segregation and delayed freezing of remanent liquid are particularly severe in copper-tin alloys, because of the wide *freezing range* (i.e. $\alpha + liquid$ region in Fig. 15.16) and because of the low melting points of tin-rich constituents.

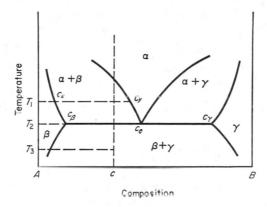

Fig. 15.17

To discuss changes which occur when an isothermal phase boundary is crossed, we use Fig. 15.17. This represents a eutectic when $T_1 > T_3$ and a peritectic when $T_1 < T_3$. At T_1 the alloy consists of the phase mixture $\alpha + \beta$ in the proportions $\alpha/\beta = (c - c_x)/(c_y - c)$. As $T \to T_2$ the proportion of β phase and the concentration of B in α both increase. The alloy arrives at T_2 containing α and β with compositions c_e and c_β and proportions $\alpha/\beta = (c - c_\beta)/(c_e - c)$. At T_2 the α phase becomes unstable and decomposes into β and γ. The alloy has then entered the $\beta + \gamma$ field. If the alloy is now taken from T_3 to T_1, in equilibrium, the opposite sequence of changes occur. The alloy reaches T_2 consisting of $\beta + \gamma$ with compositions c_β and c_γ and proportions $\beta/\gamma = (c_\gamma - c)/(c - c_\beta)$. At T_2 the γ *reacts* with part of the β to form the new phase α.

We have, of course, only described equilibrium behaviour, which requires the necessary changes of structure and phase compositions to occur freely by nucleation and atomic diffusion. Usually the temperature changes too quickly for these to maintain complete equilibrium. For example, on cooling through a peritectic change, i.e. $T_3 \to T_1$, in Fig. 15.17, the α phase product of the peritectic reaction forms as a boundary layer between the reacting β and γ phases. The reactants are thus separated from each other by the reaction product, which tends to prevent the reaction from going to completion. The microstructure at T_1 may then contain grains of unreacted, metastable γ phase encased in shells of α and embedded in a matrix of β.

Regarding Fig. 15.17 now as a eutectic or eutectoid system, we notice that, on cooling the alloy c from the α phase field to the $\beta + \gamma$ field, the β separates out in two distinct stages. First, as *primary* β during the passage through the $\alpha + \beta$ region; second, during the decomposition $\alpha \to \beta + \gamma$ when the isothermal phase boundary is crossed. These two forms of β

(a)   (b)

Fig. 15.18 Typical microstructures of two-phase alloys (a) Widmanstätten structure, (b) primary dendrites and eutectic infilling

can usually be distinguished metallographically. In eutectic freezing, the primary β usually separates out as well-defined, often large, crystals or dendrites. In a eutectoid change the primary β usually starts from nuclei in the grain boundaries of the parent α phase and then grows outwards, as small grains, by converting the nearby α-phase material to its own composition and crystal structure. Under certain conditions the β crystals grow as fairly equiaxed grains with irregular α/β boundaries. Often, however, they grow in regular sheet-like or lath-like shapes along certain crystallographic planes in the α crystals. The β crystals then have a precise crystallographic orientation in relation to the α, such that the atoms of the two phases fit together well along their common crystal boundary. This is called a *Widmanstätten structure* (Fig. 15.18).

When the isothermal phase boundary is reached, the remaining α between the β crystals decomposes to a *eutectic mixture* of $\beta + \gamma$. Two phases have to form and grow *simultaneously* at this stage. The growth of either acts as an inducement to the growth of the other. For example, if a nucleus of β forms in this α (or a spike of β grows into it from a nearby crystal of primary

β), the B atoms rejected by this particle of β, due to the change of composition from c_e to c_β, locally enrich the surrounding α with B atoms, bringing its composition into the range of c_γ, which encourages the nucleation of particles of γ phase alongside this β particle. There is therefore a strong tendency for the β and γ in the eutectic structure to separate out in a finely dispersed mixture. Such eutectic structures are commonly observed (Fig. 15.18). They may consist of regularly alternating *lamellae* of the two phases, like a multiple sandwich, or of rods or spheroids of one phase dispersed in a matrix of the other (Plate 2B). The best known example of such a structure is *pearlite*, a lamellar mixture of ferrite and cementite formed in carbon steels by the eutectoidal decomposition of austenite at 0·8 wt per cent C and 723°C (Fig. 15.13 and Plate 3A).

The amount of eutectic structure formed in an alloy is, of course, equal to the amount of the phase which decomposed to give the structure. For example, alloy c in Fig. 15.17 forms the proportion *eutectic/alloy* $= (c - c_\beta)/(c_e - c_\beta)$. The amount of β phase in the eutectic is given by $\beta/eutectic = (c_\gamma - c_e)/(c_\gamma - c_\beta)$ and so is independent of c.

The structures of eutectic alloys can be modified by chill casting or by adding traces of certain impurities. Advantage can be taken of this *modification* to improve the properties of such alloys. For example, in cast aluminium-silicon alloys the silicon normally separates as coarse needles (*acicular* structure) which mechanically weaken it. When 0·01 per cent sodium is added to the alloy, however, the silicon separates in a finer, globular form and the alloy is then much stronger. It is believed that the nuclei or growing particles of silicon are 'poisoned' by a concentration of the modifying agent at the solid–liquid interface. Their growth is then restricted, which gives increased opportunity for further nuclei to form in the liquid, so producing a finer structure. A similar improvement can be made in grey cast irons. Normally, large thin flakes of graphite are formed and are a source of mechanical weakness. By adding a trace of magnesium, calcium or cerium to the casting, however, the graphite can be made to precipitate out in a compact *nodular* form (*spherulitic* graphite). The metal is then stronger and more ductile. Much commercial cast iron is now improved in this way by the addition of magnesium.

15.7   Zone refining

The non-equilibrium freezing of a solid solution, discussed in § 13.8 and § 15.6, has an important practical application in the process of *zone-refining* (*W. G. Pfann*, 1952, *Transactions, A.I.M.E.*, **194**, 747). In a system with a phase diagram, near the solvent, like that of Fig. 15.19 the *distribution coefficient k*, where

$$k = \frac{\text{concn. of solute in solid}}{\text{concn. of solute in liquid}} = \frac{c_0}{c} = \frac{c}{c_1} \qquad 15.1$$

is less than unity so that the solid is always purer than the liquid, in equilibrium. Suppose then that, with a suitable travelling furnace, we pass a short molten zone along a long bar of such an alloy, initially of uniform concentration c. Since the concentration of solute in the liquid is higher than that in the solid which freezes from it, some solute is transported along the bar in the molten zone. The starting end of the bar is thus purified, by a factor of about k, the middle region is unaffected and the finishing end is increased in impurity content. By repeatedly passing such zones along the bar, all in the same direction, the purity and length of the pure end of the bar can both be increased substantially. The technique is used for the purification of germanium and silicon, as used in transistors and other electronic devices, and is closely related to the crystal growing and floating zone procedures discussed in § 13.11.

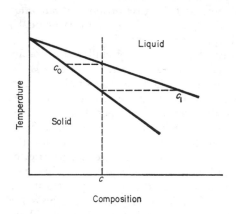

Fig. 15.19 A phase diagram suitable for zone refining

Extreme purity can be obtained in this way. For example, the arsenic content of germanium can be reduced to about one part in 10^{10} by passing six zones along the bar. Zone-refining is now regularly used for making super-pure metals, mainly for research purposes.

15.8 The determination of phase diagrams

Many techniques are used for the experimental determination of phase diagrams. Optical microscopy is one of the most important, since it allows the phases formed at various compositions to be directly recognized. By rapidly quenching an alloy from various high temperatures, so as to preserve for metallographic examination at room temperature the high-temperature structures in either a metastable form or an easily recognizable transformed form, we can use optical microscopy to determine the temperatures of phase boundaries.

Apart from x-ray analysis, which we shall discuss in Chapter 17, the other major technique is *thermal analysis*. This is a simplified form of calorimetry, based on the idea that, if a hot system is allowed to cool freely in a constant environment, any marked change in its rate of cooling, at some temperature, indicates an evolution of heat due to a change in the system itself. In a simple form of the technique a thermally-lagged crucible of the alloy, often in a vacuum or an atmosphere of inert gas, is allowed to cool slowly and its temperature is measured at frequent and regular intervals of time by means of a sensitive pyrometer, e.g. a calibrated *thermocouple* (platinum-rhodium, chromel-alumel, copper-constantan or silver-constantan) with its hot junction protected by a thin refractory sheath and immersed in the metal and its cold junctions immersed in melting ice and connected to a potentiometer. In this way cooling curves can be determined such as those of Fig. 15.20.

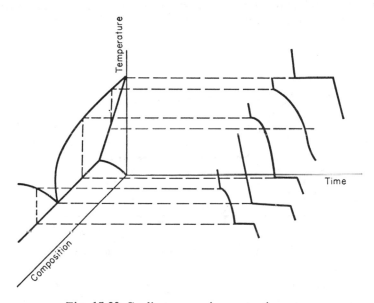

Fig. 15.20 Cooling curves in a eutectic system

The arrival at the liquidus is usually clearly indicated by a marked reduction in the rate of cooling. Sometimes, in fact, there may be a small temporary *rise* of temperature at the start of freezing. This is an indication of *undercooling*, which must be minimized by slow cooling and continual stirring if accurate results are to be obtained. At certain compositions, e.g. pure metal, eutectic alloy, intermediate phases which form directly from the melt, there is a clear *arrest* in the cooling curve when the latent heat of freezing is released at constant temperature. In most alloys, however, freezing occurs over a range of temperature, which produces a rounded cooling curve with a reduced slope. The final stage of freezing, when the phase diagram is like that of Fig. 15.15, is usually rather poorly marked on the cooling curve; the solidus in this case is better determined by other methods, e.g. by heating curves or by quenching and microscopical examination. Where freezing is completed at a eutectic line, however, there is a sharp arrest when the eutectic part of the system freezes at constant temperature. The time of this arrest is proportional to the fraction of eutectic structure in the alloy and so, by plotting it against composition and extrapolating to zero time, the compositions of the solubility limits at the eutectic temperature can be found.

Instead of plotting the time t directly against the temperature T, *inverse-rate* curves are often constructed, in which dt/dT is plotted against T, as in Fig. 15.21. These indicate the change points more sensitively. They can be constructed directly by plotting against temperature the times required for each successive unit drop in temperature.

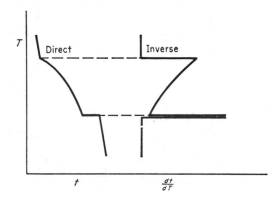

Fig. 15.21 Direct and inverse-rate cooling curves

Phase changes in the solid state are often studied by *dilatometry*, the measurement of changes of size. In a simple dilatometer the specimen in the form of a short bar is placed in a loosely fitting vitreous silica tube and pressed against an end stop by a silica rod which can slide freely along the tube. Changes in the length of the bar are transmitted by the rod to a sensitive dial gauge fixed to the other end of the tube. A thermocouple is embedded in a hole in the bar to allow the temperature to be taken. Fig. 15.22 shows typical dilatometer curves when carbon steel is heated through the *ferrite + pearlite → austenite* eutectoid change. There is a sharp contraction as the F.C.C. structure, in which the iron atoms are more densely packed, spreads through the metal. A familiar demonstration of this effect is provided by slinging an iron wire horizontally between two fixed electrodes and then passing an increasingly large current to raise its temperature

PLATE 1

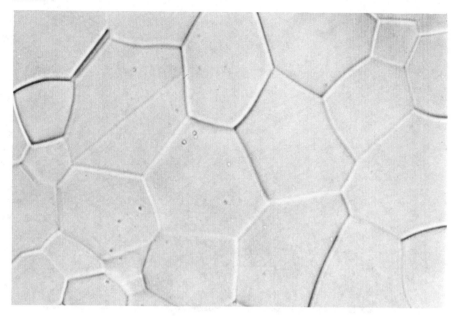

A. Grains in lightly-etched aluminium, ×100 (*courtesy of G. C. Smith*)

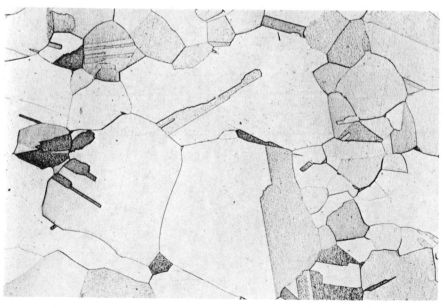

B. Etched grains in nickel, ×300 (*courtesy of G. C. Smith*)

PLATE 2

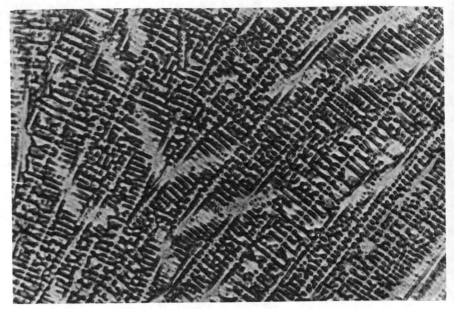

A. Cored dendritic structure in chill cast 70–30 brass, ×100 (*courtesy of G. C. Smith*)

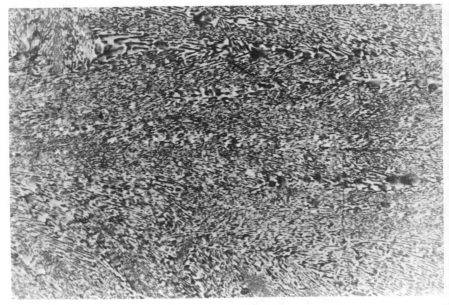

B. Structure of cadmium-bismuth eutectic, ×100 (*courtesy of G. C. Smith*)

gradually to a bright red heat. The sag of the wire, due to normal thermal expansion, is strikingly reversed when the phase change begins.

Solid-state transformations are also studied by the measurement of other physical properties. Electrical conductivity is particularly useful for determining limits of solid solubility. In a disordered solid solution the solute atoms strongly scatter the flowing electrons (cf. § 24.4) and the conductivity of a good metallic conductor falls rapidly with increase in solute concentration. In a phase mixture on the other hand, where the amounts but not the compositions of the phases vary with the alloy content, the conglomerate of grains acts as a network of interconnected metallic conductors and, provided the grains are not too finely dispersed (i.e. grain size $> 10^{-8}$ m) the conductivity simply varies linearly with the proportions of the phase mixture, as in Fig. 14.9. The usual technique is to measure the conductivities of

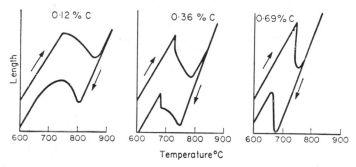

Fig. 15.22 Dilatometer curves on carbon steels heated through the $\alpha \rightarrow \gamma$ change

a series of wires of different compositions on a potentiometer, after prolonged heating at a single temperature T to establish phase equilibrium there, and then assemble the results in a diagram as in Fig. 15.23.

Electrical conductivity is also useful for studying order-disorder changes in solid solutions. The scattering power of foreign atoms for conduction electrons depends largely upon the randomness of the solution and in a well-ordered alloy the electrical conductivity becomes more nearly like that of a pure metal.

Several techniques have been developed for determining the free energies of alloy phases and are used where information is needed beyond that contained in the phase diagram. The vapour pressure of a volatile metal in solution in an alloy can be measured and from this its activity and free energy deduced by standard thermodynamical arguments (cf. § 7.6). Loss of the more volatile component at the surface of the alloy is a major source of error and it is important that diffusion in the alloy should be rapid, to maintain the surface concentration. The free energies of alloys can also be obtained from measurements of electromotive force, as discussed in § 9.1.

High temperatures are needed to enable diffusion to maintain the surface concentration and fused salts are mostly used as electrolytes.

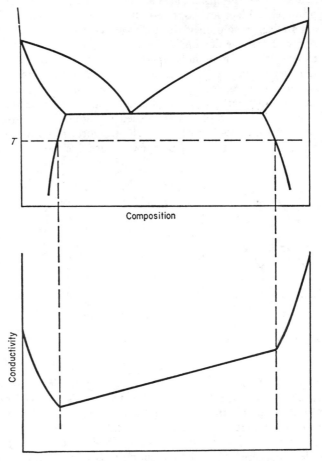

Fig. 15.23 Variation of electrical conductivity at temperature T across a simple phase diagram

15.9   Gas-metal systems

The phase equilibrium of gas-metal systems is important practically, mainly because metals so often become contaminated by gases during reduction, melting and heat-treatment, but also because gaseous elements are sometimes intentionally used as alloying elements, e.g. nitrogen in certain steels. Solutions of gaseous elements in metals are usually dilute and the simplified theory of solutions is applicable (cf. § 7.6). Experiments show that elements which normally exist as *diatomic* gases, e.g. H_2, N_2, O_2, dissolve in metals in proportion to the square root of the partial pressure in the gas phase (*Sievert's law*). This follows directly from the mass action law applied to dilute solutions and from the fact that such gases dissociate when they dissolve in metals and form *monatomic* solutions. The equilibrium of hydrogen, for example, between the gas phase and solution in a metal can be represented by the equations

$$H_2 \text{ (gas)} = 2H \text{ (metal)} \qquad \textbf{15.2}$$

$$K = c_H^2 / p_{H_2} \qquad \textbf{15.3}$$

where K is the equilibrium constant, c_H is the concentration of atomic hydrogen in solution and p_{H_2} is the partial pressure of molecular hydrogen, so that Sievert's law follows directly,

$$c_H = (K p_{H_2})^{1/2} \qquad \textbf{15.4}$$

The solubility varies with temperature in the way discussed in § 13.9.

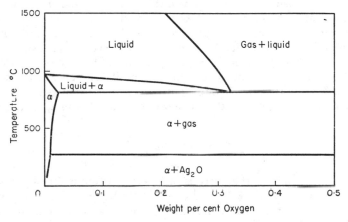

Fig. 15.24 Part of the silver-oxygen phase diagram at 1 atm pressure

To show pressure as a variable, as well as temperature and composition, we need a *three-dimensional* phase diagram. It is usually simpler to construct a series of temperature-composition phase diagrams at various constant pressures. Fig. 15.24 shows the silver-oxygen diagram at 1 atm pressure. We see in this that molten silver dissolves up to about 0·3 per cent oxygen. When it freezes, the oxygen solubility falls drastically and the alloy enters a two-phase *silver + gas* field. The evolution of gas bubbles can cause solidifying silver to 'spit' out of an open crucible.

If the gas pressure is raised above a certain level, the concentration of the element in solution in the metal may become high enough to cause a *compound* to form, e.g. Ag_2O, Cu_2O, FeO, ZrH_2, in equilibrium with the primary solid solution. Where the metal has a high chemical affinity for the gas, e.g. reactive metals and oxygen, this of course happens in the ordinary atmosphere and even in partial vacuum (cf. § 7.3). The gaseous component of the phase diagram is usually then ignored and only the equilibrium with the compound is shown, as in the iron/oxide, copper/oxide and iron/nitride diagrams.

15.10   The Chemical Potential

The concept of *chemical potential* is often used in the discussion of phase equilibria. Suppose that we have a phase at temperature T and pressure P which is composed of n_A moles of component A, n_B of B, etc. We then write its free energy G in the form

$$G = \mu_A n_A + \mu_B n_B + \cdots \qquad \textbf{15.5}$$

where μ_A is called the *chemical potential* of A, etc. This expression is suggested naturally by the fact that free energy is a capacity property, which invites us to regard it as made up *additively* from 'partial' free energies of its components (although μ_A is of course *not* the chemical potential of the pure component A, but refers specifically to the particular condition of A in this phase). It follows that $\mu_A = \partial G/\partial n_A$ = the rate of change of G with n_A when the other variables, T, P, n_B, $n_C \ldots$ are all held constant. The useful feature of the concept is that in an equilibrium between any set of phases α, β, $\gamma \ldots$ etc. the chemical potential of any component is the *same in every phase*. Thus, if μ_A^α is the chemical potential of A in α, etc., then

$$
\begin{aligned}
\mu_A^\alpha &= \mu_A^\beta = \mu_A^\gamma = \cdots \\
\mu_B^\alpha &= \mu_B^\beta = \mu_B^\gamma = \cdots \\
&\text{etc.}
\end{aligned}
\qquad \textbf{15.6}
$$

Chemical potentials in the gaseous state can be found directly from the general considerations in § 7.3 and § 7.4. At the temperatures and pressures at which they become equal to the chemical potentials in various compounds and other condensed phases, these phases then appear in the system.

FURTHER READING

As for Chapter 14; also:

BRICK, R. M. and PHILLIPS, A. (1949). *Structure and Properties of Alloys*. McGraw-Hill, New York and Maidenhead.

CHADWICK, G. A. (1963). *Progress in Materials Science*, Vol. 12: *Eutectic Alloy Solidification*. Pergamon Press, Oxford.

CHALMERS, B. (1964). *Principles of Solidification*. John Wiley, New York and London.

HANSEN, M. and ANDERKO, K. (1958). *Constitution of Binary Alloys*. McGraw-Hill, New York and Maidenhead.

PFANN, W. G. (1958). *Zone Melting*. John Wiley, New York and London.

PRINCE, A. (1965). *Alloy Phase Equilibria*. Elsevier, Holland: Cleaver-Hume, London.

RHINES, F. N. (1956). *Phase Diagrams in Metallurgy*. McGraw-Hill, New York and Maidenhead.

WINEGARD, W. C. (1964). *An Introduction to the Solidification of Metals*. Institute of Metals.

Chapter 16

Ternary Phase Diagrams

16.1 Representation of the phase diagram

When three components are mixed together to make a *ternary* alloy we need a three-dimensional space in which to construct the phase diagram, since there are two independent concentration variables and one temperature variable (ignoring pressure as a variable). Such diagrams can be represented as solid figures. Quaternary and more complex systems require four- and higher-dimensional spaces and so cannot be represented diagrammatically except by a series of diagrams in which some variables are held constant. The difficulty of visualizing phase equilibria in higher-dimensional spaces has undoubtedly hindered the rational development of the more complex alloys.

The sum of the concentrations of all components in a system is always unity. In binary alloys this allows us to represent the concentrations c and $1 - c$ of the two components by a point c on a line of unit total length. All the phase diagrams of Chapter 15 are constructed on this basis. To represent compositions of ternary alloys we need a two-dimensional figure in place of the unit line. A *triangle* has the necessary property. For convenience we use an equilateral triangle as in Fig. 16.1. The three corners represent the pure components A, B and C. The three sides, each of unit length, represent the three binary systems AB, BC and CA, so that the point P, for example, represents a binary alloy in which $c_A = PB/AB$ and $c_B = PA/AB$.

Points inside the triangle represent ternary alloys.

Consider for example the point O. We draw the three lines PQ, LM and XY through it, parallel to the sides of the triangle, and take the concentrations of the components in the alloy O to be

$$c_A = PB \quad (= OL = LY = OY = QC)$$
$$c_B = AX \quad (= OM = MQ = OQ = YC) \qquad \textbf{16.1}$$
$$c_C = XP \quad (= LB = OP = OX = AM)$$

from which it follows that

$$c_A + c_B + c_C = PB + AX + XP = AB = 1 \qquad \textbf{16.2}$$

as required. We notice two useful features of this representation:

(1) points on a line parallel to a side of the triangle represent alloys with a fixed content of the component in the opposite corner (e.g. c_A is constant along PQ);

(2) points on a straight line through one corner represent alloys in which the components at the other two corners are in constant proportions (e.g. c_B/c_C is constant along AF).

With the triangle as base we plot temperature vertically, to form a triangular prism. Since the ternary system must merge smoothly into each of the corresponding binaries as one or other of the three components is reduced to zero, the three binary phase diagrams are constructed on the three sides of this triangular prism. This is the framework in which we construct the ternary phase diagram.

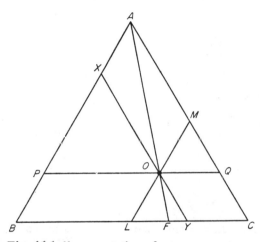

Fig. 16.1 Representation of a ternary system

Ternary phase equilibria are, of course, closely similar to binary equilibria, although the extra degree of freedom adds an extra dimension to their representation. Thus, free energy curves become surfaces, usually basin-shaped, which rest on free energy tangential planes. A phase field, which is a surface in the binary diagram, becomes a *volume* in the ternary. A phase boundary line becomes a boundary *surface* between such volumes. A singular point such as a binary eutectic point becomes drawn out into a *line* in the ternary. A singular *point* in the ternary represents a new feature, e.g. a ternary eutectic, where *four* phases coexist in equilibrium (with no degrees of freedom) which is of course impossible in a binary system.

16.2 Simple eutectic system

We shall first construct the ternary diagram equivalent to the eutectic diagram of Fig. 15.5(*f*). For simplicity of construction we shall assume initially that the three primary solid solubilities are negligibly small. The lead-bismuth-tin system is of this type. The three binary diagrams are then as shown round the sides of Fig. 16.2(a). For the moment we consider only the liquidus of the ternary. This forms three distinct surfaces, extending inwards and downwards from each corner, as indicated by the isothermal 'contour' lines shown in the diagram. The surfaces meet along three valley lines which run inwards and downwards from each binary eutectic point E_1, E_2, E_3, and converge at a central *ternary eutectic* point E.

Fig. 16.2(b) shows a projection of this liquidus surface on to the base. Consider the freezing of an alloy of composition X, which starts at the temperature T_1. Pure solid A separates out and the liquid becomes richer in B and C. Since B and C are completely conserved in the liquid, their ratio remains fixed and the composition of the liquid moves down the line

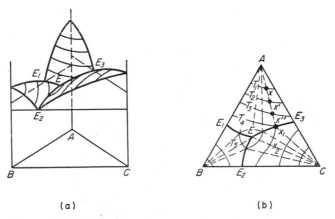

(a)　　　　　　　　　　　　　　　(b)

Fig. 16.2 (a) Liquidus surface of a eutectic system; (b) projected on to base

AX_1 as the temperature falls to T_4. At T_3, for example, the alloy contains pure solid A and liquid of composition X'' in the proportions *solid/liquid* $= XX''/AX$.

Similarly, at T_4 the pure solid A is in equilibrium with liquid X_1 in the proportions *solid/liquid* $= XX_1/AX$. However, this liquid is also now in equilibrium with pure solid C since any alloy on the line CX_1 would have produced this same liquid at this temperature. Further freezing must preserve this equilibrium with both A and C. The composition of the liquid thus moves down the line X_1E since only those points on this line represent liquid in equilibrium with both A and C. During this second stage of freezing, both pure metals A and C therefore separate out. This phase mixture is called the *binary eutectic complex* since it is an obvious generalization of the binary eutectic E_3. The liquid phase in alloys on the B-rich side of the diagram would similarly find its way to one of the valley lines E_1E or E_2E and, by depositing mixtures of A and B, or B and C, would move eventually to the point E. This is the *ternary eutectic point* at which the liquid is in equilibrium with all three solids A, B and C. At the fixed temperature corresponding to E, this liquid then freezes to a three-phase mixture of A, B and C, the *ternary eutectic structure*. We notice that a binary eutectic complex does not have a fixed composition except when its valley line accidentally points straight at the opposite corner of the diagram.

There are thus three distinct stages of freezing, *primary*, *secondary* and *tertiary*, according as one, two or three solids are separating out. Beneath the liquidus there must therefore exist secondary and tertiary phase boundary surfaces, corresponding to the various phase fields.

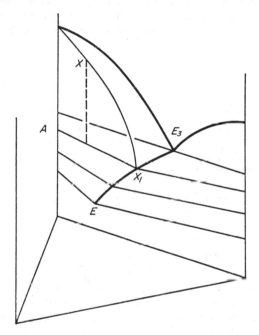

Fig. 16.3 Construction of a secondary surface

The construction of part of the secondary surface is shown in Fig. 16.3. An alloy X starts to freeze at the appropriate temperature on the liquidus, by depositing pure solid A, and the composition of its liquid runs down to X_1 at the temperature where the secondary stage begins. Any other alloy with composition on the straight line AX_1 also enters the secondary stage with liquid of the same composition X_1 and hence at the same temperature. Thus, all such alloys on this line enter the secondary stage at the same temperature (T_4 in Fig. 16.2). This isothermal line AX_1 is therefore a line in the secondary surface. The whole surface is generated by a family of such lines, as shown in Fig. 16.3.

9*

The final stage of freezing, at E, is the same for every ternary alloy whatever its composition. Hence the tertiary surface is a horizontal plane across the whole diagram through the point E. The complete phase diagram can thus be represented in section by Fig. 16.2(b). The full contour lines denote the liquidus surface. The broken lines are contour lines in the secondary surface, at the temperatures indicated by the liquidus contours which they meet at the valley lines. The position of the tertiary surface is defined by the temperature of the ternary eutectic point.

Fig. 16.4 shows some cooling curves, based on this diagram. Alloy X shows a first arrest at T_1 due to the separation of A, then a second arrest at T_4 due to the freezing of the binary eutectic complex, and a third arrest at the ternary eutectic temperature T_E. In alloy X_1 the first stage is missing because this alloy has a composition on the binary eutectic complex line.

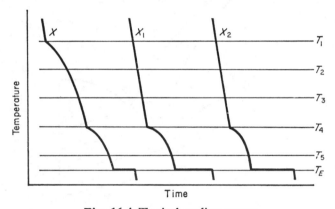

Fig. 16.4 Typical cooling curves

The cooling curve of alloy X_2 is similar but the course of freezing is quite different since no binary eutectic complex is formed in this case; as pure solid C separates out, starting at T_4, the composition of the liquid moves down the line CX_2E straight to the ternary eutectic point.

16.3   Horizontal sections

A convenient way of showing the phases formed in ternary systems is by means of *horizontal sections* through the phase diagram at various selected temperatures. From such a section we can find the compositions and proportions of the phases present in any alloy in equilibrium at the temperature concerned.

Fig. 16.5 shows some horizontal sections through the diagram of Fig. 16.2. At T_2 most alloys are still wholly liquid (L), but some solid pure A or C is also present in the A-rich or C-rich alloys. The alloy X for example at T_2 contains solid pure A and liquid of composition X' in the proportions *solid/liquid* $= XX'/AX$. The horizontal section at T_3 is similar, except that some solid B is also present, in B-rich alloys. The sections at T_4 and T_5 show that many alloys have entered the secondary stage of freezing and contain two solid pure metals as well as liquid. These three-phase regions are necessarily bounded by three straight lines.

To find the constitution of an alloy X in a three-phase field we consider the general case shown in Fig. 16.6. All alloys within the triangle PQR are mixtures of three phases whose compositions are represented by the points

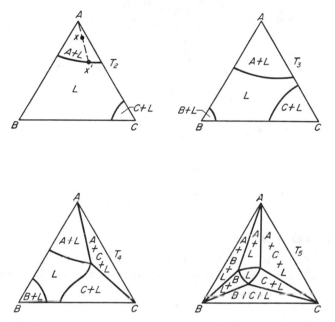

Fig. 16.5 Horizontal sections at four temperatures through the phase diagram of Fig. 16.2

P, Q and R. Since these compositions are *fixed* whatever the position of X in the triangle and only the proportions of the phases vary with X, we can think of PQR as a complete ternary system within itself, with P, Q and R as the three components. Even though the triangle PQR is not in general equilateral we can still use the method of Fig. 16.1 to determine the proportions of P, Q and R in the alloy X. This follows because, if we distort the triangle of Fig. 16.1 uniformly into the shape of the triangle PQR (e.g. by marking it out, together with all its construction lines, on to a sheet of rubber and then stretching this uniformly into the required shape), some proportional relations remain. Thus, in Fig. 16.6 we construct lines through X parallel to the sides of the triangle PQR, as shown, and deduce the proportions of the phases in X from the intercepts of these lines on any one side of the triangle, e.g.

proportion of P : *proportion of* Q : *proportion of* R

$$= LM : RL : MQ \qquad\qquad 16.3$$

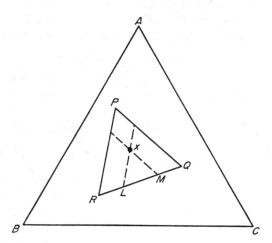

Fig. 16.6 Constitution in a three-phase field

16.4   Vertical sections

Fig. 16.7 shows two *vertical sections* taken through the phase diagram of Fig. 16.2 at 60 per cent A and 20 per cent A respectively. The intersections

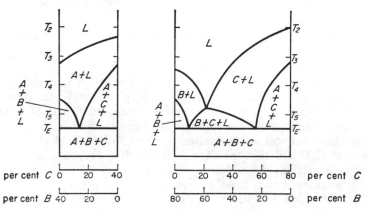

Fig. 16.7 Vertical sections at 60 per cent A and 20 per cent A through the phase diagram of Fig. 16.2

with the primary, secondary and tertiary surfaces are clearly seen. The disadvantage here however is that we cannot deduce, from such a section, the *compositions* of the phases present.

It is sometimes convenient to take vertical sections along lines through the corners of the phase diagram, since these contain the straight isothermal lines of the secondary stage of freezing (cf. Fig. 16.3). From several such sections we obtain a picture of the secondary surface. Fig. 16.8 shows such a section taken along the continuation of the line AX in Fig. 16.2(b).

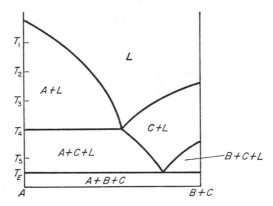

Fig. 16.8 Vertical section along the line AX through the phase diagram of Fig. 16.2

16.5   Ternary solid solutions

An obvious generalization of Fig. 15.2(f) into a lens-shaped figure spanning the ternary triangle gives the simplest form of phase diagram when the three components are completely soluble in both the liquid and solid states, as in the Ag-Au-Pd system.

We shall now consider the modification of Fig. 16.2 due to some primary solid solubility. The liquidus surface has the same form as before but beneath it, near each corner of the diagram, is a new phase boundary surface representing the limit of solid solubility in each of the three pure metals. Fig. 16.9 shows an example.

Fig. 16.10(a) shows a horizontal section through one corner of such a diagram at a temperature in the range of primary solidification. The alloy X, for example, contains a phase mixture of liquid (L) and solid solution (α) at this temperature. We meet a problem here which does not exist in binary diagrams. Because this is a two-phase structure in a three-component system at a fixed temperature, there is still one degree of freedom left for variation of composition. So far as the phase boundary lines are concerned, *any* solid solution of composition lying along PQ might come into equilibrium, in the alloy X, with liquid at a corresponding point along RS. To

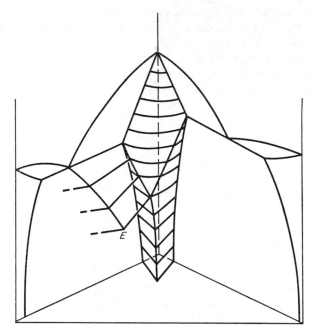

Fig. 16.9 Ternary eutectic diagram, cut away to show a surface of primary solid solution and part of the associated secondary surface leading down to the ternary eutectic point E

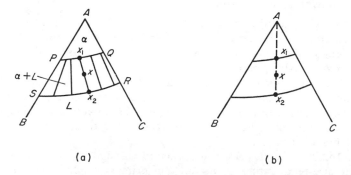

(a)   (b)

Fig. 16.10 Horizontal section through liquidus and solidus, showing tie lines

indicate which particular compositions do come into equilibrium we have to put in some additional information, in the form of experimentally determined *tie lines*, as shown in Fig. 16.10(a). We can then say, for example, that the alloy X contains α of composition X_1 and *conjugate liquid* of composition X_2 in the proportions $\alpha/liquid = XX_2/XX_1$.

When the tie lines are not known, a fair approximation to the equilibrium constitution can usually be obtained by assuming that the *partition law* is valid, i.e. that a solute dissolved in the same atomic or molecular form in two solvents is distributed between them in a constant ratio at constant temperature. Referring to Fig. 16.10, if B_α and C_α are the respective concentrations of B and C in the solid solution X_1, regarding them as solutes in

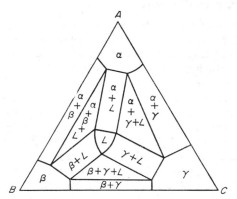

Fig. 16.11 Horizontal section just above ternary eutectic point

A, and B_l and C_l are their respective concentrations in the liquid X_2, then the partition law states that

$$B_\alpha/B_l = K_B \quad \text{and} \quad C_\alpha/C_l = K_C \qquad \textbf{16.4}$$

where K_B and K_C are constants. It follows from this that

$$B_\alpha/C_\alpha \propto B_l/C_l \qquad \textbf{16.5}$$

and hence that in both solid and liquid solutions the ratio is the same as in the alloy as a whole. Thus when the partition law is valid, X_1 and X_2 lie on the straight line AX as shown in Fig. 16.10(b).

The secondary surface of solidification develops as shown in Fig. 16.9. A typical horizontal section at a temperature just above the ternary eutectic point is shown in Fig. 16.11. We notice that alloys with compositions near

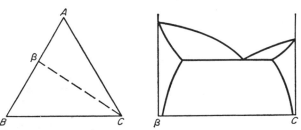

Fig. 16.12 Pseudo-binary between an intermediate phase β formed from two metals A and B, as in Fig. 15.9, and a third metal C

the corners of the diagram freeze, in equilibrium, in one stage only to primary solid solutions, that those near the sides of the triangle do not have a tertiary stage of freezing since no ternary eutectic is formed in them, and that the ternary eutectic structure itself is a phase mixture of the three solid solutions α, β and γ.

16.6   More complex diagrams

The ternary phase diagram of a system in which intermediate phases are formed and solid state transformations take place is generally very complex. Fortunately, it can largely be broken down into an assembly of ternary sub-diagrams each of which is of a simple standard type. As an example, suppose that in the system A, B, C, an intermetallic compound β is formed between components A and B, giving a binary phase diagram such as that of Fig. 15.9(f). Then, just as β interacts with A and B as if it were a separate component, splitting the AB binary diagram into two simpler and individually complete diagrams $A\beta$ and βB, so in many cases it also interacts with C similarly and splits the ternary ABC diagram into two simpler ones, $A\beta C$ and $C\beta B$. We may for example, have a *pseudo-binary eutectic* formed between β and C, as in Fig. 16.12, and this may then lead to two ternary systems $A\beta C$ and $C\beta B$ like that of Fig. 16.9.

Similar break-downs of other and more complex diagrams can generally be made. Another simplifying feature sometimes appears with secondary solid solutions. Suppose, for example, that metal A forms A-rich *secondary* solutions of the same crystal structure with both B and C. Then, in effect, both B and C are dissolved in metal A but this A is in a crystalline form different from that of the primary solution. Since the solvent and crystal structure of the secondary solution are the same in both binaries then, fairly commonly, the same phase extends as a ternary solution right across the triangle from the one binary phase field to the other.

FURTHER READING

As for Chapter 14 and 15; also:

MASING, G. (1944). *Ternary Systems*. Reinhold, New York.

Chapter 17

Metal Crystals—I Periodicity

17.1 Introduction

We turn now to the structures and properties of metal crystals. This is the starting point for many important topics in metallurgy; for example, plastic deformation, mechanical strength, heat-treatment, diffusion, sintering, radiation damage, properties of metals and alloys.

A crystal structure has three general properties; *periodicity, directionality* and *completeness*. Periodicity is the regular repetition, in space, of the atomic unit of the crystal. It is the basis of the remarkable plastic properties of crystals and of the close connection between crystals and waves. Even at the rare places where a crystal structure goes wrong, for example at a *stacking fault* (cf. § 13.1) or a *dislocation* (cf. § 17.4), it can go wrong only in certain well-defined ways. Even the faults in a crystal thus have regular and reproducible structures and properties.

The fact that properties such as conductivity and elasticity vary with the direction of measurement through a crystal follows obviously from the *directionality* of the crystal structure itself. *Completeness* is simply the filling of all crystal sites defined by the periodic structure with the required atoms. When there are too few or too many atoms for this, in a given region of the crystal, some sites may remain empty (*vacancies*) or the surplus atoms may have to lodge themselves in the interstices between the atoms of the crystalline array (*interstitials*). These *point defects* are important in diffusion and radiation damage.

17.2 Periodicity

The periodicity of a crystal is like that of a wallpaper pattern. There is a *motif*, the elementary unit of the structure, and this motif is replicated throughout the structure according to a certain pattern. The pattern itself is most simply displayed as a pattern of points. We choose any one point in a motif and mark out the pattern formed by this point and its identical replica from every other motif. This array of *lattice points* is the *space lattice* of the crystal. It is important to notice that a lattice point is not an atomic site. In certain simple crystal structures (e.g. F.C.C. and B.C.C. metals) the pattern of atomic sites happens also to form a space lattice, but in many other structures (e.g. C.P. Hex. metals) there is more than one atom in the motif.

In a space lattice, which repeats to infinity in all directions, the pattern of lattice points is exactly the same round every lattice point. Hence, if we draw a vector between any two lattice points and then repeat it, starting from a third point, it will always end at a fourth point. If we test the arrays of Fig. 13.2 in this way we find that the F.C.C. and B.C.C. are space lattices but the C.P. Hex. is not. To construct a lattice for the C.P. Hex. structure we take *two* atomic sites in the motif, chosen one from each of two neighbouring planes, and then replicate this same motif and mark out its pattern.

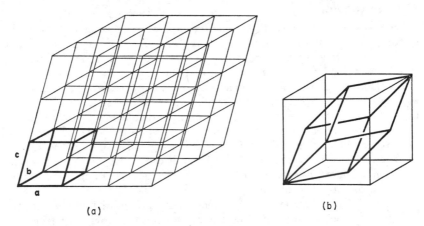

(a)   (b)

Fig. 17.1 (a) a space lattice, (b) a unit cell in the F.C.C. structure cell

Fig. 17.1(a) shows a space lattice. The lattice points lie at the intersections of three families of planes. All planes in one family are parallel and regularly spaced, but only in lattices of high symmetry (e.g. cubic) do the different families have the same or related spacings, or intersect at simple angles (e.g. 90° or 60°). Fig. 17.1(a) also shows the three *basic vectors*, **a**, **b** and **c**, of the lattice. We can conveniently choose **a** as the shortest vector between lattice points; **b** as the shortest not parallel to **a**; and **c** as the shortest not in the plane of **a** and **b**. Since we can then always reach any point of the lattice by repeated movements, i.e. *translations*, along these vectors, the space lattice is the array of points generated by the ends of a *set* of vectors **r**,

$$\mathbf{r} = l\mathbf{a} + m\mathbf{b} + n\mathbf{c} \qquad\qquad \textbf{17.1}$$

obtained by letting *l*, *m* and *n* each have, independently of the others, every value in the whole set of integers.

The intersections of the lattice planes divide the space into a set of identical *unit cells* or *primitive cells*, as shown, with one lattice point per cell since the point at each of the eight cell corners is shared between eight adjoining cells. The *structure cells* of Fig. 13.2 are not unit cells. The B.C.C. structure cell has two lattice points and the F.C.C. has four. Fig. 17.1(b) shows a unit cell of the F.C.C. lattice inscribed in the structure cell. A structure cell does not have the fundamental significance of a unit cell but is useful for displaying simplifying features of the structure.

17.3   Translational symmetry and crystal plasticity

Because of its periodicity, a crystal has *translational symmetry*. Let us take a space lattice and somewhere in this infinite framework mount the crystal which it represents. We now slide the whole crystal, without rotation, to another part of the lattice by moving it along lattice vectors, as defined in eqn. 17.1. Then the association of the crystal sites with lattice points is exactly the same as before. This translational symmetry appears here as a purely mathematical property but it also has great practical importance in connection with the plastic properties of crystals.

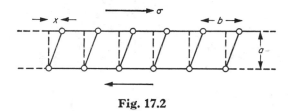

Fig. 17.2

Let us try, by applying suitable forces, to slide one part of a crystal over another, along some surface and in a certain direction. Fig. 17.2 shows a simple example of this, a sectional view of a simple cubic crystal sheared by a shear stress σ (tangential force on unit area of the plane of shear). Whatever the nature of the cohesive forces in this crystal they must, because of the periodicity of the structure, result in a resistance to the shearing process that varies *periodically* with the shear displacement x, with period b, as in the example of Fig. 17.3.

As the atoms on one side begin to slide past their neighbours on the other side their energy must rise at first, this rise being responsible for the mechanical stability and elastic resistance of the crystal. For sufficiently small stresses the displacement x is so small compared with b that *Hooke's law* is obeyed, i.e.

$$\sigma = \mu \frac{x}{a} \qquad\qquad 17.2$$

where x/a is the *shear strain* and μ is the *shear modulus* or *modulus of rigidity*.

As the atoms slide further, however, they eventually, because of the periodicity of structure, approach the next set of crystal sites in which the original crystal structure is re-formed perfectly. The energy must therefore fall again as x approaches b. This is shown in Fig. 17.3. *Because of translational symmetry*, all the original properties of the crystal are perfectly restored when $x = b$ (or any multiple of b) apart from the usually trivial effect of the surface steps at the ends. If the applied force is removed at this stage, for example, the crystal will remain stable in this displaced configuration, with a perfect crystal structure, and would oppose any subsequent distortion with its full elastic resistance.

This is a simple form of *plastic shear*. The restoration of structure and properties in the plastically deformed state is due entirely to the translational symmetry of the crystal and not, as in the viscous flow of liquids and gases,

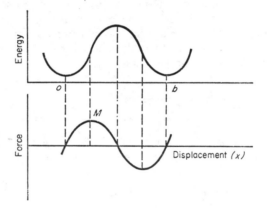

Fig. 17.3

to the effect of thermal agitation in randomizing the positions of the atoms and molecules. Some of the most important processes in metal and alloy crystals occur by plastic shear. We classify them as follows:

(1) *Slip*. In slip, the lattice of the sheared region is in every way identical with the original lattice; i.e. each atom in the slipped part of the crystal makes the same movement and this takes it forward by an integral number of basic lattice vectors. This is the common mode of plastic deformation in metals and the sheared regions in these are often only one or a few crystal layers thick. They are often parallel to a principal crystal plane and are then called *slip planes*. The steps which they produce on the surface of the crystal are *slip bands*.

(2) *Deformation twinning*. In this the sheared region has the same lattice as the parent crystal, but oriented in *twin* relationship to it, i.e. the lattice of the twin is a mirror image of that of the parent crystal. This difference between slip and twinning is shown in Fig. 17.4. The vectors which describe

the twinning displacement are *not* basic lattice vectors, since they convert the lattice to its mirror image. A twinned region is usually fairly thick, e.g. 10^3 to 10^5 atomic layers, each layer being sheared by the same amount over the next one below; this difference from slip occurs because the twinning shear can be applied only once to a given atomic layer; any further displacement of this plane in the same direction leads eventually to a slip process again. The *mechanical shear* of a twinned region can be measured (as can that of slip) from the change in shape of the crystal, or by the displacement of the parts of a line scribed on the surface, or from the angles of steps and tilts on the surface. It must be distinguished from the *crystallographic shear*, i.e. the difference in crystal orientation between the twinned and untwinned lattices. In the simplest types of deformation twinning (e.g. in B.C.C. iron) the mechanical and crystallographic shears are the same.

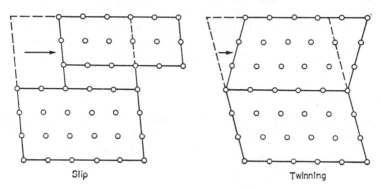

Slip   Twinning

Fig. 17.4 Slip and twinning

In other types however (e.g. in C.P. Hex. metals such as zinc) localized shuffling movements of the atom pairs in each motif also occur, in addition to the overall mechanical shear, and these shuffles drastically change the crystal orientation.

(3) *Shear transformations.* These are like deformation twinning but generate a *new* lattice. The sheared region is then, crystallographically and thermodynamically, a new *phase* and its boundary with the unsheared region is a *phase interface*. Although called shear transformations, because of the mechanical shear, the deformation is often more than a shear. For example, the new phase may have a different density from the old one. There is then some *dilatation*, because pure shear alone cannot produce a volume change. An important example of a shear transformation occurs when steel is quench-hardened. The F.C.C. austenitic structure then transforms by shear, with some dilatation, into *martensite*, the distorted form of B.C.C. iron (cf. § 14.5). In recognition of this, shear transformations are often referred to as *martensitic transformations*.

In slip, twinning and many shear transformations, there are no movements of atoms as separate individuals, only the collective movement of entire sheared regions. Such processes are described as *diffusionless*. Being independent of atomic diffusion they often require little or no activation energy, i.e. are *athermal* processes, and so are able to propagate through the crystal at great speed, comparable with the speed of sound waves in the crystal. Rapid slip enables ductile metals to absorb the shock of sudden impacts without splintering and is also essential for the high-speed machining and explosive forming of metals (cf. Chapter 22, p. 449). Rapid twinning is responsible for the 'crying' sound heard when a piece of tin is bent. Rapid martensitic transformation, even at very low temperatures, is exploited as a means of hardening certain stainless steels by plunging them into liquid nitrogen to convert austenite to martensite.

Not all shear transformations are diffusionless, however. At temperatures of about 300 to 400°C, austenite in many steels decomposes to *lower bainite*, a type of B.C.C. iron with finely dispersed cementite particles in it, by a transformation similar to martensitic but which spreads slowly due to a necessary accompanying diffusion of carbon to form the cementite precipitate.

17.4 Dislocations

Let us consider further the slip process of Fig. 17.3. The equilibrium state of the stressed crystal can be decided only by including the *source* of the applied σ, with the crystal, in the thermodynamic system. We must add

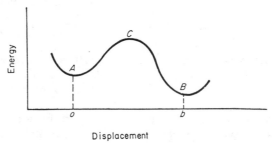

Displacement

Fig. 17.5

the energy of the source to that of the crystal. It may, for example, be elastic energy in a spring which is pulling on the crystal, or gravitational energy belonging to a weight suspended from the crystal. The inclusion of this energy modifies the energy diagram of Fig. 17.3 to that of Fig. 17.5.

This is a situation like that of Fig. 6.3 and discussed in § 12.6. There is a thermodynamic tendency for the transition, from the unslipped to the slipped state, $A \to B$, to occur but this is opposed by the energy barrier C. As in § 12.6, we conclude that the material will not all make the transition simultaneously. Instead, to minimize the amount of material in the high-energy configuration, C, at any instant, the transition occurs by nucleation and growth. We expect to find a fully slipped region growing at the expense of an unslipped region by the advance of an interfacial region, as in Fig. 17.6. We also expect the interface to be narrow, to minimize the amount of material in the high-energy transitional state.

This interfacial region is called a *dislocation*. This particular example is a *slip dislocation*; had the material in B sheared into a twin orientation or into a new crystal structure, it would have been a *twinning dislocation* or a *transformation dislocation*. The thickness C of the interfacial region is called the *width* of the dislocation. The narrower the dislocation, the smaller is the interfacial energy, for the reason already mentioned above. The wider the dislocation, the smaller is the *elastic energy* of the crystal because then

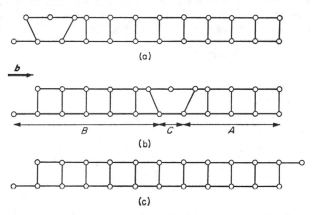

Fig. 17.6 (a), (b), (c) Stages in the nucleation and growth of a unit of slip deformation; A denotes the unslipped region, B the slipped region and C the interfacial region

the atomic spacing along the direction of slip is nearer its ideal value. The equilibrium width is determined by the balance between these two opposing changes of energy.

The elastic energy is proportional to the *elastic constants* of the material and these in turn are proportional to the sharpness of curvature of the energy curve near the minima A and B in Fig. 17.5. The interfacial energy, per unit width of dislocation, is proportional to the height C of the maximum in the curve. It then follows that, in Fig. 17.7, the energy curve 1 would give a narrow dislocation and curve 2 a wide one.

The width is important because it determines the *mobility* of the dislocation, i.e. the ability of the dislocation to *glide* through the crystal under a small driving stress, so allowing the slipped region to grow. Fig. 17.6 shows the glide process. We see that the dislocation preserves its form as it moves from site to site; this is a direct consequence of the translational symmetry of the crystal structure. Because of this there is no *progressive* change in the energy of the dislocation as it moves. At this stage in the argument, *no* applied force is required to move the dislocation. Because of the periodicity of the structure, however, the dislocation can still alter its form—and hence energy—*periodically* as it moves from site to site, provided that it returns to exactly the same form at each new site. An applied stress, known as the *Peierls-Nabarro* stress, is needed to overcome this periodic change of energy. *This must however be very small when the dislocation is wide because the highly distorted region at the centre of the dislocation is then not sharply localized on any one single central crystal site.* The practical importance of this can hardly be over-emphasized. It is the basis of the remarkable *malleability* that enables metals to be shaped by mechanical working; and of the great mechanical *toughness* that enables them to resist strong forces and absorb sudden blows without shattering. Without it, there would be little point in metallurgical technology and we would still be living in a stone age.

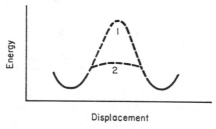

Fig. 17.7

The width of a dislocation depends on the nature of the interatomic forces and on the crystal structure. With *directional covalent* bonds, as in diamond (cf. § 4.3), the energy increases when the *angle* of a bond is altered, as well as the length. The interfacial energy is then high and the dislocations are narrow and immobile (except at high temperatures where thermal energy can help a dislocation overcome its energy barriers). Atoms in metals are held together mainly by the free electron gas (cf. § 4.6) and the bond energy is then insensitive to angle, so long as the interatomic spacing is preserved. It then becomes possible, in a suitable crystal structure, to have an energy curve like that of curve 2 in Fig. 17.7, which gives wide and mobile dislocations. In ductile metals such as copper, silver and gold, dislocation widths of order ten atomic spacings are expected.

The requirement to preserve the interatomic spacing as far as possible, during the slip process, is best satisfied when slip occurs along *close-packed directions* on *close-packed slip planes*. Close-packing in the direction of slip means that the distance b from A to B in Fig. 17.5 is small, which keeps down the height of C. Close-packing in the plane of slip means that such planes are spaced as far apart as possible, which allows them to slide over one another with minimal changes of interplanar spacing. The advantage of simple metal crystal structures, such as the F.C.C., is that they provide such close-packed slip systems (e.g. octahedral planes in F.C.C. metals) in which dislocations are highly mobile. On the other hand, when the crystal structure is complex and does not contain any highly close-packed planes and directions, dislocations are usually fairly immobile. The hardness and brittleness of many intermetallic compounds are largely due to this.

Since it runs along the boundary of a sheared *surface* in the crystal, a dislocation is a *line* imperfection and Fig. 17.6 represents a sectional view through this line. We must consider the orientation of the dislocation line in relation to that of the vector, called the *Burgers vector*, which defines the plastic displacement produced by the dislocation (e.g. **b** in Fig. 17.6).

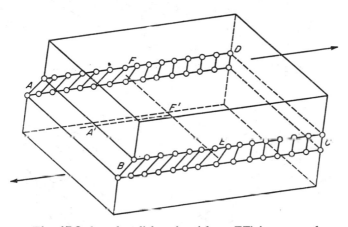

Fig. 17.8 An edge dislocation (along *EF*) in a crystal

Figs. 17.8 and 17.9 show *edge* and *screw* dislocations, in which the dislocation line *EF* is respectively perpendicular or parallel to the Burgers vector. *ABCD* is a slip plane, *ABEF* is the slipped region and *FECD* is unslipped. *Mixed* or *compound* dislocations, in which the line is oblique to the Burgers vector, are also possible. We can describe them in terms of edge and screw dislocations by resolving their Burgers vectors into an edge component, perpendicular to the line, and a screw component, parallel to the line.

Fig. 17.10 shows the structure round a screw dislocation in a simple crystal. The striking feature is that the 'planes' intersected by the dislocation are no longer a family of separate parallel planes; they are all joined up into a single *helicoidal* surface, from which the screw dislocation gets its name. For instance, if we start on the top face at the top of the cliff *BC* and go round the dislocation, staying always in this face, we arrive at the bottom of the cliff, i.e. one layer down. The same effect is found, but less obviously, in edge dislocations; planes intersected by *both* the dislocation line and the

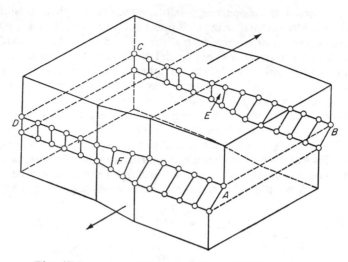

Fig. 17.9 A screw dislocation (along *EF*) in a crystal

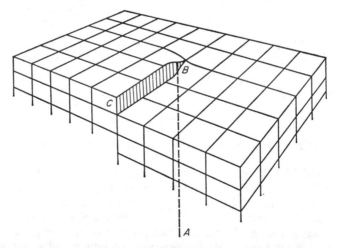

Fig. 17.10 A screw dislocation *AB* which ends at a point *B* on a crystal face

Burgers vector are converted into spirals. In fact, one general way to define dislocations would be to replace one or more of the families of parallel planes in Fig. 17.1(a) by helicoids.

This helicoidal crystal structure is important in the growth of crystals from dilute solutions and vapours (*F. C. Frank*, 1952, *Adv. Phys.*, **1**, 91). Atoms or molecules depositing on such a growing crystal need a ledge, such as *BC*, against which to fasten themselves permanently to the crystal. With-

out this, crystal growth is very difficult at low supersaturations. The helicoidal structure provides a self-perpetuating growth step. The observation
of such growth steps, usually in the form of spirals, on crystals has provided
a striking confirmation of Frank's theory.

This means that most crystals, so grown, are necessarily *imperfect*. They
contain dislocations *ab initio*, before any plastic deformation. Crystals
growing from a *melt*, as in Chapter 13, do not have the same great need for
dislocations in order to grow. Nevertheless they usually contain many dislocations in the as-grown condition, due to various irregularities in the
crystallization process. Small, flat contraction cavities may collapse to
form dislocation *loops* round their perimeters. Gradients of temperature
and composition may produce misalignments between neighbouring
dendritic arms growing from the same nucleus. Various methods of observation (electron microscopy, x-ray microscopy, field-ion microscopy,
etch-pitting, decoration of dislocations with precipitates) have in fact
shown that crystals ordinarily contain large numbers of dislocation lines,
either in the form of *networks*, as in Fig. 17.11, or in grain boundaries.

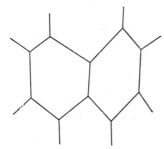

Fig. 17.11 A dislocation network

Because of the presence of this initial dislocation structure, there is
usually no nucleation of dislocations in the process of plastic deformation,
only the glide of those already present. Furthermore, because of the
interconnected nature of dislocation networks, plastic deformation in fact
generally *increases* the dislocation content of a crystal, so that the capacity
of a crystal for plastic deformation is in this sense *never exhausted*.

Fig. 17.12 shows a simple configuration for multiplying dislocations,
known as the *Frank-Read source*. We imagine that the diagram is in the slip
plane of a dislocation line *AB* which is linked by its *nodes* at *A* and *B* to
immobile branches of the dislocation network. The applied stress pushes
the centre of the line forward into a loop (b) which eventually folds round
the nodes (c) and reunites with itself behind them (d) to form a closed
dislocation ring, that can expand outwards across the slip plane, together
with a reconstituted line between the nodes from which the process can be
repeated. An almost unlimited amount of slip can thus be generated on this
plane by the repeated creation of dislocation loops.

Fig. 17.13 shows another method of multiplication. Part of a dislocation line, *AB* in the lower slip plane, of screw orientation has transferred itself by *cross-slip* on the plane *ABCD* to the upper slip plane and has then bowed itself out in this plane to the form *CEFD*. The configuration in this upper slip plane is similar to that of Fig. 17.12 and is geometrically suitable for multiplication. This process of multiplication by cross-slip can be seen particularly clearly in ionic crystals such as LiF and NaCl. It has also been observed in B.C.C. iron.

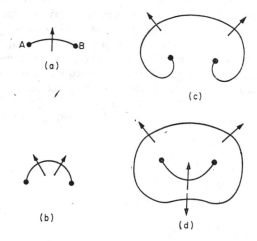

Fig. 17.12 Stages (a), (b), (c) and (d) in the formation of a dislocation ring from a Frank-Read source

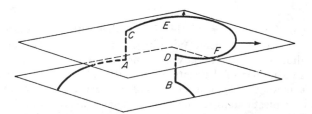

Fig. 17.13 Multiplication of dislocation by cross-slip

17.5   Miller indices

For the more precise study of crystal structure a notation for crystal planes and directions is necessary. We set up coordinate axes *X*, *Y*, *Z*, parallel to the edges of a unit cell, or of a structure cell where this is more symmetrical, and graduate each axis in units equal to the length of its own cell edge, as in Fig. 17.14.

To specify a crystal plane we count its intercepts on the three axes and take their reciprocals. These reciprocal intercepts can always be written as h/n, k/n and l/n, where n is the common denominator and h, k and l are integers. We write these as (hkl) and call them the *Miller indices* of the plane. For example, in Fig. 17.14 we have $OP = 2$, $OQ = 2$, $OR = 3$; their reciprocals $1/2$, $1/2$, $1/3$, become $3/6$, $3/6$, $2/6$, which identifies the plane as a (332) plane.

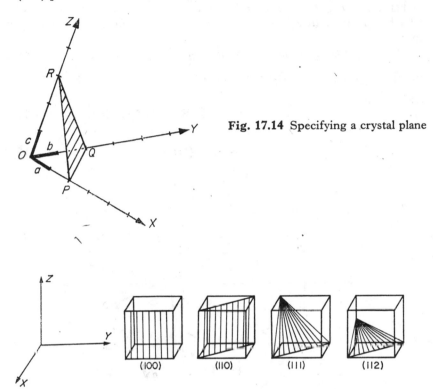

Fig. 17.14 Specifying a crystal plane

Fig. 17.15 Miller indices of planes in cubic crystals

Fig. 17.15 shows important planes in cubic crystals, referred to the structure cell. A plane parallel to an axis intercepts that axis at infinity and has a zero index. A *negative* intercept is indicated by a bar over the corresponding index, e.g. $(\bar{h}kl)$.

The symbol (hkl) refers to the entire family of *parallel hkl* planes. Planes of equivalent crystallographic type, but in different orientations, are denoted as $\{hkl\}$. For instance, $\{111\}$ stands for all octahedral planes in the cubic lattice, i.e. for (111), (11$\bar{1}$), (1$\bar{1}$1), ($\bar{1}$11), ($\bar{1}\bar{1}\bar{1}$), ($\bar{1}\bar{1}$1), ($\bar{1}$1$\bar{1}$), and (1$\bar{1}\bar{1}$).

To define a *direction* in a crystal we draw a line through the origin in Fig. 17.14, parallel to it, and give the coordinates of a point on it, measured in

cell edge lengths and converted to a set of smallest integers, written as
[*hkl*]. For example, if the coordinates are $3a$, $-b$, $c/2$, the line is a [6$\bar{2}$1]
direction. A family of differently oriented but crystallographically equivalent directions is written as $\langle hkl \rangle$.

In *cubic lattices* Miller indices are proportional to *direction cosines*. Several
results follow from this. Let OP be the [*hkl*] direction, in Fig. 17.16(a),
so that $OA = ah$, $OB = ak$, and $OC = al$, where a is the length of the cube
cell edge, i.e. the *lattice parameter*. Since $\cos \alpha = OA/OP$, etc. the direction
cosines are

$$\cos \alpha = mh, \qquad \cos \beta = mk, \qquad \cos \gamma = ml \qquad \textbf{17.3}$$

where $m = a/OP = $ constant. We note that

$$\cos^2 \alpha + \cos^2 \beta + \cos^2 \gamma$$
$$= [(OA)^2 + (OB)^2 + (OC)^2]/(OP)^2 = 1 \qquad \textbf{17.4}$$

and that

$$h^2 + k^2 + l^2 = (OP)^2/a^2 \qquad \textbf{17.5}$$

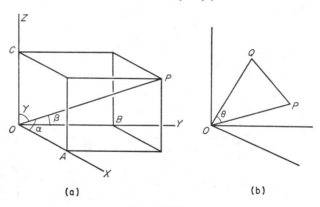

(a)   (b)

Fig. 17.16

In Fig. 17.16(b) let P and Q have coordinates ah, ak, al, and au, av, aw,
respectively. From the properties of triangles,

$$(PQ)^2 = (OP)^2 + (OQ)^2 - 2(OP)(OQ) \cos \theta \qquad \textbf{17.6}$$

but

$$(PQ)^2 = a^2[(h - u)^2 + (k - v)^2 + (l - w)^2]$$
$$= a^2(h^2 + k^2 + l^2) + a^2(u^2 + v^2 + w^2) - 2a^2(hu + kv + lw)$$
$$= (OP)^2 + (OQ)^2 - 2a^2(hu + kv + lw)$$

and hence

$$\cos \theta = \frac{hu + kv + lw}{(h^2 + k^2 + l^2)^{1/2}(u^2 + v^2 + w^2)^{1/2}} \qquad \textbf{17.7}$$

which gives the angle between directions in terms of Miller indices. For example, the angle between [100] and [111] is given by

$$\cos \theta = (1 + 0 + 0)/(1)^{1/2}(3)^{1/2},$$

i.e. $\theta = 54° 44'$.

In Fig. 17.17 let the plane $A'B'C'$ be (hkl). Then $OA' = a/h$, etc. Let OP be the normal to the plane and draw PA perpendicular to OA'. In the triangle OPA' we have $\cos \alpha = OP/OA' = OA/OP$, so that $OA = (OP)^2/OA' = nh$, where $n = (OP)^2/a$. The other coordinates of P are similarly nk and nl. Hence there is a point on OP that satisfies eqns. 17.3, i.e. the direction OP is $[hkl]$. Thus, in *cubic lattices*, $[hkl]$ is perpendicular to (hkl) and eqn. 17.7 also gives the angle between planes (hkl) and (uvw).

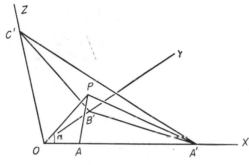

Fig. 17.17

It then follows, by setting $\cos \theta = 0$ in eqn. 17.7, that the direction $[uvw]$ lies in the plane (hkl) when

$$hu + kv + lw = 0 \qquad\qquad \textbf{17.8}$$

Consider now the spacing of planes (hkl) in a *simple cubic* lattice, i.e. the length d of OP in Fig. 17.17. Substituting $\cos \alpha = d/(a/h)$, etc. in eqn. 17.4, we obtain

$$d = \frac{a}{(h^2 + k^2 + l^2)^{1/2}} \qquad\qquad \textbf{17.9}$$

We see that *low-index* planes have the widest spacings. For example, the spacings of (100), (110), (111), (210) are respectively a, $a/\sqrt{2}$, $a/\sqrt{3}$, $a/\sqrt{5}$. Since there is a fixed total number of lattice points in a given volume of lattice and since every point belongs to a lattice plane, the planes of widest spacing are also the planes of *closest packing*.

Eqn. 17.9 has to be modified for B.C.C. and F.C.C. lattices, because the planes in these are *interleaved*. The B.C.C. lattice, for example, can be regarded as two interpenetrating simple cubic lattices, as in Fig. 14.18, so that the spacing of (100) planes is $a/2$, not a. Fig. 17.18 shows, two-dimensionally, how the spacing of some planes is affected by the interleaving, but not others. Because of this, (110) are the most closely packed planes in B.C.C. crystals and (111) in F.C.C.

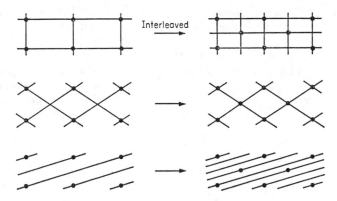

Fig. 17.18 Interleaving reduces the spacing of some planes but not others

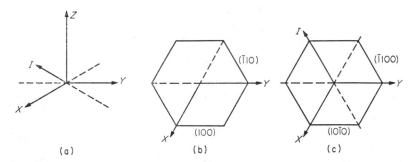

Fig. 17.19 (a) Miller-Bravais axes, (b) and (c) prismatic planes in Miller and Miller-Bravais notation

In hexagonal crystals, *Miller-Bravais* indices are commonly used. Four axes are chosen, Z perpendicular to the basal plane and X, Y, I in the basal plane along three close-packed directions as in Fig. 17.19(a). The intercepts of a plane on these axes are measured, in the order X, Y, I, Z, and converted to indices exactly as before. The positive axes are as shown. The Miller-Bravais indices of the plane are then $(hkil)$ where i is obtained from the I axis. By geometry, $i = -(h + k)$ and so is redundant. The advantage of the notation can be seen from Fig. 17.19 where we index the prismatic faces of a hexagonal prism. Miller indices have different forms, e.g. (100) and ($\bar{1}$10), for different planes of the same crystallographic type. In Miller-

PLATE 3

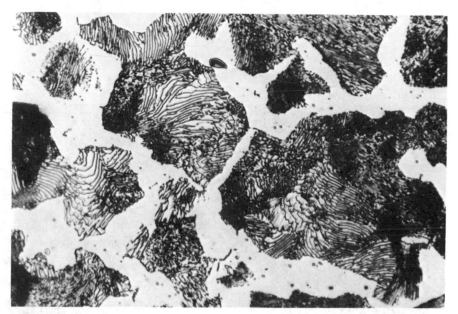

A. Ferrite and pearlite in a slowly cooled 0·55 per cent carbon steel, ×1000
(*courtesy of G. C. Smith*)

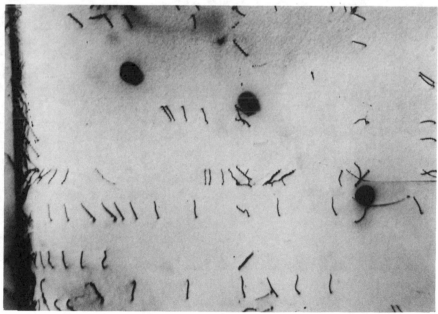

B. Electron microscope picture of dislocations in two sets of slip lines near a
grain boundary in an Ni-Fe-Cr alloy, ×35,000 (*courtesy of R. B. Nicholson*)

PLATE 4

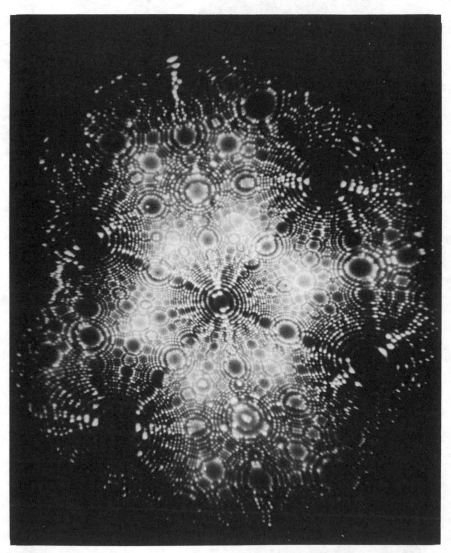

Helium field-ion microscope picture of a tungsten tip, showing atomic structure
(*courtesy of B. Ralph*)

Bravais notation these planes become (10$\bar{1}$0) and ($\bar{1}$100), respectively, and their crystallographic equivalence, as members of the {10$\bar{1}$0} family, is then plain.

A direction in this notation is indicated by constructing four vectors, in order along the X, Y, I, Z, axes, which combine to give a vector in the required direction. These four must be chosen so that the length of the third, in cell edge units, is always equal to minus the sum of the first two. The four vector lengths are then reduced to smallest integers and written as [*hkil*], which is of course *not* in general perpendicular to (*hkil*). Fig. 17.20 shows examples of directions in the basal plane.

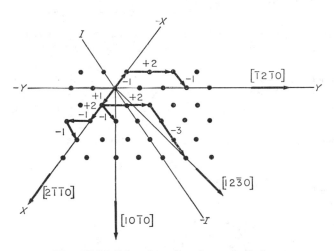

Fig. 17.20 Directions in a hexagonal plane

17.6   Burgers vectors and glide systems in metals

We introduced the Burgers vector in § 17.4 as the plastic displacement produced by a dislocation. It follows that all parts of a dislocation line have the same Burgers vector and hence that a dislocation line cannot end inside a crystal since this would involve a change of Burgers vector, to zero, at the end point. Furthermore if two dislocations join to form a third—either in time as when one glides into another; or in space at a *node* in a dislocation network (Fig. 17.11)—the Burgers vector of the third dislocation is the vector sum of the other two.

The Burgers vector is usually defined from a *Burgers circuit*, shown in Fig. 17.21. This is a sequence of lattice vectors, from one lattice point to the next, which forms a closed clockwise ring round the dislocation. *Exactly* the same set of lattice vectors is then used to make a second circuit, in a 'good' region of the crystal. An extra vector is then needed to join the end point of this circuit to the starting point. This is the Burgers vector.

In Miller index notation a Burgers vector is written as $na[hkl]$ and has components nah, nak, nal, along the three axes, where a is the length of the cell edge and n is a number. F.C.C. metals slip along $\langle 110 \rangle$ close-packed directions and mostly on $\{111\}$ close-packed slip planes. A perfect unit dis-

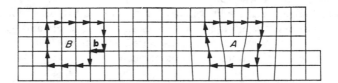

Fig. 17.21 A Burgers circuit (A) round a dislocation; when repeated in a perfect crystal (B) it fails to close unless completed by the Burgers vector **b**

location in this case has a Burgers vector such as $(a/2)[110]$, of length $a/\sqrt{2}$, which connects a cube corner to a nearby face centre. In B.C.C. metals the Burgers vector for slip from a cube corner to cube centre, along the close-packed $\langle 111 \rangle$ direction, is $(a/2)[111]$, of length $a\sqrt{3}/2$. Several slip planes, e.g. $\{110\}$, $\{112\}$, $\{123\}$, may be used in this case, giving corru-

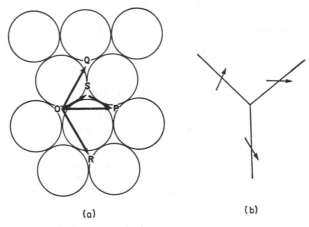

(a)   (b)

Fig. 17.22 Burgers vectors (a) on a close-packed plane, (b) at a dislocation node

gated slip surfaces known as *pencil glide*. C.P. Hex. metals normally slip along a direction of $[2\bar{1}\bar{1}0]$ type and the Burgers vector is then $(a/3)[2\bar{1}\bar{1}0]$, i.e. has components $2a/3$, $-a/3$, $-a/3$, 0 (cf. Fig. 17.20) and is of length a, where a is the lattice spacing in the (0001) plane. The favoured slip plane in metals such as zinc and cadmium is the (0001) basal plane; this is also the close-packed plane when the axial ratio c/a is greater than ideal. High temperatures and smaller axial ratios (e.g. Mg, Be, Ti, Re) favour other planes for slip, particularly $\{10\bar{1}1\}$ pyramidal planes and $\{10\bar{1}0\}$ prism planes.

Fig. 17.22(a) shows a close-packed plane of atoms in an F.C.C. or C.P. Hex. crystal. An atom sitting at O in the next layer above can make a perfect slip displacement to P, Q or R. Hence if this atom is moved $O \rightarrow Q \rightarrow P$ by two successive dislocations with Burgers vectors OQ and OR (equivalent to QP), respectively, the effect is as if one dislocation with vector OP had passed through. It is thus possible for three such dislocations, with these Burgers vectors, to meet at a node, as in diagram (b), and so form networks such as that of Fig. 17.11. In Miller index notation this can be represented as, for example

$$\frac{a}{2}[10\bar{1}] + \frac{a}{2}[0\bar{1}1] = \frac{a}{2}[1\bar{1}0] \qquad \textbf{17.10}$$

A billiard-ball model of close-packed slip planes immediately shows that the 'atoms' prefer to slide, not along perfect lattice vectors such as OP, but along zig-zag paths such as $O \rightarrow S \rightarrow P$. The same effect appears in metals, probably because the zig-zag path involves less change of volume. To produce the zig-zag slip, a perfect dislocation with Burgers vector OP splits into two separate *partial* dislocations with Burgers vectors OS and SP respectively; for example, on a (111) plane in an F.C.C. metal such a dissociation is

$$\frac{a}{2}[1\bar{1}0] = \frac{a}{6}[1\bar{2}1] + \frac{a}{6}[2\bar{1}\bar{1}] \qquad \textbf{17.11}$$

The displacement OS takes the atoms into twinning positions and so creates a *stacking fault* (cf. § 13.1), as in Fig. 17.23, which can be thought of as a thin sheet of C.P. Hex. structure or of twinned F.C.C.

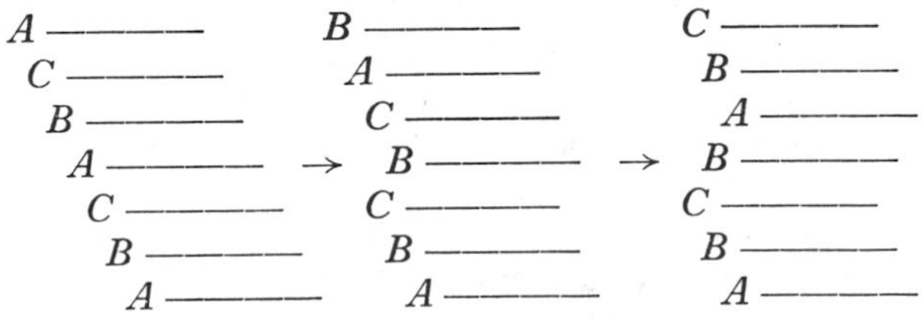

Fig. 17.23 Formation of stacking faults and twins by partial slip on single and successive planes

The second displacement $S \rightarrow P$ in Fig. 17.22 corrects the fault and completes the perfect slip $O \rightarrow P$. The two partial dislocations thus form a pair of parallel dislocation lines bounding a *ribbon* of stacking fault, in the close-packed plane which contains their Burgers vectors. As experiments with a billiard-ball model immediately show, the 'atoms' fit together well only on this particular plane. The severe misfit developed by faulting on other planes leads to a prohibitively high misfit energy, i.e. *high stacking fault energy*, and so prevents dissociation on them. The fact that dissociation can occur easily only when the dislocations lie in {111} planes is probably why these are the strongly preferred slip planes in F.C.C. metals.

Dissociation also restricts the cross-slipping of screw dislocations (cf. Fig. 17.13). Geometrically, a screw dislocation can glide in any plane that contains its line, since every such plane is parallel to its Burgers vector. However, if the screw dislocation is dissociated in one of these planes it must then glide in that plane, otherwise a band of badly misfitting material would be created from the stacking fault. Cross-slip can still occur if the

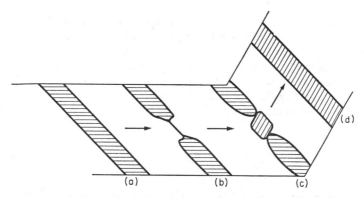

Fig. 17.24 Stages (a) to (d) in the cross-slip of a screw dislocation by the formation of a constriction

dissociated dislocation contracts to a single line, forming a *constriction*, over part of its length, as in Fig. 17.24. This constricted part can then dissociate in the cross-slip plane and grow, at the expense of the neighbouring parts, into a long dissociated dislocation in the cross-slip plane.

The more widely the partials are separated initially, the more the energy required to bring them together to form the constriction and the more rare is the cross-slip. Some metals cross-slip rarely and straight slip bands parallel to {111} planes are formed (e.g. cobalt, gold, α-brass, austenitic stainless steel). Others cross-slip abundantly and form wavy interlaced slip bands (e.g. aluminium, lead) except at very low temperatures where the necessary thermal activation energy is not available.

The width w of a stacking fault ribbon in a dissociated dislocation is set by a balance between the mutual repulsion of the partial dislocations, due to elastic forces (cf. § 19.3) and the surface energy of the stacking fault which acts like a surface tension in holding the dislocations together. It can be shown that

$$w = \frac{\mu a^2}{24\pi\gamma}$$ 17.12

where μ is the shear modulus, a is the lattice parameter and γ is the energy per unit area of stacking fault. For example, if $\mu = 4 \times 10^{10}$ N m^{-2} and $a^2 = 10^{-19}$ m^2, a stacking fault energy $\gamma = 0.05$ J m^{-2} gives $w = 10$ Å. Various methods for estimating w and γ exist. In metals of low stacking fault energy, w can sometimes be measured directly in the electron microscope. The abundance of *annealing twins* in a polycrystal (cf. § 19.5), which are bounded by interfaces similar to a stacking fault, can sometimes be used to infer a value of γ. Other evidence can be obtained from mechanical properties, cross-slip behaviour and dislocation structures produced by quenching. Although some values are still controversial it is now clear that cobalt, gold, austenitic steel and many F.C.C. alloys based on copper, silver or gold have low stacking fault energies (below 0.03 J m^{-2}), that aluminium and nickel have high values (above 0.15 J m^{-2}), and that copper and silver have intermediate values.

It is sometimes possible to drive one partial dislocation away from the other and so produce a wide stacking fault extending right across the crystal. This happens fairly readily in cobalt, a metal in which F.C.C. and C.P. Hex. stacking are almost equally preferred. In other metals it usually occurs only after severe plastic working, e.g. in beaten gold leaf. If the same fault is produced repeatedly on each layer, by the same Burgers vector displacement, a *deformation twin* is formed, as in Fig. 17.23. Deformation twins form rarely in F.C.C. metals because slip usually prevents the shear stress from rising to the high levels needed for twinning. B.C.C. transition metals twin at low temperatures, or under impact stresses at room temperature, because the stress for slip becomes large under these conditions. The twinning shear in this case occurs in the [11$\bar{1}$] direction on a (112) plane. The necessary partial dislocation, $(a/6)[11\bar{1}]$, may be formed by the decomposition of a perfect slip dislocation, i.e.

$$\frac{a}{2}[11\bar{1}] \rightarrow \frac{a}{6}[11\bar{1}] + \frac{a}{3}[11\bar{1}]$$ 17.13

Another possibility is that the perfect $(a/2)[11\bar{1}]$ dislocation may split into *three* $(a/6)[11\bar{1}]$ twinning dislocations, one on each of three neighbouring layers. Twinning in C.P. Hex. metals, which occurs on the $(10\bar{1}2)$ plane in the $[10\bar{1}1]$ direction (or its reverse, depending on the axial ratio), is a more complicated process. The mechanical shear is small, but accompanying local readjustments of atomic positions produce a large change of crystallographic orientation.

When the twinning shear is small it is not difficult to nucleate twinning dislocations within the perfect crystal, and there is evidence in zinc that this can occur, although in general we expect the crystal to take advantage of any existing dislocations or other imperfections to make the twinning process easier. Where the twinning shear is large it is reasonable to expect that twins grow by a process similar to the Frank-Read mechanism for slip.

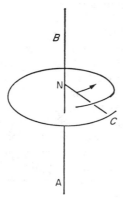

Fig. 17.25 Pole mechanism for twinning

Fig. 17.25 shows a possible twinning mechanism. The *pole* dislocation *AB* has a component of its Burgers vector perpendicular to the twinning plane, which is thus converted into a helicoidal surface as shown. A twinning dislocation *NC*, attached to it, can then rotate in the twinning plane about the node *N*, changing by one layer each time round and so producing homogeneous twinning on a thick stack of planes.

Certain types of partial dislocations are incapable of gliding. Fig. 17.26

Fig. 17.26 A simple sessile dislocation

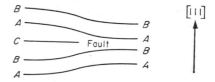

shows a simple example of such a *sessile* dislocation, formed in an F.C.C. crystal. One of the close-packed layers is incomplete. The dislocation is formed along the internal edge of this layer; and a stacking fault in the missing region. The dislocation is an *edge*, with Burgers vector $(a/3)$ [111]. It can glide only in the direction of its Burgers vector and only in the plane of its stacking fault. But the Burgers vector does not lie in this plane. Hence it cannot glide. We shall see in § 19.7 how such dislocations may be formed.

Fig. 17.27 shows another sessile configuration, a dislocation *lock* formed from two dissociated glide dislocations, moving on (111) and (11$\bar{1}$) planes respectively. The Burgers vectors are as shown. The two dislocations run together at the line of intersection of their slip planes, where the two leading partials unite by the reaction

$$\frac{a}{6}[\bar{1}21] + \frac{a}{6}[2\bar{1}\bar{1}] = \frac{a}{6}[110] \qquad \qquad \textbf{17.14}$$

so forming a *stair-rod* dislocation joined by stacking faults on two different planes. This obviously cannot glide. Dislocation locks, the formation of which has been observed in the electron microscope, are important because they are readily created at the intersections of active glide planes and then block the traffic of other, mobile, dislocations on or near those planes, so *hardening* the metal.

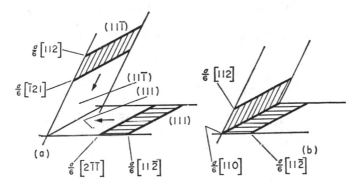

Fig. 17.27 Formation of a dislocation lock (b) from two dissociated glide dislocations (a)

17.7 Diffraction of x-rays

We turn now to another aspect of the periodicity of crystals, the *diffraction of waves* by a lattice. Consider, in particular, *x-rays*. These are produced by bombarding a metal target with fast electrons in a vacuum tube. Below a certain voltage for the acceleration of these electrons a smooth spectrum of 'white' x-rays is produced over a wide range of wavelengths. Above about 50,000 volts, however, intense 'monochromatic' x-rays are emitted at particular wavelengths, characteristic of the target metal. They are produced when the bombarding electrons have enough energy to knock electrons out of the low energy levels (e.g. $1s$; cf. § 3.6) in the atoms of the target. Other electrons in higher energy levels in these atoms then drop down into the empty quantum states so created and, in so doing, release energy in the form of x-rays according to eqn. 3.3. The radiation produced by electrons dropping from the second ($n = 2$) and third electronic shells is denoted as K_α and K_β respectively. The K_α radiation is emitted at two slightly different wavelengths and is referred to as a *doublet*.

X-rays are electromagnetic waves, the same as light or radio waves but with wavelengths of order 1 Å. Their wavelengths are often measured in kX units, where $1kX = 1.00202$ Å. Fig. 17.28 shows such a wave moving along the x axis. We shall take it to be a *plane wave*, i.e. as a sequence of plane wave fronts, alternately crests and troughs, perpendicular to the x axis. The wave is compounded of two sinusoidal waves, electric and magnetic, only the first of which is shown in the diagram. The electric wave field is given by

$$E = A \sin 2\pi[(x/\lambda) - \nu t] \qquad\qquad \textbf{17.15}$$

meaning that, if a particle of electrical charge e is situated at a point x, it is acted on by a force Ee, in the direction of the E axis in the diagram, at an instant of time t; the wave has amplitude A, wavelength λ and frequency ν, where $\lambda\nu = c =$ velocity of light $= 3 \times 10^8$ m s^{-1}. The magnetic wave similarly provides a magnetic field H, perpendicular to both axes x and E.

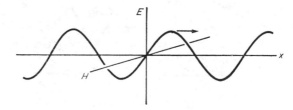

Fig. 17.28 An electromagnetic wave

An electrically charged particle, such as an electron in an atom, oscillates under the periodic force Ee and as a result emits electromagnetic waves of the same frequency which spread out symmetrically from it as a series of concentric, expanding, spherical wave fronts. When a piece of crystal is bathed in the x-ray beam, all its atoms emit spherical waves. These waves expand and overlap one another. Where wave crests coincide, the amplitude of the total scattered wave is increased. Where crests and troughs coincide there is mutual cancellation and no scattered wave. The original analysis of this, due to *Laue*, considered all the atoms individually and then added their contributions, allowing for the phase differences in the superposed waves. Later it was shown by *Bragg* that the analysis is much simpler if whole *planes* of atoms are considered and their contributions then added.

To derive the *Bragg law* we consider in Fig. 17.29 plane wave fronts (ACE) incident upon crystal planes at an angle θ. Clearly, if the angle of emergence of the scattered rays, forming the diffracted wave fronts (DBF), is the same as the angle of incidence, then all path lengths from the *same* crystal plane are equal, e.g. $ANB = CMD$. Thus if the diffracted beam is regarded as simply *reflected* off a plane, all the individual waves from the atoms in that plane scatter in phase with one another. However, the reflected beams off *different* planes have also to be in phase. The condition for this is that the path length EOF must be an integral number n of wavelengths longer than the path length ANB. Since $EOF - ANB = OP + OQ = 2d \sin \theta$, where d is the spacing of the planes, the diffraction condition is

$$n\lambda = 2d \sin \theta \qquad \qquad \textbf{17.16}$$

where λ is the wavelength. This is the *Bragg law* and the values of θ which satisfy it are *Bragg angles*. The integer n is the 'order of reflection'. For cubic crystals, where d is given by eqn. 17.9, it is convenient to take n into the Miller index, i.e.

$$\lambda = \frac{2a \sin \theta}{(n^2h^2 + n^2k^2 + n^2l^2)^{1/2}} = \frac{2a \sin \theta}{N^{1/2}} \qquad \textbf{17.17}$$

where N is the 'reflection number' or 'line number'. For example, the $n = 2$ reflection from the (100) planes gives 'line 4' or 'the 200 reflection'.

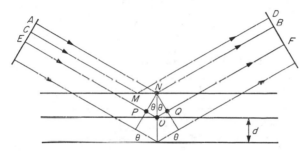

Fig. 17.29 Bragg reflection

We note from eqn. 17.16 that Bragg reflection cannot occur unless $\lambda \leqslant 2d$. For most crystal planes, d is not more than about 3 Å, so that λ has to be less than about 6 Å.

If a crystal is set up at some arbitrary angle in a beam of monochromatic x-rays, in general the Bragg law will not be satisfied and there will be no reflections from it. To produce reflections we may either provide a range of wavelengths, using white x-rays, as in the *Laue method*; or, with monochromatic rays, provide a range of orientations by rotating the crystal in the beam or by using a polycrystalline aggregate, as in the *Bragg method* (or *powder method*).

Fig. 17.30 shows the powder method. A thin rod specimen, made from powdered crystal cemented on a thread or from polycrystalline wire, is suspended and rotated vertically at the centre of a circular camera of radius R, round which is wrapped a strip of photographic film. The diffracted beams, at the appropriate Bragg angles (θ), leave the specimen at angles 2θ

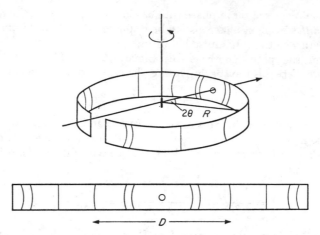

Fig. 17.30 The powder method

TABLE 17.1.   REFLECTIONS FROM CUBIC LATTICES

Simple cubic		B.C.C.	F.C.C.
Reflection	N		
100	1	*m*	*m*
110	2		*m*
111	3	*m*	
200	4		
210	5	*m*	*m*
211	6		*m*
220	8		
300, 221	9	*m*	*m*
310	10		*m*
311	11	*m*	
222	12		
320	13	*m*	*m*
321	14		*m*
400	16		
410, 322	17	*m*	*m*
411, 330	18		*m*
331	19	*m*	
420	20		
421	21	*m*	*m*
332	22		*m*

m = missing reflection

to the incident x-ray beam and produce lines, in pairs, spaced at distances
D apart when the film is laid flat, where $\theta = D/4R$.

We then have the problem of *indexing* these lines, i.e. identifying each
with a certain reflection from a certain crystal plane. For a *simple cubic*
lattice, the first line, at the smallest value of θ, is the first order (100)
reflection, at $\sin\theta = \lambda/2a$. The other lines then follow in sequence, as in
Table 17.1, with increasing values of N in eqn. 17.17.

In B.C.C. and F.C.C. crystals, the *interleaving* of the crystal planes causes
some of these reflections to be *missing*. For example, a B.C.C. crystal does
not give a 100 reflection because the (100) plane through the centre of the
cubic cell gives a reflection exactly out of phase with those from the (100)
faces of the cell.

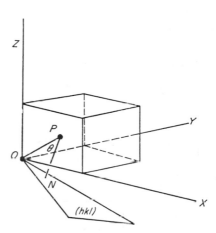

Fig. 17.31

To analyse this, we regard the B.C.C. or F.C.C. lattice as a simple cubic
lattice with a *basis*, i.e. with more than one atom per lattice point. Fig.
17.31 shows an example (e.g. B.C.C.). We construct a simple cubic lattice
with one of the atoms O as origin. Let the coordinates of a second atom P,
within the cell, which also belongs to this lattice point, be $ma[uvw]$ in Miller
index notation. The length OP is then $ma(u^2 + v^2 + w^2)^{1/2}$. In the B.C.C.
cell, for example, $m = \frac{1}{2}$, $u = v = w = 1$. Consider the possibility of
x-ray reflections from planes (hkl). One such plane passes through O,
another through P. We need the distance d' between them. Construct the
normal PN from P on to the plane through O, as shown. The angle θ is
then given by eqn. 17.7. We also have $d' = PN = OP \cos\theta$. Hence

$$d' = \frac{ma(hu + kv + lw)}{(h^2 + k^2 + l^2)^{1/2}} = m(hu + kv + lw)d \qquad \textbf{17.18}$$

where d is the spacing of the *simple cubic* lattice planes (hkl). Consider the
nth order reflection from the planes. The path difference between reflections

from successive simple cubic lattice planes is then $n\lambda$. The corresponding path difference between reflections from the planes through O and P is given by

$$n\lambda d'/d = m(nhu + nkv + nlw)\lambda = n'\lambda \qquad \textbf{17.19}$$

If this is a half-integral number of wavelengths the waves are out of phase and the reflection is missing. For B.C.C. crystals we have

$$n' = \tfrac{1}{2}(nh + nk + nl)$$

i.e. missing reflections when $nh + nk + nl$ is *odd*, as in Table 17.1.

The F.C.C. analysis is more complicated. We have $m = \tfrac{1}{2}$ and, in addition to the origin position O, three other atomic positions corresponding to the

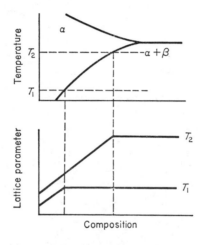

Fig. 17.32 Determination of solubility from lattice parameter measurements

u, v, w values 110, 101, 011, respectively, which we shall denote as A, B and C. We have then to consider whether

$$\begin{aligned} nh + nk \quad &(OA) \\ nh + nl \quad &(OB) \qquad\qquad \textbf{17.20} \\ nk + nl \quad &(OC) \end{aligned}$$

are even or odd. If nh, nk and nl are *all* even or *all* odd, these sums are even and the reflections are present, as in Table 17.1. Suppose however that nh is odd and the others are even. We then have four waves, from O, A, B, C, respectively; O and C form one pair, in phase with each other; A and B form another pair, out of phase with O (and C) and hence in phase with each other. The OC and AB pairs mutually cancel and the reflection is missing. Exactly the same argument holds if nh is even and the others are odd. Hence *mixed* even and odd values of nh, nk and nl give missing reflections, as in Table 17.1.

These characteristic missing reflections enable us to identify the lattice type and label the reflections. Knowing these, we can then apply Bragg's law to the measured angles of reflection to find the lattice parameter. The measurement of lattice parameter as a function of composition provides a valuable method for determining the limits of solubility in solid solutions, as shown in Fig. 17.32.

The mutual cancellation of certain reflections from interleaved planes occurs completely only if the planes have the same composition. When different kinds of atoms separate on alternate planes, as in *superlattices*, the cancellation is usually incomplete because the x-rays reflected from one layer may be more intense than those (of opposite phase) reflected from the alternate layer. In place of missing reflections there are then residual reflections, *superlattice reflections*, weaker than the main lattice reflections. This effect is invaluable for the study of order in alloys.

17.8 Electron and neutron diffraction

Because of their wave properties, beams of particles such as electrons or neutrons can be used to study crystals by diffraction methods. An electron beam is produced by accelerating electrons, emitted from a hot filament, with a voltage V (volts) of order 100 000 which gives a wavelength (Ångströms) of order 0·05, according to eqn. 3.4. The relation is

$$\lambda = (150/V)^{1/2} \qquad\qquad 17.21$$

neglecting relativistic effects. These waves can be reflected from crystals in accordance with Bragg's law. The low-order reflections are the strongest and therefore the most used. Because of the small wavelength, the Bragg angles for these are only of order 1 degree, so that the important reflections take place at glancing incidence to the crystal planes. The electrons are strongly scattered by matter, through their electrostatic interactions, and so are much less penetrating than x-rays. It is thus necessary to enclose the electron source, the specimen and photographic plate in a vacuum chamber. The strong scattering makes electrons very suitable for examining surface films (oxide layers, corrosion products, etc.) by glancing incidence from the surface, and for examining thin foils ($\simeq 1000$ Å) by transmission through the material.

Neutrons are scattered mainly by interaction with the atomic nuclei. This process does not depend particularly on the atomic number Z. As a result, neutron diffraction has certain advantages over x-ray and electron diffraction. Light atoms such as hydrogen and carbon scatter neutrons strongly, which enables their positions in a lattice to be determined. X-ray and electron diffraction suffer here from the fact that light atoms, having only a few electrons, are weak scatterers of these radiations. Neutrons are also useful for determining the positions of atoms with similar atomic numbers (e.g. Fe and Co; Cu and Zn) whose scattering powers for x-rays and electrons are nearly equal. A third advantage is that neutrons interact magnetically with the spinning electrons in atoms and so can be used to examine the electronic structure of magnetic materials.

A neutron beam is produced from a hole in a nuclear reactor. The neutrons released by uranium fission are slowed down, i.e. *moderated*, within the reactor by collisions with the nuclei of a substance such as graphite or heavy water. They stream through the hole with energies of order kT (T = reactor temperature, k = Boltzmann's constant) and wavelengths,

$$\lambda = h/(2mkT)^{1/2} \qquad\qquad \textbf{17.22}$$

in the range 1 to 2 Å. The beam can be 'monochromatized' by reflecting it off a crystal set at the Bragg angle for the required wavelength.

17.9 Electron microscopy

Many of the principles of electron diffraction have been exploited to make the *electron microscope* one of the most valuable instruments available to the metallurgist. The microscope can be used in many different ways (e.g. for

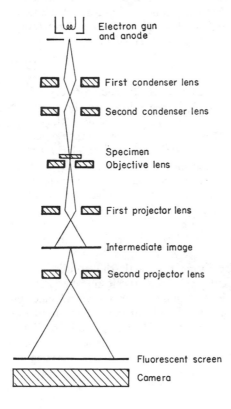

Fig. 17.33 Electron microscope

obtaining electron diffraction patterns from selected small areas of specimens; for examining surface structures as imprinted in replica films of plastic, stripped off the surface) but the most useful general technique involves the *transmission* of the electron beam through a thin foil of the specimen, usually prepared by electrolytic polishing. Structural details down to a size of about 10 Å can be observed and the microscope is able to provide direct observation of dislocations, stacking faults, grain boundary structures, fine precipitates, centres of strain and small clusters of point defects. Under special conditions, e.g. by superposing a film of one crystal lattice on another to obtain a moiré effect, it is even possible to observe the lattice structure.

Fig. 17.33 shows the arrangement inside the vacuum chamber of an electron microscope. The lenses are magnetic, i.e. D.C. coils in soft iron, and their focal lengths can be changed by regulating the current through the coil. The two condenser lenses collimate the beam, which passes through

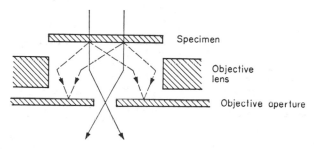

Fig. 17.34 Removal of diffracted beams to form a bright-field image

the specimen and the objective lens to form a first image, magnified about × 40, in the object plane of the first projector lens. A small area from this image is then projected as an intermediate image, again magnified about × 40, by the first projector lens. A small area of this is then projected on to the fluorescent viewing screen, with a further magnification of about × 50, by the second projector lens. Total magnifications in the range × 20 000 to × 100 000 are attained.

The most useful of the several techniques of thin-film transmission electron microscopy is the *bright-field diffraction contrast technique* shown in Fig. 17.34. The incident beam passes through the specimen and on to the viewing screen but is weakened in intensity by the transfer of some of its electrons to the diffracted beams, shown as broken lines in the diagram. A small aperture at the back focal plane of the objective lens removes these diffracted beams. The amount of intensity lost to the diffracted beams increases sharply as the orientation of the specimen, relative to the beam, approaches the Bragg angle for a prominent set of reflecting planes in the crystal. It follows that when the specimen is set near a Bragg angle, the brightness of its image on the screen varies very sensitively with orientation.

If there are regions in it of slightly different orientations, they will show up dark or bright according as they lie nearer to the Bragg angle or not. It is this *diffraction contrast* that gives the picture. A simple effect seen in this way is *bending* of the foil (e.g. due to buckling by heating in the electron beam). This produces *extinction contours*, i.e. dark bands across the image which indicate regions in the specimen where the orientation of the bent crystal passes through a Bragg angle.

Dislocation lines are made visible by this effect. Fig. 17.35(a) shows a simple edge dislocation. The rotation of the crystal on either side of it provides the necessary local variations of orientation for diffraction contrast and in a critically oriented foil such a dislocation becomes clearly visible in the image (Plate 3B). The line seen in the image generally lies to one side or the other of the dislocation in the crystal, according to which side of the dislocation is providing the diffraction contrast. In a bright field the dis-

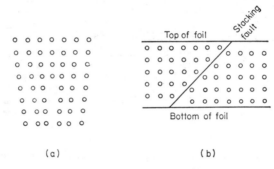

(a)   (b)

Fig. 17.35 Atomic arrangement round an edge dislocation (a) and a stacking fault (b)

location appears as a dark line. Where the dislocation line in the crystal crosses an extinction contour the dislocation line in the image then appears bright.

Diffraction contrast also enables *stacking faults* to be seen. The atoms on one side of such a fault are offset, relative to those on the other side, and the mutual interference of rays reflected from the two sides produces the contrast in the image. A fault which is inclined to the plane of the foil, as in Fig. 17.35(b), shows itself as a set of alternating light and dark *fringes* which run parallel to the line of intersection of the fault with the surface of the foil.

A dislocation or stacking fault cannot always be seen in the image. Certain conditions have to be satisfied between the Burgers vector of the dislocation, or the vector which describes the displacement across the stacking fault, and the vectors which describe the diffraction process. A detailed theory of diffraction contrast from lattice defects has now been built up, with the aid of which it is possible to gain detailed information about dislocations, stacking faults, etc. from electron microscope pictures.

17.10   Field-ion microscopy

An entirely different technique, *field-ion microscopy*, has recently been developed for looking at crystal structures. It has the advantage of being able to resolve down to 2 or 3 Å, so that the individual atoms themselves can be clearly seen.

Fig. 17.36 shows a diagram of the microscope. The specimen consists of a short fine wire, electrolytically polished at one end to a sharp tip about 100 atoms radius. It is mounted on a filament attached to electrodes, which are cooled by liquid or solid N_2, H_2 or He, in a vacuum chamber, with its sharp tip pointing towards a fluorescent viewing screen a few centimetres

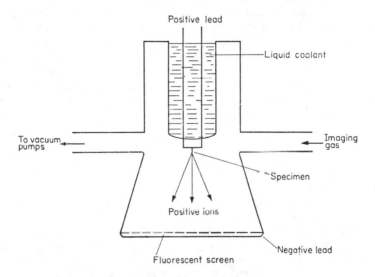

Fig. 17.36 Field-ion microscope

away. The specimen is positively charged, relative to the screen, to about 10 000 volts. Lines of electrostatic force thus radiate out from the atoms of the tip to corresponding points on the screen. These lines radiate out in a regular manner (approximately as straight lines) so that the pattern of points where they reach the screen corresponds closely to the pattern of atoms on the tip. The geometry is such that lines of force which start from neighbouring atoms are diverged to a spacing of about 1 to 2 mm where they reach the screen. This divergence is responsible for the millionfold magnification which is achieved by the microscope.

The image of the tip is carried to the screen by charged ions which travel along the lines of force and activate the phosphor when they reach the screen. These ions are produced from an *imaging gas*, such as helium or neon, which is allowed to leak into the vacuum chamber in small amounts. These gas atoms, while still electrically neutral, are *polarized* by the strong electrical field near the tip and are pulled down on to the metal surface there, where they bounce about, eventually with a thermal energy kT corresponding to the temperature of the tip. When such an atom happens to attain a critical distance, just above one of the metal atoms of the tip which is in a rather exposed position and hence carries a particularly strong positive charge, this gas atom becomes ionized by giving up one of its electrons to the tip. Now a positive ion, it is immediately repelled from the tip and flies down its line of force to the screen. The total image produced continuously by such ionized gas atoms streaming down the various lines of force is faintly visible on the screen and can be photographed.

What we see on the screen is a pattern of bright spots (Plate 4) each of which is the continuous image produced by all the gas atoms ionized from a single atom of the metal tip and so, in a sense, is a 'picture' of this single atom. There are two features to notice about the pattern. First, not all of the atoms of the tip give bright images. Those in a close-packed surface layer are crowded together too closely to act as strong ionization centres. Such regions of the tip appear dark. The atoms in exposed positions, particularly at the edges of crystal planes, ionize best and give the brightest images. Second, the image is a flat projection of a hemispherical surface cut through the crystal. In some places this surface is built up from a stack of parallel planes, roughly as concentric circular discs parallel to the surface, one disc for each atomic plane cut by the hemisphere. In other places the surface appears as a mosaic of small facets of crystal planes which happen to lie locally parallel to the surface. The atomic structure of the crystal can be clearly seen in some of these facets.

Field-ion microscopy is particularly valuable for observing lattice defects and for studying their atomic structures. Vacant atomic sites, grain boundaries and dislocations have all been studied and information gained about their detailed structures (Plate 5). The development of superlattice structures in alloys is also a very suitable subject for the technique.

FURTHER READING

AMELINCKX, S. (1964). *The Direct Observation of Dislocations*. Academic Press London.

BACON, G. E. (1955). *Neutron Diffraction*. Oxford University Press, London.

CHARLES BARRETT and MASSALSKI, T. B. (1966). *Structure of Metals*. McGraw-Hill, New York and Maidenhead.

BRAGG, W. H. and W. L. (1933). *The Crystalline State*. Bell, London.

BROWN, P. J. and FORSYTH, J. B. (1973). *The Crystal Structure of Solids*. Edward Arnold, London.

BUERGER, M. J. (1956). *Elementary Crystallography*. John Wiley, New York and London.

CAHN, R. W. (1965). *Physical Metallurgy*. North Holland Publishing Co., Amsterdam.

COTTRELL, A. H. (1953). *Dislocations and Plastic Flow in Crystals*. Oxford University Press, London.

CULLITY, B. D. (1956). *Elements of X-ray Diffraction*. Addison-Wesley, London.

FRIEDEL, J. (1964). *Dislocations*. Pergamon Press, Oxford.

GOMER, R. (1961). *Field Emission and Field Ionization*. Oxford University Press, London.

GRUNDY, P. T. and JONES, G. A. (1975), *Electron Microscopy in the Study of Materials*, Edward Arnold, London.

HIRSCH, P. B., HOWIE, A., NICHOLSON, R. B., PASHLEY, D. W. and WHELAN, M. J. (1965). *Electron Microscopy of Thin Crystals*. Butterworth, London.

HULL, D. (1965). *Introduction to Dislocations*. Pergamon, Oxford.

MCKIE, D. and MCKIE, C. (1974). *Crystalline Solids*. Nelson, London.

PHILLIPS, F. C. (1949). *Introduction to Crystallography*. Longmans Green, London.

PINSKER, Z. G. (1953). *Electron Diffraction*. Butterworth, London.

READ, W. T. (1953). *Dislocations in Crystals*. McGraw-Hill, New York and Maidenhead.

SCHMID, E. and BOAS, W. (1950). *Plasticity of Crystals*. Hughes, Wrexham.

SMALLMAN, R. (1970). *Modern Physical Metallurgy*. Butterworth, London.

Chapter 18

Metal Crystals—II Directionality

18.1 Rotational symmetry

Many properties of crystals are *directional* in character and vary with direction of measurement in accordance with the directionality of the crystal structure itself. The extent of this *anisotropy* is limited, partly by the nature of the particular property concerned and partly by the symmetry of the crystal structure. Some properties are necessarily *isotropic* in cubic crystals. Others vary within limits set by the necessity to repeat the same values along all directions of the same crystallographic type.

We are now concerned not with the translational symmetry of a crystal but its *rotational symmetry*. Fig. 18.1 shows some symmetry axes in a cube. The *tetrad* axis has *four-fold* symmetry, i.e. when the cube is rotated about this axis it coincides with its initial self every 90°. There are three tetrad axes, along the three ⟨100⟩ directions. The cube also has four *triad* axes, along ⟨111⟩ directions, which have *three-fold* symmetry, and six *diad* axes, along ⟨110⟩ directions, which have *two-fold* symmetry. Fig. 18.1 also shows symmetry axes in a hexagon. The *hexad* axis, perpendicular to the basal plane, has *six-fold* symmetry. The diad axes fall into two groups of three each, through the centres and edges of the side faces.

18.2 Stereographic projection

It is useful to have a systematic method for displaying the planes, directions and rotational symmetry of a crystal in a flat diagram. A field-ion picture of course does this naturally, though with some distortion. The most useful geometrical method is by *stereographic projection*. An infinitesimal crystal is assumed to lie at the centre of a sphere, as in Fig. 18.2(a). A plane in the crystal can then be represented on the sphere, either by the *great circle C* which traces the intersection of the extended plane with the sphere, or by the *pole P* where the perpendicular from the centre of the plane meets the sphere. The angle between two planes or directions can be measured in terms of the distance along a great circle between the corresponding poles.

The information on the reference sphere is then projected on to a *basic circle*, as in diagram (b). For this we may think of the sphere as transparent, with a source of light S at the 'south pole' which casts shadows P' of points such as P on to a tangent plane at the 'north pole' N. The stereographic projection is the information thus presented within the basic circle.

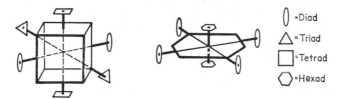

() = Diad

△ = Triad

□ = Tetrad

⟨⟩ = Hexad

Fig. 18.1 Diad, triad and tetrad axes of a cube and diad and hexad axes of a hexagon

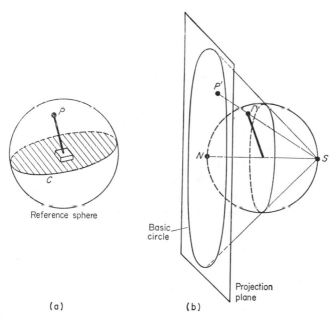

(a) (b)

Fig. 18.2 Stereographic projection (a) reference sphere showing a great circle C which traces a crystal plane, together with the pole P of the plane; (b) projection of the reference sphere on to a basic circle, with corresponding projection of the pole P to P'

This projection has some important geometrical properties. It preserves angular truth; the angle at which two great circles intersect on the reference sphere is preserved between the traces of these circles on the projection plane. Small circles on the sphere project as circles on the stereogram (although with radially displaced centres). Great circles project as circles which cut the basic circle at opposite ends of a diameter. Great circles through the north pole N project of course as diameters.

To measure the angular 'distance' between two points on the stereogram, we construct the great circle through them and then measure the distance between them along this circle. This is conveniently done with the help of a *Wulff net*, i.e. a transparent stereogram of all great circles (e.g. at 2° intervals) which pass through a given pair of poles at opposite ends of a diameter, each circle being marked at suitable intervals (e.g. 2°) along its length to indicate the angular distance. The net is laid on the stereogram, centre to centre, and then rotated about the centre, round the basic circle,

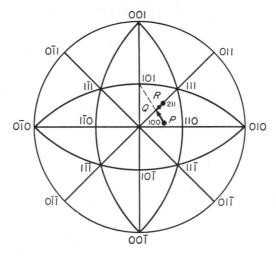

Fig. 18.3 Standard projection of a cubic crystal

until the two points under study fall on the *same* great circle of the net. The angle between them is then read off directly from the graduations along this circle.

Various geometrical methods have been developed for the determination, on the stereogram, of the crystallographic orientations of the planes of twin crystals, Widmanstätten structures, slip bands, stacking faults, etc. from measured traces of these features on metallographic specimens. These methods are very useful in advanced metallographic work. For details, the reader should consult standard texts (e.g. *C. S. Barrett, Structure of Metals*). In many problems, *standard projections* are used.

Fig. 18.3 shows the standard projection of a cubic crystal, viewed along the [100] axis. The plane of the basic circle is parallel to (100) and the vertical and horizontal diameters are traces of the (010) and (001) planes, respectively, at an angular distance of 90° from the [010] and [001] poles.

The inclined diameters similarly represent the (011) and (0$\bar{1}$1) planes and the four remaining great circles denote the other {110} planes.

The rotational symmetry of the crystal lattice is clear from this standard projection. Because of this symmetry the diagram is divided into three-sided figures, called *unit 'triangles'*, which are all equivalent except for their different orientations, each being bounded by one {100} line and two {110} lines. In many problems it is sufficient to construct and use only one unit triangle, e.g. that with [100], [110] and [111] poles at its corners. If we were considering, for example, the orientation of a metal single crystal in the form of a wire, we could indicate the axis of the wire in this unit triangle by the position of a point such as P in Fig. 18.3. We note from eqn. 17.7 that the angular distances along the three sides of this triangle are [100]⌒[110] = 45°, [100]⌒[111] = 54° 44' and [110]⌒[111] = 35° 16'.

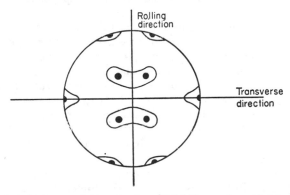

Fig. 18.4 Pole figure of distribution of [111] axes in rolled brass sheet

To show any *preferred orientations or textures* which may exist in a polycrystalline metal it is usual to use not a standard projection but a *pole figure*. In this the stereogram is aligned to the overall geometrical features of the specimen (e.g. basic circle parallel to the plane of a rolled metal sheet) and the numbers of grains in various ranges of orientation are then indicated in it by a series of contour lines or by density of shading. The experimental information is usually obtained from the relative intensities of x-ray reflections from the polycrystal at various angular settings. Fig. 18.4 shows an example, the distribution of [111] crystal axes in heavily rolled alpha-brass sheet. This corresponds approximately to a (110) [112] texture, i.e. with (110) planes parallel to the rolling plane and [112] directions parallel to the rolling direction.

18.3 Anisotropy

The extent to which a property such as electrical conductivity varies with the direction of measurement in a crystal depends on the nature of the property itself and on the rotational symmetry of the crystal. Some properties, called *scalars*, are independent of direction, e.g. mass, volume, density, enthalpy, entropy. Others are more complicated, because they are related to quantities that are in themselves directional.

Many properties are measured as *coefficients*. We impose some condition on a crystal, e.g. apply a load, a voltage difference or a temperature difference, and then measure the response of the crystal to this stimulus, e.g. deformation, or flow of electricity or heat. In many cases the intensity of the response is directly proportional to that of the stimulus, as for example in Hooke's law, Ohm's law and Fourier's law. *A coefficient of proportionality* is then defined through a relation of the type

$$\text{response} = \text{coefficient} \times \text{stimulus} \qquad \textbf{18.1}$$

Many important properties, e.g. elastic moduli, electrical and thermal conductivity, coefficients of diffusion and thermal expansion, are defined in this way.

There are then various possibilities, according to the scalar or directional characters of the stimulus and the response. In the simplest cases they are both scalars, e.g. heat and temperature, and the coefficient which connects them, e.g. specific heat, is then also a scalar.

In many cases the stimulus and response are both vector quantities, e.g. temperature gradient with components $\partial T/\partial x$, $\partial T/\partial y$, $\partial T/\partial z$, along the three coordinate axes, and heat flow also with components along these axes. The link between them is then of a *vector-vector* type and the coefficient expressing this link has to take this into account. Let us consider *linear flows*, e.g. flows of electricity, heat or matter (i.e. diffusion) down gradients of potential, temperature or concentration. Let these gradients be $\partial\theta/\partial x_1$, $\partial\theta/\partial x_2$, $\partial\theta/\partial x_3$, where θ stands for potential, temperature or concentration, and x_1, x_2, x_3 stand for the coordinates x, y, z; and let I_1, I_2, I_3 be the components of *current density* (of electricity, heat or diffusion) along these axes. To link the current to the gradient we have to write three equations, one for each component of current, and in each equation we must make the current proportional to all three components of the gradient. These equations are

$$I_1 = k_{11}\frac{\partial\theta}{\partial x_1} + k_{12}\frac{\partial\theta}{\partial x_2} + k_{13}\frac{\partial\theta}{\partial x_3}$$

$$I_2 = k_{21}\frac{\partial\theta}{\partial x_1} + k_{22}\frac{\partial\theta}{\partial x_2} + k_{23}\frac{\partial\theta}{\partial x_3} \qquad \textbf{18.2}$$

$$I_3 = k_{31}\frac{\partial\theta}{\partial x_1} + k_{32}\frac{\partial\theta}{\partial x_2} + k_{33}\frac{\partial\theta}{\partial x_3}$$

The coefficient of proportionality thus has nine distinct components, $k_{11}\ldots k_{33}$, referred to collectively as the *conductivity tensor* k_{ij}, where $i = 1, 2, 3$ and $j = 1, 2, 3$. The three equations can then be written as

$$I_i = \sum_j k_{ij}\frac{\partial\theta}{\partial x_j} \qquad \textbf{18.3}$$

or simply as

$$I_i = k_{ij} \frac{\partial \theta}{\partial x_j} \qquad\qquad \text{18.4}$$

if we make the rule that, when a suffix (i.e. j) appears twice in one term (in this case once on k and once on x), that term stands for the sum of all terms produced by substituting all the various values that the suffix can have. Thus, when $i = 1$ eqn. 18.4 represents the first of eqns. 18.2.

Fortunately, crystal symmetry and various other effects make many of the tensor components either equal to one another, or to zero, so that the measurement and specification of such coefficients is usually simpler than we might expect at first sight. For example, it can be proved that the conductivity tensor is *symmetrical*, i.e.

$$k_{ij} = k_{ji} \qquad\qquad \text{18.5}$$

which reduces the number of independent components to six. In cubic crystals, symmetry requires that the physical properties of the material shall be the same along the three cubic axes. The effect of this is to make $k_{ij} = $ constant when $i = j$ and $k_{ij} = 0$ when $i \neq j$. Such crystals are thus *isotropic* in their conductivities and diffusivity. Only one value is required for each conductivity coefficient. This does not apply, however, to crystals of lower symmetry. A hexagonal crystal, for example, has two conductivity coefficients, one for directions in the basal plane, the other for conduction along the hexagonal axis.

In a more general use of the word *tensor*, a scalar is a *zero-order tensor*, a vector is a *first-order* tensor and the conductivity tensor is *second-order*. Tensors of higher order are also possible; we shall see an example in § 18.6. The number of components in an nth order tensor is 3^n although, as we have seen in the case of the conductivity tensor, various factors can reduce the actual number of independent components to much less than this. We notice that the second-order conductivity tensor is the link between two first-order tensors. This is an example of a general effect; the link between an mth-order tensor and an nth-order tensor is an $(m + n)$th-order tensor.

18.4   Strain

Stimuli and responses may themselves be tensors of higher order than first. An important example of a response that is a symmetrical second-order tensor is *strain*. Consider first the problem in two dimensions. Suppose that we have a flat thin sheet of rubber and inscribe on it a grid of small unit squares. Then deform the sheet by pulling its edges. We may then find one of these squares typically displaced and deformed, as in Fig. 18.5, from $OABC$ to $PQRS$.

We can think of the overall change as occurring in three stages. First, a pure *translation* of the square, as a rigid body, by the vector OP. There is no strain in the square at this stage. Hence, so far as strain is concerned, we can redefine the coordinate axes as X_1' and X_2', to allow for the effect of this rigid-body translation. In the second stage, the square *rotates* as a rigid body about the new origin P. Again there is no strain from this and so we rotate the axes accordingly, to X_1'' and X_2'', to allow for this rigid-body rotation. Finally, the square in this position deforms to the shape $PQRS$. Because the square is small the deformation in it is *uniform* and $PQRS$ is a parallelogram. Because there is no rigid-body rotation in this final stage, then $\widehat{UPQ} = \widehat{SPV}$.

We shall suppose that the strains are small enough to be regarded as *infinitesimal*. This simplifies the definition of the strain components. For the stretching of the material along the coordinate axes we define, remembering that $OA = OC = 1$, the *tensile strains* or *unit elongations* $\epsilon_{11} = PU$

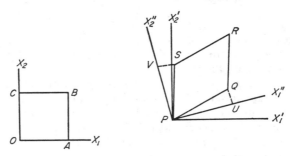

Fig. 18.5 Displacement and deformation of a square

$- OA$ and $\epsilon_{22} = PV - OC$, where QU and SV are perpendiculars to the coordinate axes. For the *shearing* of the material we define the *tensor shear strains* $\epsilon_{12} = \widehat{SPV}$ and $\epsilon_{21} = \widehat{UPQ}$, and also their sum, the *angle of shear*

$$\gamma_{12} = \gamma_{21} = \epsilon_{12} + \epsilon_{21} = 90° - \widehat{QPS} \qquad \textbf{18.6}$$

In all then, the two-dimensional strain tensor has the components

$$\begin{matrix} \epsilon_{11}, & \epsilon_{12} \\ \epsilon_{21}, & \epsilon_{22} \end{matrix} \qquad \textbf{18.7}$$

and may be written as ϵ_{ij} where $i = 1, 2$, and $j = 1, 2$. It is symmetrical, i.e. $\epsilon_{12} = \epsilon_{21}$ and so has three independent components. We note (a) that a component ϵ_{ij} refers to the displacement in the direction parallel to axis i of a point initially on the axis j; (b) that a strain is a *pure number* so that there is no need to write units such as 'inches per inch' after it.

The extension of this definition of strain to three dimensions is now straightforward. The strain tensor has nine components

$$\begin{matrix} \epsilon_{11}, & \epsilon_{12}, & \epsilon_{13} \\ \epsilon_{21}, & \epsilon_{22}, & \epsilon_{23} \\ \epsilon_{31}, & \epsilon_{32}, & \epsilon_{33} \end{matrix} \qquad \textbf{18.8}$$

but only six are independent because it is symmetrical, i.e.

$$\epsilon_{ij} = \epsilon_{ji} \qquad\qquad \textbf{18.9}$$

If we give different non-zero values to all six components we have produced the most general kind of uniform strain that is possible. Such a deformation changes the volume of a unit cube and also changes the shape to that of a parallelepiped. It can be proved that there is always a simple representation of such a strain in *principal axes*. Suppose, for example, that we had scribed small *circles* on the rubber sheet. These would be deformed into ellipses. The minor and major axes of these ellipses are then the principal axes. A small square parallel to these axes deforms to a rectangle, without rotation. Similarly, in three dimensions a cube parallel to the principal axes deforms to a rectangular prism, without rotation. In principal axes there are only three strain components to consider, ϵ_{11}, ϵ_{22}, ϵ_{33}, the others being zero. This does not mean of course that there is no shear strain in the material. The shear is easily seen from the distortion of a diamond inscribed in the unit square or cube.

All the various simpler types of strain can be produced from the general strain tensor by setting some of the components either to zero or equal to one another. Suppose that all are zero except ϵ_{11}. This represents a simple tensile strain along the X_1 axis. It is *not* what we observe, however, when we stretch a metal wire by pulling at its ends because in this case there is usually also a *lateral* contraction (the *Poisson's ratio* effect). In a mechanically isotropic material the strain tensor in this latter case is of the form

$$\begin{matrix} e & 0 & 0 \\ 0 & -ve & 0 \\ 0 & 0 & -ve \end{matrix} \qquad\qquad \textbf{18.10}$$

where e is the tensile strain along the X_1 axis, $-ve$ is the lateral strain along each transverse axis X_2 and X_3) and v is *Poisson's ratio*.

Consider next the deformation $\epsilon_{11} = \epsilon_{22} = \epsilon_{33}$, with all other components equal to zero. This is a *pure dilatation*, i.e. change of volume without change of shape. Since the strains are infinitesimal we can write this as

$$\Theta = \frac{\Delta V}{V} = \epsilon_{11} + \epsilon_{22} + \epsilon_{33} \qquad\qquad \textbf{18.11}$$

This also defines the dilatational component of the general strain tensor ϵ_{ij}. On average, a part $\frac{1}{3}\Theta$ of each tensile strain component thus belongs to the dilatation. The remaining strain components, i.e.

$$\begin{matrix} \epsilon_{11} - \tfrac{1}{3}\Theta & \epsilon_{12} & \epsilon_{13} \\ \epsilon_{21} & \epsilon_{22} - \tfrac{1}{3}\Theta & \epsilon_{23} \\ \epsilon_{31} & \epsilon_{32} & \epsilon_{33} - \tfrac{1}{3}\Theta \end{matrix} \qquad\qquad \textbf{18.12}$$

thus represent the pure shear strains in the material. They are referred to collectively as the *deviatoric strain*.

The simplest stimuli which produce strain are the scalars: *temperature T* and *hydrostatic pressure p*. Strictly, hydrostatic pressure is a particular scalar form of the general stress tensor (cf. § 18.5), just as dilatation is a particular scalar form of the general strain tensor.

Strain is related to temperature by the *coefficient of thermal expansion* α_{ij},

$$\epsilon_{ij} = \alpha_{ij}(T - T_0)   \qquad 18.13$$

and this is a second-order tensor since it links a second-order tensor to a scalar. A sphere inscribed in a body would thus in general change to an ellipsoid when the temperature is changed. However, in a cubic crystal the symmetry requires the thermal strain to be the same along the three cubic axes. Since an ellipsoid with three equal axes is a sphere, thermal expansion is isotropic in cubic crystals. This is of great practical importance for the behaviour of such materials when heated and cooled as polycrystals. Since all the grains expand or contract isotropically and equally, no *thermal stresses* are developed between them when the temperature is uniform. In crystals of lower symmetry, e.g. hexagonal or tetragonal, the thermal expansion no longer need be isotropic. In general, severe misfits develop between the grains due to changes of shape as they are heated or cooled. These misfit strains weaken the material by producing internal stresses, localized plastic deformations and cracks.

The strains produced by hydrostatic pressure have the same tensor characteristics as thermal strains. Cubic crystals deform isotropically, but not others. The simplest response to pressure is a dilatation, through the relation

$$p = -K\Theta   \qquad 18.14$$

where K is the *bulk modulus of elasticity*. This is a particular form of Hooke's law.

18.5 Stress

The most important stimulus of strain in engineering metals is of course *stress*, the distribution of internal forces produced through a material by applied forces. To define the state of stress at a point in a body we take an infinitesimal cube of material enclosing that point and consider the forces applied to the faces of this cube by the material round it. We need consider only three perpendicular faces since the forces on the other three are equal and opposite. The force F on a face is then divided by the area A of the face to convert it to a stress F/A. This stress can be represented by a vector which points out of the face, usually at some oblique angle. We then resolve this vector into its components parallel to the edges of the cube, as in Fig. 18.6. These are the *stress components*.

The stress at a point is in fact a second-order tensor with nine components, as in the Figure. The *normal stress* components σ_{11}, σ_{22} and σ_{33} are *tensile* (positive) or *compressive* (negative) stresses. The others are shear stresses. The shear stresses exert couples on the cube and, in order that these couples should not rotate the cube about its axes, it is necessary that the couples on conjugate pairs of faces should balance, i.e.

$$\sigma_{ij} = \sigma_{ji} \qquad\qquad\qquad \textbf{18.15}$$

so that the stress tensor, like that of strain, is symmetrical and has only six independent components.

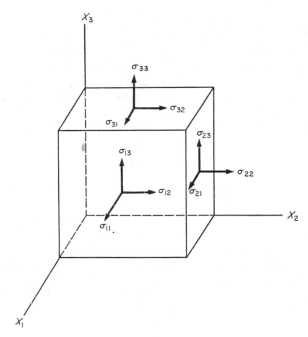

Fig. 18.6 Stress components

There are in fact many similarities with strain. Principal axes can be found (which coincide with those of strain in mechanically isotropic materials) in which only the normal stresses appear, those with $i \neq j$ then vanishing. Simple uniaxial tensile stress is obtained by setting to zero all components except σ_{11} and this stimulates the strain response 18.10. Pure hydrostatic pressure p is represented by

$$\sigma_{11} = \sigma_{22} = \sigma_{33} = -p \qquad\qquad \textbf{18.16}$$

with all other components zero. In a general state of stress we define the hydrostatic component as

$$p = -\tfrac{1}{3}(\sigma_{11} + \sigma_{22} + \sigma_{33}) \qquad\qquad \textbf{18.17}$$

and hence the deviatoric stresses as

$$
\begin{matrix}
\sigma_{11} + p & \sigma_{12} & \sigma_{13} \\
\sigma_{21} & \sigma_{22} + p & \sigma_{23} \\
\sigma_{31} & \sigma_{32} & \sigma_{33} + p
\end{matrix}
\qquad \textbf{18.18}
$$

We notice that the sum of the deviatoric normal stresses is always zero.

18.6   Hooke's law

In the limit of low stresses, the displacements of the atoms with respect to their neighbours are so small that the strain response satisfies eqn. 18.1. In this case the equation becomes *Hooke's law of elasticity*. To express this law in complete generality we have to write each strain component as a linear function of all six stress components. To define the *elastic constants* of the material we shall in fact do the reverse, i.e. write each stress component in terms of all the strain components. To avoid unwieldy expressions the notation used in eqn. 18.4 now becomes necessary. Hooke's law is then the *six* equations

$$
\sigma_{pq} = c_{ijpq}\epsilon_{ij}
\qquad \textbf{18.19}
$$

given by $p = 1, 2, 3$ and $q = 1, 2, 3$ (with $\sigma_{pq} = \sigma_{qp}$) with the rule that we must sum all the terms generated by the repeated suffixes i and j. The 81 coefficients c_{ijpq} represent the *elastic constant tensor*, which is a fourth-order tensor since it links two second-order tensors. The number of independent components is however much less than this. The symmetry of the stress and strain tensors immediately reduces the 81 to 36. Considerations of elastic energy reduce this further to 21 and crystal symmetry then brings in other substantial reductions. These symmetry effects allow us to simplify the notation as follows:

Old: σ_{11} σ_{22} σ_{33} σ_{23} σ_{31} σ_{12} ϵ_{11} ϵ_{22} ϵ_{33} $2\epsilon_{23}$ $2\epsilon_{31}$ $2\epsilon_{12}$

New: σ_1 σ_2 σ_3 σ_4 σ_5 σ_6 ϵ_1 ϵ_2 ϵ_3 γ_4 γ_5 γ_6

In terms of this, Hooke's law becomes

$$
\begin{aligned}
\sigma_1 &= c_{11}\epsilon_1 + c_{12}\epsilon_2 + c_{13}\epsilon_3 + c_{14}\gamma_4 + c_{15}\gamma_5 + c_{16}\gamma_6 \\
\sigma_2 &= c_{21}\epsilon_1 + c_{22}\epsilon_2 + c_{23}\epsilon_3 + c_{24}\gamma_4 + c_{25}\gamma_5 + c_{26}\gamma_6 \\
\sigma_3 &= c_{31}\epsilon_1 + c_{32}\epsilon_2 + c_{33}\epsilon_3 + c_{34}\gamma_4 + c_{35}\gamma_5 + c_{36}\gamma_6 \\
\sigma_4 &= c_{41}\epsilon_1 + c_{42}\epsilon_2 + c_{43}\epsilon_3 + c_{44}\gamma_4 + c_{45}\gamma_5 + c_{46}\gamma_6 \\
\sigma_5 &= c_{51}\epsilon_1 + c_{52}\epsilon_2 + c_{53}\epsilon_3 + c_{54}\gamma_4 + c_{55}\gamma_5 + c_{56}\gamma_6 \\
\sigma_6 &= c_{61}\epsilon_1 + c_{62}\epsilon_2 + c_{63}\epsilon_3 + c_{64}\gamma_4 + c_{65}\gamma_5 + c_{66}\gamma_6
\end{aligned}
\qquad \textbf{18.20}
$$

When the symmetry of the crystal is taken into account, many of these elastic constants become either equal or zero. In *cubic crystals* only c_{11}, c_{12} and c_{44} need to be known because

$$c_{11} = c_{22} = c_{33}$$

$$c_{12} = c_{21} = c_{13} = c_{31} = c_{23} = c_{32}$$

$$c_{44} = c_{55} = c_{66}$$

18.21

and all others are zero. In *hexagonal crystals* five constants are needed, c_{11}, c_{12}, c_{13}, c_{33} and c_{44}. The elastic constants of many crystals have now been measured. Some values are given in Table 18.1.

TABLE 18.1.   ELASTIC CONSTANTS OF CUBIC CRYSTALS
(10^{11} N m^{-2})*

	c_{11}	c_{12}	c_{44}	$2c_{44}/(c_{11}-c_{12})$
Aluminium	1·08	0·622	0·284	1·23
Copper	1·70	1·23	0·753	3·3
Gold	1·86	1·57	0·42	3·9
Iron (B.C.C.)	2·37	1·41	1·16	2·4
Lead	0·48	0·41	0·144	3·9
Molybdenum	4·6	1·79	1·09	0·77
Nickel	2·50	1·60	1·185	2·6
Potassium	0·046	0·037	0·026	6·3
Sodium	0·06	0·046	0·059	7·5
Silver	1·20	0·897	0·436	2·9
Tungsten	5·01	1·98	1·51	1·0
β-brass	0·52	0·275	1·73	18·7
Diamond	9·2	3·9	4·3	1·6
Sodium chloride	0·49	0·124	0·126	0·7

(* 10^5 N m^{-2} = 14·5 pound inch^{-2})

A cubic crystal is *elastically isotropic* when

$$2c_{44} = c_{11} - c_{12}$$

18.22

We see from the Table that most metals are far from isotropic. This in fact is a natural consequence of their free electron structure. The most 'ideal' metals from the free electron point of view, i.e. the alkali metals, are generally the most anisotropic. The exceptional anisotropy of β-brass is related to the fact that the B.C.C. structure in this alloy is almost mechanically unstable.

Most problems of engineering elasticity are tackled on the assumption that the material is elastically isotropic, which is not unreasonable for polycrystalline metals, provided these are not strongly textured, since the individual anisotropies of the grains then average out in the bulk. Various other elastic constants are then used, in place of the c_{ij}. These are: *Young's modulus E*, the ratio of tensile stress to tensile strain in uniaxial tension;

the *shear* or *rigidity modulus* μ, the ratio of shear stress to angle of shear, cf. eqn. 17.2, usually measured in a torsion experiment; the *bulk modulus* K, defined in eqn. 18.14; and *Poisson's ratio* ν, the ratio of lateral contraction to longitudinal extension in uniaxial tension. These constants and the c_{ij} (*for isotropic crystals*) are all inter-related, e.g.

$$K = \frac{E}{2(1-2\nu)}, \qquad \mu = \frac{E}{2(1+\nu)}, \qquad E = \frac{9K\mu}{3K+\mu} \qquad \textbf{18.23}$$

$$c_{11} = K + \tfrac{4}{3}\mu, \qquad c_{12} = K - \tfrac{2}{3}\mu, \qquad c_{44} = \mu$$

Table 18.1 shows that in most anisotropic cubic metals $2c_{44} > (c_{11} - c_{12})$. Young's modulus is then greatest along $\langle 111 \rangle$ axes. Tungsten is isotropic and aluminium nearly so. Molybdenum is unusual in having the reverse anisotropy; its Young's modulus is greatest along $\langle 100 \rangle$ directions. Table 18.2 gives average elastic constants of quasi-isotropic polycrystalline metals.

TABLE 18.2. AVERAGE ELASTIC CONSTANTS OF POLYCRYSTALLINE METALS
(10^{11} N m^{-2})

	Ag	Al	Au	Be	Bi	Cd	Cu	Fe
E:	0·79	0·71	0·80	3·0	0·32	0·5	1·23	2·10
μ:	0·29	0·26	0·28	1·5	0·12	0·2	0·45	0·83
K:	1·09	0·75	1·66	1·5	0·31	0·4	1·31	1·60
ν:	0·38	0·34	0·42	0·08	0·33	0·29	0·35	0·28

	Mg	Ni	Pb	Pd	Pt	Sn	W	Zn
E:	0·45	2·0	0·16	1·13	1·68	0·54	3·9	0·9
μ:	0·17	0·77	0·06	0·51	0·61	0·20	1·5	0·4
K:	0·45	1·76	0·50	1·76	2·47	0·52	3·0	0·9
ν:	0·33	0·31	0·45	0·39	0·38	0·36	0·28	0·35

18.7 Plastic glide of metal crystals

The plastic properties of metal crystals are markedly anisotropic. Slip occurs along close-packed directions on (mainly) close-packed planes. A given slip system, i.e. plane and direction, becomes active when the resolved shear stress on it reaches a critical value. This is the *law of critical resolved shear stress*. The anisotropy appears in the markedly different values of this stress on different systems.

The usual experimental arrangement is to pull a rod-shaped single crystal of cross-sectional area A, as in Fig. 18.7, and to determine the force F at which slip begins. This can be detected either by a sensitive *extensometer*, to measure the elongation and indicate the deviation from Hooke's law, or by observing the formation of *glide ellipses*, i.e. traces of the slip *steps* formed where the active slip planes meet the cylindrical surface of the specimen. Suppose that the inclined plane in Figure 18.7 is such a plane, of area $A/\cos\theta$, where θ is as shown. The tensile stress F/A due to F can be regarded, in its effect on this plane, as a normal stress along ON of magnitude

Normal stress along $ON = (F\cos\theta)/(A/\cos\theta) = (F/A)\cos^2\theta$ **18.24**

and a shear stress along OS,

Shear stress on $OS = (F\sin\theta)/(A/\cos\theta) = (F/A)\sin\theta\cos\theta$ **18.25**

In general the crystallographic direction of slip does not lie along OS, which is in the plane of ON and OF, but along some direction such as OR. We then have

Shear stress on $OR = (F/A)\sin\theta\cos\theta\cos\psi$ **18.26**

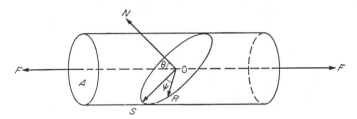

Fig. 18.7 Resolution of stress on an inclined plane

It follows from the law of critical resolved shear stress that the *tensile yield stress* F/A varies with θ and ψ in such a way as to make the shear stress on this particular slip system *independent* of θ and ψ when slip starts on this system.

This has been well established experimentally. Theoretically it means that the dislocations, in their *glide* movements, are driven along only by the resolved shear stress component of the applied stress and that they react to obstacles in their path as if these were regions of unfavourable internal shear stress. This comes from the *work* done through the plastic displacement produced by a moving dislocation. It is convenient to interpret this effect in terms of a certain force acting on the dislocation line and pushing it forward. The formula for this force can be derived quite generally but we shall obtain it only for a geometrically simple situation.

In Fig. 18.8 suppose that a crystal of rectilinear dimensions L_1, L_2, L_3 is subjected to an applied stress with a resolved component σ in the direction of the Burgers vector, of length b, of an edge dislocation line that is moving across the crystal as shown. The applied force on the top face of the crystal is $\sigma L_1 L_2$. A complete traverse by the dislocation moves the top face a distance b in the direction of this force. The work done is then $\sigma b L_1 L_2$. Let us define a force f per unit length, acting on the dislocation line in its direction of motion in such a way that this force does the same work. The dislocation line has a length L_2 so that the total force on it is fL_2. The dislocation moves a distance L_1 in crossing the crystal, so that the work done is fL_1L_2. Hence

$$f = \sigma b \qquad\qquad \textbf{18.27}$$

This is the formula of *Mott and Nabarro*. It is very useful in the dislocation theory of the strength of solids (Chapter 21). The law of resolved shear stress comes from the fact that f depends only on the resolved component σ of the total stress system because no work is got from the other components through the glide of the dislocation.

An important effect of the anisotropy of the slip process is *lattice rotation*. Consider a crystal, e.g. of a hexagonal metal such as zinc, which is slipping

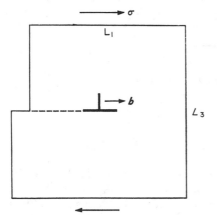

Fig. 18.8 An edge dislocation ($\perp$) driven forward by a shear stress

on only one system. Then, as shown in Fig. 18.9, both the direction and plane of slip rotate towards the tensile axis as the crystal is stretched. This has several effects. A round bar becomes elliptical in cross-section, and eventually ribbon-shaped, since there is no thinning down of diameters perpendicular to the slip direction. Second, the resolved shear stress (for a constant applied load) on the active slip system continually increases as the crystal extends. This is because the area of the slip plane remains fixed at its original value (apart from the usually small effect of the slip steps at the ends of the plane), whereas the resolved component of the applied load continually increases as the plane rotates towards the tensile axis. We note that eqn. 18.26 cannot be used here because it refers only to the *initial* values of θ and ψ. If the critical resolved shear stress of the crystal remains constant during the deformation, then the crystal becomes *softer* in the sense of supporting less tensile load as it stretches. This *geometric softening* leads to a *plastic instability*, i.e. if one region of the bar begins to deform first, all subsequent deformation then concentrates into this region with the result that one part of the bar is severely stretched while the others remain practically undeformed. This happens less often in practice than we might expect because *work hardening* usually occurs and increases the c.r.s.s. sufficiently rapidly to suppress geometrical softening.

The third effect of lattice rotation is the development of preferred orientations in heavily worked metals. Suppose that we have a number of single crystal rods, of initially random orientations, which we then extend as in Fig. 18.9. Their orientations are afterwards no longer random since, in every case, the crystal axis and plane which formed the slip system have become more nearly parallel to the axis of the rod. The development of *deformation textures* in practice is more complicated than this, because of the simultaneous operation of more than one slip system in a grain, but the effect is due essentially to this lattice rotation.

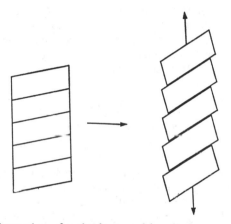

Fig. 18.9 Deformation of a single crystal by slip on one glide system

Consider now the extension of a single crystal rod of an F.C.C. metal of initial orientation P in Fig. 18.3. There are twelve possibilities for slip on {111} planes in ⟨110⟩ directions. Application of eqn. 18.26 to these shows that the one that first becomes active, i.e. the *primary* system, is (11$\bar{1}$) [101]. The (11$\bar{1}$) plane forms a great circle through the [01$\bar{1}$], [101], [1$\bar{1}$0] and [0$\bar{1}\bar{1}$] poles in the stereogram. Other slip systems on which, for the orientation P, the resolved shear stress is fairly high, though lower than that on the primary system, are the *cross-slip system* (1$\bar{1}\bar{1}$) [101], the *conjugate system* (1$\bar{1}$1) [110] and the *critical system* (1$\bar{1}$1) [10$\bar{1}$].

As the crystal extends, the lattice rotation due to slip on the primary system causes the tensile axis to move along a great circle towards the [101] pole, i.e. $P \rightarrow Q$ in Fig. 18.3. At Q however the resolved stresses on the primary and conjugate systems become equal and, if these systems are equally work hardened at this stage, *double slip* then occurs equally and simultaneously on both. In practice this usually happens, at least approximately, which shows that there has been a *latent* work hardening on the inactive system comparable with the work hardening on the active system. Because of the double slip, the tensile axis then moves down the symmetry line along the side of the unit triangle towards the [211] pole, i.e. $Q \rightarrow R$ in Fig. 18.3. In metals and alloys such as α-brass, which form stacking faults fairly easily, latent hardening often occurs so strongly that the full-scale operation of the second slip system is delayed and the tensile axis *overshoots* the symmetry line and then zigzags back and forth across it.

18.8   Plasticity of polycrystals

The ability to deform on several differently oriented systems is vital to the plasticity of polycrystals. Only a limited set of shapes can be produced by the single glide of Fig. 18.9, whereas if a grain in a polycrystal is to remain joined to its neighbours it must be able to change its shape *arbitrarily*. An arbitrary deformation is provided by giving six independent and arbitrary values to the six components of the strain tensor. However, when a metal deforms plastically there is no volume change, to a good approximation, so that

$$\epsilon_{11} + \epsilon_{22} + \epsilon_{33} = 0 \qquad\qquad \textbf{18.28}$$

and hence there are only *five* independent strain components. A condition for polycrystalline plasticity therefore is that the grains should possess at least five independent glide systems. These must be truly independent; the three slip directions in a (111) plane in an F.C.C. crystal, for example, provide only two independent systems since the strain produced by a slip on one of them can also be produced by appropriate slips on the other two.

The cubic metals fully satisfy this five-shear condition and this is the basis of their remarkable, seemingly isotropic, plasticity in polycrystalline form (Plate 6A). The importance of this property can hardly be overemphasized. It has enabled plastic working (cf. Chapter 22) to be developed as a major process for shaping metals and improving their mechanical properties. Without it, metals would have much less value as engineering materials.

It should be noticed that the grain boundaries *per se* play little part in all this. They are merely the interfaces along which the grains meet and mutually influence one another's modes of deformation. Provided that the grains are deforming compatibly with one another, then the dislocations running into a boundary, from the grains on either side of it, mutually annihilate one another within the boundary. Ideally, this annihilation occurs perfectly but in practice, because the dislocations tend to be bunched together in the active slip planes, the annihilation is not perfect and narrow zones of complex deformation, containing tangled dislocations (i.e. *turbulent* deformation) develop near the boundaries.

In hexagonal and other metal crystals of lower symmetry, there is real difficulty because the number of independent glide systems with low critical resolved shear stresses is usually too small to satisfy the five-shear condition. Some help is provided by sliding along the grain boundaries themselves, and by the development of non-uniformities in the deformations of the grains, but there is nevertheless a real danger of brittleness in such metals due to insufficient glide systems at low stresses. The mechanical working of metals such as zinc, magnesium, beryllium, titanium and uranium is made difficult through this effect. At high temperatures however, many additional glide systems become active at low stresses because the thermal agitation enables dislocations in them to move easily despite their large Peierls-Nabarro stresses (cf. § 17.4). This is one of the reasons for *hot working* such metals.

When the five-shear condition is satisfied, each grain deforms to a first approximation like the specimen as a whole, so that the grains become elongated into long thin threads in drawn wires, or flattened into long ribbons in rolled sheet, etc. Without their mutual constraints, however, the grains would nevertheless preferably deform anisotropically. There is a little freedom for them to do this at the surface of the metal since the constraints there are absent on one side. Local differences of deformation, from one grain to another, may then appear on the surface and can give a worked metal an undesirable 'orange peel' appearance if the grain size is large.

Table 18.3 gives some deformation textures commonly developed in

TABLE 18.3.   DEFORMATION TEXTURES IN METALS

	Wire (Fibre Texture)	Sheet (Rolling Texture)
B.C.C.	[110]	(100)[011]
F.C.C.	[111] + [100]	(110)[$\bar{1}$12] + (112)[11$\bar{1}$]
C.P.Hex	[10$\bar{1}$0]	(0001)[11$\bar{2}$0]

heavily worked wires and sheets. In wires, of course, there is only a unique axis but in sheet there is alignment of crystal planes to the plane of rolling and of crystal axes to the direction of rolling. When the stacking fault energy is small, in F.C.C. metals and alloys such as α-brass, the 'double' rolling texture gives way to a 'pure' or 'single' texture of the (110) [$\bar{1}$12] type. It seems that this may be due to difficulty of cross-slip when the dislocations are widely extended (cf. § 17.4).

18.9   Coherent and non-coherent crystal boundaries

The boundary or interface between adjoining crystals may be *coherent*, i.e. there is a continuity of crystal structure across it; or *semi-coherent*, when there is partial continuity; or *non-coherent*, when there is none. The twin interface shown in Fig. 17.4 is an example of a coherent boundary; all the atomic sites in it are common to both crystals joined by it.

Because the atoms in them fit together well, coherent boundaries have low energies. Thus, all boundaries tend to become as coherent as possible. The structures of grain boundaries are largely governed by this principle. Where there is some freedom of choice, a system will generally arrange for the mutual orientations of neighbouring crystals (and even the crystal structures themselves in some cases) to be such as to provide the greatest possible coherency between these crystals. We see this in twinned crystals, in Widmanstätten structures, in martensitic transformations, and in precipitates from supersaturated solid solutions. The *orientation relationships* and *habit planes* of many phases in multi-phase alloys appear to satisfy the principle that the atoms in the adjoining crystals must, at the common boundary, have closely similar patterns and spacings. These crystallographic features of alloy phases are specified by two pieces of information:

(1) A statement of the relative orientations of the crystallographic axes in the adjoining crystals. These orientation relationships are often very simple, certain important crystal planes and directions in the two structures being parallel to each other. For example, when crystals of α-brass form within a β-brass crystal, a (111) plane of α is parallel to a (110) plane of β and a [110] direction of α is parallel to a [111] of β. These relations are usually written in the form:

$$(111)_\alpha \| (110)_\beta \quad \text{and} \quad [110]_\alpha \| [111]_\beta$$

(2) A statement of the *habit* of a crystal, i.e. the shape of the particle and, for plate-shaped or needle-shaped particles, of the crystal planes and axes in the adjoining crystal which are parallel to the plate or needle. The regular pattern of plate-shaped particles of a second phase on certain crystal planes in a matrix crystal gives a Widmanstätten structure. Where particles have plate-shaped habit, the habit plane is always one of good fit between the adjoining crystals.

In polycrystalline single-phase materials we have to consider only differences of orientation, not those of crystal structure. We specify the difference in orientation by the *rotations* which we would have to give to one crystal to bring it into the same orientation as the other. There are three degrees of freedom available for such rotations, corresponding to the three perpendicular axes about which rotations can be made. Having settled the orientation relationship, there are two further degrees of freedom then available for choosing the orientation of the grain boundary itself, regarding this as a plane interface between the crystals.

We can classify the rotations in increasing complexity. The simplest are the *rotational symmetry operations* of § 18.1. These bring the crystal lattice into *self-coincidence*, so that there is no orientation difference and no boundary. Next are *twinning rotations* which bring the crystals into twinned orientations in relation to each other. As we have seen in Fig. 17.4, fully coherent boundaries are possible when the habit plane of the twin coincides with the *twinning plane* (i.e. the plane of mirror reflection between the twinned lattices). Next are the rotations associated with *coincidence lattices*. If we rotate a crystal lattice through itself, about a simple crystal axis, we find that some of the lattice sites come into coincidence with those of the unrotated lattice, at certain angles of rotation, and that the pattern of coinci-

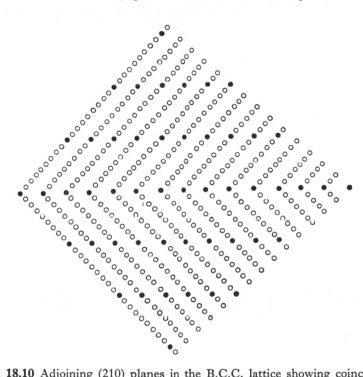

Fig. 18.10 Adjoining (210) planes in the B.C.C. lattice showing coincidence lattice sites (black circles) obtained from a rotation of 96·4°

dent sites at any one of these angles is itself a lattice, the *coincidence lattice*, of large cell dimensions. A simple way to demonstrate this is to make duplicate lattices of holes in black paper, pin the sheets together through one common lattice point, hold them against the light and rotate into positions in which various other points are seen to come into coincidence.

Most coincidence lattices do not have high densities of coincident sites so that coherent boundaries between crystals in such relative orientations are not possible. However in certain cases the coincident sites lie in moderately close-packed planes, as in the example shown in Fig. 18.10. A fairly coherent boundary is then possible, along such a plane. Examples of such boundaries have been observed in field-ion microscope studies of bi-crystals (*D. G. Brandon et al.*, 1964, *Acta metall.*, **12**, 813).

Consider next small deviations from orientations that give fully coherent boundaries. The simplest deviation of crystal orientation produces the *symmetrical small-angle tilt boundary*, shown in Fig. 18.11. The maximum coherence is obtained when the two slightly tilted crystals are *stepped*, as shown, and these steps become edge dislocations when the crystals join together. This semi-coherent grain boundary is thus formed from a 'wall'

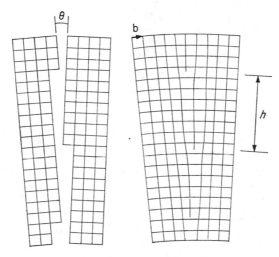

Fig. 18.11 A symmetrical small-angle tilt boundary formed from edge dis-locations which derive from steps on the adjoining crystal faces

of parallel edge dislocation lines, vertically above each other at a spacing h, given by

$$b/h = 2 \sin \tfrac{1}{2}\theta \simeq \theta \qquad\qquad \textbf{18.29}$$

where b is the length of the Burgers vector and θ is the angle of misorientation between the crystals. Such boundaries have been observed by electron microscope and etch-pitting methods and eqn. 18.29 has been verified.

Other small-angle boundaries are produced by rotation about other axes. The *twist boundary*, formed by rotation about an axis perpendicular to the boundary, contains dislocations which are at least partly of screw type, usually assembled as a hexagonal network of dislocation lines.

Fig. 18.12 shows a simple example of a small deviation of a boundary from its symmetrical (and most coherent) orientation. The twin interface shown there lies at a small angle to the twinning plane. The angle is obtained by the interface stepping from one atomic layer of the twinning plane to the next, at suitable intervals, with a *twinning dislocation* T at each step. Similar steps have been observed in semi-coherent boundaries between crystals in coincidence-lattice orientations. Additional dislocations also appear in small-angle tilt and twist grain boundaries when these are rotated away from symmetrical orientations.

By suitably and steadily increasing the density of dislocations in a boundary, to represent increasing rotations of the crystals or the boundary, or both, from orientations that give coherent boundaries, a gradual transition from coherent to non-coherent boundaries can be obtained. In the limit, the dislocations become too densely packed to be easily resolved and such large-angle non-coherent boundaries are then better described

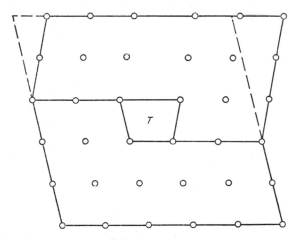

Fig. 18.12 A twinning dislocation T at a step in a twin interface

in other ways (cf. Fig. 13.10). It seems likely that even in boundaries of this type there are 'islands' where the atoms fit together well, surrounded by non-coherent regions.

When second phases form in alloy crystals the problem generally becomes one of matching two *different* crystal structures. However, if the particles of one of the phases are extremely small they may adopt a structure that allows them to be coherent with the matrix crystal, even if this involves considerable elastic distortion in them and their immediate surroundings. This is because the free energy associated with the crystal structure and its elastic strain field goes as the *volume* of the particle, whereas the energy of the interface goes as the *surface area*. In the limit of small thin particles the surface predominates and the particle becomes fully coherent. It can then be thought of as a solute-rich *cluster* in the parent solution. In quenched and aged Al + 4 per cent Cu alloys, for example, such particles consist of copper-rich discs, about 100 Å diameter and 4 Å thick, on {100} planes of the aluminium matrix crystal.

When such a particle grows larger, due to the acquisition of more solute atoms, the volume becomes more important relative to the surface and the particle may then 'break-away' from its coherent attachment to the matrix in order to reduce its elastic distortion and adopt a more stable crystal structure. The early stages of this break-away to a non-coherent structure are accomplished by the formation of dislocations in the interface, as shown in Fig. 18.13.

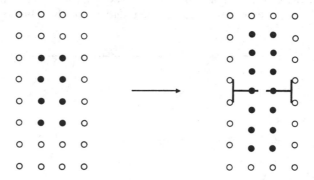

Fig. 18.13 A particle which becomes semi-coherent by the formation of dislocations in its interface

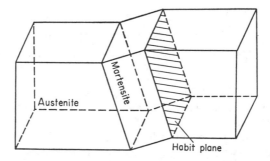

Fig. 18.14 Idealized form of martensite plate in a crystal of austenite

Martensitic transformations are similar to deformation twinning in many respects, but involve a change of crystal structure and (usually) a change of volume as well as a shear. The driving force is, of course, the free energy released by the change of crystal structure. The character of martensite in quenched steel varies considerably with carbon and alloy content and with rate of quenching, but commonly in steels containing 0·5 to 1·4 wt per cent carbon it forms *plates* of the type idealized in Fig. 18.14 (cf. Plate 6B).

This martensite has a body-centred tetragonal structure with an axial ratio that depends on the carbon content (cf. Fig. 14.4), an orientation that satisfies the *Kurdjumov-Sachs* relationships,

$$(111)_A \| (101)_M \quad \text{and} \quad [1\bar{1}0]_A \| [11\bar{1}]_M \qquad \textbf{18.30}$$

where A and M refer to austenite and martensite respectively, and a habit plane parallel to $\{225\}_A$. The coherent nature of the transformation is demonstrated by the regular *tilts*, i.e. inclined steps, produced on the free surface of the transformed plate.

It was first recognized by *E. C. Bain* in 1924 that a simple and small set of movements could convert an austenitic crystal cell into a martensitic one.

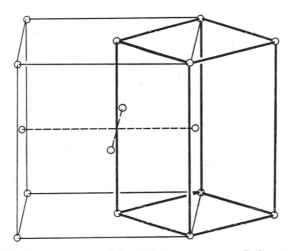

Fig. 18.15 Interpretation of the F.C.C. structure as B.C. tetragonal

As Fig. 18.15 shows, the F.C.C. cell can be regarded as a B.C. tetragonal cell with an axial ratio of $\sqrt{2}$. A compression of the long axis by about 17 per cent and an expansion of the short axes by about 12 per cent could thus produce the required structure. There can be little doubt that the martensite cells are created in this way. However, this process without additional movements cannot produce the observed orientation relationships. We have not allowed for the possibility that the cells *rotate* as they change their axial ratio.

Clearly, the rotations must be such that, combined with the Bain transformation, they leave the habit plane *unrotated* and, as far as possible, *undistorted* during the transformation; otherwise there would be gross misfit between the plate and its matrix. We express this by saying that the habit plane must be an *invariant* plane of the transformation. Looked at from this point of view, the transformation as shown in Fig. 18.14 bears a resemblance to deformation twinning, although it is not a simple shear on the habit plane. Measurements of the change of shape produced across the region of transformation, show two distinct differences. First, the gross deformation is one in which successive layers of the habit plane move slightly away from each other, uniformly, as well as sliding over each other. This is referred to as an *invariant plane strain*. Second, the gross deformation does *not* produce the martensitic structure if applied to the austenitic lattice. It was deduced from this that the invariant plane strain (together with a small additional dilatation), which *homogeneously* distorts the transforming plate to its observed final shape, must be supplemented by an additional *inhomogeneous* shear strain which produces the correct lattice in the correct orientation without any further change of shape. Electron microscopy has shown

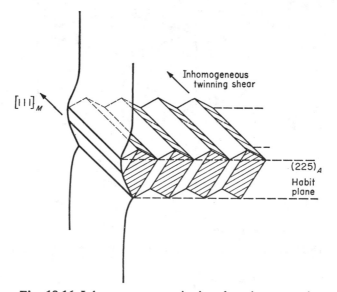

Fig. 18.16 Inhomogeneous twinning shear in martensite

that this inhomogeneous shear is produced by deformation twinning on (112) planes in the martensite, as shown in Fig. 18.16. The twin bands are extremely thin ($\simeq 20$ Å). The system thus chooses a relatively complex mode of transformation and a high-index habit plane, all for the purpose of obtaining an interfacial plane which remains as invariant as possible.

At high carbon content (1·5 to 1·8 wt per cent) the habit plane of martensite changes to $\{259\}_A$. The B.C.C. martensite formed at low carbon contents (see § 14.14), for example in Fe-Ni alloys, does not have an acicular habit. Instead, it forms *massive martensite*, i.e. irregular slabs of B.C.C. iron with jagged boundaries. Each slab consists of parallel thin plates of sheared material, with a $\{111\}_A$ habit, and the jagged outline is due to the impingements of other nearby packets of such plates.

FURTHER READING

As for Chapter 17; also:

CHARLES BARRETT and MASSALSKI, T. B. (1966). *Structure of Metals*. McGraw-Hill, New York and Maidenhead.

CHRISTIAN, J. W. (1965). *The Theory of Transformations in Metals*. Pergamon Press, Oxford.

MCLEAN, D. (1957). *Grain Boundaries in Metals*. Oxford University Press, London.

NYE, J. F. (1957). *Physical Properties of Crystals*. Oxford University Press, London.

REED-HILL, R. E. (1973). *Physical Metallurgy Principles*. Van Nostrand, New York and London.

WANNIER, G. H. (1959). *Elements of Solid State Theory*. Cambridge University Press, London.

Chapter 19

Metal Crystals—III Energies and Processes

19.1 Cohesion

In this chapter we shall look at the structures and properties of metals from the point of view of energy. Innumerable topics of physical metallurgy come up for discussion under this heading; but we must start first with *cohesion*. The practical value of metals is due almost entirely to the nature of the cohesive forces between metal atoms. Large forces give large elastic constants, high melting-points and small thermal expansions. The absence of saturation and directionality in the metallic bond leads to close-packing and hence to simple crystal structures in which dislocations can be highly mobile (cf. § 17.4) and this is the basis of the outstanding plasticity and toughness of metals.

The simplest metals to discuss are the alkalies. They have only one valency electron per atom and these valency electrons behave almost ideally as free electrons in the metallic state (cf. § 4.6). Moreover, the ions are small compared with the atomic spacing in these metals, as the following comparison with copper, silver and gold shows:

	Li	Na	K	Cu	Ag	Au
$\dfrac{\text{ionic radius}}{\text{atomic radius}}$	0·39	0·51	0·58	0·75	0·88	0·95

The ions in alkali metals are thus 'not in contact with each other'. These are '*open metals*', whereas copper, silver and gold are '*full metals*'. In open metals we need consider only the valency electrons, for the electrons in the ionic shells are negligibly disturbed by the metallic binding. We thus picture such a metal as an array of positively charged spheres—the ions—dispersed evenly in a uniform negatively charged fluid—the free electron gas.

What holds such a structure together? It is the fact that, when one of the atoms joins the metal, its valency electron is no longer confined to the small volume of the atom itself but is free to move through the entire metal. We saw in Chapters 3 and 4 that the structures of atoms and molecules are largely determined by opposing trends in the potential and kinetic energies of the electrons. The potential energy is lowest when an electron remains close to the centre of positive charge. In that case, however, the wave function of the electron has to bend to and fro very sharply to localize itself in a small volume near the nucleus. This sharp bending implies a high kinetic energy, as can be seen, for example, in eqn. 3.9 when L is made small. The kinetic energy of the valency electron in a free atom remains large because the electron cannot move far from the positive ion without its potential energy becoming high. But when the atom joins the metal this difficulty disappears since, as the electron moves away, it continually meets other ions which keep its potential energy as low as that in its parent ion. The electron can thus reduce its *kinetic energy of localization.*

It is not difficult to estimate this reduction, at least approximately. We can appreciate the order of magnitude from the following argument. By substituting $r = a$ in the formula $-e^2/4\pi\,\varepsilon_0 r$ for the potential energy, where a is given by eqn. 3.14, and then comparing with eqn. 3.15, we find that the potential energy (negative) has twice the magnitude of the kinetic energy (positive) for an electron in the ground state of the hydrogen atom. The same remains approximately true of a valency electron in an alkali metal atom. The loss of the kinetic energy of localization will thus roughly double the binding energy of the electron, which amounts to a few electron volts. A precise calculation, based on the fact that the slope of the wave function $d\psi/dr$ must be zero at the mid-point between neighbouring ions, gives the value 3.13 eV for sodium.

Most of the valency electrons do not do as well as this, however. They now exist throughout the metal in quantum states like those of Fig. 3.2. In the ground state ($\lambda = 2L$) the kinetic energy is negligibly small (for large L) but, because of the Pauli exclusion principle, only two electrons (with opposite spins) can go in this state. All others must go into higher states (i.e. $n > 1$) in which they have higher kinetic energies, as given by eqn. 3.9 and as represented by a *conduction band* like that of Fig. 4.5.

To evaluate this effect, consider a cube of metal of sides L. Eqn. 3.9 then generalizes into

$$E = \frac{h^2}{8mL^2}\,(n_x^2 + n_y^2 + n_z^2) \qquad\qquad \textbf{19.1}$$

where n_x, n_y, n_z are the quantum numbers (1, 2, 3 ... etc.) associated with components of the motion along the three axes of the cubical block. Each quantum state is defined by three particular values for these three quantum numbers. At 0 K all these quantum states are filled, with two electrons in each, up to a certain energy level E_F, called the *Fermi energy*.

This is usually represented by a *Fermi distribution* curve as in Fig. 19.1 which gives the probability that a quantum state is occupied. Unit probability means that it is always full, zero probability means that it is always empty, and a fractional probability means that it is occupied for that fraction of the time. We draw the curve as continuous because in a large piece of

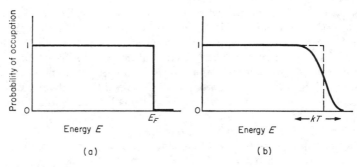

Fig. 19.1 The Fermi distribution curve, (a) at 0 K, (b) at T K

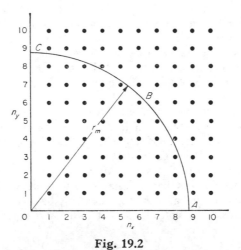

Fig. 19.2

metal the individual quantum states are too numerous and too closely spaced in energy to be marked separately on the diagram.

To calculate E_F consider first a 'two-dimensional' metal, by dropping n_z from eqn. 19.1, and let it contain N free electrons. We can represent the various possible values of n_x and n_y by points in a diagram, as in Fig. 19.2. Suppose that the scale is such that there is one point per unit area in the diagram. The distance of a point from the origin is r, where $r^2 = n_x^2 + n_y^2$, and all states with energies up to E are represented by points in the quadrant $OABC$ of radius r_m, where $r_m^2 = 8mL^2E/h^2$. The area of the quadrant is $\pi r_m^2/4$. Since each point has unit area and there are $N/2$ occupied points up to E_F, then

$$\frac{N}{2} = \frac{\pi r_m^2}{4} = \frac{\pi}{4}\left(\frac{8mL^2E_F}{h^2}\right) \qquad \textbf{19.2}$$

which gives E_F. In three dimensions we replace the quadrant by an octant (one eighth of a sphere) of volume $\pi r_m^3/6$. Eqn. 19.2 then changes to

$$\frac{N}{2} = \frac{\pi r_m^3}{6} = \frac{\pi}{6}\left(\frac{8mL^2E_F}{h^2}\right)^{3/2} \qquad \textbf{19.3}$$

which gives

$$E_F = \frac{h^2}{8m}\left(\frac{3N}{\pi L^3}\right)^{2/3} = \frac{h^2}{8m}\left(\frac{3N}{\pi V}\right)^{2/3} \qquad \textbf{19.4}$$

where $V = L^3$. We notice that E_F depends on N/V, i.e. on the *density* of free electrons in the metal; it thus does not alter when we join together two pieces of the same metal. Values of E_F for monovalent metals are as follows:

	Li	Na	K	Cu	Ag	Au
E_F (eV)	4·7	3·2	2·1	7·1	5·5	5·6

For the cohesion we need, not the maximum kinetic energy E_F, but the *average*. To obtain this we need to know the *density of states*, i.e. the number of quantum states $N(E)\,dE$ per unit volume of metal with energies in the range E to $E + dE$. The number n of occupied quantum states per unit volume is $N/2V$. From eqn. 19.4 we have

$$n = AE^{3/2} \qquad \textbf{19.5}$$

where $A = (\pi/6)(8m/h^3)^{3/2}$, and hence the density of states is

$$N(E) = \frac{dn}{dE} = \frac{3}{2}AE^{1/2} \qquad \textbf{19.6}$$

The average kinetic energy of an electron is thus

$$E = \frac{\int E\,dn}{\int dn} = \frac{\int E^{3/2}\,dE}{\int E^{1/2}\,dE} = \frac{3}{5}E_F \qquad \textbf{19.7}$$

For sodium this is $1·92$ eV. The expected binding energy is thus $3·13 - 1·92 = 1·21$ eV, which is close to the observed evaporation energy of sodium, i.e. $1·13$ eV per atom or $109\,000$ J mol^{-1}.

In this calculation we have neglected the electrostatic interactions of the free electrons between themselves, which reduce the cohesion. The contribution due to these turns out to be quite small however because the electrons are able to correlate their motions so as largely to keep out of one another's way, as in Fig. 4.6.

The next simplest metals to consider are the alkaline earths (Be, Mg, Ca, etc.). The effect of delocalizing the wave functions is now larger than in the alkalies but this is offset by the increased Fermi energy due to the doubling of the number of electrons in the Fermi distribution, i.e. $N \to 2N$ in eqn. 19.4. In fact, if the density of states in these metals continued to follow eqn. 19.6, the cohesion would be weak. In practice it is, of course, much stronger than that of the alkali metals. The reason is that there is a large increase in the density of states because many additional conduction band states are formed from p-states (and to a lesser extent d-states) which lie just above the valency electron s-states in the free atom. These extra states in the conduction band prevent the Fermi energy from rising unduly high. The cohesion is then increased.

A similar effect, involving the partly-filled d-states, is mainly responsible for the cohesion of the transitional metals. The great cohesion of metals such as tungsten and molybdenum is connected with the fact that the band of d-states in the Fermi distribution has room for lots of electrons in low-lying energy levels. It remains difficult to explain the fairly strong cohesion of copper, silver and gold, and the metals immediately before them in the periodic table, because in these the band of d-states is full, or nearly so. It has been suggested that in these metals there is a strong *van der Waals* attractive force (cf. § 4.2) between the d-shell electrons of neighbouring ions but the general problem of developing the theory for such electrons, which are in states transitional between those of free electrons and those of ionic core electrons, is very difficult.

The strong cohesive forces in the copper group pull the ions together until they are 'in contact' and begin to repel by the overlap forces discussed in § 4.2. We can regard such metals approximately as hard spheres held in contact by non-directional attractive forces. In contrast to the alkali metals, the equilibrium atomic spacing here is greater that that at which the valency electrons have lowest energy. However, the energy of valency electrons depends not on atomic *spacing* directly but on atomic *volume*. To achieve minimum volume for a given spacing the structure must be close-packed. It is thus not surprising that the copper group and the metals before them in the periodic table are all close-packed.

A structure of hard spheres held together by non-directional attractive forces would not in fact be mechanically stable in B.C.C. form. This is easily demonstrated by trying to stack a B.C.C. crystal from ping-pong or billiard balls. Fig. 19.3 shows what happens. Here we have a side view of a (110) plane with one atom of the next layer also shown. This atom can slide along the [$1\bar{1}0$] direction, as shown, and 'roll down the hill' into a more close-packed configuration.

B.C.C. structures are in fact found only in metals not too near the copper group in the periodic table. An exception is β-brass and here it is found that the shear modulus for (110) [1$\bar{1}$0] shear is very small, only one-eighteenth that of (100) [010] shear (cf. Table 18.1).

Apart from a few qualitative arguments such as the above, it has not so far been possible to explain the particular crystal structures of the various metals from the theory of cohesion. Allotropy is very common amongst the metals and the heats of transformation from one crystal structure to another are quite small, so that metals are fairly indifferent to their precise structure. This is just what we would expect from the theory, which shows that the energies of free electrons are sensitive to the *volume* in which they are packed and not much to the (uniform) arrangement of the atoms within a given volume. This makes the calculation of the relative cohesive energies of different crystal structures extremely difficult, since such calculations have to be done very exactly to determine reliably the small differences between nearly equal quantities.

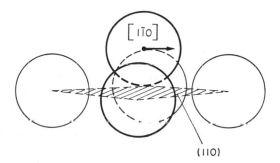

Fig. 19.3 Unstable shear of (110) planes in a B.C.C. lattice of hard spheres

The calculation of the energies of surfaces, grain boundaries and crystal imperfections from the quantum-mechanical theory of metallic cohesion is generally extremely difficult and has been done for only one or two special cases. In general we have to find alternative simpler methods. Elasticity theory has been a great help here because it allows energies associated with small deviations from the ideal structure to be calculated rather easily. Many energies of features in metals have been estimated in this way, some accurately, some approximately. When all else fails, we can still gain an idea of some energies by semi-geometric methods, e.g. by counting numbers of 'broken bonds' or by comparing the structures of different defects. These methods usually give only order-of-magnitude values.

19.2 Thermal properties

When a metal is heated, disorder of various kinds sets in. Each involves absorption of thermal energy and so contributes to the specific heat. In principle, the nuclei and the electrons can all separately absorb such energy. In practice, the contribution of the electrons to the specific heat is small at ordinary temperatures. The quantum of energy needed to excite an electron in the ionic core into a more energetic state of motion is generally so large, compared with the available thermal energy per particle ($\simeq kT$), that this thermal excitation practically never happens. We can think of a nucleus and its core electrons as a single particle when discussing specific heat.

The free electrons also absorb little thermal energy at ordinary temperatures. The specific heat of an ordinary gas at constant volume is $3k/2$ per particle (k = Boltzmann's constant). At room temperature $kT \simeq 1/40$ eV. The Fermi energy in, say, copper is 7·1 eV. Electrons in this Fermi level thus move about at 0 K with kinetic energies equivalent to those of the particles in an ordinary gas at about 50 000 K. It is thus not surprising that the classical theory of specific heats does not apply to the free electron gas. The Pauli principle rules that an electron can enter only an empty quantum state, so that thermally excited electrons must go into states above E_F. Thermodynamics shows that the average thermal energy allowable to a particle is of order kT in a system at a temperature T K. Only electrons within an energy interval kT from E_F, approximately, can then absorb thermal energy. Those further down the Fermi distribution are prevented from doing so by the absence of empty quantum levels there. At room temperature kT is not more than about $0.01E_F$ so that only about 1 per cent of the electrons can take up their kT of thermal energy. The specific heat of the electron gas is thus very small at such temperatures and the Fermi distribution curve is hardly altered. The effect of temperature on this curve is shown with exaggeration in Fig. 19.1(b).

We turn to the contribution made by atomic vibrations to the specific heat. The atoms in a crystal are never at rest. They vibrate rapidly about their equilibrium sites, even at 0 K. As the temperature is raised they vibrate with increased amplitudes but almost unchanged frequencies. Increase in amplitude involves increase in the kinetic and potential energies of vibration. The amount of vibrational energy absorbed per 1 degree rise in temperature is the *vibrational heat capacity*; or, for unit mass, the *vibrational specific heat*.

To a good approximation we can regard the vibrations as elastic oscillations with a restoring force proportional to displacement, from Hooke's law. We can thus picture the vibrating atoms as *simple harmonic oscillators*. It can be shown from quantum mechanics that a particle vibrating up and down a line in simple harmonic motion, i.e. a *linear oscillator*, can have a vibrational energy given by the formula $(n + \frac{1}{2})h\nu$ where $n = 0, 1, 2 \ldots$ etc., h = Planck's constant and ν = frequency of vibration. Even in the lowest level, $n = 0$, the particle vibrates with the *zero-point energy* $\frac{1}{2}h\nu$. This zero-point motion, which exists at 0 K, is analogous to the zero-point translational motion of the free electron gas (cf. Fig. 19.1).

For a simple approximate theory, we assume that the N atoms of the crystal (monatomic, with a simple lattice) vibrate as if they were $3N$ independent linear oscillators, each with frequency ν. The factor 3 allows for the three axes of oscillation available to each atom.

How do the selected values of n depend on temperature? Suppose that, in warming up from 0 to T K, the system takes in M quanta. The disorder and entropy are then increased through the various ways in which these quanta can be shared out amongst the $3N$ oscillators. To count these distributions let us imagine that we can label the quanta $a_1, a_2 \ldots a_M$, and the oscillators $z_1, z_2 \ldots z_{3N}$. To represent one arrangement we write down the a's and z's in arbitrary order, starting with a z, e.g.

$$z_6 a_2 a_4 z_2 a_7 a_1 a_5 z_8 z_3 a_9 a_6 \ldots$$

and suppose that the a's between two successive z's belong to the z on their immediate left, e.g. oscillator z_6 has the quanta a_2 and a_4, z_2 has a_7, a_1, a_5, z_8 has none, and so on. We can thus represent every possible arrangement by writing down all the variants of this sequence. The first z can be chosen in $3N$ ways and the remaining symbols can then be written down in $(M + 3N - 1)!$ ways, so that the total number is $(M + 3N - 1)!3N$. However, many of these are indistinguishable. All which differ by mere interchange of one z label with another, or of one a label with another, represent the *same* physical distribution of the quanta amongst the atoms. The number of such permutations is $M!3N!$. Hence the number w of distinguishable distributions is

$$w = \frac{(M + 3N - 1)!3N}{M!3N!} = \frac{(M + 3N - 1)!}{M!(3N - 1)!} \qquad 19.8$$

The entropy is $S = k \ln w$. Using Stirling's approximation, neglecting the 1 in comparison with the large numbers $3N$ and $M + 3N$, and differentiating with respect to M, we obtain

$$\frac{dS}{dM} = k \ln \left[\frac{M + 3N}{M} \right] \qquad 19.9$$

The internal energy is $E = E_0 + Mh\nu$, where E_0 is the zero-point internal energy. Hence $dE/dM = h\nu$. The equilibrium condition is $dF/dM = 0$ at constant temperature T, where $F = E - TS$. Substituting, we obtain

$$M = \frac{3N}{e^{h\nu/kT} - 1} \qquad 19.10$$

The specific heat at constant volume, C_V, is dE/dT. This is the *atomic heat* when $N = N_0 =$ Avogadro's number. Thus

$$C_V = \frac{d}{dT} \left(\frac{3Nh\nu}{e^{h\nu/kT} - 1} \right) = 3R \left(\frac{\theta}{T} \right)^2 \frac{e^{\theta/T}}{(e^{\theta/T} - 1)^2} \qquad 19.11$$

where $R = N_0 k =$ gas constant, and $\theta = h\nu/k =$ the *Einstein characteristic temperature*. This is Einstein's formula for the specific heat, obtained in 1907.

Fig. 19.4 shows the form of C_V as a function of T/θ. The specific heat starts at zero at 0 K and rises to a limiting value $3R$ ($= 25$ J mol^{-1} K^{-1}) at temperatures well above θ. Many solids behave in this manner and the theory undoubtedly provides a sound first approximation to the thermal behaviour of crystals.

However, the specific heat in practice diminishes at low temperatures less sharply than eqn. 19.11 suggests. The reason is that the atoms do not vibrate independently. They are all coupled together by their cohesive forces and do not vibrate about fixed sites but about positions defined by their neighbours. We can liken these coupled vibrations to the standing vibrations of a stretched rubber string along which are mounted equal masses at equal intervals. Various modes of vibration can be set up in the string, some of

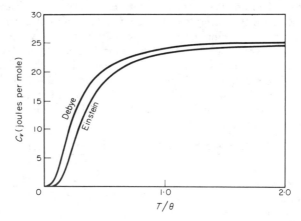

Fig. 19.4 The atomic heat at constant volume according to Einstein's and Debye's theories

long wavelengths in which many masses move in unison, some of short wavelengths in which neighbouring masses move in opposition. Returning to the crystal of N atoms, we find a similar spectrum of $3N$ different modes of vibration of the whole crystal, ranging from long wavelengths in which the whole crystal vibrates elastically like a tuning fork, to short wavelengths in which neighbouring atoms vibrate against one another. The analysis leading to eqn. 19.10 can be applied to each mode and the specific heat is then given by

$$C_V = \frac{d}{dT}\left(\sum_{r=1}^{3N}\frac{h\nu_r}{e^{h\nu_r/kT}-1}\right) \qquad \textbf{19.12}$$

where the sum is taken over all the frequencies ν_r of the various modes ($r = 1, 2 \ldots 3N$). The problem then is to find the *vibrational spectrum*, i.e. the number of modes in each frequency range.

This is a difficult problem in the mechanics of lattice vibrations. In 1912 two independent attempts were made. The first, by *Born and Kármán*, was a rigorous atomistic treatment but limited to a one-dimensional chain of atoms. The second, by Debye, was three-dimensional but ignored the atomic structure apart from fixing the number of modes at $3N$ and the highest frequency at v_m. *Debye* pictured the vibrating crystal as a structureless elastic 'jelly', i.e. an *elastic continuum*, vibrating in its various natural modes of elastic vibration. This led him to the formula

$$C_V = 3R \left[\frac{12T^3}{\theta^3} \int_0^{\theta/T} \frac{x^3 \, dx}{e^x - 1} - \frac{3(\theta/T)}{e^{\theta/T} - 1} \right] \qquad \textbf{19.13}$$

for the specific heat, where $\theta = hv_m/k =$ the *Debye characteristic temperature* and $x = hv/kT$. Fig. 19.4 shows the specific heat curve. It is broadly similar to Einstein's theory but at low temperatures it gives $C_V \propto T^3$, in much better agreement with experiment. This is because the long-wave vibrations, having low hv values, are able to absorb thermal energy at correspondingly low temperatures.

The characteristic temperature is determined experimentally by fitting the theoretical curve to the observed specific heats. Some values of the Debye θ (K) are as follows:

Na	Al	Cu	Ag	Fe	Pb	Be	Diamond
150	385	315	215	420	88	1000	2000

Because of its dependence on vibrational frequency, the characteristic temperature is a measure of the elastic properties of the crystal. To gain an idea of v, let us think of a vibrating atom as a simple harmonic oscillator of mass m. If a displacement x produces a restoring force $f = \alpha x$, the frequency of harmonic oscillation is $(1/2\pi)\sqrt{\alpha/m}$. If the atomic spacing is a the restoring force f per atom corresponds to a stress of about f/a^2 per unit area and the displacement x corresponds to a strain x/a. If Y is the elastic constant corresponding to this deformation, then $Y = \alpha/a$. The atomic vibrational frequency is thus given in order of magnitude by

$$v \simeq \frac{1}{2\pi} \sqrt{\frac{Ya}{m}} \qquad \textbf{19.14}$$

Since, typically, $Y \simeq 10^{11}$ N m^{-2}, $a \simeq 3 \times 10^{-10}$ m, and $m \simeq 10^{-26}$ kg, then $v \simeq 10^{13}$ s^{-1}. We thus expect characteristic temperatures to be a few hundred deg K, as observed. We see also that strongly bonded crystals of light atoms (e.g. diamond) have high v and θ values; and weakly bonded crystals of heavy atoms (e.g. lead) have low values.

Detailed comparison with experiment has shown that the Debye theory does not perfectly fit the experimental facts. The main error comes from the type of vibrational spectrum used in the theory. A consequence of the assumption of an elastic continuum is that the spectrum should have the simple parabolic shape shown in Fig. 19.5. More detailed work which takes the atomicity into account has shown, however, that a spectrum more like that of the full line in Fig. 19.5 is to be expected, with a strong peak at a frequency well below ν_m in addition to a main peak nearly at ν_m. Neutron scattering experiments are useful for determining the actual shapes of lattice vibrational spectra.

All these theories give the specific heat at constant volume but if the heating is done at constant pressure there is also the effect of *thermal expansion* to consider. We saw in Fig. 4.1 that the potential energy rises

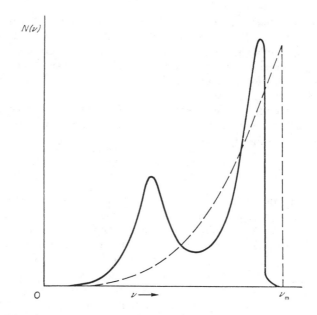

Fig. 19.5 The frequency spectrum of a crystal according to Debye (dotted line) and Blackman (full line)

more sharply when the atoms swing close together than when they swing apart, so that there is in practice some deviation from simple harmonic behaviour. The effect is that the vibrational frequency falls as the lattice expands. This decrease reduces the size of the energy quantum, so that more quanta can be absorbed for a given total energy and hence more entropy gained. This provides the necessary thermodynamic driving force for thermal expansion, which is of course opposed by the cohesive forces of the crystal (i.e. the bulk modulus). The specific heat usually measured experimentally is C_p and this is slightly greater than C_V (by a few per cent at room temperature) because of work done against the cohesive forces when the crystal is allowed to expand thermally. It can be shown from thermo-dynamics that

$$C_p - C_V = \alpha^2 KVT \qquad\qquad\qquad 19.15$$

where α is the volume coefficient of thermal expansion, V is the volume per mole, K is the bulk modulus, and T the absolute temperature.

19.3   Energies of dislocations

A dislocation is a centre of internal strain in a crystal. In the core of the dislocation the atoms are too misaligned for their energies to be estimated

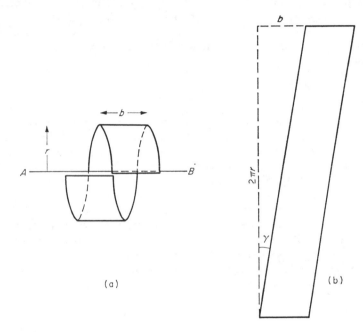

Fig. 19.6

accurately by elasticity theory. The strains become weaker at larger distances from the dislocation and, outside the central core, they can be analysed reliably by elasticity theory. Fortunately, many important properties of dislocations depend mainly on this long-range strain field and so can be calculated accurately.

The strain field of a screw dislocation can be found very simply. Consider in Fig. 19.6 a thin ring of material, of radius r and width b, round the dis-location AB, with Burgers vector of length b. The ring is distorted by the dislocation into a screw, as shown. If we open out this ring without releasing the shear in it, we obtain a flat strip of length $2\pi r$ which is sheared by a displacement b at one end. If the material is elastically isotropic this shear must be distributed evenly along the strip, i.e. the shear strain γ at a distance r from the dislocation is given by

$$\gamma = \frac{b}{2\pi r} \qquad\qquad 19.16$$

and so falls rather slowly with increasing distance.

The shear stress is $\mu\gamma$, where μ is the shear modulus. Hooke's law shows that the strain energy is $\frac{1}{2} \times$ *stress* $\times$ *strain*. Consider now a cylinder round the dislocation, of radius r, wall thickness dr and unit length. The strain energy per unit volume in it is then $(1/2)\mu\gamma^2$. The cylinder contains a volume $2\pi r\, dr$ and hence a strain energy $(\mu b^2/4\pi)(dr/r)$. The sum of this, over all cylinders from a 'core radius' r_0 to an outer 'crystal radius' r_1, is then the elastic energy U of the dislocation in the crystal, i.e.

$$U = \frac{\mu b^2}{4\pi} \int_{r_0}^{r_1} \frac{dr}{r} = \frac{\mu b^2}{4\pi} \ln\left(\frac{r_1}{r_0}\right) \qquad\qquad 19.17$$

The fact that the crystal may not be cylindrical in shape is of small consequence because the energy depends only logarithmically on r_1. We can take r_1 as the (narrowest) width of the crystal. For a unit dislocation we expect r_0 to be of the order of one atomic spacing. In general, there is so much elastic energy outside the core that the additional contribution of the highly strained core region itself is small by comparison. For example, if the length of dislocation line $= r_1 = 10^{-2}$ m, $b = 3 \times 10^{-10}$ m, $r_0 = 10^{-9}$ m, then $U \simeq 5 \times 10^7 \mu b^3$. Usually $\mu b^3 \simeq 5$ eV so that $U \simeq 250$ MeV. Clearly there can be no thermal creation of defects with such large energies as this and the crystal will try to rid itself of dislocations. This is the basis of most of the annealing processes which occur when plastically deformed metals are heated (cf. § 21.7) although the extent to which dislocations can be removed from metals is limited by the formation of metastable networks, as in Fig. 17.11.

The logarithmic factor in eqn. 19.17 should be noted. It implies that the energy is by no means localized in the centre of the dislocation, even though the *density* of energy is highest there. In the above example, most of the energy lies *outside* the radius $r = 10^{-6}$ m. This is a result of the long-range character of the strain field.

Eqn. 19.17 also shows that the strain energy goes as the *square* of the length of the Burgers vector. It follows that a large dislocation, with a vector $2b$ can reduce its energy, in effect from $4b^2$ to $2b^2$, if it splits into two well-separated unit dislocations. This is why large dislocations tend to break down into smaller ones and why dissociation into partial dislocations occurs, despite the surface tension of the stacking fault between them (cf. eqn. 17.12). To estimate the repulsion between the two dislocations b and b when they are at a distance r apart, we note that the strain field at much farther distances is still very like that of a *single* dislocation $2b$; and that within the distance r from the dislocations the strain field resolves itself into that from *two* distinct unit dislocations. Hence, at least approximately, the strain energy of the pair at the separation r is given by

$$U = \frac{\mu b^2}{4\pi} \left[4 \ln \left(\frac{r_1}{r} \right) + 2 \ln \left(\frac{r}{r_0} \right) \right] \qquad\qquad \textbf{19.18}$$

The change of U with r then gives the force per unit length exerted by one dislocation on the other, i.e.

$$f = -\frac{dU}{dr} = \frac{\mu b^2}{2\pi r} \qquad\qquad \textbf{19.19}$$

From the Mott and Nabarro formula, eqn. 18.27, this is equivalent to a stress $\mu b/2\pi r$. But this is precisely the stress which each dislocation exerts, at the position of the other. We may thus think of each dislocation as sitting

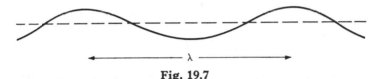

λ

Fig. 19.7

in the stress field of the other and responding to this stress, at its centre, as if this were a uniformly applied stress of the same magnitude. By the same argument, if the two dislocations have Burgers vectors of opposite signs they are attracted together by a force given by eqn. 19.19. When they coalesce, they annihilate each other and their strain energy becomes zero. Because stresses become zero at a *free surface*, such a surface acts rather like a dislocation of opposite sign in attracting dislocations towards itself.

A dislocation network such as that of Fig. 17.11 is mechanically stable because of *line tension* in its dislocation lines, analogous to the surface tension which makes soap froths stable. To estimate line tension, consider in Fig. 19.7 a dislocation line which, instead of being straight (dotted line) has taken up the sinuous form shown, with wavelength λ. At distances large compared with λ from the dislocation, the strain field and strain energy are unaltered. Near the dislocation line we have, roughly, the ordinary strain field of a dislocation in a sinuous tube which follows the dislocation line and so has a length L_1 which is greater than that, L_0, of a straight tube between the same end points. The dislocation line thus has an *extra* energy, approximately $(L_1 - L_0)(\mu b^2/4\pi) \ln (\lambda/r_0)$ due to its sinuous form. We know from soap films that surface tension (N m^{-1}) is numerically equal to surface energy (J m^{-2}). The same argument applies to the line tension of a dislocation. Hence we deduce the line tension τ of the dislocation as

$$\tau \simeq \frac{\mu b^2}{4\pi} \ln \left(\frac{\lambda}{r_0} \right) \qquad\qquad \textbf{19.20}$$

In many practical problems of the strength of metals and alloys, λ is 100 to 1000 Å. For these we can take

$$\tau \simeq \tfrac{1}{2}\mu b^2 \qquad\qquad \textbf{19.21}$$

as an order-of-magnitude value.

Line tension is important for all processes of slip in which dislocation *loops* have to be formed. Examples are the operation of Frank-Read sources (Fig. 17.12) and the looping of dislocation lines between foreign particles which act as obstacles in the slip plane. Consider a Frank-Read source between two nodes spaced a distance l apart. The dislocation has to pass through a semi-circular form (at least approximately) before it can spread out into the slip plane. In this semi-circular form it is pushed forward by the Mott and Nabarro force σbl, from the resolved shear stress σ, and held back by the line tensions 2τ on its ends. The minimum applied stress required to operate the source is thus

$$\sigma \simeq \frac{2\tau}{bl} \simeq \frac{\mu b}{l} \qquad\qquad \textbf{19.22}$$

This formula is the starting point of the theory of the *real* strengths of ductile metals. In Chapter 17 we met some ideal strengths, the shear strength of a perfect lattice (Figs. 17.2 and 17.3) and the Peierls-Nabarro force, i.e. the resistance offered by a perfect lattice to a straight dislocation. The first of these is very high. It is obvious geometrically that the atoms in Fig. 17.2 must slide forward by an appreciable fraction of an atomic spacing, e.g. about $b/4$, before running on to the next equilibrium site. Hence the ideal shear strength cannot be much smaller than about $\mu/10$, even allowing for the softening of the law of force at large shear strains. Ordinary samples of metals slip at much lower stresses than this (e.g. $10^{-4}\mu$ to $10^{-2}\mu$) because they contain dislocations. The Peierls-Nabarro force, although important in hard non-metallic crystals (e.g. silicon, diamond, alumina) and probably also in B.C.C. transition metals at temperatures below about 150 K, is normally negligibly small in ductile metals, e.g. F.C.C. metals, at room temperature. The practical *yield strength* of such metals, when pure, is then determined largely by the characteristic scale, l, of the links in dislocation networks in the crystal, through eqn. 19.22. It is convenient to define a *dislocation density* ρ as the total length of dislocation lines in unit volume. Then $l \simeq \rho^{-1/2}$ for networks like that of Fig. 17.11 and the shear yield stress is given by

$$\sigma = \alpha \mu b \rho^{1/2} \qquad\qquad \textbf{19.23}$$

where α is a number, of order unity, whose precise value depends on the specific geometrical structure of the dislocation lines involved in the yielding process. Most specific models give $\alpha \simeq 0\cdot2$ to $0\cdot3$. On this basis a pure metal with small Peierls-Nabarro force has a shear yield stress of order $10^{-4}\mu$ when annealed (i.e. $\rho \simeq 10^{12}$ m^{-2}) and of order $10^{-2}\mu$ when severely distorted by plastic working ($\rho \simeq 10^{16}$ m^{-2}). This agrees broadly with what is observed.

We now consider *edge* dislocations. Many of the above arguments apply also to these. For example, in a ring of radius r round the dislocation line, a displacement b has to be accommodated in a length $2\pi r$ so that the elastic strain must be of order $b/2\pi r$. The order-of-magnitude results, eqns. 19.17 to 19.23, thus apply also to edge dislocations.

There is, however, an important difference. The strain field round the edge dislocation does not have such simple symmetry as that round the screw, because the atoms on one side of the slip plane are compressed and those on the other side are expanded, as seen in Fig. 19.8. The strain components at a point with polar coordinates r, θ thus change sign as θ increases from 0 to 2π, and so contain a factor such as sin θ or cos θ. Another difference is

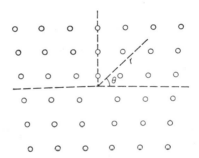

Fig. 19.8 Strain field round an edge dislocation

that, because dilatational and shear strains are both present, Poisson's ratio ν appears in the formulae. For example, the dilatation Θ at the point r, θ for the dislocation of Fig. 19.8 in an elastically isotropic medium is given by

$$\Theta = -\frac{b}{2\pi r}\left[\frac{(1-2\nu)}{(1-\nu)}\sin \theta\right] \qquad\qquad \textbf{19.24}$$

where positive and negative values indicate expansion and compression respectively.

This angular dependence of the strain field affects the forces exerted between edge dislocations. Like dislocations still repel and unlike attract along the line between them, but the resolved components of the forces along the slip planes of the dislocations behave in a more complicated way because of the contribution of the forces in the tangential direction which come from the angular dependence. In Fig. 19.9 we show one dislocation at the origin O and another, like, dislocation at D. Then D is *attracted* along its glide plane EF towards E, when $OC < CD$, and is repelled in the direction of F when $OC > CD$. The shear stress exerted at D along EF by the dislocation at O is in fact given by

$$\sigma = \frac{\mu b}{2\pi(1 - \nu)} \frac{x(x^2 - y^2)}{(x^2 + y^2)^2}$$ **19.25**

where $x = OC$ and $y = CD$.

Parallel edge dislocations of the same sign thus tend to arrange themselves vertically above one another. This striking conclusion from the elasticity theory of dislocations is confirmed by the effect known as *polygonization*. If a crystal is plastically bent to introduce into it parallel edge dislocations of the same sign, it is then observed, particularly if the crystal is afterwards annealed, that these dislocations arrange themselves into small-angle tilt boundaries of the type shown in Fig. 18.10.

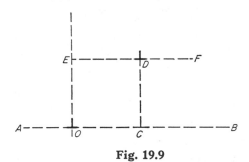

Fig. 19.9

19.4 Energies of surfaces

The energies of surfaces and interfaces play an extremely important part in determining the microstructures and properties of metals and alloys.

The atoms in the free surface of a solid or liquid have higher free energy than those inside because they have no neighbours, and hence no cohesive bonds, on one side. We can thus very roughly estimate surface energies from the numbers of 'broken bonds' at the surface. An atom in the surface of a cubic crystal generally has about one-quarter of nearest neighbours missing, compared with one inside. We thus expect the surface energy to be about one-quarter of the cohesive energy, per atom; e.g. about 1 eV for a metal such as iron. Typically, there are about 10^{19} atoms m^{-2} in a crystal surface, so that the surface energy should be of order 10^{19} eV m^{-2}, i.e. $1\cdot6$ J m^{-2}.

It is difficult to make much more accurate estimates, because the electron theory of metallic surfaces is not well developed. There are several experimental approaches. A simple one is to take the value for a liquid metal, measured by a standard capillarity method at the melting point, and add a factor, e.g. 30 per cent, to allow for the extra cohesion of the solid. A direct method is to suspend small weights from a thin wire or foil held at high temperatures, in vacuum or inert gas, and determine the value at which the sample neither shortens nor lengthens. The surface tension is then in balance with the weight. Typical values of surface free energies measured in this way are as follows:

	Ag	Au	Cu	Fe	Ni	Pt	Sn
γ, J m^{-2}	1·13	1·35	1·73	1·915	1·725	2·1	0·685

We expect, of course, the surface energy to be related to other cohesive properties of solids. It is found in fact that

$$\gamma \simeq 0·05Eb \qquad\qquad 19.26$$

where E is Young's modulus and b is the atomic spacing.

The experimental determination of surface energies is made difficult by the tendency for clean surfaces to become contaminated by foreign atoms and molecules attracted there by cohesive forces. Such contamination can reduce the measured values by an order of magnitude. The general effects of adsorption are of course of wide importance. The value of metals as *catalysts* in chemical processes is well known. Adsorption is also important in metallography; etch-pits become sharp and deep when adsorbed impurities protect their sides from chemical attack. The study of foreign atoms on metal surfaces has been greatly helped by the development of field-ion microscopy and by low-energy electron diffraction analysis. The atomic structure of a clean metal surface is found to be very similar to that of a parallel plane of atoms inside the metal but when a small amount of a substance such as oxygen is allowed to collect, the structure is drastically changed, even before one monolayer is absorbed. Oxygen atoms on nickel, for example, become incorporated in the surface itself and form a kind of two-dimensional superlattice with the metal atoms there.

Because of the unselective nature of the free electron bond, any two metals with clean surfaces will adhere well. It follows that a liquid metal of low cohesive energy will generally spread freely over the clean surface of a metal of high cohesive energy, since in this way a surface of high energy can be replaced by a low-energy surface and a low-energy metal-metal interface. The wetting action of *solders* and *brazes* depends on this.

Surface energy is also important in the *fracture* of solids, since fracture is a process in which new surfaces are created. In an ideally simple process of tensile fracture we uniformly stretch all the interatomic bonds between two adjoining cleavage planes, by an increasing tensile stress, until they all simultaneously break. If Hooke's law were obeyed right up to the fracture stress σ_t, we could equate the elastic energy, $a\sigma_t^2/2E$ per unit area, where a is the initial spacing of the planes, to the energy 2γ of the two surfaces so created, i.e. $\sigma_t = \sqrt{4E\gamma/a}$. However, this overlooks the softening of the law of force at large strains. Much of the contribution to the surface energy in fact occurs after the maximum in the force-displacement curve has been passed. The estimate of this effect depends on the detailed law of force between the atoms. Most calculations lead to the order-of-magnitude value

$$\sigma_t \simeq \sqrt{\frac{E\gamma}{a}} \qquad \qquad \textbf{19.27}$$

The very high strengths given by this formula, e.g. 4×10^{10} N m^{-2} ($\simeq 6 \times 10^6$ psi) for iron, represent an ideal upper limit to the strength of a solid. The highest observed values, obtained from crystals in the form of fine whiskers, always fall below this, e.g. $1{\cdot}3 \times 10^{10}$ N m^{-2} for iron. There are several reasons for this. The natural vibrations of the crystal must help to break the bonds. In metals at least, the ideal shear strength is significantly smaller than σ_t, even when considered as tensile yield stress, so that shear plays a major part in the process of fracture. Most solids in bulk have much lower strengths than these (e.g. $< 0{\cdot}01E$) because of the weakening effects of dislocations and cracks in them.

19.5   Grain boundaries and interfaces

The elasticity theory of dislocations enables the energy of a low-angle grain boundary to be estimated quite simply. We saw that in the boundary of Fig. 18.11 the dislocations are spaced at distances $h = b/\theta$. Because of the mutual tilt of the two adjoining half-crystals, the strain field of each dislocation hardly extends beyond a distance h from the boundary. The energy of each dislocation can then be written as

$$U = \frac{\mu b^2}{4\pi(1 - \nu)} \ln \left(\frac{h}{r_0}\right) + C \qquad \qquad \textbf{19.28}$$

where C is the energy of the core. There are θ/b dislocations in unit (vertical) length of the boundary. Hence the energy per unit area γ of the boundary is $U\theta/b$, i.e.

$$\gamma = A\theta(B - \ln \theta) \qquad \qquad \textbf{19.29}$$

where $A = \mu b/4\pi(1 - \nu)$. This is the formula of *Read and Shockley*. It is hard to evaluate B theoretically, but when a value to suit the experimental results is chosen, eqn. 19.29 then fits the observations very well. What is remarkable is that such agreement continues up to large-angle boundaries, e.g. $\theta \simeq 25°$, for which $\gamma \simeq 0{\cdot}02Eb$, i.e. about one-third of the surface energy.

The energies of large-angle boundaries are usually measured by *thermal etching*. When a polycrystal is heated at high temperatures, grooves are developed in its free surface along lines where grain boundaries meet that surface. This is an example of the general effect of balancing the forces of surface tension at an interfacial triple junction, as shown in Fig. 19.10, and its importance in influencing the microstructures of metals and alloys at high temperatures, where the atoms are sufficiently mobile (by diffusion and evaporation) to enable the necessary movements of boundaries to take place, was pointed out in a classic paper by *C. S. Smith* (1948, *Transactions A.I.M.E.*, **175**, 15). The condition for equilibrium is that the three surface tensions should form a triangle of forces. Thus, in Fig. 19.10

$$\frac{\gamma_A}{\sin A} = \frac{\gamma_B}{\sin B} = \frac{\gamma_C}{\sin C} \qquad\qquad \textbf{19.30}$$

where A, B and C are the interfacial angles. If C represents the *dihedral angle* of the surface groove and if $\gamma_A = \gamma_B = \gamma_S$ = the surface energy, and $\gamma_C = \gamma_{GB}$ = the grain boundary energy, then

$$\gamma_{GB} = 2\gamma_S \cos \frac{C}{2} \qquad\qquad \textbf{19.31}$$

so that γ_{GB} can be determined from γ_S and C.

This method gives the *free energy* of a boundary. The total energy is larger

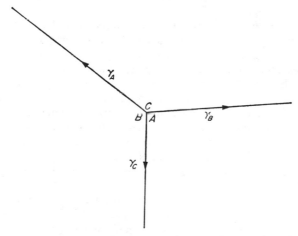

Fig. 19.10 Relation between surface tensions and angles

than this because of a contribution associated mainly with the vibrational entropy of the atoms in the boundary. Total energies can be measured directly, with some difficulty, by calorimetry. Some typical values of free energies of large-angle boundaries are as follows:

	Au	Cu	Fe	Ni	Sn
γ_{GB} J m^{-2}	0·36	0·60	0·78	0·69	0·16

We see that $\gamma_{GB} \simeq \frac{1}{3}\gamma_S$.

Eqn. 19.30 assumes that the energy is independent of the orientation of the boundary relative to the crystal axes of the adjoining grains. For large-angle grain boundaries in random polycrystalline metals this is usually a fair assumption. If we apply Fig. 19.10 now to the junction between three such large-angle grain boundaries, inside a polycrystal, then in general $\gamma_A \simeq \gamma_B \simeq \gamma_C$ so that $A \simeq B \simeq C \simeq 120°$. This limitation upon equilibrium angles between boundaries generally conflicts with geometrical requirements for the numbers of faces and edges on grains. In two dimensions, where the grains can be represented as polygons, only a six-sided polygon can have straight sides and 120° angles at its corners. Polygons with either more or fewer sides than six cannot be in equilibrium, since they must either have curved sides or angles other than 120°. At temperatures where the atoms are mobile, curved boundaries tend to straighten out, which in general alters the angles from 120°; and triple junctions where the angles are not 120° tend to move in an attempt to restore the correct angles, which produces curves boundaries again. As a result, polygons with fewer than six sides tend to shrink and disappear, while those with more than six tend to grow. This instability in a random array of grains is one cause of *grain growth* which occurs when polycrystals are heated, particularly after plastic deformation, at high temperatures.

In zone-refined metals grain boundaries migrate readily, e.g. at 10^{-5} m s^{-1} at temperatures near the melting point T_m, and some movement can be detected even at $0·25T_m$. The activation energy is only about $5RT_m$ per mole, which suggests that atoms break away singly from the surface of the shrinking grain to join the boundary before being deposited in turn on the growing grain. This high rate of grain growth is rarely observed in less pure metals because impurities obstruct the boundaries. Soluble impurities which segregate strongly to boundaries, for example to reduce their elastic misfit energies (cf. §19.6), are dragged along to some extent on the trailing sides of the boundaries and this involves activation energies more comparable with those of lattice diffusion. The boundaries may then not be detectably mobile at temperatures below $0·4T_m$.

Insoluble impurities have an even stronger effect. The boundaries become attached to inclusion particles to economize the incoherent interfacial area and have then to be dragged away from them by the driving forces for grain growth. As the grain size increases, the driving forces per unit area decrease, due to the reduction in the number of triple junctions, and eventually a stable grain size is reached at which the anchoring forces balance the driving forces. From an estimate of these forces *Zener* (cf. *C. S. Smith*, 1948, *Transactions A.I.M.E.*, **175**, 15) has deduced that this stable grain size is about $4r/3\alpha$ in a metal which contains a volume fraction α of inclusions of radius r. For example, if $\alpha = 10^{-4}$ and $r = 10^{-8}$ m the stable size is $1·3 \times 10^{-4}$ m. The grain size of many industrial metals is controlled in this way (e.g. Al-killed steels).

The measured activation energies for grain growth in impure metals are often anomalously large because, as the temperature is raised, increased solubility and coarsening of inclusions releases the boundaries for faster movement. A striking effect often occurs when a fine-grained metal is heated just below the solubility temperature of a precipitate the particles of which are anchoring the boundaries. At first there is normal grain growth. Then, as the particles dissolve and coarsen, a few boundaries are released and sweep quickly through the metal, giving a very coarse grain size. This is *abnormal grain growth*. By contrast, if the same fine-grained metal is quickly heated to a temperature above the solubility limit, so that all grain boundaries are released together, only normal grain growth occurs.

Surface energies influence the structures of multi-phase alloys. Small amounts of a second phase tend to form in the grain boundaries of a primary phase, because the total interfacial energy can then be reduced, parts of the large-angle grain boundaries there becoming the (incoherent) boundary of the second phase. Fig. 19.11 shows a typical arrangement, a piece of second phase β in a triple junction between three grains of the primary

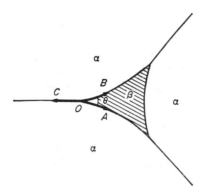

Fig. 19.11 A second phase at the junction of three grains of a phase

phase α. At high temperatures the dihedral angle or *contact angle* θ adjusts itself to bring the forces $\gamma_{\alpha\alpha}$ and $\gamma_{\alpha\beta}$ (i.e. OC, and OA and OB) into balance, giving

$$\gamma_{\alpha\alpha} = 2\gamma_{\alpha\beta} \cos \frac{\theta}{2} \qquad\qquad 19.32$$

The equilibrium shapes of the included phase for various dihedral angles are shown in Fig. 19.12. This angle can range from 0° to 180°. When $2\gamma_{\alpha\beta} \leqslant \gamma_{\alpha\alpha}$ it is zero and the second phase can spread as a continuous film between the grain boundaries of the main phase. When $2\gamma_{\alpha\beta} > \gamma_{\alpha\alpha}$ the dihedral angle is not zero and complete spreading cannot occur.

These conclusions have great practical importance when the second phase has a low melting point. If the contact angle is zero, heating above this melting point will cause the alloy to fall to pieces along its grain boundaries, since each of these is continuously enveloped in a film of liquid. This is *hot-shortness*, a problem in metals which contain trace impurities that produce liquid phases that 'wet' the grain boundaries. A good example is provided by bismuth and lead in copper. Both are insoluble and a liquid phase is formed at high temperatures. A trace of bismuth makes copper extremely brittle by spreading round its grain boundaries. The addition of an equal amount of lead makes the copper ductile again by increasing the interfacial energy between the copper grains and the Bi-Pb liquid phase, so preventing complete spreading.

Intergranular penetration by liquid metals also raises problems in the use of containers for liquid metals, e.g. in liquid-metal cooled heat exchangers. On the other hand, wetting of grains by a liquid metal is essential to the production of useful *cermets* (ceramic grains bonded by metal films) by *liquid-phase sintering*. The best example here is *cemented tungsten carbide*, for high-speed cutting tools, made by pressing together mixed powders of tungsten carbide and cobalt and then heating this *powder compact* at such a temperature that the cobalt melts and binds together the carbide grains.

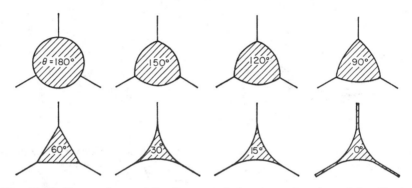

Fig. 19.12 Shape of a particle of a second phase as a function of the dihedral angle (after C. S. Smith)

We saw in § 18.9 that when two adjoining grains bear a certain orientation relationship to each other and their boundary is symmetrically oriented, the structure in the boundary is coherent, or partly so. Measurements of interfacial energy γ as a function of orientation θ have shown that coherent boundaries do indeed have low energies. The curve drops sharply, in a cusp, as the ideal orientation for coherence is approached, where it reaches a value comparable with that of a stacking fault (cf. § 17.6). It is difficult to estimate the energies of coherent boundaries theoretically, because there are no long-range elastic fields and the elementary electron theory, which ignores crystal structure, is of no help.

Interfacial energy is responsible for the formation of *annealing twins* (cf. Fig. 19.13) which are a common feature of certain metals and alloys with low stacking fault energies (e.g. α-brass) in the worked and annealed condition. Annealing twins are not deformation twins. They form, not by mechanical shear, but as part of the process of grain growth, which involves localized disordering and re-ordering of the crystal structure. As a grain grows, it runs into other grains and forms boundaries with them. It sometimes happens that such a boundary has high energy but can have lower energy if this part of the growing grain takes up a twinned configuration relative to its other parts. Such a twin, once started, then continues to grow until a further change of the grain boundary orientation relationships requires this twin also to change. The number of annealing twins in a grain thus increases in proportion with the number of new grains encountered, which has been confirmed experimentally, and also with decrease in the stacking fault energy of the material, since such a fault is structurally similar to a coherent twin interface. Thus α-brass is profusely twinned, in the worked and annealed condition, whereas aluminium is not. The abundance of annealing twins in such metals has in fact been used as a means of estimating the energies of twin interfaces.

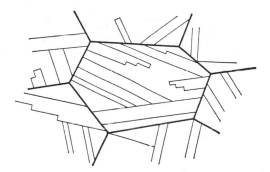

Fig. 19.13 Annealing twins formed during grain growth

It is important to notice that eqn. 19.30 does *not* apply to triple junctions which involve coherent interfaces. The reason is that if such an interface deviates slightly from its symmetrical orientation interfacial dislocations must then appear in it (cf. Fig. 18.12) and these increase its energy sharply. A simple triangle of forces no longer represents the equilibrium and a more general formula is then required, which takes account of the variation of energy with orientation of the interface.

19.6 Misfit energies of solute atoms and inclusions

The size factor rule of Hume-Rothery shows clearly that misfit strains round solute atoms are an important factor in solid solutions. Although elasticity theory cannot be strictly applied on the atomic scale it does nevertheless give a simple rough guide to the energies involved. It can be shown that the elastic energy of an isotropic elastic medium, strained by putting a rigid sphere of radius r_1 into a hole in it of unstrained radius r_0, is given by

$$U = 8\pi\mu r_0^3\epsilon^2 \qquad\qquad \textbf{19.33}$$

where $\varepsilon = (r_1 - r_0)/r_0$ and μ is the shear modulus. If $\mu = 4 \times 10^{10}$ N m^{-2}, $r_0 = 1.5 \times 10^{-10}$ m, and $\varepsilon = 0.14$ (corresponding to Hume-Rothery's rule; cf. §14.3), this gives an elastic energy of order 40 000 J mol^{-1} of solute. *Darken and Gurry* (*Physical Chemistry of Metals*) have noted that primary solid solubility is severely limited (e.g. < 1 atomic per cent) if the free energy of solution exceeds about $4RT$ at the temperature concerned. For example, the average heat of mixing in Fig. 14.17 is about 40 000 J mol^{-1} and the average solubility is about 1 per cent at 1000 K. We can thus roughly explain the Hume-Rothery 14 per cent rule.

The same argument could be applied to particles of second phases in alloys. The elastic energies of such phases can, however, be substantially

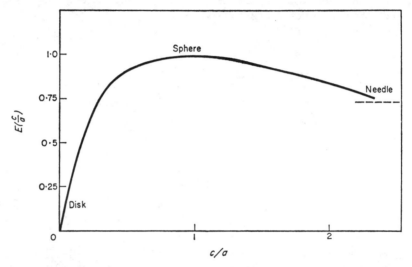

Fig. 19.14 The strain energy E of an ellipsoidal particle as a function of its shape; a is the equatorial diameter and c the polar diameter of the ellipsoid (after Nabarro)

reduced by other effects. If such a particle has an incoherent interface with the matrix, the size of hole in which it fits can be freely adjusted towards the ideal value by the removal or addition of material at the interface in a process of internal 'crystal growth' brought about by *vacancy creep* (cf. § 19.9). Secondly, the elastic energy can be reduced, even when the interface is coherent, if the particle grows into the shape of a flat disc rather than a sphere. Fig. 19.14 shows the results of the calculations of *Nabarro* (1940, *Proc. Phys. Soc.*, **52**, 90). He considered an ellipsoidal hole, with semi-axes, a, a and c, which contains a particle of different natural volume. When $c \ll a$ the particle is disc-shaped, when $c = a$ it is spherical and when $c \gg a$ it is needle-shaped. The strain energy clearly favours the disc. The reason is that the sides of the disc, near the centre, can move in or out elastically rather easily, rather as the faces of a long crack can be prised apart by a small force (cf. § 21.10). A third method of relieving elastic energy round a misfitting particle is by localized plastic deformation. This occurs mainly during heat-treatment processes as a means of accommodating strains due to differences in the coefficients of thermal expansion between the particle and its matrix. The particle in effect makes room for itself by 'punching' out edge dislocation loops into the surrounding crystal. Tangled clusters of dislocations round such particles are frequently observed in electron microscope studies.

A misfitting solute atom can reduce its elastic energy by migrating to some other irregularity in the material where it may find sites more suitable to its own size. Incoherent boundaries and the cores of dislocations provide such sites and the solute atoms may become elastically 'bound' in such sites by a binding energy which is an appreciable fraction of the misfit energy calculated above. It is obvious in Fig. 19.8, for example, that a solute atom which expands the lattice will be attracted to the expanded region of the dislocation, below the half-plane and one that contracts the lattice will be attracted similarly to the contracted side. This attraction and migration of foreign atoms to dislocations and grain boundaries is responsible for certain *ageing* effects when metals and alloys are rested at room temperature or heated at moderate temperatures (cf. § 21.2).

19.7 Energies of point defects

The most important point defect in a metal is the *vacancy* or *Schottky defect* (cf. Fig. 19.15), i.e. the hole left by a missing atom in a crystal. The

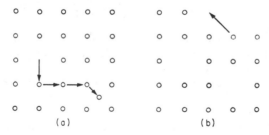

Fig. 19.15 (a) A Frenkel defect and (b) Schottky defects in a crystal

other simple point defect, the *interstitial*, i.e. an extra atom inside a crystal, has too high an energy of formation to appear in most metals except in special circumstances. For the same reason a vacancy is hardly ever created inside a perfect crystal because this would require the simultaneous creation of an interstitial; only vacancy-interstitial *pairs*, called *Frenkel defects*, can be created inside a perfect crystal (cf. Fig. 19.15).

To avoid this difficulty, the vacancies must be created at discontinuities such as free surfaces, grain boundaries and dislocations, where an atom can vacate its hole without having to become an interstitial. The process is obvious at a free surface (cf. Fig. 19.15(b)). At an incoherent grain boundary atoms can leave the crystals to become part of the disorganized material in the boundary and this material can subsequently crystallize on to these crystals to prevent the boundary becoming too thick. Experiments have proved in fact that such boundaries are as good as free surfaces for both creating ('*sources*') and annihilating ('*sinks*') vacancies. Vacancies can also be created or annihilated at *jogs* on edge dislocations, as shown in Fig. 19.16. The corresponding addition or removal of a row of atoms along the edge of the half-plane 'lifts' the dislocation from one slip plane to the next. This process of *climb* is very important for the mechanical behaviour of metals at high temperatures (cf. § 21.9).

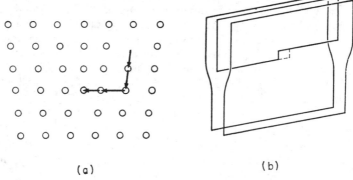

(a)   (b)

Fig. 19.16 (a) Climb of an edge dislocation by the creation (or annihilation) of vacancies; (b) a jog, at which this process takes place

Vacancies occur *naturally* in crystals at high temperatures. This is because their energies of formation are relatively small. The argument at the start of § 12.6 is directly applicable here. The formation of one vacancy, by a thermodynamically reversible process, at a given site in a crystal requires a work w, essentially to take the atom from this site and place it on the surface of the crystal. The chance of this site being empty is thus $\exp(-w/kT)$. A crystal of N atoms provides N sites where this can happen. Hence the number n of vacancies in the crystal is given by

$$n = Ne^{-(w/kT)} = Ne^{-(W/RT)} \qquad\qquad 19.34$$

provided $n \ll N$, where W is the energy per mole. As usual, we can write $W = E - TS$ where E is the energy of formation and S is the entropy of formation of the vacancy at a given site. This entropy comes from the slower vibrations of the atoms next to the vacancy, due to the lower cohesion there (cf. eqn. 19.14).

The thermal formation of vacancies can be detected from small changes in several physical properties, e.g. electrical resistivity, lattice parameter, density, thermal expansion and specific heat. Such changes can be observed directly at temperatures near the melting point or, in some cases, preserved by quenching rapidly from such temperatures. Vacancies can also be studied in the field-ion microscope and electron microscope. Such experiments have shown that about 1 site in 10^4 becomes vacant, near the melting point, and that the energy of formation is typically of order $10kT_m$ and the entropy of order k.

The energy of formation is obviously related to the cohesive energy of the crystal. When an atom is taken from the inside to the surface of the crystal, some of its bonds are broken. When the atom then evaporates the remaining ones are broken. Hence the energy of formation should be comparable with the energy of vaporization. In practice it is about one-quarter to one-half of it, because slight rearrangements of atomic and electronic structure round the vacancy allow the energy to be reduced. A fair estimate of the energy of a vacancy in an electronically simple metal can be made from the free electron theory. We have seen in § 19.1 that the energies of free electrons are sensitive to the specific volume of the metal. The calculated energy of a vacancy, due to this effect, is of order 1 eV.

An interstitial atom must make room for itself by pushing its neighbours away. It is thus a centre of intense local strain and elasticity theory can obviously be used to estimate its energy roughly. For an order of magnitude value we substitute $\epsilon \simeq 0.5$ in eqn. 19.33 to obtain, with the other values given there, $U \simeq 6$ eV. This is probably an overestimate but it is clear from eqn. 19.34 that with values of this order very few interstitials could exist in thermal equilibrium at temperatures in the range 1000 K.

Large formation energies similarly forbid the presence, except rarely, of most other defects as equilibrium features of crystals. Next to the single vacancy, the *divacancy* is probably the most abundant. When two vacancies come together there is some economy in the number of broken bonds, which may lead to a binding energy between them of order 0·3 eV. This would typically allow something like 1 per cent of the vacancies to exist in pairs, near the melting point. Divacancies and larger clusters can, of course, exist more abundantly as *non-equilibrium* defects produced by quenching, irradiation or plastic deformation. In plastic deformation the defects tend to be produced in lines, due to the coming together of opposite dislocation lines on neighbouring parallel slip planes. An opportunity for this is provided by screw dislocations which intersect those planes. When a dislocation, gliding in such a plane, meets and cuts through an intersecting screw dislocation, the part of this dislocation that glides round one side of the screw is raised or lowered by one layer relative to the part that glides round the other side of it.

Point defects can move about inside crystals by the thermally activated jumping of atoms into them, as shown in Fig. 19.15(a) and 19.16(a). This mobility enables the excess vacancies, temporarily retained in a metal by fast quenching from a high temperature, to disappear by migrating to free surfaces and other sinks. The accompanying changes of physical properties such as electrical resistivity enable the process to be followed experimentally and, from measurements of the rate of decay at various ageing temperatures, the *activation energy of migration J* of vacancies can be determined. The value is usually comparable with that of the energy of formation. An interstitial, on the other hand, is usually much more mobile because the crystal round it is too badly distorted to hold it firmly in place.

Quenched vacancies near a free surface or a grain boundary generally disappear there but those further inside the crystal, away from any large sinks, more often run into one another and form *clusters* instead. If there are atoms of volatile impurity elements in the crystal, which try to separate out in the form of gas 'bubbles', the vacancy clusters generally develop as small cavities containing this gas. In the absence of volatile elements, however, the clusters usually develop in 'collapsed' forms to avoid the surface energy associated with a cavity. The most common collapsed form is a *dislocation ring*. If a number of vacancies gather together in one atomic layer, usually a close-packed plane, they form a disc-shaped hole in this sheet of atoms. Above and below this hole, the neighbouring sheets can collapse inwards, into the hole, so forming a stacking fault enclosed by a ring of edge dislocation along the rim of the disc. This is an imperfect dislocation but it can be converted to a perfect one by sliding the collapsed sheets along each other, in the disc, to eliminate the stacking fault. These dislocation rings are commonly seen in quenched and aged metals under the electron microscope (Plate 7A).

19.8 Diffusion

Diffusion in the solid state is important in metallurgy because many processes for improving metals and alloys by *heat-treatment* depend on it. The general equations of diffusion were discussed in § 12.4. We shall now consider the atomic processes. The frequency of jumping f, in eqn. 12.10, is the vibrational frequency (in the required direction) multiplied by the probability that the vibrating atom has the activation energy to make the jump. We can thus write

$$f \simeq \tfrac{1}{3}\nu e^{-(F_D/RT)} \qquad\qquad \textbf{19.35}$$

where ν is the vibrational frequency; the $\tfrac{1}{3}$ allows roughly for the fact that only about $\tfrac{1}{3}$ of randomly directed vibrations are in the required direction and F_D is the free energy of activation. Writing $F_D = E_D - TS_D$ and substituting into eqn. 12.10, we obtain eqn. 12.22 with

$$D_0 \simeq \tfrac{1}{6}b^2\nu e^{S_D/R} \qquad\qquad \textbf{19.36}$$

If $b^2 = 10^{-19}\,\text{m}^{-2}$ and $\nu = 10^{13}\,\text{s}^{-1}$, then $b^2\nu/6 \simeq 1\cdot5 \times 10^{-7}\,\text{m}^2\,\text{s}^{-1}$. The most reliable experimental measurements on the diffusion of lattice atoms in metal crystals give D_0 in the range 10^{-5} to $10^{-3}\,\text{m}^2\,\text{s}^{-1}$. C. *Zener*

(Chapter 11, *Imperfections in Nearly Perfect Crystals*, edited by W. Shockley) has emphasized that an appreciable entropy contribution is expected because the surrounding atoms temporarily suffer a severe elastic strain while the jump is taking place. He has estimated that

$$S_D \simeq \beta E_D / T_m \qquad\qquad \mathbf{19.37}$$

where T_m is the absolute melting point and β ($\simeq$ 0·25 to 0·45) is a parameter that depends on the elastic constants.

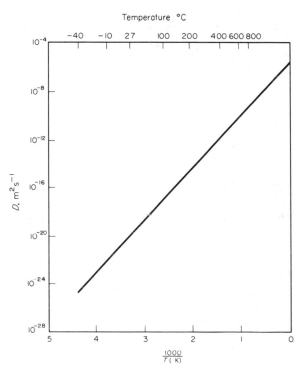

Fig. 19.17 Diffusion coefficient of carbon in ferrite (after *C. Wert*, 1950, *Phys. Rev.*, **79**, 601)

A good example of the application of eqn. 19.36 is in the diffusion of interstitially dissolved carbon and nitrogen atoms in B.C.C. iron (cf. § 14.5), for which very accurate data are available, summarized in Fig. 19.17. The atomic process here is simply the jumping of such an atom from one interstitial site to a neighbouring one. The activation energy E_D is then required mainly as strain energy, for pulling the neighbouring iron atoms apart to let the jumping atom through. This strain energy leads to a temporary lowering of elastic constants and vibrational frequency (cf. eqn. 19.14) and hence to an increased vibrational entropy. Zener deduces for carbon in ferrite that $\beta = 0.43$, that exp $(S_D/R) = 12$, and that $D_0 = 2.6 \times 10^{-6}$, in good agreement with the experimental values, $D_0 = 2 \times 10^{-6}$ m²s⁻¹ and $E_D = 84\,100$ J mol⁻¹.

The small atoms of C, N and O, interstitially dissolved in B.C.C. transition metals such as Fe, Nb and Ta, are able to diffuse rapidly along the interstitial channels, with an activation energy usually of order $5RT_m$. Hydrogen atoms diffuse even more rapidly, with an $E_D \simeq 30\,000$ J mol⁻¹ in several metals. We have to distinguish here between diffusion and *permeation*. Molecular hydrogen gas does not leak appreciably out of a sealed metal container because the molecules have to dissociate to atoms (e.g. on the surface) before the hydrogen can go into the metal. The rapidity of the diffusion becomes more evident when the hydrogen is brought to the metal in atomic form, e.g. through electrolysis or chemical reaction.

For many years the atomic mechanism of self-diffusion and diffusion of substitutional alloy atoms in metals was controversial, but it has now become clear that the main process—certainly in the full metals—is the migration of vacancies. There is the following evidence:

(1) *The activation energy.* An atom of the crystal lattice can jump into an adjacent lattice site if (i) this site is vacant, the probability of which goes as exp $(-W/RT)$, (ii) the atom has the activation energy for the jump, the probability of which goes as exp $(-J/RT)$. The total activation energy for diffusion of the atom, i.e. $E_D \, (\simeq F_D)$, should thus equal $W + J$. Experiments have shown that this is generally so in practice. Most other possible processes of self-diffusion require the atoms to pass through interstitial-like configurations of much higher energy. Some experimental values of E_D for self-diffusion in cubic metals, obtained from the migration of radio-isotopes, are given in Table 19.1. In crystals of lower symmetry diffusion is generally anisotropic.

TABLE 19.1.   SELF-DIFFUSION DATA IN METALS

Metal	E_D (kJ mol⁻¹)	E_D/RT_m
Pb	100	20
Ag	184	18
Au	176	16
Cu	197	17·5
Co	280	19
γ-Fe	310	20
α-Fe	280	18·5
Nb	439	19
W	549	19·5

(2) *The Kirkendall effect.* In binary substitutional alloys it is possible for one species of atom to diffuse *faster* than the other. This would be impossible if the atoms moved by simply exchanging places, since every jump of one

atom would require a compensatory jump of the other atom in the opposite direction, but it can happen in vacancy diffusion since we then have in effect a 'ternary' alloy; atoms of species A, atoms of species B, and vacancies V. A rapid flow of A in one direction can then be compensated by an equal flow of $B + V$ in the reverse direction, with the V flow compensating for the slowness of the B flow.

This unequal diffusion of alloys is demonstrated by the *Kirkendall effect*. Fig. 19.18 shows a typical specimen for this experiment. A block of α-brass, with fine molybdenum wires to mark the original position of its surface, is mounted in a block of copper and annealed. As the copper and zinc interdiffuse across the interface, the markers move together by a distance greater than can be accounted for by the change of lattice parameter. More atoms have left the brass than have entered it, i.e. the zinc atoms move faster than the copper and the centre block has shrunk in consequence. In effect, vacancies have entered the centre block and disappeared on sinks in it.

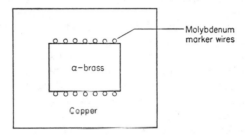

Fig. 19.18 A specimen for demonstrating the Kirkendall effect

In a binary alloy of A and B we must give each component its own diffusion coefficient, D_A and D_B. The Kirkendall effect occurs when $D_A \neq D_B$. If no porosity is generated by the unequal migrations, an *overall chemical diffusion coefficient* D can be defined,

$$D = c_B D_A + c_A D_B \qquad\qquad 19.38$$

as has been shown by *L. S. Darken* (1948, *Transactions A.I.M.E.*, **175**, 184), where c_A and c_B are the respective atomic concentrations. When $c_A \ll 1$ then $D \simeq D_A$. In practice, porosity is often produced by the Kirkendall effect, particularly when the counter-migrating vacancies condense on foreign inclusions and form cavities round them.

Measured diffusion curves in alloys rarely have the perfectly symmetrical form of Fig. 12.2. The diffusion coefficient is generally a function of composition, e.g. increasing by an order of magnitude in many copper alloys as the composition increases from pure copper up to the solubility limit. This is not surprising since the cohesive forces in alloys generally vary with composition and these govern the activation energy for diffusion. Important additional effects are caused by interactions between vacancies and solute atoms. Both electronic and elastic effects can lead to an association between a vacancy and a solute atom, with a binding energy that may amount to the order of $\frac{1}{3}W$. The movements of the atom and its vacancy are then *correlated*. The pair cannot migrate purely by the jumping of the solute atom into the vacancy, since such jumps merely shuttle this atom between the same two adjoining sites. Migration through the crystal can occur only if the vacancy changes places also with the solvent atoms next to the solute atom.

In general the diffusion coefficients of substitutional solute atoms lie within a factor 10 of that of the parent metal. Solutes of less cohesive elements tend to migrate faster than that of the solvent (e.g. $D_{arsenic} = 7 \cdot 6 \, D_{copper}$ in copper at 1000 K) and those of more cohesive elements migrate more slowly (e.g. $D_{nickel} = 0 \cdot 12 \, D_{copper}$ in copper at 1000 K).

In substitutional diffusion, values of D_0 orders of magnitude smaller than the theoretical value are often observed, always accompanied by low values of E_D. These anomalies are particularly noticeable at low temperatures and are mainly due to rapid diffusion along a few special channels through the solid which act as 'short-circuiting' paths, the structure of which allow atoms to move easily, with low activation energies. The small D_0 then follows from the small 'cross-section' of these conducting channels. At high temperatures, where diffusion is rapid everywhere, the main flow goes through the normal regions of the solid simply because the entire cross-section of the specimen is open to this type of diffusion; normal values of D_0 and E_D are then obtained. At low temperatures, however, this diffusion practically ceases because RT is too small in relation to the E_D of normal diffusion. Flow is then restricted to the special channels where E_D is small. At such temperatures diffusion is *structure-sensitive*, i.e. it can be increased by increasing the content of dislocations, grain boundaries and other faults in the material.

Three possibilities have been considered for the conducting channels: grain boundaries, dislocation lines and chemical inhomogeneities. The idea about chemical channels is that, in brass for example, a vacancy changes places with zinc atoms more often than with copper atoms and so confines its movements mainly to the network of adjoining zinc atoms in the alloy. In fact if zinc is removed from the surface by a suitable solvent (e.g. sea water, on brass condenser tubes in marine engines) it is possible to *dezincify* the brass, leaving behind a porous skeleton of copper.

Fig. 19.19 shows the effect of diffusion along a grain boundary. Although the atoms migrate down the boundary at great speed, their concentration in the nearby lattice rises almost as fast because sooner or later most of those atoms wander off the boundary into the neighbouring crystal. The boundary thus acts as an 'irrigation channel' which feeds the adjoining crystals with atoms and so brings these atoms to regions that would be quite inaccessible by direct lattice diffusion in the time allowed. From a mathematical analysis of the shapes of concentration contours near grain boundaries it is possible to determine the ratio D_b/D_l of diffusion coefficients in the boundary and lattice, where D_b is based on the assumption of a certain definite thickness, e.g. 5 Å, for the boundary (*J. C. Fisher*, 1951, *J. appl. Phys.*, **22**, 74). Energies of activation for grain boundary diffusion are usually about $10RT_m$; and $D_b \simeq 10^5 D_l$ at $T \simeq 0.6T_m$.

Experiments have shown that diffusion is faster in incoherent large-angle boundaries than in small-angle ones. As the orientation difference across a small-angle boundary is increased, the D_0 for boundary diffusion correspondingly increases but E_D stays constant (at about $10\ RT_m$). This suggests that the dislocations in these boundaries act as 'pipes' along which the atoms can diffuse rapidly (*D. Turnbull*, 1954, *Bristol Conference, Physical Society*). The properties of such a pipe, e.g. its E_D, remain constant as the orientation of the small-angle boundary is increased but the number of such pipes, and hence the D_0 of the whole boundary, increases in accordance with eqn. 18.29.

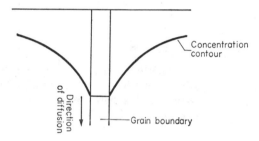

Fig. 19.19 Diffusion at a grain boundary

Diffusion on a free surface is generally faster than that inside the crystal but it can vary greatly with the structure of the surface. An atom adsorbed on a perfect close-packed face is weakly bound to the crystal and can clearly move about on it easily. On the other hand, to lift an atom out of a site in that close-packed face and move it to a site on the open surface, leaving behind a surface vacancy, is a process which requires a large activation energy. The steps along the edges of incomplete layers on close-packed faces are more suitable for surface diffusion. Experimentally, surface diffusion has been studied through the rate of formation of grooves at grain boundaries (cf. Fig. 19.10), from the blunting of tips in the field microscope, and from the mobility of small gas bubbles inside crystals. Values of E_D for surface diffusion range from about $10\ RT_m$ (tungsten) to $20\ RT_m$ (copper).

19.9 Vacancy creep and sintering

Under non-equilibrium conditions more vacancies may be generated than annihilated, or vice-versa, at a given source such as a free surface. We saw an example of this in the Kirkendall effect. As a result atoms are taken from one part of the material and deposited at another part, so that in general there is a change of shape. This slow deformation is called *vacancy creep* and plays a part in many metallurgical processes.

The systematic drift of the vacancies in one direction, superposed on their otherwise random movements, represents a response of the system to a force. This force may have any origin, e.g. a simple load on the material, or a chemical potential due to a concentration gradient, but is always defined in terms of the free energy change which results from the deformation produced by the drift flow.

This is a special case of the general theory of diffusion in force fields. Consider any system of randomly migrating particles. Let a force F act on each particle along the x axis, so that Fx is the work done when the particle moves a distance x. Given a certain concentration gradient, this system can be in statistical equilibrium, cf. the atmosphere in the earth's gravitational field. From § 6.6, the equilibrium concentration c is given by

$$c = c_0 e^{Fx/kT} \qquad \textbf{19.39}$$

i.e.

$$\frac{dc}{dx} = \frac{Fc}{kT} \qquad \textbf{19.40}$$

The drift flow due to F then exactly balances the diffusion flow due to the gradient. If there is no gradient, the drift flow due to F remains the same. Let v be the velocity of a particle in this drift flow. The *flux*, i.e. the number which pass through a plane of unit area in unit time, is then vc. Hence, from eqns. 12.9 and 19.40,

$$v = \frac{DF}{kT} \qquad \textbf{19.41}$$

This is *Einstein's formula*. Applied to vacancy creep, D is then the diffusion coefficient of vacancies. It is convenient however to identify D with the coefficient of self-diffusion; the diffusion coefficient of vacancies is then D/c_v where c_v is the equilibrium concentration of vacancies, i.e. n/N in eqn. 19.34. For vacancy creep then we write eqn. 19.41 as

$$v = \frac{DF}{c_v kT} \qquad \textbf{19.42}$$

where D is the coefficient of self-diffusion.

Fig. 19.20 shows a simple arrangement for vacancy creep. The faces move in the directions of the applied stress σ, acting on them, by the creation of vacancies on the tensile faces, which then migrate in the directions shown, to be annihilated on the compressive faces. Let b^3 be the atomic volume. The movement of one atom, outwards or inwards on a face to create or eliminate a vacancy, then causes the applied force on that face to do the work σb^3. Let L be the linear size of the solid. The average path length from source to sink is then about $\frac{1}{2}L$, so that the force F on a vacancy is about $4\sigma b^3/L$. Vacancies occur at average distances b/c_v along this path. Hence the atom at an end of the path moves once during the time $b/c_V v$. Its displacement b in the direction of the stress contributes an increment $2b/L$ to the creep strain. The creep rate is thus given approximately by

$$\dot{\epsilon} = \frac{(2b/L)}{(b/c_V v)} = \frac{8\sigma b^3 D}{L^2 kT} \qquad\qquad 19.43$$

Apart from the precise value of the numerical factor, which depends on the geometry of the model and the rigour of the calculation, this is the *Nabarro-Herring* equation of vacancy creep.

Vacancy creep is a slow process but, because the creep rate varies linearly with the stress, it becomes significant at very low stresses where plastic flow ceases. The presence of the factor D/L^2 shows that it is particularly

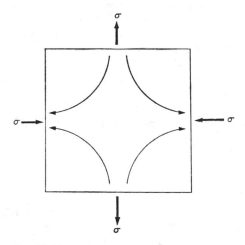

Fig. 19.20 Deformation by vacancy creep

important at high temperatures in fine-grained polycrystals, where the grain boundaries act as sources and sinks. For example, if $b^3 = 10^{-29}$ m^3, $D = 10^{-13}$ m^2 s^{-1}, $L = 10^{-3}$ m, and $kT = 10^{-20}$ J (i.e. $T \simeq 750$ K), then a stress $\sigma = 10^7$ N m$^{-2} \simeq 1500$ psi produces a creep rate $\dot{\epsilon} \simeq 10^{-8}$ s^{-1}, i.e. the material extends about 0·5 per cent per week. Vacancy creep in fact is an important mode of deformation in certain practical situations, e.g. in magnesium components used at high temperatures in nuclear reactors; see also § 22.13.

Vacancy creep contributes to the process of *sintering* by which powders become consolidated into dense solids when heated at temperatures just below their melting points, which is important for making solid bodies from powders of substances such as tungsten and alumina that are too refractory for melting and casting. The driving force for sintering comes of course from the surface energy, which diminishes with the surface area as the particles join together. In the early stages of sintering the pores are inter-connected and the compacting of the material can then occur by surface diffusion and also to some extent by plastic deformation. The final stage of sintering, however, involves the filling up of isolated spheroidal pores in the body, and this is usually done by the migration of atoms into them by vacancy creep. The effective stress for this creep, due to the surface tension γ on a pore of radius r, is $2\gamma/r$, as in the elementary theory of pressure in bubbles. These pores are sources of vacancies and for rapid sintering it is necessary to have some nearby sinks. Normally these are provided by grain boundaries since each powder particle becomes one grain and its pores are situated round its boundary. The drift flow thus takes place between the pore and its adjoining grain boundary. In the very late stages of sintering, however, grain growth may occur. Some pores are then left 'stranded' inside the growing grains and they then sinter extremely slowly. To obtain full densification by sintering it is thus necessary to prevent this grain growth, which can be done by including certain impurity particles in the powder compact to 'anchor' the grain boundaries.

Negative sintering, i.e. expansion of cavities, can also occur by vacancy creep in solids which contain dissolved gases, so that gas pressure in the cavities provides the driving force for their expansion, or by applied tensile stress. These effects are important in certain types of radiation damage (i.e. *swelling*; cf. § 19.10) and in *cavitation creep failure* in stressed materials at high temperatures (cf. § 21.13).

19.10 Radiation damage

Vacancies and interstitials are produced in large numbers in materials irradiated in nuclear reactors. We saw in § 2.5 that when a fast particle collides with a lattice atom it may give to this atom an energy up to E_m, as in eqn. 2.13, and that a fast neutron ($E \simeq 2$ MeV) typically gives an atom of medium weight some 10^5 eV in such a collision. In a reactor with a fast neutron flux of 10^{16} m^{-2} s^{-1} about 1 atom in 10^4 would be so hit during one year.

The energy E_d required to knock an atom irreversibly out of its site inside a crystal is clearly comparable with, but larger than, the binding energy. Both theory and experiment have shown that, typically, $E_d \simeq 25$ eV. The average energy received by a struck atom, about $\frac{1}{2}E_m$, is far higher than this and most of it is taken up as kinetic energy, i.e. the atom becomes a fast moving ion in the crystal. Unlike the neutron, however, it interacts very strongly, through its electrically charged nucleus and electrons, with the electrons of the atoms through which it passes. Most of its energy is imparted to these electrons and in a metal this is rapidly converted into heat with little or no permanent damage. Metals are thus, because of their free electron structure, much less sensitive to radiation damage than most other solids.

The fast moving ion is quickly slowed down, by this loss of energy, to a speed comparable with that of a free electron. It is then an almost neutral atom and can no longer give up much of its energy to the electrons. There is therefore a 'cut-off' energy L_c for the ion, below which the electrons are not appreciably excited. A rough estimate of this is

$$L_c \simeq ME_F/16m \qquad\qquad 19.44$$

where M and m are the respective masses of the ion and an electron and E_F is the Fermi energy. Typically, L_c is 10^4 to 10^5 eV. The remaining process for dissipating the energy is by atom-atom collisions. While the kinetic energy is still high this may occur by direct confrontation of the two nuclei, when the centre of the moving atom happens to pass closely to that of a lattice atom. This produces a *Rutherford collision*, in which the struck atom usually acquires sufficient energy to become itself a fast moving particle,

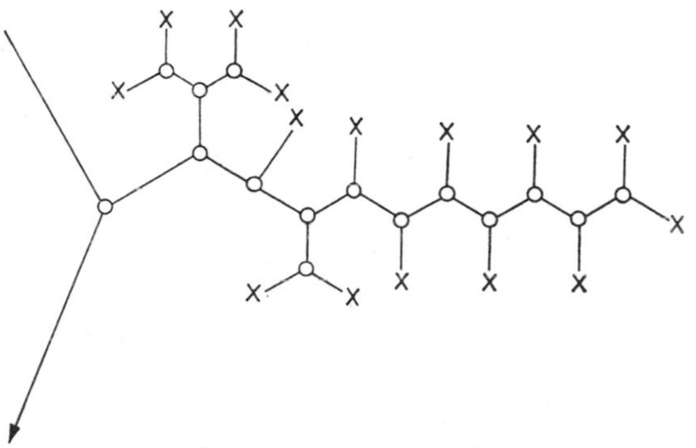

Fig. 19.21 A spike of displaced atoms (x) and vacancies (o)

capable of producing secondary knock-on damage. As the kinetic energy drops lower, the electron clouds of the atoms prevent the deep penetration of one atom by another. In effect, the electron clouds bounce off one another and the atoms collide rather like hard spheres at this stage. The collision cross-section is now very large, of atomic rather than nuclear dimensions, so that in this final stage the damage develops as a highly concentrated cluster, or *displacement spike*, of vacancies surrounded by interstitials, as in Fig. 19.21.

The cascade of displaced atoms increases until no moving atom has enough energy to displace another one. The lowest kinetic energy at which a moving atom can displace another, without itself getting trapped in the vacancy so created, must be about $2E_d$. Each branch of the cascade then ends at a displaced atom with an energy of $2E_d$ or less. The number of such branches is thus about $E/2E_d$, where E is energy of the primary atom of the cascade.

In most situations, where $E_m < L_c$, we can take $E \simeq \frac{1}{2}E_m$. The number of vacancy–interstitial pairs produced from one primary is then about

$$n \simeq \frac{E_m}{4E_D} \simeq \frac{E_m}{100} \qquad\qquad 19.45$$

For copper, $n \simeq 1000$.

Slow neutrons sometimes cause severe radiation damage, by *transmutation* (cf. § 2.5). Not only does this produce foreign atoms such as helium in the material; these atoms are often created with great kinetic energy (e.g. as alpha particles) and are primary sources of knock-on damage. The extreme example is the fission of uranium in which two nuclei of elements such as Sr, Kr, I, Xe are thrown apart with a total kinetic energy of order 200 MeV. Most of their energy is immediately converted to heat, in a metal, but nevertheless some 10^4 to 10^5 atoms are displaced by one fission event. As a result every atom in natural uranium is displaced about once a week in a highly-rated power reactor.

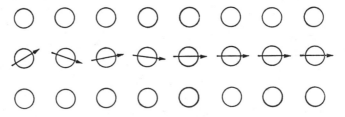

Fig. 19.22 Sequence of focussed collisions

The picture outlined above is complicated by several additional effects. First, the heat released in the damaged region enables a great deal of atomic rearrangement and elimination of point defects, by recombination, to take place before it is conducted away. Only a minority of defects survive this *thermal spike annealing*. Second, much of the kinetic energy is dispersed in *replacement collisions* and *focussed collisions*. In a replacement collision the incident atom enters and stays in the site vacated by the atom it strikes. Although they do not increase the number of vacancies and interstitials, replacement collisions do contribute to radiation-induced diffusion in metals and to the disordering of ordered alloys. Fig. 19.22 shows a focussed collision. Each struck atom is hit by a neighbour which comes from a particular crystallographic direction, so that there is a tendency for it to be knocked forward in that direction. Some of the initial momentum thus tends to be transmitted along close-packed rows of atoms by a sequence of well-directed replacement collisions. When such a sequence reaches a free surface, the last atoms may be thrown off in this same crystallographic direction, which has been observed experimentally; in some cases, as field-ion microscopy has shown, a *surface vacancy* may then be created.

Many of the point defects which survive the thermal spike are rapidly eliminated by ordinary thermal annealing. Interstitials are particularly mobile and many fall back into nearby vacancies, even at temperatures as low as 40 K in metals such as copper. Others disappear at sinks (surfaces, grain boundaries, dislocations) or join with one another to form plate-like clusters. In most metals these processes are very active at temperatures well below room temperature. In alloys the recovery from radiation damage tends to be spread out to somewhat higher temperatures because the point defects get trapped at foreign atoms. The vacancies which survive these annealing stages also become mobile at temperatures of the order $T_m/4$ and disappear at sinks (including clustered interstitials) or form clusters themselves. At progressively higher temperatures the numerous small clusters are gradually replaced by a few larger ones, which are more stable and do not finally disappear until temperatures are reached where self-diffusion and vacancy creep are rapid (e.g. $T_m/2$).

Radiation damage can affect metals and alloys in many ways. We have already mentioned the disordering of superlattices and enhancement of diffusion. Somewhat related is the dispersal of fine precipitates, particularly by fission spikes in uranium alloys. In some materials, notably graphite, the *stored energy* associated with the point defects can be important. As the defects anneal away this energy is released as heat, which reduces the apparent specific heat of the material. After severe irradiation at temperatures below about 150°C, enough energy can be stored to raise the temperature of the graphite spontaneously when released. The energy release is then auto-catalytic since the initial temperature rise speeds up the rate of release which enables the temperature to rise still further. This self-heating in graphite-moderated reactors is overcome in practice, either by periodically annealing the reactor before the stored energy reaches the unstable level or by operating the reactor at an elevated temperature where extensive annealing takes place during normal operation.

In cubic metals, the main practical effect is *radiation hardening* (sometimes accompanied by *embrittlement*). This is brought about mostly by the defect clusters which obstruct the motion of dislocations in a manner very similar to that of fine alloy clusters in precipitation-hardened alloys. In non-cubic materials there is an additional effect due to the tendency of the point defects to gather in certain crystal planes which produces an anisotropic dimensional change, called *growth*. In graphite, for example, interstitials gather between the widely spaced basal planes and cluster together there to form small new basal layers between the existing layers. As a result, the crystal grows along the basal axis and shrinks along axes in the basal plane. Most graphite blocks used in reactors are made of textured polycrystalline material and so exhibit a bulk growth in accordance with their texture. In building such a reactor it is necessary to stack and key the blocks together in such a way as to avoid gross distortion due to growth.

Very large dimensional changes occur in textured α-uranium under irradiation. The crystal structure, orthorhombic, is reminiscent of a stack of corrugated sheets laid with their corrugations parallel and resting in one another. On irradiation, new corrugated sheets grow between the existing ones and the crystal correspondingly elongates in the direction perpendicular to the sheets and contracts, by an equal amount, along the axis of corrugation. It is quite possible for a crystal to double its length in this way. The effect leads to three practical problems, *growth*, *wrinkling* and *creep*. The large overall growth of a textured bar is quite unacceptable in a reactor fuel element. This problem is solved by producing a random texture in the bar, so that different grains grow in different directions and their overall effect cancels out. Wrinkling occurs when the grain size is large. Some grains grow out of the surface more than others and distort the wall of the fuel can. This is solved by heat treating the metal to refine the grain size.

The third problem, *irradiation creep*, is less tractable. As the individual grains grow anisotropically in different directions, they press against one another and produce internal stresses. These stresses rise to the plastic yield strength of the metal and the grains then become spontaneously plastic. In this condition they are incapable of supporting an external load, however small, without slowly yielding to it. The time required for the growth strain to reach the yield strength is a characteristic *time of relaxation* in the material. An initial state of self-stress in the material would be largely relaxed during this time. Let the rate of crystal growth be $\dot{\epsilon}_g$, the tensile yield strength be σ_y and Young's modulus be E. Then the time of relaxation t_m is given by

$$t_m = \frac{\sigma_y}{E\dot{\epsilon}_g} \qquad\qquad \textbf{19.46}$$

Typically, in a power reactor, $\sigma_y \simeq E/500$, and $\dot{\epsilon}_g \simeq 3 \times 10^{-8}$ s^{-1}. Then $t_m \simeq 1$ day. An externally applied stress σ produces an elastic strain σ/E. This strain is converted to plastic strain in a time t_m, but is also continuously renewed to support the continuing applied stress. The steady rate of creep $\dot{\epsilon}$ under the applied stress σ is then given by

$$\dot{\epsilon} \simeq \frac{\sigma}{Et_m} \simeq \frac{\sigma}{\sigma_y \dot{\epsilon}_g} \qquad\qquad \textbf{19.47}$$

For example, if $\sigma \simeq E/250\ 000$ (a stress often provided by the weight of the metal) then $\dot{\epsilon} \simeq 6 \times 10^{-11}$. This creep rate is sufficient to produce a large bend in a bar in a few weeks' irradiation. This problem cannot be solved metallurgically except by converting the uranium to a cubic crystal structure. In practice, fuel element cans are often braced to support the α-uranium fuel.

The main problem of radiation damage due to slow neutrons is *fission gas swelling*. Uranium produces krypton and xenon, about one atom for every five uranium atoms destroyed. At high temperatures these atoms gather in small pockets and form gas 'bubbles' which expand by vacancy creep or, when large, by plastically deforming or fracturing the surrounding metal. There is about $2 \, m^3$ (s.t.p.) of gas per m^3 of metal after a 'burn-up' of $0 \cdot 3$ per cent of the uranium atoms. Thus, for a 3 per cent increase in volume at 800°C a gas pressure of about 270 atm. (nearly $3 \times 10^7 \, Nm^{-2}$) is expected in the gas bubbles.

The practical problem is to keep the swelling small. Electron-microscopy studies show that initially the gas is finely dispersed in very small bubbles, only about 10^{-7}m across. In this condition there is little swelling because the surface tension of the metal keeps the gas in the bubbles at high pressure. For example, if $\gamma \simeq 1 \, N \, m^{-1}$ and $r \simeq 10^{-7} \, m$, then $2\gamma/r \simeq 2 \times 10^7 \, N \, m^{-2} \simeq 200$ atm. The problem then is to keep the gas in *small* bubbles. At high temperatures the bubbles can move through the metal by the surface migration of metal atoms from one side of a bubble to the other and if, they thus meet and coalesce, particularly on grain boundaries, the swelling can become much larger. This effect is aggravated by cracking which can occur along the grain boundaries and very large swellings (e.g. 100 per cent) may then result. To keep the bubbles fine and stable, a fine precipitate of nearly insoluble iron and aluminium intermetallic compounds is produced in the metal by heat treatment. These provide nucleation and anchoring centres for the bubbles and the swelling is then restricted to a few per cent. Difficulties can still arise if the temperature of the uranium is allowed to rise into the range where phase transformations occur (above 660°C) but these effects are not primarily due to radiation damage.

Gas-producing nuclear reactions (mainly helium) can cause difficulties in other reactor materials, e.g. materials that contain boron (even as an impurity), beryllium and lithium. Traces of helium can embrittle metals by forming fine bubbles along their grain boundaries.

FURTHER READING

As for Chapters 17 and 18; also:

BUEREN, H. G. VAN (1960). *Imperfections in Crystals*. North-Holland Publishing Co., Amsterdam.

COCHRAN, W. (1973). *The Dynamics of Atoms in Crystals*. Edward Arnold, London.

DAMASK, A. C. and DIENES, G. J. (1963). *Point Defects in Metals*. Gordon and Breach, New York.

DARKEN, L. S. and GURRY, R. W. (1953). *Physical Chemistry of Metals*. McGraw-Hill, New York and Maidenhead.

DIENES, G. J. and VINEYARD, G. H. (1957). *Radiation Effects in Solids*. Interscience London.

HENDERSON, B. (1972). *Defects in Crystalline Solids*. Edward Arnold, London.

JASWON, M. A. (1954). *The Theory of Cohesion*. Pergamon Press, Oxford.

KITTEL, C. (1953). *Introduction to Solid State Physics*. John Wiley, New York and London.

MOTT, N. F. and JONES, H. (1936). *The Theory of the Properties of Metals and Alloys*. Oxford University Press, London.

PEIERLS, R. E. (1955). *Quantum Theory of Solids*. Oxford University Press, London.

SEITZ, F. (1940). *Modern Theory of Solids*. McGraw-Hill, New York and Maidenhead.

SHEWMON, P. G. (1963). *Diffusion in Solids*. McGraw-Hill, New York and Maidenhead.

SHOCKLEY, W., Editor (1952). *Imperfections in Nearly Perfect Crystals*. John Wiley, New York and London.

THOMPSON, M. W. (1969). *Defects and Radiation Damage in Metals*. Cambridge University Press.

WILKES, P. (1973). *Solid State Theory in Metallurgy*. Cambridge University Press.

ZIMAN, J. M. (1964). *The Theory of Solids*. Cambridge University Press, London.

Chapter 20

Heat-Treatment of Alloys

20.1 Introduction

The classical alloy for heat-treatment is, of course, medium and high carbon steel. The hardening of steel, by plunging the metal red-hot into water (so producing martensite), and its toughening, by tempering the quench-hardened metal at a moderate temperature, have been known empirically and used for thousands of years. Until this century it was not realized that other alloys could be substantially hardened by heat-treatment. *Wilm's* discovery that aluminium alloys which contain copper could be hardened by quenching and ageing or tempering came in 1906 and this was the starting point for the development of duralumin and innumerable other heat-treatable alloys. It was the realization in 1920, by *Merica, Waltenberg and Scott*, that this form of hardening, which appears only during the ageing or tempering treatment, is due to the decomposition of a supersaturated solid solution produced by the quench, that opened the way to the large-scale development of age-hardening and temper-hardening alloys. Once this scientific principle was known it became a simple matter to choose, from the phase diagram, likely alloy compositions and heat-treatment temperatures; and, since the essential thing required—a solubility that decreases with decreasing temperature—is a simple and common feature of phase diagrams, there was no shortage of potentially heat-treatable alloys.

Even pure metals can in fact be hardened by quenching and ageing or tempering. The quench, from a temperature near the melting point, produces a supersaturated solution of vacancies and, on subsequent ageing or annealing at a moderate temperature, these vacancies agglomerate to form dislocation rings or other defects which harden the metal by acting as obstacles to gliding dislocations. However, because vacancies exist normally in low concentrations and are also highly mobile, this form of hardening is usually rather small and difficult to produce. To obtain large, easily managed, effects alloying is essential for two reasons. The first is thermodynamic. By greatly increasing the number of phases possible in the solid state, and by providing sloping solubility lines and eutectoid transformations in the phase diagram, alloying enables major changes of constitution to be produced in the solid by changes of temperature. Hardening can also be induced through the order-disorder transformation, by a heat-treatment process which produces a fine-scale domain structure; this *order-hardening* is not widely used, although important in some alloys of the noble metals.

Fig. 20.1 shows the type of phase diagram generally required for hardening by quenching and tempering. We first homogenize an alloy, of composition c, by heating it at a temperature T_1 long enough to take all the solute into solid solution and to disperse it uniformly (as indicated by the diffusion formula $x \simeq \sqrt{Dt}$ of § 12.4, where x is the scale of the microstructure to be homogenized). After this *solution treatment* we then cool the alloy to some lower temperature T_2, usually room temperature, quickly enough to prevent any separation of the solute. This gives supersaturated α solid solution. The hardening is then produced by allowing the solute to separate

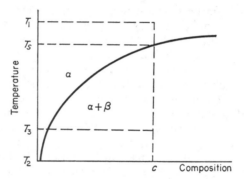

Fig. 20.1 Solid solubility line suitable for a precipitation hardening alloy

in a very finely dispersed form, to provide numerous obstacles to dislocations (cf. § 21.5), by either *ageing* the alloy at T_2 or, if the separation occurs too slowly at this temperature, by *tempering* the alloy at some slightly higher temperature T_3.

The second reason why alloying is important is kinetic. In both precipitation hardening and eutectoid transformation systems the decomposition of the high-temperature phase, during or after quenching, to a duplex structure is generally retarded by nucleation difficulties and by the need to bring the solute atoms together, through diffusion, to form the nuclei. This slowing down enables quenching to be done rather more leisurely without becoming ineffective, which is important when large industrial objects are heat-treated, and it enables the metastable states to be usefully long-lived at T_2.

Fig. 20.2 shows qualitatively the basis of the kinetic effect. The creation of a particle of a new phase in the parent solution generally requires an activation energy A which, as in the simple nucleation formula of eqn. 12.35, is extremely large at the equilibrium temperature T_e (e.g. the eutectoid temperature; or T_S in Fig. 20.1), and decreases rapidly with decreasing temperature. The total activation energy for the decomposition is then $A + E_D$, where E_D is the activation energy for diffusion, since the atoms have to migrate together to form the particle. The characteristic time of decomposition t at any temperature T, below T_e, then varies as

$$t = t_0 e^{(A + E_D)/kT} \qquad\qquad \textbf{20.1}$$

At temperatures near T_e the decrease of A with T is usually so sharp that t also decreases sharply, as shown in Fig. 20.2. At lower temperatures where $A \ll E_D$, however, the exponential term then approximates to $\exp(E_D/kT)$ and t then lengthens out with decreasing T in the usual way. The time of

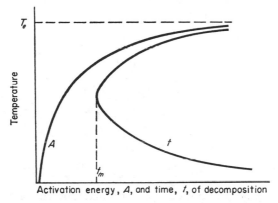

Fig. 20.2 Change of activation energy and time of decomposition with increasing undercooling below the equilibrium temperature

decomposition thus follows a C curve, as shown, with a minimum t_m. To be effective, a quench must be completed in a time short compared with t_m. Water-quenching is usually sufficiently fast for this, particularly when E_D is the value for substitutional atomic diffusion.

20.2   Decomposition of supersaturated solid solutions

A solubility line in a phase diagram refers of course to the equilibrium second phase (β in Fig. 20.1) and it is natural to expect this phase to form when the quenched solution decomposes. However, because the solution is so extensively undercooled it is often also unstable with respect to other types of precipitates, representing phases that may not exist in the equilibrium phase diagram. For example, in Fig. 20.3 the alloy c could precipitate either β or β' from homogeneous supersaturated α solution.

A more extreme possibility exists when the ageing or tempering temperature lies below the *spinodal line* of the supersaturated solution (cf. § 14.11) because this solution is then unstable with respect to *composition fluctuations within itself*. Regions which happen to be richer in solute than average tend to become still richer; and poorer regions tend to become still poorer. The driving force for this intensification of composition fluctuations is of course the free energy change in the composition range where d^2f/dc^2 is negative, as discussed in § 14.11. The rich and poor zones of the supersaturated solution develop on a very fine scale, since chance groupings of small numbers of nearby atoms can change the local composition quickly and radically; and since the nucleation difficulties which oppose fine-scale changes are absent (or nearly so).

Both the intermediate non-equilibrium precipitates and the zones of clustered solute are of course transitory structures since, although their formation releases free energy, they have higher free energy than the final

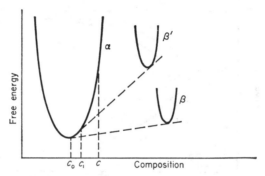

Fig. 20.3 Arrangement of free energy curves that allows more than one decomposition product from a supersaturated solution

equilibrium $\alpha + \beta$ structure. Their importance lies in the facts that they can often form more quickly than the equilibrium structure, and so can form before it when the quenched solution is aged or lightly tempered; that they are often extremely long-lived at the ageing temperature; and that, when they form on an extremely fine scale, they often produce great hardening (cf. § 21.5). Their formation in preference to the equilibrium structure is an illustration of the general principle that a reacting system follows the path of lowest activation energy.

X-ray diffraction and electron microscope observations, and the measurement of physical properties such as electrical resistivity and heat content, have shown that in supersaturated solid solutions there often develops a fine-scale clustering into a zone structure. Although not completely proved, it seems most likely that these *G.P. zones*, so-named after their discoverers, *A. Guinier* and *G. D. Preston*, are produced by spinodal decomposition. They form homogeneously on a fine scale throughout the matrix crystal, there is little or no nucleation barrier, and they are fully coherent with the lattice of the matrix, i.e. can be regarded as small regions of the parent lattice in which the solvent atoms have been removed and solute atoms put in their places, one for one. Where there is little size difference between the solute and solvent, as in Ag and Zn zones in Al, the zones are spherical (Plate 7B). Where there is a large size difference the zones form as thin flat plates, presumably for the reason discussed in § 19.6 (cf. Fig. 19.14). In Al + 4 wt per cent Cu alloys, for example, the G.P. zones are discs of copper on {100} planes of the Al matrix, about one atom thick and about 30 atoms in diameter. Some 10^{23} to 10^{24} are formed per m^3.

G.P. zones form at a rate that depends on the quenching temperature and the quenching rate, but which is usually strikingly high. In pure Al-Cu alloys quenched from 550°C they form within one hour at 20°C. If each copper atom moves some three to four atom distances to join a zone, this implies a diffusion coefficient of about 10^{-20} m^2 s^{-1} which is some 10^{10} times larger than that expected for Cu in Al at room temperature. Other alloys behave similarly and there is good evidence that this rapid diffusion is caused by excess vacancies, retained from the solution temperature by the rapid quench. If the quench is interrupted for a few seconds at 200°C to eliminate these vacancies, the zones then form much less rapidly.

The further evolution of the microstructure, particularly at higher ageing or tempering temperatures, varies greatly from system to system. Most systems develop *intermediate* forms of precipitate, between the G.P. zone stage and that of the final equilibrium structure, and the greatest precipitation hardening is usually associated with the most finely dispersed form of intermediate precipitate.

In the Al-Cu system the quenched solution evolves through the following stages: *G.P. zones* → θ'' → θ' → θ. The θ precipitate is the equilibrium $CuAl_2$ intermetallic compound. The intermediate precipitates θ'' and θ' have been identified as distinct crystal structures from their diffraction patterns. Both form on {100} planes of the aluminium matrix, the structure of θ'' being nearer to that of the matrix and that of θ' nearer to that of $CuAl_2$. The θ'' is formed, at temperatures of about 150°C, as plates some 20 Å thick and 400 Å diameter. It forms homogeneously throughout the crystal, probably by the growth of some G.P. zones at the expense of others, and gives the alloy its greatest hardness. The θ' phase forms on a much coarser scale, at about 200°C, generally heterogeneously on dislocations, boundaries and active slip lines. Unlike the earlier precipitates, the θ' plates are partially incoherent with the matrix and there is little or no elastic distortion of the matrix round them. Careful correlation of their orientations with those of edge dislocation lines on which they have formed (also on screw dislocations which are turned into spiral lines by the absorption of vacancies) has shown that they form only in those orientations which allows them to relieve elastic stresses in the field of the dislocations (cf. § 19.6).

Because of their higher free energies, G.P. zones and intermediate precipitates are more soluble in the matrix than the equilibrium precipitate.

For example, in Fig. 20.3 the solubility limit for β' is c_1 and that for β is c_0. It is thus possible to construct metastable solubility limits for equilibrium with these precipitates in the phase diagram. Fig. 20.4 shows an example, for G.P. zones and θ'' in Al-Cu alloys. An interesting consequence of these metastable limits is the effect known as *reversion*. If, for example, a quenched Al + 4 per cent Cu alloy is quenched, aged at room temperature to harden it with G.P. zones, then heated for a few seconds at 200°C and re-quenched, it is softened again because the copper in the G.P. zones

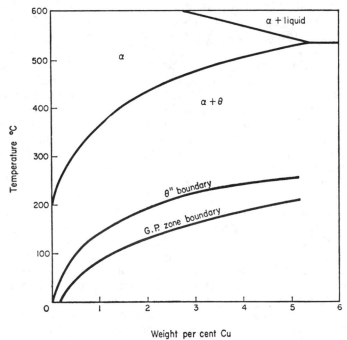

Fig. 20.4 Metastable phase boundaries for G.P. zones and θ'' in the Al-Cu system (after R. H. Beton and E. C. Rollason, 1957–8, *J. Inst. Metals*, **86**, 77)

dissolves at temperatures above the G.P. zone solubility line. The hardening returns on ageing again at room temperature, but more slowly because there are now fewer vacancies to help the copper atoms to migrate to zones.

The equilibrium precipitate generally forms on a coarse scale (10^3 to 10^4 Å) and is associated with *overageing* in which the hardness falls back towards that of a slowly cooled alloy. The precipitate is incoherent and, having a large interfacial energy, tends to be nucleated at incoherent interfaces. The $CuAl_2$ phase, for example, forms along grain boundaries in polycrystalline Al-Cu at temperatures above about 250°C. It also sometimes forms by *discontinuous precipitation*. This is a process of grain boundary migration driven along by the free energy released by precipitation. As a grain boundary moves through a crystal it sweeps up solute, brings this together by diffusion along the boundary and then deposits it as coarse particles of equilibrium precipitate in its rear.

Grain boundaries, by providing interfaces in which diffusion and incoherent nucleation can occur readily, generally speed up the overageing stage and diminish the age-hardening effect. They can also make the alloy very weak, brittle and susceptible to localized corrosion damage (cf. § 23.9). In some cases a continuous film of brittle second phase is formed along a boundary, which gives an easy path for fracture. In others, the boundary with its coarse precipitates is surrounded by adjoining thin regions of the crytals which are free from precipitates. In heavily overaged alloys this is due to the draining of solute from these regions to feed the coarse precipitates in the boundary (*denudation*). At lower ageing temperatures there seems to be no removal of solute, but precipitation cannot readily occur there because the boundary has removed the quenched-in vacancies from these regions. The weakness in such cases is due to the softness of the precipitation-free zones. All the plastic deformation tends to concentrate there and produce cracks at the triple junctions (cf. § 21.12). Several techniques have been developed to prevent the formation of these precipitate-free regions near grain boundaries, including interruption of cooling during the quench (*step-quenching*). Plastic deformation is sometimes helpful in promoting nuclei for precipitation in the regions. Trace amounts of certain substances may also be added to modify the precipitation process (see § 20.3).

The final stage in the sequence of changes, as the quenched alloy is heated at successively higher temperatures, is *coarsening*, in which the particles of equilibrium precipitate become larger and fewer. This is due to *competitive growth*, driven along by the large interfacial energies of these incoherent particles. The smaller of these particles have higher free energies per atom than the larger ones, because of their higher ratios of surface to volume (cf. eqn. 12.32). The concentration of solute in the matrix is thus higher near the small particles than the large ones. This sets up a diffusion flow, from the small particles which thereby dissolve and disappear, to the large ones which thereby grow still larger.

20.3   Development of commercial age-hardening alloys

For high strength, an age-hardened alloy should contain a large volume fraction ($\gg 0\cdot01$) of fine particles (10–100 Å) distributed uniformly without precipitate-free regions. The most direct way to obtain a large volume fraction is to have a high solubility at high temperatures which falls to a very low value at the ageing temperature, as in the successful Cu-Be and Ni-(Ti, Al) alloys. Another approach is to develop complex alloys, containing several solutes which may precipitate independently or in combination. Thus magnesium and silicon may be added to aluminium to form zones and intermediate precipitates leading towards the very stable Mg_2Si

compound. The greatest strength in any known aluminium alloy is obtained by the addition of zinc and magnesium (e.g. 8 wt per cent Zn, 1 wt per cent Mg), which form zones and intermediate precipitates leading towards the stable $MgZn_2$ compound.

For the necessary fineness of particles, it is essential that G.P. zones, or coherent or semi-coherent intermediate precipitates, should be formed. The good age-hardening properties of the Al-Cu system, which is the basis of duralumin alloys, follow from the ready formation of G.P. zones and the θ'' precipitate. The binary Al-Mg and Al-Si systems both show poor age hardening properties because intermediate precipitates are not readily formed in them and the equilibrium precipitate forms heterogeneously on a coarse scale. The binary Al-Zn system has favourable solubility and zone and intermediate precipitate characteristics but is not a good age-hardening system, at room temperature, because the zinc atom is too mobile and the coarse equilibrium precipitate (Zn) forms at quite low temperatures by continuous and discontinuous precipitation.

Many commercial alloys are greatly improved by the addition of various elements in trace amounts, which are able to enhance or retard the formation of various structures. For example, Cu-Be alloys soften rapidly by discontinuous precipitation at temperatures above about 300°C, but this can be prevented by the addition of about 0·4 wt per cent cobalt. A useful addition to Al-Cu alloys is 0·1 wt per cent cadmium. This trace element retards the formation of G.P. zones and so delays the age-hardening process at room temperature, which gives more time for mechanically fabricating the quenched alloy before it becomes too hard (otherwise the quenched alloy has to be refrigerated to keep it soft); and it speeds up the formation of θ' and also leads to a greater hardness from this precipitate. It is thought that the cadmium atoms trap the excess vacancies, retained in the quench, and so reduce the rate of formation of G.P. zones; and that they help the formation of θ' particles by segregating to the θ'-*matrix* interfaces and reducing the interfacial energy.

Much effort has gone into the study of trace elements in Al-Zn-Mg alloys. Although very hard, the basic alloy is plagued by grain boundary weakness due to precipitate-free regions. Small additions of silver have a very beneficial effect in refining the precipitate structure and removing the precipitate-free regions.

20.4 Heat-treatment of steels

A plain carbon steel is usually heat-treated by raising it through the eutectoid transformation to a temperature some 50°C into the single-phase austenitic field (cf. Fig. 15.13), holding it there long enough to dissolve the cementite and disperse the carbon uniformly, and then cooling it to room temperature. Slow cooling in a furnace is referred to as *annealing* (a term also widely used for any softening treatment by heat); somewhat faster cooling in air is *normalizing*; and fast cooling in a liquid bath (water or

PLATE 5

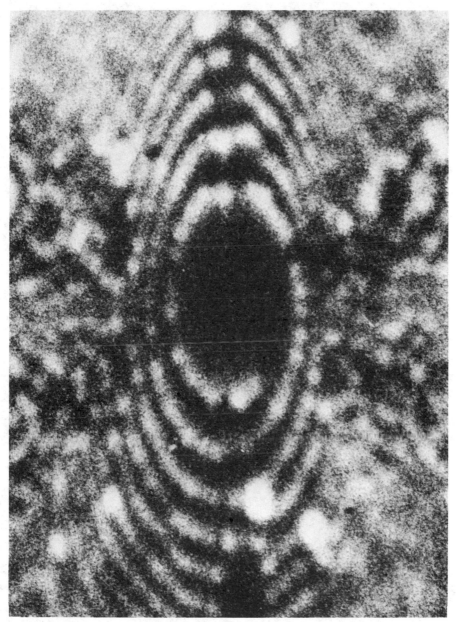

Helium field-ion microscope picture of a dislocation in iridium, showing the helicoidal structure of atomic layers round the dislocation (*courtesy of S. Ranganathan*)

PLATE 6

A. Slip bands on differently oriented glide systems in polycrystalline aluminium, ×500 (*courtesy of G. C. Smith*)

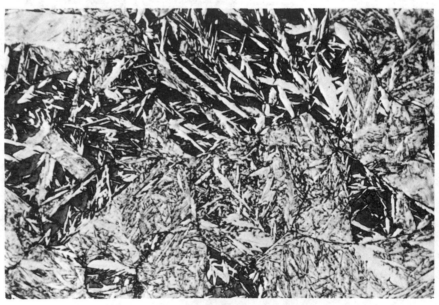

B. Structure of martensite in quenched 1·2 per cent carbon steel, ×750 (*courtesy of G. C. Smith*)

brine, for plain carbon steels) is of course *quenching*, which is followed up afterwards by a *tempering* treatment.

Early researches on the heat-treatment of steel attempted to follow the transformations during continuous cooling but the effects observed were complicated by the continual change of temperature. Decisive progress came only after the development of the technique of *isothermal transformation* (*E. S. Davenport and E. C. Bain*, 1930, *Transactions A.I.M.E.*, **90**,

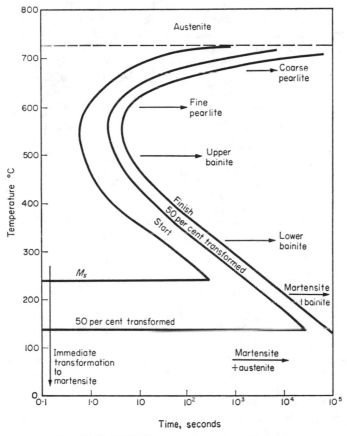

Fig. 20.5 *TTT* diagram of a eutectoid steel

117) in which specimens are quenched into a bath of molten lead or salt at some pre-determined temperature and the course of their transformation at this fixed temperature is then determined, either by measuring a physical property such as length (*dilatometry*) or by removing them at intervals, quenching into water, and then determining the extent of the isothermal transformation metallographically, from the non-martensitic regions.

Fig. 20.5 shows the isothermal decomposition of a eutectoid steel (0·8

13 + I.T.M.

wt per cent C) presented as a time, temperature, transformation (TTT) diagram. Over the range in which the austenite decomposes isothermally the time of transformation follows a C-curve like that of Fig. 20.2. Cutting across this curve at low temperatures are horizontal lines denoting the formation of martensite; M_S denotes the start and M_F the finish. We see that martensite forms almost instantaneously, by change of temperature, below M_S, and not by time. In steels which contain more than 0·7 per cent C the M_F temperature lies below 0°C so that there is some *retained austenite* usually present in the quenched structure, unless the metal is chilled to sub-zero temperatures. Retained austenite, being soft, is an undesirable constituent of quench-hardened steel.

The diagram of Fig. 20.5 is particularly simple because the isothermal decomposition product, *pearlite* (or a modified structure, *bainite*, at lower

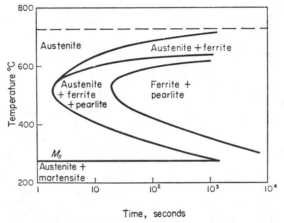

Fig. 20.6 *TTT* diagram of a 0·6 wt per cent steel

temperatures, as discussed below) has the same composition as the austenite from which it is formed. In hypo-eutectoid and hyper-eutectoid steels an extra line appears in the high-temperature part of the diagram, representing the preliminary formation of a *pro-eutectoid* constituent, *ferrite* or *cementite* respectively. When the carbon content of the untransformed austenite reaches the eutectoid value, due to the abstraction of the pro-eutectoid constituent, the pearlite reaction begins. Fig. 20.6 shows an example.

How do we relate a *TTT* diagram to the transformation behaviour on continuous cooling? We can draw the cooling curve as in the full line of Fig. 20.7 and consider its intersection with the isothermal curve but this is misleading because it fails to integrate the effects of time spent in different temperature ranges. An approximate allowance for this can be made by dividing the temperature range into steps ΔT, determining the time of transit Δt through each step, dividing it by the time t of isothermal transformation at the mean temperature in this step and so, by adding the fractions of time 'used up' in successive steps, building up a *cumulative fraction* α, i.e.

$$\alpha(T) = \sum_{T_e}^{T} \frac{\Delta t}{t} \qquad\qquad 20.2$$

of isothermal transformation time already used up at each temperature T reached during continuous cooling. The transformation begins when $\alpha \simeq 1$. We can in fact construct an 'effective cooling curve' by plotting αt against T, as in Fig. 20.7. The method is not accurate but broadly indicates the expected behaviour.

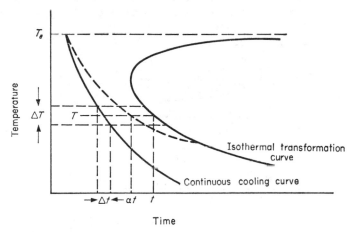

Fig. 20.7 Relation of transformations under isothermal and continuous cooling conditions

In plain carbon steels the reaction is rapid near the *pearlite nose* of the TTT curve (about 550°C; see Fig. 20.5). Fast cooling is necessary to produce an effective quench. This has practical disadvantages. It is not possible to quench the interior regions of thick sections; and the large temperature gradients produced by fast quenching set up thermal and transformation stresses in the material (cf. Fig. 15.22) which distort it and sometimes give *quenching cracks*. This can be overcome by adding small amounts of certain alloy elements (Mn, Ni, Cr, Mo, V) to the steel, which shift the pearlite nose to the right, so that effective quenching can be got by cooling more slowly (e.g. oil bath quenching, even air quenching in some more highly alloyed *air-hardening* steels). Fig. 20.8 shows a TTT diagram of an alloy steel. One feature which appears in this and many such diagrams is the separation of the pearlite and bainite reactions into two distinct C-curves with a temperature range between in which the austenite decomposes very slowly.

In many steels the rate of decomposition to bainite, just above M_S, is fairly slow. Advantage can be taken of this to produce deep hardening without quench cracking by quenching rapidly to a temperature just above M_S, holding there for long enough to equalize the temperature (without forming bainite) and then cooling slowly through the $M_S \rightarrow M_F$ range. This is called *martempering*. Alternatively, the metal may be quenched rapidly to a somewhat higher temperature above M_S and then decomposed isothermally to *lower bainite*. This is *austempering* and gives a reaction product similar to quenched and tempered martensite.

The *hardenability* of a steel is not the hardness of its martensite but the depth to which martensite can be produced in thick sections. It is measured by the *Jominy end-quench test* in which a 1 inch diameter bar is quenched by a jet of water directed against one end. The hardenability is the length

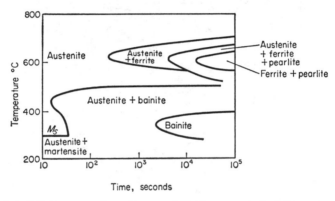

Fig. 20.8 *TTT* diagram of an alloy steel (0·42 per cent C, 0·78 per cent Mn, 1·78 per cent Ni, 0·8 per cent Cr, 0·33 per cent Mo)

of the martensite zone along the bar, as indicated by hardness measurements, and it varies from a fraction of an inch for plain carbon steels to several inches in some alloy steels.

20.5   Formation of pearlite

Experimental work mainly by *Mehl* (*R. F. Mehl and W. C. Hagel*, 1956, *Progress in Metal Physics*, Pergamon Press, **6**, 74) has shown that pearlite nucleates heterogeneously. In a homogeneous eutectoid steel the pearlite nuclei form in the austenitic grain boundaries and then grow into the adjoining grains as *nodules*, as shown in Fig. 20.9. Each nodule contains one or more *colonies*. A colony consists of alternate sheets of ferrite and cementite, the boundary of the colony corresponding to a change in the crystallographic orientation of the ferrite and to a change in the direction of the cementite sheets.

The pearlite-austenite interface is incoherent and advances radially outwards from its nucleus into the untransformed austenite, with a constant speed at constant temperature. At the interface the lamellae tend to grow perpendicularly to the interface, to minimize the ferrite-cementite distance along the interface over which the carbon must diffuse to separate from the austenite. This conflict with the tendency for the lamellae to grow as parallel plates is resolved by the occasional *branching* of a plate to start off a new colony at an angle to the parent colony.

The rate of decomposition to pearlite is determined by two factors: the *rate of nucleation* $\dot{N}$, i.e. the number of nuclei formed per unit time per unit

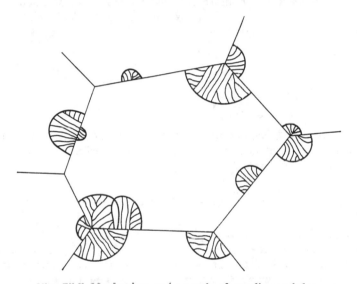

Fig. 20.9 Nucleation and growth of pearlite nodules

volume of untransformed phase, and the *rate of growth G*, i.e. the speed of advance of the pearlite-austenite interface. The rate of nucleation is proportional to the area of austenitic grain boundary and hence is greatest in fine-grained austenite. It follows that increasing the austenitizing temperature, to enlarge the austenite grains, increases the hardenability of a steel but this is not a practical technique because coarse-grained martensite is fragile.

Let X be the volume fraction of austenite decomposed in time t at a constant temperature T. The rate of formation of pearlite is then

$$\frac{dX}{dt} = GA_f \qquad\qquad 20.3$$

where A_f is the area of austenite-pearlite interface, at this time, in unit volume. At high temperatures (near T_e) nucleation is difficult, for the usual reasons, but growth is rapid. A few nuclei only are then formed and grow into very large pearlite nodules. Frequently, a single nodule becomes much bigger than its parent austenite grain, having crossed many grain boundaries during its growth. Under these circumstances we can assume that $\dot{N}$ is constant and that the distribution of the nodules relative to one another in space is random. It is helpful to define an *extended volume fraction* X_x and *extended interfacial area* A_x. These are the hypothetical volumes and areas the nodules would have if nucleation and growth had continued with all the original austenite remaining everywhere still available to them. The fraction of A_x not *inside* other nodules is equal to the untransformed volume, $1 - X$, so that $A_f = (1 - X)A_x$. We also have $dX_x/dt = GA_x$. Hence, combining these with eqn. 20.3,

$$\frac{dX}{dX_x} = 1 - X \qquad\qquad 20.4$$

which integrates to

$$X = 1 - e^{-X_x} \qquad\qquad 20.5$$

The number of nodules, per unit volume of austenite, nucleated in the time interval t' to $t' + dt'$ is $\dot{N} dt'$. If these grow spherically they have radii $G(t - t')$ at a later time t. Hence

$$X_x = \int_0^t \frac{4}{3} \pi(t - t')^3 G^3 \dot{N} \, dt' = \frac{\pi}{3} \dot{N} G^3 t^4 \qquad\qquad 20.6$$

Eqn. 20.5, with this value of X_x, is the *Johnson-Mehl equation*. It gives a characteristic *sigmoidal* curve for pearlite formation, as shown in Fig. 20.10. The reaction rate is slow at first, because A_f is small, is a maximum at about half-transformation, when A_f is large but the growing nodules have not yet begun seriously to run into one another, and becomes slow again at the end as A_f shrinks with the remaining small regions of austenite.

At temperatures near the pearlite nose, where $\dot{N}$ is large and G is small, we enter a different regime of behaviour. The entire austenitic grain boundary area quickly becomes covered with small pearlite colonies. These impinge on one another and form a roughly plane reaction front which advances into the grain from the boundary. From the point of view of hardenability the important parameters then are G and the austenitic grain diameter d, since the time to complete the transformation is given essentially by

$$t_f \simeq d/2G \qquad\qquad 20.7$$

This is usually the decisive equation in practice.

There is experimental evidence that both cementite and ferrite can nucleate pearlite. It is significant that a pearlite colony frequently grows on only one side of the austenitic grain boundary at which it is nucleated. It has been suggested (*C. S. Smith*, 1953, *Trans. Am. Soc. Metals*, **45**, 533) that a nucleus first forms in a coherent orientation with one of the adjoining grains, to avoid a high nucleation energy, and then grows into the *other* grain by the advancement of the incoherent interface provided by the original austenitic grain boundary, because incoherent interfaces are relatively mobile at such temperatures. The growth is then similar to discontinuous precipitation with grain growth, discussed in § 20.2.

The sideways growth of a pearlite nodule appears to occur by the branching of fingers from the plates to form two interlaced crystals, one of ferrite,

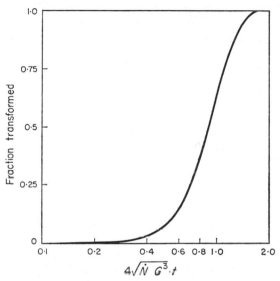

Fig. 20.10 Reaction curve for transformation of austenite to pearlite according to the Johnson-Mehl equation

one of cementite. This avoids the necessity for repeated nucleation. The edgeways growth occurs by the advancement of the pearlite-austenite interface, as in Fig. 20.11, and the separation of carbon to the cementite plates. It has often been assumed that diffusion of carbon is the rate-determining process governing the edgeways growth of pearlite. For steady-state growth the situation in Fig. 20.11 provides a well-defined diffusion problem for which solutions have been obtained. There are difficulties however because several possible diffusion paths exist, i.e. through the austenite, through the ferrite or along the austenitic-pearlite interface. Experiments have shown that in commercial plain-carbon steels the pearlite grows at about the rate expected from carbon segregation by diffusion in austenite. However, in high-purity steels the pearlite grows some 50 times faster than this, for a similar ferrite-cementite spacing. Carbon diffuses some 100 times faster in ferrite at such temperatures but its solubility in ferrite is so low that diffusion in ferrite also could not explain the speed of growth. It seems that there must be some highly conducting paths available at or near the interface for rapid transport of the carbon. However, there is no strong evidence for rapid carbon diffusion along grain boundaries.

There are also difficulties in understanding the spacing of the ferrite and cementite plates in pearlite. This spacing is coarse near T_e and becomes progressively finer as the transformation temperature is lowered towards the pearlite nose. An attractive suggestion (*C. Zener*, 1946, *Transactions A.I.M.E.*, **167**, 550) is that the spacing adjusts itself to give the maximum rate of growth. If the spacing is too large, the length L of the diffusion path is correspondingly large and the carbon thereby separates slowly. If the spacing is too small, the area and hence energy of the ferrite-cementite

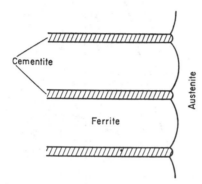

Fig. 20.11 Growth of pearlite colony

interfaces become large and this reduces the overall free energy change driving the reaction forward. The argument leads to the conclusion that the optimum spacing should be twice that at which the ferrite-cementite interfacial energy would exactly cancel the free energy released by the austenite → pearlite reaction. However, in plain-carbon steels the observed spacings are an order of magnitude larger than this so that some other factor, perhaps related to the failure of some plates to keep up with the advancing pearlite interface and to the difficulty of replacing them by the branching process, may be operative.

The effect of some alloy elements in retarding the growth of pearlite is more clearly understood. In general the alloy atoms tend to segregate to one of the constituent phases of the pearlite; carbide-forming elements such as chromium, for example, prefer the cementite. If they remain randomly dispersed, as in the austenite, the free energy of the austenite → pearlite reaction is decreased, so much so that in many cases the reaction cannot go forward without segregation occurring. The rate of pearlite growth is then limited by the slow diffusion of these substitutional atoms. Even though the incoherent austenite-pearlite interface is available for more rapid diffusion, the segregation of alloy atoms generally still remains the rate-determining step and makes the reaction slow.

20.6 Pro-eutectoid reactions

In a steel not of eutectoid composition the transformation at temperatures above about 550°C begins with the separation of pro-eutectoid ferrite or cementite, usually at the austenitic grain boundaries. The composition of the separating structure is now different from that of the parent austenite so that long-range migration of carbon is necessary to change the composition in the two regions of the decomposing grain. If the grain size is large the central region of a grain is so far from the boundary that the necessary migration cannot occur unless the rate of cooling is very slow. Usually some pro-eutectoid constituent then nucleates inside the grain as well as at the boundary.

Nucleation at a boundary seems to occur coherently with one of the adjoining grains and hence, through the original grain boundary, incoherently with the other. Pro-eutectoid ferrite tends to grow rapidly along the boundary and slowly into one of the grains. If the incoherent side grows we get a *massive transformation* in which large irregular-shaped nodules of ferrite are formed, rather like pearlite nodules. This happens in normalized hypo-eutectoid steel. If the coherent side grows we get an elongated structure, such as the Widmanstätten structure, along close-packed planes in the austenite. The factors governing the choice and precise form of structure are complex and involve climb of interfacial dislocations in semi-coherent interfaces, as well as surface energy and diffusion factors.

20.7 Formation of bainite

As the transformation temperature is lowered, below the pearlite nose, and the free energy change driving the F.C.C. → B.C.C. transformation increases, ferrite plays an increasingly predominant part in the transformation. The pro-eutectoid ferrite and ferrite + cementite curves in the *TTT* diagram run together and the transformation is nucleated by the formation of Widmanstätten plates of ferrite. The transformation interface is mainly austenite-ferrite and any cementite plates there tend to become separated from the austenite by the ferrite growing round them. The continuous lamellar structure of pearlite is no longer possible. Instead, ferrite plates grow out into the austenite, pushing carbon away from them until carbide particles are nucleated in the carbon-rich regions between the ferrite plates. This is *upper bainite*.

At lower reaction temperatures (< 400°C) the bainite appears increasingly like tempered martensite. It forms as well-defined acicular plates, crystallographically oriented to the austenitic matrix, speckled with fine carbide particles. This is *lower bainite*. There is clear evidence from relief effects seen on metallographic surfaces that lower bainite forms by a shear process, similar to martensite (*T. Ko and S. A. Cottrell*, 1952, *J. Iron Steel Inst.*, **172**, 307). It seems probable that this is crystallographically a martensitic transformation, the progress of which is delayed by diffusion of carbon. Carbides nucleate and form in the (slightly supersaturated) ferrite behind the interface and, as they draw carbon away from the adjoining austenite, allow this interface to move forward by the crystallographic shear mechanism.

20.8   Formation of martensite

A martensite plate forms rapidly, in about 10^{-7} sec, by the glide of transformation dislocations which shear the material (or, more strictly, produce an invariant plane strain; cf. § 18.9), as can be seen by the tilt produced on a free surface intersected by the plate. Inside the metal the plate grows until stopped by an obstacle such as a grain boundary or another martensite plate lying across its path. The elastic constraint of the surroundings causes the plate to take on a thin lens-like shape, parallel to the habit plane along which the main shear takes place.

Because of the speed of the transformation to martensite there is no time for separation of atoms from solid solution and so the composition of the martensite is identical with that of the austenite from which it is formed. There is thus an ideal temperature T_0 for the formation of martensite, i.e. the point at which curves of bulk free energy versus temperature for austenitic and martensitic phases of the same composition cross. Martensite actually first appears in quenched steel at the M_S temperature, some 200°C lower than T_0 in plain carbon steels, this undercooling being due to the need to acquire extra free energy (about 1200 J mol^{-1}) to provide the driving force necessary to expand loops of transformation dislocations. By applying stress to the quenched austenite, however, the necessary dislocation movements can be started at higher temperatures and in this way martensite can be created at temperatures much nearer to T_0. Both T_0 and M_S fall, equally and linearly, with increasing carbon content; M_S for example starts at about 600°C for zero content and reaches room temperature at about 1·6 wt per cent C.

Because martensite forms so rapidly, the transformation is commonly described as *athermal*, implying that no activation energy is needed. Certainly, the rapid growth of the plates (even at 4 K) suggests a process of pure dislocation glide, without thermal activation. The nucleation process is however thermally activated and careful experiments have shown that martensite can be produced isothermally in some iron alloys. The reason why the whole process appears usually to depend only on temperature and not on time probably lies in the fact that the formation of martensite is a special kind of plastic deformation in which the driving force is provided by internal release of free energy instead of external stress. We then have a close comparison with processes of plastic yielding generally, which set in rather sharply at a given yield stress that is itself usually fairly independent of temperature. The peculiar role of temperature in the martensitic transformation lies in the fact that, with increasing undercooling, the free energy release and hence the 'effective applied stress' due to it increases.

Processes of plastic yielding tend mostly to be rather insensitive to temperature because the elementary step, the creation of a plastic nucleus, which in its simplest form is a single dislocation loop large enough to be expanded by the applied stress, involves many atoms. The work to make it, which is proportional to the number of atoms involved, is then much too large for thermal activation unless the applied stress is practically large enough to overcome by itself the resistance to the creation of this nucleus.

We expect a similar situation in the formation of martensite, with temperature playing a double role: a major one of providing, with increasing undercooling, an increasing driving force analogous to increasing applied stress; and a minor one of providing thermal activation for nuclei when the driving force is *almost* equal to the resisting force. As with other plastic processes, we do not expect the loops of transformation dislocations, which are the boundaries of the martensitic nuclei, to be created *ab initio* in the perfect lattice but to grow out of pre-existing dislocations in the austenite. The transformation is then heterogeneously nucleated and we can suppose that each potential nucleus has its own characteristic driving force (cf. the Frank-Read stress for slip, § 21.2) and so can become active at a characteristic degree of undercooling. The activation energy for the formation of a given nucleus remains prohibitively large until the particular characteristic temperature for this nucleus is almost reached, on cooling, when it drops to a negligible value and the nucleus can then form at the high speed associated with a process of dislocation glide.

There are many other striking analogies between the martensite transformation and plastic processes. There is a *burst phenomenon*, analogous to deformation twinning, in which the mechanical disturbance set up by the formation of one martensite plate triggers off the formation of other plates, so that a zigzag sequence of several plates often forms in one 'burst'. There is also a *stabilization* effect. If the quench is interrupted at some temperature in the martensite range and the metal held there for a period, the transformation is then delayed on subsequent cooling until a certain degree of undercooling is achieved, when it resumes again with a burst of new martensite plates. This is strongly reminiscent of the effect of *strain ageing* on the process of plastic yielding (cf. § 21.3) and may have a similar cause, i.e. the immobilization of dislocations by the migration of foreign atoms to them, which have a pinning effect.

20.9 Tempering of quenched steel

In the as-quenched condition martensite is too brittle for practical use. The metal must be tempered to improve its toughness. At the higher temperatures (above 400°C) this tempering treatment also substantially softens

plain carbon steel and in practice a balance has to be struck between requirements of toughness and hardness. For treatments of about 1 hour duration
the first stage of tempering is the precipitation at about 100°C of ε-*carbide*,
a C.P. Hex. structure of composition about Fe_5C_2, with an accompanying
drop in the carbon content of the martensite to about 0·25 wt per cent C.
The hardness actually increases slightly during this stage, due to the fine-
scale of the precipitate (so that razor blades do not soften in boiling water!).
At about 250°C retained austenite decomposes and at about 350°C particles
of cementite (Fe_3C) begin to form, replacing the ε-carbide and reducing the
low-carbon martensite to ferrite. Above about 500°C the cementite particles
grow competitively, as discussed at the end of § 20.2, into larger rounded
particles dispersed through the B.C.C. iron matrix, giving a *spheroidized*
structure (*sorbite*). There is also considerable relief of quenching stresses at
these temperatures.

Carbide-forming alloy elements profoundly affect the tempering reactions and are used, in addition to improving hardenability, to delay the
softening stage of tempering so that higher tempering temperatures may be
used, thereby improving the toughness of the steel without seriously softening it. This is important also for the use of strong steels at elevated temperatures, e.g. high-speed tool steels (which contain tungsten) that retain their
hardness and cutting edge when running at red-heat in machine tools. These
alloys act by forming stable carbides, at the expense of cementite, on tempering.

If plain carbon or low-alloy steels are tempered below about 250°C they
usually remain somewhat brittle. Above 350°C they become too soft. Unfortunately a particular form of brittleness appears on tempering between
these two temperatures, believed to be due to the formation of thin films
of cementite along the martensite boundaries. It is therefore necessary to
add sufficient alloys to prevent undue softening at temperatures of 350°C
and above. There are many possibilities (Mo, W, V, Si, Co) and the choice
has to be based on requirements of hardenability, avoidance of a low M_s
temperature (otherwise quench cracking is troublesome), and cost, as
well as on resistance to tempering.

Some alloy additions, e.g. Mo, V, even reverse the temper-softening,
producing a *secondary hardening* at temperatures ($\simeq 550$°C) at which
they diffuse freely in ferrite and form fine carbide precipitates. These
carbides, e.g. Mo_2C and V_4C_3, are highly resistant to overageing. Medium-
alloy creep-resistant steels owe their high-temperature strength to this
effect.

Some alloy steels, particularly deep-hardening Ni-Cr steels, become
embrittled after tempering at high temperatures unless cooled very quickly
through the temperature range 550 to 500°C. This *temper-brittleness*, which
can be suppressed by the addition of 0·25 wt per cent Mo, is associated with
fracture along grain boundaries and is believed to be due to the segregation
of traces of phosphorus to these boundaries.

20.10 Case hardening of steel

For many engineering components such as gears, shafts and valves it is an advantage to use a steel with a hard surface and a softer, tough interior. This is usually done by one of several *case hardening* treatments in which the surface layer of a soft steel (e.g. 0·2 wt per cent C) is enriched with a hardening alloy element, usually carbon or nitrogen, by a diffusion treatment. In *carburizing*, the steel is heated for some hours at about 900°C in a carbon-rich atmosphere provided by gaseous, liquid or solid carbonaceous substances, to produce a carburized layer by diffusion to a depth of about 10^{-3} m. The surface is then hardened by quenching from about 750°C, a treatment that leaves the interior soft. In *nitriding*, the metal contains about 1 per cent aluminium and is heated at 500–550°C in ammonia for some hours. Nitrogen atoms diffuse into the surface and form fine stable nitride precipitates with the aluminium, so that the metal is precipitation-hardened in its surface. No subsequent heat-treatment is needed. In *cyaniding*, a steel containing about 0·2 per cent C is heated at 850–900°C for about 1 hour in molten sodium cyanide. Carbon and nitrogen atoms diffuse into the metal and the surface is then hardened by quenching in oil or water. *Carbonitriding* is similar, except that the metal is heated in a gas of hydrocarbon and ammonia for some hours at 700–850°C.

FURTHER READING

AMERICAN SOCIETY FOR METALS (1964). *Principles of Heat-Treatment.*
BRICK, R. M. and PHILLIPS, A. (1949). *Structure and Properties of Alloys.* McGraw-Hill, New York and Maidenhead.
BULLENS, D. K. (1948). *Steel and Its Heat Treatment.* John Wiley, New York and London.
CHRISTIAN, J. W. (1965). *The Theory of Transformations in Metals.* Pergamon Press, Oxford.
GUY, A. G. (1959). *Elements of Physical Metallurgy.* Addison-Wesley, London.
KELLY, A. and NICHOLSON, R. B. (1963). 'Precipitation Hardening', *Progress in Materials Science*, 10. Pergamon Press, Oxford.
SMOLUCHOWSKI, R., MAYER, J. E. and WEYL, W. A. (1951). *Phase Transformations in Solids.* John Wiley, New York and London.
ZACKAY, Z. V. and AARONSON, H. I. (1962). *Decomposition of Austenite by Diffusional Processes.* Interscience, London.

Chapter 21

Mechanical Properties

21.1 The tensile test

The mechanical properties of a metal are usually measured in a *tensile test*. A straight rod or strip is gripped at its ends, which are usually enlarged into *shoulders* to strengthen them, and it is then pulled axially by a gradually increasing tensile force. The shoulders taper smoothly from the grips to the geometrically uniform central section of the test bar, along which is marked a *gauge length* over which the tensile elongation is measured by an extensometer or strain gauge.

Measurements are made of the increments Δl of elongation over the gauge length which are produced by successive additions to the total tensile load F. We can then define the tensile stress F/A and the increment of tensile strain $\Delta l/l$ either in terms of the *initial* values of the cross-sectional area A and gauge length l, in which case they are called *nominal* or *engineering* stress and strain; or in terms of the *current* values of A and l at the particular stage in the test at which these measurements of F and Δl have been taken. They are then called *true* stress and strain. During the elastic stage of deformation in most metals it matters little which is used because the elastic strain is usually too small to change A and l appreciably from their initial values. During the plastic stage, the material elongates, not by any additional stretching of the atomic bonds but by increasing the number of atoms along the gauge length. In this real physical sense, each increment of plastic elongation *increases* the gauge length over which the next increment is to be measured and, since the total number of atoms in the test bar remains constant, correspondingly decreases the cross-sectional area. *True* stress and strain are then the proper quantities to use.

The total true strain ϵ_t is obtained by summing the incremental strains. Hence

$$\epsilon_t = \int_{l_0}^{l_1} \frac{dl}{l} = \ln\left(\frac{l_1}{l_0}\right) \qquad \qquad 21.1$$

where l_0 and l_1 are the initial and final gauge lengths. Because of its form,

ϵ_t is often referred to as the *logarithmic* strain. We notice that at small strains, when $(l_1 - l_0) \ll l_0$,

$$\epsilon_t = \ln \left[1 + \frac{l_1 - l_0}{l_0} \right] \simeq \frac{l_1 - l_0}{l_0} = \epsilon_n \qquad 21.2$$

where ϵ_n is the nominal strain.

Fig. 21.1 shows some typical true stress-strain curves given by polycrystal-line metals. The initial *yield stress* is often not clearly marked and in such cases is conveniently defined by the *proof stress*, the stress at which a standard small plastic strain (e.g. 0·001) is first produced. Beyond the yield

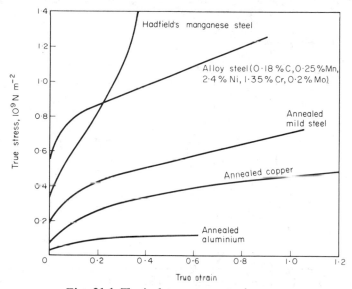

Fig. 21.1 Typical true stress-strain curves

stress, the stress-strain curve usually rises through *work hardening*, according to a roughly parabolic curve,

$$\sigma_t = c\epsilon_t^n \qquad 21.3$$

where the *strain-hardening exponent* n usually lies between 0·1 and 0·3.

As the test bar extends, it is strengthened by work hardening but also weakened by reduction of cross-sectional area. At first the work hardening effect usually predominates and the deformation is then *stable*, i.e. all parts of the gauge length deform equally, since if one part deformed excessively it would become stronger than the rest. Beyond a certain stage, however, the effect of reduction of cross-sectional area predominates and this leads to a *plastic instability* in which all subsequent deformation becomes concentrated in one short section of the gauge length which stretches excessively and forms a narrow *neck*, as n Fig. 21.2. The local reduction of cross-section in the neck causes the tensile load F to fall, even though the material in the neck *continues to work harden*. This deformation of the neck under a falling load continues until either the remaining cross-section breaks or the neck draws down to a point or chisel-edge, as in Fig. 21.2(b).

Because of necking, the capacity of the material for plastic deformation in tension, i.e. the *ductility*, is not truly indicated by the strain to fracture measured along the gauge length. A longer gauge length would give a smaller value. A better measure of ductility is the reduction of area at the narrowest part of the neck. This can be expressed as a logarithmic strain in the form $\ln(A_0/A_1)$ where A_0 and A_1 are the initial and final areas. In an

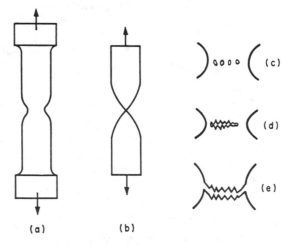

Fig. 21.2 (a) Formation of a tensile neck; (b) point rupture or chisel-edge rupture; (c) and (d) rupture by the growth and coalescence of holes in the neck, leading to (e) fibrous cup-and-cone rupture

ideally ductile material a total reduction of area occurs, down to a point or chisel-edge rupture. Many soft metals approximate to this ideal ductility when very pure. In most metals however, small holes usually form at weak or weakly adherent foreign inclusions (Plate 8A). These holes then expand by plastic deformation and coalesce by localized necking of the 'bridges' of metal between them, to form a *ductile fibrous fracture* by *internal necking*, as shown in Fig. 21.2. The circumferential rim of metal usually finally separates along a surface of localized shear deformation, giving a characteristic *cup-and-cone* fracture.

A tensile instability occurs when the load F supportable by the specimen fails to rise during an increment of tensile strain, i.e. when $dF/d\epsilon \leqslant 0$. Since $F = \sigma_t A$ we have $dF = d(\sigma_t A) = A\,d\sigma_t + \sigma_t\,dA = 0$ as the limiting condition for the instability. Plastic deformation in metals occurs at almost constant volume, so that $dV = d(Al) = A\,dl + l\,dA = 0$. Combining these, the condition is

$$\frac{d\sigma_t}{\sigma_t} = \frac{dl}{l} = d\epsilon_t \qquad\qquad \textbf{21.4}$$

Thus, when eqn. 21.3 is obeyed, the instability occurs at the strain $\epsilon_t = n$. Alternatively, we can write eqn. 21.4 in the form

$$\frac{d\sigma_t}{\sigma_t} = \frac{dl}{l_0}\frac{l_0}{l} = \frac{d\epsilon_n}{1 + \epsilon_n} \qquad\qquad \textbf{21.5}$$

and then plot σ_t against ϵ_n and locate the instability at the point where a tangent, through the point $\sigma_t = 0$, $\epsilon_n = -1$, touches the curve, as in Fig. 21.3. This is *Considère's construction*.

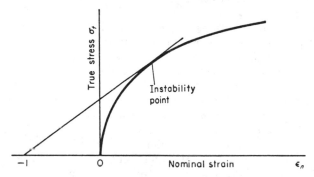

Fig. 21.3 Considère's construction

The nominal stress at the instability is the *ultimate tensile stress* or *tensile strength* of the metal. It is important in structural engineering because it determines the highest load which a tensile strut of a given cross-sectional area can support. It is clear from Considère's construction that if a material has a high yield stress and then work hardens only slightly, so that in place of eqn. 21.3 there is a sharp change of slope at the yield stress, the ultimate tensile stress can coincide with the yield stress. The instability then sets in as soon as the metal becomes plastic. There is no uniform plastic deformation, only necking. This condition, which can exist in metals excessively hardened by cold working or precipitation hardening, is dangerous in thin tensile members used in structural engineering because plastic deformation then leads immediately to localized failure at very low overall elongations even though the metal may rupture in a fully ductile manner. The danger is aggravated in some cases by effects of high rates of strain which usually develop in such regions of localized failure and by the tendency of some metals in such circumstances to change to a brittle type of failure. A high ratio of yield stress to ultimate tensile stress is thus to be avoided in metals for structural engineering.

21.2 The yield stress

The typical commercial metal contains inside each grain a network of dislocation lines and a host of other irregularities, e.g. foreign atoms, precipitates, vacancies, small-angle boundaries, that can act as obstacles to dislocations. There are also often places on the grain boundaries from which dislocations can be released into the grain fairly readily. There may additionally be local groupings of dislocations created round large foreign inclusions by thermal shrinkage strains.

Above a certain level of applied stress, some of these various dislocations are able to move between obstacles to an extent which is visible in, for example, the electron microscope. This does not represent general yielding, however, but a stage of 'pre-yield micro-strain' in which certain dislocations are able to move, either because the applied stress is enhanced by local self-stresses or by stress-concentrating features such as notches and sharp edged foreign inclusions, or because such dislocations happen to be unusually free from obstacles or lie in very favourably oriented grains. The overall plastic strain produced by these localized movements remains smaller than or comparable with the elastic strain, however, until the general yield stress is reached. Since the applied stress has to climb from its 'pre-yield' value to the yield stress within this narrow range of plastic strain, this preliminary plastic stage seems, on the stress-strain curve, to be a stage of very rapid work hardening.

General yielding is certainly possible at a stress at which average dislocation sources (e.g. Frank-Read sources) can create slip bands in average regions of the metal. We may thus expect the general yield stress σ_y to be of the form

$$\sigma_y = \sigma_s + \sigma_i \qquad\qquad 21.6$$

where σ_s is the stress to operate such a source (cf. eqn. 19.22) and σ_i is the 'friction stress' which represents the combined effect of all the obstacles opposing the motion of a dislocation coming out of a source. For commercial metals σ_y usually lies in the range $10^{-3}\mu$ to $10^{-2}\mu$, where μ is the shear modulus. As mentioned previously, this general yield stress is often very ill-defined, partly for the reason discussed above, partly because the work-hardening which follows general yielding is often rapid, and partly because, in polycrystals, sources on variously oriented planes have to be brought into operation before the strain compatibility condition between the grains, discussed in § 18.8, can be satisfied.

In some materials, however, general yielding can begin in a very striking manner with a *yield drop* in which the applied stress falls, during yielding, from an *upper yield point* to a *lower yield point*. This effect is frequently seen in nearly perfect crystals of non-metallic materials such as silicon, germanium and sapphire, in which there is a fairly low density of dislocations (e.g. $\rho \simeq 10^7\,\mathrm{m}^{-2}$) and a large Peirels-Nabarro force (cf. §17.4). If we think of the tensile test as applying a certain overall rate of strain to the specimen, then, because ρ is small, the plastic strain rate $bv\rho$, where b is the Burgers vector length, can match the applied rate only if the speed v of the dislocations is high. With a large Peierls-Nabarro force this speed is reached only at a high stress. However, as the dislocations glide they also multiply and ρ thus increases rapidly. This introduces some work hardening but the effect of this is usually swamped by the fact that the dislocations can now move much more slowly and hence at much lower stresses. The stress supportable by the specimen thus drops during initial yielding and then rises again as work hardening comes fully into play at higher strains, cf. Fig. 21.4.

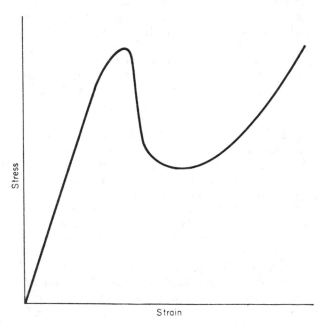

Fig. 21.4

21.3 Yield points in metals

In F.C.C. metals the Peierls-Nabarro force is extremely small and the stress to move a dislocation is almost independent of speed, at least up to the order of 10^3 m s^{-1}. B.C.C. transition metals such as iron behave at room temperature in a manner intermediate between the F.C.C. metals and the non-metals mentioned above.

It follows that the density of effective dislocations must be reduced virtually to zero if such metals are to show large yield drops. This can be done by eliminating dislocations from the crystal initially, as in some whisker crystals. Yielding then begins at the stress required to create dislocations in the perfect lattice, which is about $\mu/30$ in F.C.C. metals, and continues at the stress to multiply and glide these dislocations, well below $10^{-3}\mu$.

A more general way of producing yield drops in metals is by immobilizing the dislocations already present by *pinning* them with foreign atoms. We saw in § 19.6 that misfitting solute atoms are attracted to dislocations where they can fit with less elastic strain. The elastic interaction can be quite large. To find the order of magnitude, consider an atomic site near the centre of an edge dislocation where the stress is of order $0\cdot1\mu$. Below the half-plane, this stress is tensile. The surface area of this atomic site is of order πb^2 where b is the atomic diameter. Let us put into this site a solute atom which expands the diameter by, say, 20 per cent. The displacement of an element of surface is then about $0\cdot1b$. The work done by the tensile stress through this displacement is about $(0\cdot1\mu)(\pi b^2)(0\cdot1b) \simeq \mu b^3/30 \simeq 0\cdot2$ eV. This indicates the order of magnitude of the binding energy, B, to be expected. If, by migration of solute atoms, an equilibrium 'atmosphere' of solute builds up round a stationary dislocation, the usual Maxwell-Boltzmann formula gives the concentration c of solute near the dislocation as

$$c = c_0 \exp\left(B/kT\right) \qquad\qquad \textbf{21.7}$$

in a solid solution of concentration c_0. With $B \simeq 0\cdot2$ eV and $kT = 0\cdot025$ eV (i.e. room temperature) we obtain $c \simeq 10^3 c_0$. A limit to c is of course set by *saturation*, i.e. filling of all the attractive sites along the dislocation with solute atoms. The amount of solute absorbed in dislocations to produce saturation is very small since, in an annealed metal, only about one atomic site in 10^8 is at the centre of a dislocation; even in a heavily worked metal this only rises to one site in 10^3. Hence, badly misfitting solute atoms that are mobile at relatively low temperatures can saturate dislocations even when present in very small amounts in the metal.

These considerations underlie the dramatic effect of small amounts of carbon and nitrogen on the mechanical properties of soft iron and structural steel. Interstitial carbon and nitrogen atoms are very mobile in B.C.C. iron, even at room temperature, and they strongly distort the lattice (as is evident from Fig. 14.4), which gives them a binding energy of about $0\cdot5$ eV with a dislocation. Furthermore, this lattice distortion, being tetragonal, contains a large component of shear strain and this enables the atoms to interact strongly with screw dislocations, through their shear stress field, as well as with edge dislocations, through lattice expansion. Conditions are thus exceptionally favourable for intense segregation of these atoms to all dislocations in B.C.C. iron.

Dislocations in iron are immobilized by the strong pinning effect of the carbon and nitrogen segregated in them. The density of dislocations available for glide is thus reduced virtually to zero until the very high applied stresses are reached at which dislocations can either be released from their segregated atoms or created from points of stress concentration.

There is then a higher upper yield point, followed by a drop to a lower yield stress at which the yielding process is propagated through the specimen. Fig. 21.5 shows the type of stress-strain curve obtainable from a well-aligned specimen. In practice, the observed upper yield point is often brought down to the lower yield stress by non-uniformity in the applied stress distribution, which causes yielding to begin prematurely in high stressed regions.

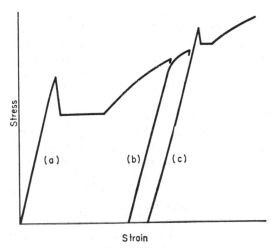

Fig. 21.5 Yielding of soft steel, (a) initial yield drop; (b) absence of yield drop on immediate re-testing; (c) return of yield drop on ageing after straining

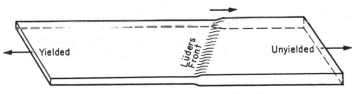

Fig. 21.6 A simple Lüders band

This yielding leads to a plastic instability known as the *yield elongation* or *Lüders strain*, in which the metal has to strain a certain amount, usually up to about 0·05, before work hardening raises its plastic strength sufficiently to overcome the initial loss due to yielding and so enables the stress-strain curve to rise. This instability propagates along the specimen as a *Lüders band* or *stretcher strain*, a simple form of which is shown in Fig. 21.6. The yielded region grows by the advance of a sharp *Lüders front*, about one grain thick, into the unyielded region. In homogeneous metal the front moves at constant speed, at constant stress. The lower yield stress thus remains roughly constant during the yield elongation, although it may jump from one level to another as new Lüders bands form in the unyielded regions. When the entire sample has yielded in this way, the stress-strain curve begins to rise through work hardening.

If the yielded sample is unloaded and immediately reloaded, it shows no yield point since it contains unpinned dislocations. However, if it is rested or mildly heated to allow carbon and nitrogen atoms to migrate to the dislocations once more, the yield point returns. These *overstrain* and *strain ageing* effects are shown in Fig. 21.5. In mild steel strain ageing usually takes a few days at room temperature, or about 30 minutes at 100°C, the rate being controlled by the diffusion of nitrogen and carbon atoms, i.e. by an activation energy of 0·8 to 0·9 eV. If the ageing treatment is more prolonged there may also be some general hardening due to precipitation of carbides and nitrides.

These effects have several important practical consequences. The pinning of dislocations makes structural steel an almost perfectly elastic solid up to a fairly high stress (e.g. $E/500$) above which the metal becomes almost perfectly plastic in the sense of deforming at constant stress. These properties are very useful in structural engineering. To set against this, strain ageing increases the temperature at which steel is brittle (see § 21.11). Also, stretcher strains are troublesome in sheet steel pressings because they give the metal an unsightly appearance. They can be avoided by *temper-rolling* or *leveller-rolling* the sheet, just before pressing (see § 22.7). The metal is then overstrained uniformly over its whole surface. Alternatively a non-ageing steel may be used. In this, some aluminium or vanadium is added to fix the nitrogen as nitride precipitates. Carbon is less troublesome as a strain-ageing element because of its low solubility at room temperature.

21.4 Effects of grain size

We saw in § 18.8 that in a polycrystalline metal the grains are forced to deform plastically on several independent glide systems. This is the main effect of polycrystallinity on the plastic properties of metals. In F.C.C. and B.C.C. metals it increases the rate of work hardening. In materials with fewer glide systems the effect is more drastic. The plastic strain is constrained to the same order as the elastic strain until the applied stress is raised sufficiently to bring the necessary additional glide systems into play. In polycrystalline zinc, for example, there is little plastic deformation until the twinning stress is reached, despite the great capacity of the individual crystals for slip on their basal planes.

Grain boundaries as such do not greatly harden a metal of normal grain size. They are too far apart. To see this, consider in Fig. 21.7 a slip band of length d across a grain. Its dislocations pile up against the grain boundaries and produce a large concentration of stress in the nearby parts of the adjoining grains. If the applied stress is σ and the 'friction stress' (see § 21.5) which opposes the glide of a dislocation in a slip band is σ_i (where $\sigma > \sigma_i$) then, when the slip band forms, a load $(\sigma - \sigma_i)d$ is transferred from it (per unit thickness) on to the material just beyond the ends of the slip band. This effect is directly similar to the stress concentrating action of a *crack*. It is known from elasticity theory that the applied stress is locally increased by a factor of about $(c/r)^{1/2}$ at a distance r ahead of a crack of length c. Similarly, at the distance r ahead of our slip band there is an additional stress of about $(\sigma - \sigma_i)(d/r)^{1/2}$. At points where $r \ll d$ this is a high concentration of stress and the neighbouring grain cannot usually withstand it.

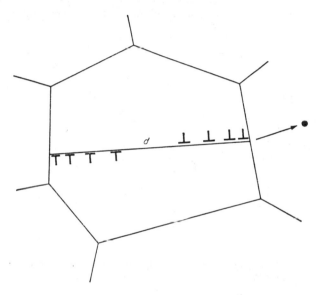

Fig. 21.7 Pile-ups of dislocations against grain boundaries at ends of a slip plane

There can be several effects of this. In a material such as zinc, where the neighbouring grains may be unfavourably oriented for relaxing the pile-up stresses, brittle cracking may occur. In metals with free dislocations which can move on many differently oriented glide systems, e.g. copper, the adjoining grains begin to yield almost as soon as some additional load is thrown on to them and there is then little effect of grain boundary pile-ups. In metals in which the dislocations are strongly pinned by impurity atoms there is an important effect because each grain has, locally, to be stressed up to its upper yield point before it can become plastic. Suppose that the central grain in Fig. 21.7 has in fact so yielded (e.g. because of the stress-concentrating effect of a foreign inclusion in it), but its neighbours are still elastic. Let the local stress needed to trigger yielding at any point be σ_d.

Then the pile-up stresses can make the neighbouring grains yield if $\sigma_d = (\sigma - \sigma_i)(d/r)^{1/2}$, i.e. at an applied stress

$$\sigma_y = \sigma_i + k_y d^{-1/2} \qquad\qquad \textbf{21.8}$$

where $\sigma_y = \sigma$ and $k_y = \sigma_d r^{1/2}$. This is the *Petch-Hall* formula for the *lower yield stress* σ_y in materials with a yield point. Experiments show that it is well obeyed in practice, cf. Fig. 21.8.

Except where impurity-pinning is weak, it seems that σ_d does not represent the stress to unpin dislocations in the adjoining grains. This is because the concentration of stress falls off as $r^{-1/2}$ from the end of the pile-up. To produce a moderate increase of stress on the nearest pinned dislocation, which may be some 10^{-6} m from the pile-up, a stress concentration is required

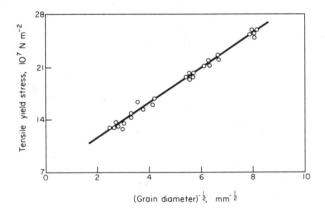

Fig. 21.8 Effect of grain size on the lower yield stress of mild steel (after Cracknell and Petch)

on the grain boundary at the tip of the pile-up which may exceed the ideal strength of the material there. In strongly pinned metals it is then easier to propagate Lüders bands by *creating* dislocations from the grain boundaries where the pile-ups occur rather than by unpinning dislocations in the next grains. Electron-microscope pictures support this conclusion. The same argument shows that even in weakly pinned metals, where unpinning is the easier process, a sizeable k_y is to be expected when the $r^{1/2}$ factor in $\sigma_d r^{1/2}$ is large. In fact even the F.C.C. metals, which do not usually give marked yield drops, generally follow eqn. 21.8 with an appreciable variation of yield stress with grain size.

21.5   Alloy hardening

Alloying is used in many different ways to strengthen metals. The most important general method is to obstruct the movements of dislocations by a fine dispersion of foreign particles distributed throughout the matrix crystal. These particles may be single atoms, as in solid solution hardening, or some larger clusters or separate phases, as in precipitation hardening and dispersion hardening. It is these dispersed obstacles that are responsible, together with any Peierls-Nabarro stress that may be present, for the 'friction stress' in eqn. 21.8.

Most of the main features of this type of alloy hardening are brought out in the study of plain carbon steels and aluminium alloys hardened by heat-treatment. In the first case the quenched supersaturated solution, i.e. martensite, is very hard and there is little further hardening when the metal is tempered to bring the carbon out of solution as a fine carbide precipitate. In the second case the quenched supersaturated solution is quite soft and intense hardening takes place on ageing or tempering, as the alloy element is allowed to cluster and then form fine precipitates. Finally, in both cases, the metal becomes soft again if tempering is carried through to the overageing stage where the precipitates become coarse.

Why is the initial behaviour so different? The answer appears to depend on several effects. Measurements of solid solution hardening have shown that solutes which produce tetragonal or other non-spherical lattice distortions harden much more strongly than those that produce spherically symmetrical distortions. The value of $d\sigma/dc$, i.e. increase in shear yield stress σ with atomic fraction c of solute, in dilute solutions, is of order 3μ for the first type and $0\cdot1\mu$ for the second, where μ is the shear modulus. Thus interstitial solutes in B.C.C. transition metals produce great solid solution hardening whereas substitutional solutes, e.g. copper in quenched aluminium alloys, do not. Probably the reason is that atoms which produce non-spherical distortions can interact with shear stress fields and so strongly obstruct both screw and edge dislocations, whereas spherically symmetrical distortions can relieve hydrostatic stresses only, which leaves the screw dislocations relatively free. An additional effect in the quench-hardening of steel is that large numbers of dislocations are produced in martensite by the transformation of the crystal structure, during the quench, and the total hardening is then a combination of solution and work hardening (cf. § 21.6 and § 21.8).

Another effect depends on the *flexibility* of a dislocation line (the *Mott-Nabarro* effect). We saw in eqn. 19.21 that a dislocation line has a certain line tension which causes it to take up a radius of curvature of about $\mu b/2\sigma$ in a shear stress field σ. Let us think of a solute atom as a centre of self-stress in the crystal. Then a dislocation line must generally change direction wherever it passes through or near such an atom, the angle of deflection increasing with the force of interaction between them. From one such atom to the next the line will loop with a curvature set by the applied stress. If there is equilibrium between the forces on the dislocation from the applied stress and from the obstacles, the dislocation will consist of a series of short loops in a macroscopically straight line. However, if the forces of interaction

are weak, as in the spherically symmetrical case, the individual loops will
be extremely shallow and the dislocation will appear as an almost rigid
straight line. Clearly if we draw a long straight line anywhere through a
random solid solution there will, in general, be as many places along it
where the forces from nearby solute atoms are pushing it backwards, on
average, as are pushing it forwards. Hence to a first approximation there is
no solid solution hardening in this case. Such hardening only becomes
possible when the dislocation line is more flexible, so that it can be pushed
against the resisting obstacles by the applied stress and, at the same time,
by looping forward, avoid those obstacles which would otherwise be pushing
it forwards. Flexibility is therefore necessary for strong hardening and, in
a solid solution where the obstacles are closely spaced, the condition for
this is that the individual obstacles should interact strongly with the dislocation.

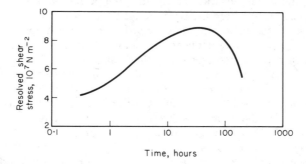

Fig. 21.9 Precipitation hardening in a single crystal of an Al-Cu alloy at 190°C
(after Dew-Hughes)

We thus expect little hardening from randomly dispersed solute atoms in
substitutional solution in aluminium. When these atoms cluster together,
or form precipitates, several factors contribute to the ensuing hardening,
of which the two most general ones are as follows. First, a moving dislocation has to *cut through* the particles lying across its path in the glide plane.
The passage of the dislocation may create a disordered layer or a stacking
fault along its path through the particle; and additional interface (which may
include interfacial dislocations) is created between the particle and the
matrix. Calculations have shown that the energy absorbed in creating these
is mainly responsible for the hardening caused by coherent precipitates.
The second general effect is that, because the obstacles are larger and more
widely spaced than in the solid solution, a dislocation line loops from one
particle to the next, so that the particles can present their full resistance to
the dislocation.

Fig. 21.9 shows the typical course of a simple precipitation-hardening
reaction, as indicated by the shear stress for slip in single crystals of an

Al-Cu alloy quenched and aged at 190°C. In commercial alloys the shear strength rises from about $10^{-3}\mu$ in the as-quenched condition to a maximum of about $10^{-2}\mu$ in the fully hardened condition. This maximum hardening corresponds to a particle spacing usually in the range 100 to 1000 Å. Particles much smaller than about 100 Å do not harden so much because thermal energy fluctuations are able to help push the dislocations through them. In such finely dispersed alloys there is thus a marked increase of yield strength at very low temperatures, where the necessary activation energy is no longer available.

Particles more widely spaced than about 1000 Å harden less because the dislocations, by forming deep loops, are able to glide through the passages between them and so by-pass them (the *Orowan effect*), as shown in Fig. 21.10. The yield strength in this case is given approximately by eqn. 19.22,

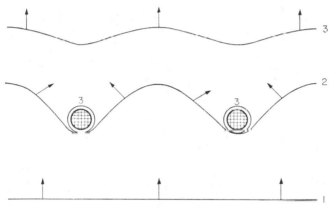

Fig. 21.10 Stages 1, 2 and 3 in the by-passing of obstacles by looping of the dislocation line between them

where l is the distance between neighbouring particles. Refined versions of this formula give good agreement with yield strengths measured on alloys containing coarse dispersions of spheroidal particles.

21.6   Work hardening

Hardening by plastic deformation, e.g. by rolling or wire drawing, is an important method for strengthening certain metals and alloys, particularly those such as copper which do not form many precipitation-hardening alloys but which are very ductile and can be heavily worked. Shear strengths up to about $\mu/300$ can be readily attained in wrought industrial metals and alloys.

Experiments on single crystals have shown that plastic deformation on two or more intersecting glide systems is necessary for rapid work hardening.

Thus in Fig. 21.11 single crystals of the hexagonal metals, slipping on basal planes only, show very little work hardening, whereas the cubic metals work harden at a much higher rate. Even in the cubic metals there are strong orientation effects. Those set for single slip, initially, show a first stage of 'easy glide' in which there is little work hardening, before the characteristic rapid rise begins. The greatest rate of work hardening in pure single crystals, measured over strains large compared with the elastic strain, is of order $d\sigma/d\epsilon \simeq \mu/150$. The stress-strain curve rises linearly at a

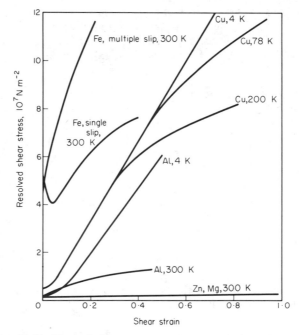

Fig. 21.11 Typical shear stress-strain curves of single crystals

rate near this value and then, above a certain stress that depends on temperature, the rate of work hardening falls off, more drastically at higher stresses, and the curve follows a roughly parabolic path.

Electron microscopy has confirmed the idea that rapid work hardening is caused by the mutual obstruction of dislocations gliding on intersecting systems. These dislocations interact in various ways. They influence one another through their stress fields. When they meet, they may combine to form sessile dislocation locks (cf. § 17.6). They may penetrate one another's glide planes, like trees in a forest, which have to be cut through for glide to continue. This can lead to the formation of dislocation jogs and dipoles and even to the combining of the glide and forest dislocations into a network. This mutual entanglement and immobilization of dislocations on different glide systems is mainly responsible for work hardening.

The rate of hardening is low in hexagonal metal crystals (e.g. $d\sigma/d\epsilon \simeq 10^{-4}\mu$) because of the difficulty of operating dislocation sources on secondary slip systems. The dislocations on the primary system then mostly run out at the free surface and are lost, so that the primary slip sources can continue to operate in an almost unchanging environment. This also happens during the easy glide stage in a cubic metal crystal oriented for single slip. The difference here however is that the secondary slip sources in this case are prevented from operating, initially, purely because they carry a smaller resolved shear component of the applied stress than the primary slip system (cf. § 18.7). It follows that these sources can be activated by locally high stresses from nearby dislocations gliding in the primary system. The process is then auto-catalytic. If a secondary source is in this way made to release a few dislocations, these interact with local dislocations of the primary system and an obstacle is formed. Other dislocations of the primary system then pile up against this obstacle which produces a local field of high stress near the obstacle, which then activates the secondary sources still further, and so on. The electron microscope has shown in fact that the transition from easy glide to rapid hardening takes place when there is marked activity on the secondary systems. Many dislocations are produced on these systems but they contribute little to the plastic strain because they move only short distances before forming obstacles with the dislocations of the primary slip system.

The density of dislocations and the shear strength build up together, during work hardening, in accordance with eqn. 19.23. The density is particularly high (e.g. $\rho \simeq 10^{16}$ m^{-2}) in the dislocation tangles which form in the obstacles themselves and these tangles gradually spread, by acquiring more dislocations from nearby slip planes, until they form a rough cell-like structure surrounding the active primary slip sources. At low temperatures the primary dislocations are confined to their own slip planes and cannot escape the continuous obstacles which encircle the sources in these planes. Work hardening then continues rapidly to high stresses.

21.7 Recovery, recrystallization and hot working

The dislocations in the active slip bands glide until they get trapped in obstacles or are pushed against them in pile-ups by the applied stress. The cold worked state is fairly stable mechanically. Apart from a small amount of running back by some of the dislocations on unloading it can persist indefinitely at low temperatures. It is of course not thermodynamically stable. Some 5 per cent of the total energy of plastic working is stored in the dislocation structure of the metal (the rest being liberated, during working, as heat) and this can amount typically to some 4×10^{7} J m^{-3}. The metal is prevented from releasing this energy so long as the dislocations are confined to their slip planes.

Thermal energy allows the dislocations to move out of their slip planes, first by the cross-slip of screw dislocations and then at higher temperatures by the climb of edge dislocations. Cross-slip enables screw dislocations to by-pass obstacles in their primary slip planes and is responsible for the diminution of the rate of work hardening in the 'parabolic' stage of the stress-strain curve.

The general effect of the extra kinematic freedom allowed the dislocations by these thermally-assisted movements is to 'tidy up' the cold worked structure. Dislocations of opposite signs come together and annihilate each other. Those of the same sign polygonize into well-defined cell walls. These *recovery* processes limit the amount of hardening attainable by cold working and also cause cold worked metals to soften when heated. The cross-slip process, being a form of glide, depends on shear stress in the material for its occurrence, irrespective of whether this stress is produced by the applied forces or by the dislocation structure. It thus tends to set in characteristically at a certain level of work hardening and can happen even at very low temperatures, e.g. $0 \cdot 1 \, T_m$, at sufficiently high stresses, e.g. $10^{-2}\mu$. By contrast, the climb process depends on the movements of vacancies (some of which may have been produced by the plastic deformation itself) and so tends to set in characteristically at a certain temperature level, about $0 \cdot 3 \, T_m$. Because it is less sensitive to the level of stress in the material it can produce much more complete softening than the cross-slip process.

A very drastic softening process, *recrystallization*, commonly occurs when cold worked metals are heated, at temperatures of about $0 \cdot 3 \, T_m$ in pure metals or about $0 \cdot 5 \, T_m$ in impure metals and alloys. Large-angle grain boundaries sweep through the metal and replace the cold worked grains by a new set of more perfect grains, giving complete softening. The annealing of cold worked metals in industry is based almost entirely on recrystallization. The process is closely related to grain growth but the driving force for moving the boundaries is provided in this case by the line tensions of the dislocations. We picture these dislocation lines as strings, attached at various points to a boundary, each string pulling the boundary into its own grain. On the one side of the boundary is the worked grain, which may have some $10^{15} \, \mathrm{m}^{-2}$ strings; on the other side is the recrystallized grain, which may have only some $10^{10} \, \mathrm{m}^{-2}$. The boundary thus moves into the worked grain at a rate which is proportional to the difference in dislocation density and which increases exponentially with temperature through the usual Maxwell-Boltzmann factor. *Primary recrystallization* is complete when all the original worked grains have been so replaced. At higher temperatures in highly textured materials this primary set of recrystallized grains may be replaced by a different set in a second wave of recrystallization. This is *secondary recrystallization* (cf. *abnormal grain growth* in § 19.5).

In industrial practice the usual aim of a recrystallization anneal is to soften a work hardened metal without producing a coarse grain size. The temperature and time of annealing have then to be chosen carefully. With thick pieces, which may require many minutes at temperature to become uniformly heated, recrystallization is usually done at the lowest temperature that will complete the process in the necessary time. Thin sheet and wire, on the other hand, may be *flash annealed* or *pulse annealed* at much higher temperatures for times of a few seconds only.

The plastic strain given to the metal before annealing has a profound effect on the recrystallized grain size. Fig. 21.12 shows schematically the effect of annealing at a given temperature after various amounts of plastic deformation. Below a critical strain ϵ_c there is no recrystallization, only recovery. At this critical strain, usually between 0·01 and 0·05, a few

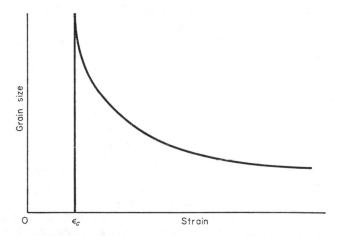

Fig. 21.12 Effect of plastic strain on recrystallized grain size

recrystallized grains form and grow extremely large, so large that this effect is sometimes exploited to grow single crystals. Where a fine grain size is required the critical range of strain must obviously be avoided but this is often difficult when the metal is non-uniformly deformed, e.g. in sheet metal pressings, and coarse-grained zones may appear in regions between deformed and undeformed parts of the metal.

The reason for the effect is that recrystallization is a process of nucleation and growth and the rate of nucleation (at a given temperature) increases far more rapidly with increasing strain than does the rate of growth. What seems to be required for nucleation is a high-angle boundary, relatively free from pinning impurities, on one side of which is a region much lower in dislocation density than elsewhere. Only a high-angle boundary has the kinematic freedom to sweep freely in all directions through the surrounding grains, as a incoherent interface, and it produces a drastic change of orientation when a highly textured structure is recrystallized. In a homogeneously deformed single crystal there may be no part of the worked structure sufficiently different in orientation from any other to allow such a boundary to form. Recrystallization then does not occur unless it is stimulated, for example by making a needle scratch on the surface to create a groove of locally disoriented material. Single crystals deformed in double or multiple slip usually recrystallize when subsequently heated. It seems that the necessary rotations of small regions in a crystal are produced in this case and that polygonization then creates large-angle boundaries round these regions. When a polycrystal is plastically worked, the mutual constraints of adjoining grains cause multiple slip; large localized lattice rotations develop at small strains, so that recrystallization becomes possible.

Experiments, mostly on F.C.C. metals such as lead, have shown that in zone-refined samples all large-angle boundaries have similar mobilities but that in less pure samples the presence of certain solute elements in trace amounts (e.g. 10^{-3} to 10^{-2} per cent) reduces the mobilities of most boundaries by two or three orders of magnitude, whereas coincidence-lattice boundaries remain fairly mobile. The effect of this selectivity on the recrystallization behaviour of highly textured commercial metals is striking. In F.C.C. crystals, boundaries which separate grains rotated by about 38° about $\langle 111 \rangle$ or 27° about $\langle 100 \rangle$, are particularly mobile; both of these are coincidence-lattice rotations. Accordingly, when the textured metal is recrystallized, it changes its orientation by one of these rotations, so forming a *recrystallization texture*. The best example of this is the *cube texture*, (100) [001], formed by the recrystallization of heavily rolled sheets of F.C.C. metals (cf. § 18.2). This texture can be extremely sharp, so that the sheet may resemble a single crystal in extreme cases.

Experiments on thin sheets have shown that the energy of the free surface plays a part in recrystallization and grain growth. Primary recrystallization is frequently brought to a halt by grooves which form by thermal etching along the lines where grain boundaries meet the surface and which then anchor the boundaries to these positions. In addition, differences of surface energy, due to different orientations of the grains, provide a driving force favouring the growth of some grains that is sometimes large enough to overcome the anchoring effect of the grain boundary grooves. This seems to be the basis of the process of secondary recrystallization in sheets and it leads to the development of specific textures in which the surface has low energy. The effect is particularly important in B.C.C. silicon-iron, used for transformer sheet. For good magnetic properties this sheet should have a cube texture. Fortunately, the (100) surface has low energy, sufficient to provide enough driving force for the growth of cube-oriented nuclei. Adsorbed impurities from the atmosphere alter the surface energies of different crystal planes and this effect has to be taken into account in the development of techniques for producing sheets with controlled recrystallization textures.

PLATE 7

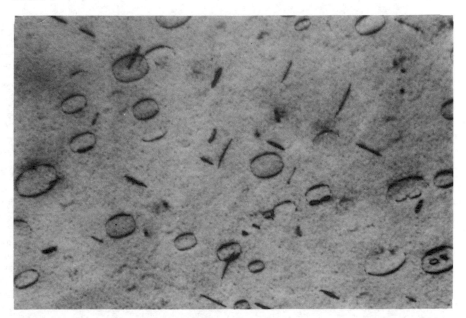

A. Electron microscope picture of dislocation rings in a quenched Al-Cu alloy, ×56,000 (*courtesy of R. B. Nicholson*)

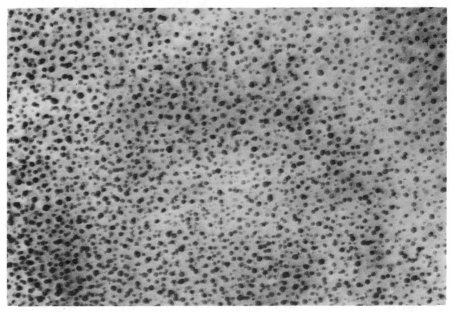

B. Electron microscope picture of spherical G.P. zones, about 90 Å diameter in a quenched and aged Al-Ag alloy, ×190,000 (*courtesy of R. B. Nicholson*)

PLATE 8

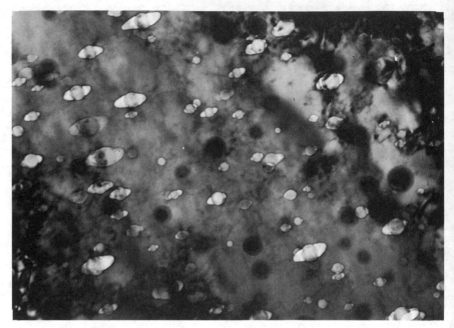

A. Electron microscope picture of holes formed by plastic deformation round spherical silica particles in copper, ×80,000 (*courtesy of G. C. Smith*)

B. Small cleavage cracks near a major crack in a silicon-killed steel broken at 10°C, ×650 (*courtesy of J. R. Griffiths*)

When metals are worked at temperatures above about $0.7\,T_m$ they recrystallize while being deformed and so remain soft. This is the basis of industrial *hot working* processes in which metal ingots or billets are given massive deformations by comparatively small applied stresses. To avoid tensile instabilities, which would result from the absence of work hardening, most hot working processes use compressive forces, e.g. rolling, forging, extrusion. The time required for recrystallization sets a limit to the speed of a hot working operation at any given temperature and very rapid operations must be done at appropriately higher temperatures. An upper limit to the hot working temperature is reached when melting occurs, usually local melting of impure regions at the grain boundaries which causes the grains to fall apart; this is *hot shortness*. In general it is necessary to work at not less than 50°C below the melting point, particularly in rapid operations where the heat of plastic deformation may raise the temperature appreciably during the working. Alloys are often much more difficult to hot work than pure metals. This is because the alloy elements usually reduce the solidus temperature and at the same time raise the recrystallization temperature, so that the hot working range between these two temperatures is narrowed. Eutectic alloys, because of their low melting points, usually cannot be hot worked. In many of these one of the phases remains hard and brittle right up to the melting point; see, however, § 22.13.

21.8 Combined dispersion hardening and work hardening

We have seen in § 21.5 that when particles obstructing dislocations become coarsely dispersed the yield stress drops because the dislocations can

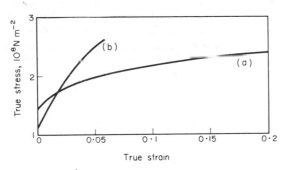

Fig. 21.13 Tensile true stress-strain curves of Al-Cu single crystals, quenched and then (a) aged and (b) over-aged at 190°C

loop between them. However, provided the particles are themselves very strong and the matrix adheres strongly to them, this low initial yield stress is generally followed by a high rate of work hardening. Fig. 21.13 shows the effect in precipitation-hardened Al-Cu crystals. After plastic working, the over-aged structure gives the stronger alloy.

14+I.T.M.

The rate of work hardening is high because the dispersed particles do not deform plastically with the matrix. As a result, dislocations of the matrix accumulate round the particles, initially as simple loops, as in Fig. 21.10, but later as dense tangles. This has two hardening effects. By promoting glide on intersecting systems near the particles, it stimulates the work hardening of the matrix. Secondly, the stresses from the dislocation loops tend to make the particles conform, by *elastic* deformation, to the plastic deformation of the matrix. This leads of course to very large elastic strains in the particles and hence to the particles supporting a large fraction of the total applied load. This effect, first recognized by *Fisher, Hart and Pry*, hence called *FHP hardening*, is undoubtedly largely responsible for the high tensile strengths of commercial metals, such as annealed carbon steels, which contain coarsely dispersed foreign particles and hard compounds.

A limit to the hardening achievable by this means is set by the fracture or yield strength of the particles, or by the matrix tearing itself away from them or flowing round them by complex three-dimensional plastic deformation. Fortunately the particles, although coarse by precipitation-hardening standards, can nevertheless still be fine enough to have ideal fracture strengths ($\simeq \mu/10$); they can also have structures that resist the passage of glide dislocations through them; and, particularly if formed by precipitation within the matrix, can adhere strongly to the surrounding crystal. The limiting factor in such cases is the level of work hardening attainable in the matrix metal since this fixes the level of stress at which the matrix can, in principle at least, flow round the particles.

There are several possibilities here. First, if the matrix metal has a high melting point compared with the temperature under consideration, there will be little recovery and work hardening can develop in it to a high level. Thus, iron is capable of being greatly strengthened at room temperature by combined dispersion and work hardening. Flow of the matrix round the particles can also be made difficult by keeping the scale of the dispersion fine, so that dislocations have to bend sharply if they are to produce complex plastic movements near particles. In this sense a fine dispersion can stabilize the work hardened structure of the matrix and suppress the recovery, polygonization and grain boundary processes that lead to softening.

The combined effects of dispersion and work hardening are frequently used to make strong steels. In the *patenting process* a medium carbon steel wire is transformed from austenite to a finely and uniformly dispersed ferrite-cementite structure in a molten lead or salt bath at 400–500°C and then hard drawn through dies. Tensile strengths of up to about $E/50$ (E = Young's Modulus) can be produced in 0·95 per cent C steel by this process, which is used for making piano wire and steel rope. In the *ausforming process* an alloy steel (e.g. > 3 per cent Cr to suppress the pearlite transformation) containing about 0·45 per cent C is converted to austenite and then cooled to about 500°C and plastically worked. This work-hardened austenite transforms to a type of tempered martensite which contains a high density of dislocations stabilized by finely dispersed carbides. Tensile strengths of up to about $E/70$, with good ductility, are then obtained.

One method of overcoming the tendency of a soft matrix to flow round coarsely dispersed particles is to use needle-shaped particles, so that distant parts of the matrix are then connected by strong elastic struts. The logical development of this idea leads to the fibre-strengthened structures discussed in § 21.15.

21.9 Creep

Metals in service are often required to support steady loads for long times. The progress of plastic deformation with time at constant stress, i.e. *creep*, is therefore an important property. Experiments show that creep occurs at all temperatures. However, in practice it is usually only creep at high temperatures, i.e. above about $0.5\,T_m$, that matters.

To see the reason for this, suppose that we load a specimen to a stress a little above its initial yield stress. It deforms rapidly, plastically, to a strain at which work hardening has raised its yield stress to the applied stress. At this stage there is a precarious balance between the driving and resisting forces on the dislocations and not much thermal activation energy is required to upset this balance and keep the dislocations moving. Rapid *transient creep* thus follows on smoothly from the initial deformation. The resisting forces, due to the continuing build-up of the work hardening obstacles, then rise above the driving force and the creep slows down. At low temperatures the only softening process is cross-slip and this can operate only so long as the applied stress remains close to the self-stresses due to the dislocation structure. The characteristic effect of temperature in this range, therefore, is to *alter the shape of the short-time stress-strain curve* rather than to produce extensive creep at constant stress. Only a rather small transient creep is observed, which increases with time t according to a logarithmic law,

$$\epsilon(t) = \epsilon_0 + \alpha \ln t \qquad \textbf{21.9}$$

At higher temperatures more general processes of thermal softening occur and allow creep to develop extensively. Recrystallization when it occurs produces a catastrophic increase in creep rate but is generally avoided under practical creep conditions. At extremely high temperatures *vacancy creep* (cf. § 19.9) can produce significant deformation under very low stresses, but this is a special case. At temperatures in the range $0.5\,T_m$ to $0.7\,T_m$, in which creep-resistant alloys are mainly used, the important softening process is *dislocation climb*. The significant point here is that mentioned in § 21.7. Climb is mainly a thermally-activated process, not critically sensitive to the stresses acting on the dislocations, and it enables the forces resisting dislocation glide, from the work hardening obstacles, to relax continually. As a result, the driving force can maintain this glide at such a rate as to build up the obstacles by work hardening as fast as they are relaxed by thermal softening. We therefore expect a continuing *steady-state* creep to develop in the material. Using σ_i to denote the self-stress from the obstacles, we can define a *coefficient of work hardening*, $h = \partial\sigma_i/\partial\epsilon$, from the change of this stress with plastic strain ϵ at constant time t, and a *coefficient of thermal softening*, $r = -\partial\sigma_i/\partial t$, from the change with time at constant strain. In steady-state creep σ_i is constant, i.e.

$$d\sigma_i = \frac{\partial\sigma_i}{\partial\epsilon}\,d\epsilon + \frac{\partial\sigma_i}{\partial t}\,dt = 0 \qquad\qquad \textbf{21.10}$$

so that the steady creep rate is given by

$$\dot\epsilon = \kappa = \frac{r}{h} \qquad\qquad \textbf{21.11}$$

There is of course a smooth transition from the rapid deformation on first loading to the slow steady creep, through a stage of transient creep which often increases as $t^{1/3}$ (*Andrade's law*), and the whole creep-time relation can be represented by

$$\epsilon(t) = \epsilon_0 + \beta t^{1/3} + \kappa t \qquad\qquad \textbf{21.12}$$

Experiments show that the rate of steady-state creep depends on temperature and applied stress (σ) through a relation of the type

$$\dot\epsilon = A\sigma^m \exp\left(-Q/kT\right) \qquad\qquad \textbf{21.13}$$

where Q is usually close to the activation energy for self-diffusion, as would be expected since the climb process is a form of localized vacancy creep between dislocations. The driving force on a vacancy in the stress fields of these dislocations is usually small so that we might expect that $m = 1$. In practice however, $m \simeq 4$. The reason appears to be that at higher applied stresses there is a greater density of dislocations in the material at any given time, so that there are larger numbers climbing and the distances they have to climb to annihilate or polygonize are smaller.

Creep tests in practice are usually made by suspending constant loads from simple tensile bars in thermostatically-controlled furnaces and recording the creep strains indicated by extensometers or strain gauges mounted on the specimens. Fig. 21.14 gives some typical creep curves. They show the features outlined above and also, at the higher temperatures, a final *accelerating* stage of creep. When creep strains are large this accelerating stage is usually due to the effect of increasing stress, since the cross-section decreases under constant load, and it can be eliminated by using a device to reduce the load in proportion to the cross-section, so maintaining constant stress. In engineering tests on creep-resistant materials, however, the strains are usually too small for this cross-section effect to be significant. The acceleration in this case is then due either to some metallurgical change in the material, leading to a softer structure, or to the formation of small cracks. The accelerating stage usually presages the fracture of the material.

Engineering creep tests are expensive. They have to be run for long times, e.g. 10^3 to 10^5 hr, with precise temperature control, and run in large numbers to explore the behaviour of materials over a range of temperatures and stresses. There is thus a great incentive to try to deduce long-term creep behaviour from short-term tests. Extrapolation by time is extremely unreliable, because the later stages of creep curves by no means follow uniquely from the earlier stages. Complex creep alloys rarely follow eqns. 21.9 or 21.12 with sufficient precision for a reliable extrapolation and it is hardly possible to predict the onset and extent of the accelerating stage. A better method is to make complete short-term creep tests at a higher temperature and then, using the activation energy formula, stretch these

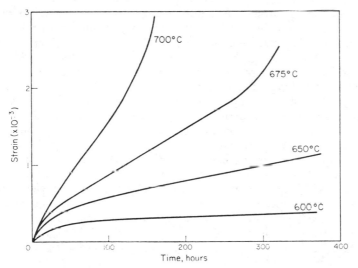

Fig. 21.14 Tensile creep-time curves of a low-alloy steel at a stress of 7×10^7 N m^{-2}

same creep curves out to longer times at the required temperature. The main risk in this method is that different processes occurring in the material may have different activation energies and so may not contribute to the same degree at different temperatures.

The development of creep-resistant metals and alloys involves two separate problems: first, the development of refractory metals such as molybdenum, niobium and chromium into practical metallurgical materials; second, the achievement of the best high temperature properties in a chosen metal or alloy. The first is mainly a problem of developing important anciliary properties such as oxidation resistance, low-temperature toughness, weldability, etc. The second is a problem of providing obstacles to dislocations that are stable at high temperatures. For this stability we depend on dispersed particles, either to stabilize a work-hardened state by pinning the dislocations through a strain ageing process, as in some low-alloy boiler tube steels, or to provide a stable dispersion-hardened state, as in nickel-base gas turbine alloys.

The obvious way to make such dispersions is by precipitation. Many creep-resistant alloys are in fact of the precipitation-hardening type. There is however a fundamental difficulty. For the solution treatment a good solubility at high temperatures is required but this very same solubility enables the alloy to over-age and soften when used for long times at moderately high temperatures. There are several ways of attacking this problem. We can use less soluble precipitates, by taking the solution temperature practically up to the melting point. This is done in some steels, hardened by alloy carbides. Secondly, we can try to minimize the driving force for coarsening by developing coherent precipitates that have low interfacial energies with the matrix, e.g. Ni_3Al in nickel-base creep-resistant alloys. Thirdly, we can make a stable dispersion by a process of diffusion and solid-state reaction, as in *internal oxidation*. In this process, a dilute solution of a highly oxidizable element in a fairly noble metal (e.g. Si in Cu) is exposed to an oxidizing atmosphere at high temperatures. Oxygen diffuses in and reacts with the oxidizable element to form very fine and extremely insoluble oxide particles. Great hardness which is stable to high temperatures can be achieved in this way. A difficulty is that grain boundary regions tend to become preferentially oxidized and greatly embrittled in consequence. This can be overcome by internally oxidizing a fine powder of the alloy and then sintering the particles together. Closely related to this are processes of dispersion hardening in which fine powders of metal and oxide are mixed and sintered together. An example is S.A.P. (sintered aluminium powder) which consists of finely dispersed Al_2O_3 in Al and has good creep strength up to about $0.7T_m$. Another is ThO_2 in nickel.

21.10   Theory of fracture

We saw in Fig. 21.2 some of the ways in which a ductile metal can break by *plastic strain concentration* when work hardening is no longer able to protect the material from plastic instability. Although usually referred to as *ductile fracture*, such a failure is better described as a *rupture*, since fracture characteristically involves the propagation of sharp *cracks*.

Fig. 21.15 shows a simple crack. We can regard the material as a sandwich made up of a thin failure zone F, along which the crack C propagates, and two elastic blocks E (the substance and structures of which may be the same as in F or may be different, according to the material). During the propagation of the crack the deformation in E is perfectly elastic, or nearly so (although it may have been plastic before the crack started) and this is what we mean when we say that such a fracture is *macroscopically brittle*.

This brittle fracture may be produced by many different processes of failure—elastic, plastic, viscous, chemical—in the layer F. At one extreme we may think of the two blocks E as held together by a single layer of atomic bonds, as shown in Fig. 21.15, which stretch and break as the crack advances through them. This is the ideal form of brittle tensile fracture, approached closely in the *cleavage* fracture of brittle crystals in which the crack runs along a crystallographic plane of low ideal tensile strength. Alternatively, the same bonds in Fig. 21.15 could represent filaments of

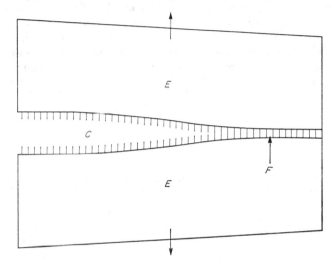

Fig. 21.15 A simple crack

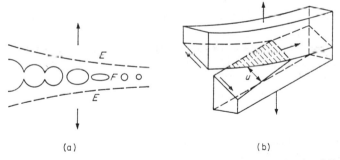

(a)   (b)

Fig. 21.16 Fracture due to (a) localized necking, (b) shear in the failure zone

adhesive in a glued joint, or 'bridges' of plastically deforming metal in a 'brittle' failure produced by localized plastic deformation. Fig. 21.16 shows two of the ways in which the latter could happen. In the first, we obtain a tensile failure by the necking down of ductile bridges between holes existing along the path of failure. It is because of this type of process that incompletely sintered powder-metal compacts have poor tensile properties; also rolled metals, containing large amounts of slag, when pulled in directions transverse to the plane of rolling. In the second example a plastic shear failure occurs along a zone at about 45° to the tensile axis. This type of failure often occurs in thin metal sheets.

In all such brittle failures the overall mechanics of the process is essentially the same. The resistance to failure comes from the work which has to be done to separate the material in the failure zone. We define a *displacement u*, from the increase in distance between corresponding points on opposite faces of the zone, e.g. the increase in distance between the broken lines in diagram (a) of Fig. 21.16, or the shear displacement shown in diagram (b). The stress $\sigma(u)$ supported by any length of the failure zone then varies in a manner such as that of Fig. 21.17 as the displacement u of this section

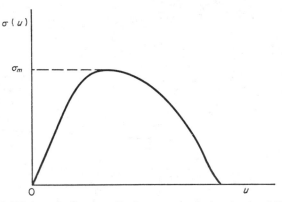

Fig. 21.17 A typical stress-displacement relation in the failure zone

increases from zero. The work done is the integral of force times displacement over the whole process of failure in this section, i.e. the *work of fracture* 2γ, per unit area, is given by

$$2\gamma = \int_0^\infty \sigma(u) \, du \qquad \text{21.14}$$

We write it as 2γ since in the particular case of ideal cleavage γ is then the surface energy of the crystal plane exposed by the crack. In processes such as those of Fig. 21.16 plastic work is the main contributor to 2γ. It is sometimes convenient in engineering problems to combine 2γ with Young's modulus E, in the form

$$K_c = (2\gamma E/\pi)^{1/2} \qquad \text{21.15}$$

and to call K_c the *fracture toughness* of the material.

What applied stress is required to propagate such a crack? This is not a matter of concentrating enough stress at the tip to break the bonds there because by the very nature of the crack there are bonds at its tip at all stages of deformation and failure, including some at the maximum stress σ_m of Fig. 21.17. It is essentially a thermodynamic question, as *A. A. Griffith* first pointed out. Enough energy has to be released by the growth of the crack to supply the work of fracture 2γ. For a simplified calculation, we consider in Fig. 21.18 a large body, of unit thickness in the direction normal to the diagram, which is stressed to an average tensile stress σ by being held at constant strain between fixed grips at its ends. Let it contain a small crack of length $2c$, as shown. Then, approximately, a circular region of radius c, round the crack, is relieved of stress σ and hence of elastic energy $(\sigma^2/2E)\pi c^2$ by the presence of the crack. If the crack grows infini-

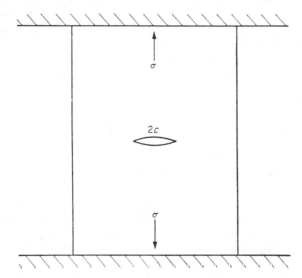

Fig. 21.18 A crack in a body held between fixed grips

tesimally, to $2(c + dc)$, the additional elastic energy released is $(\sigma^2/2E)2\pi c\,dc$. An exact calculation gives just twice this value. The energy absorbed as work of fracture due to this infinitesimal growth is $4\gamma\,dc$. Equating these, we have

$$\sigma \simeq \sqrt{\frac{\gamma E}{c}} \qquad\qquad \textbf{21.16}$$

as the approximate value of the breaking strength of the body. This is the *Griffith formula* or, when γ contains plastic work, the *Griffith-Orowan-Irwin* formula for brittle fracture. It is valid when $\sigma \ll \sigma_m$, i.e. when c is large compared with the length of the failure zone at the tip of the crack. Its validity is not limited to the fixed-grip situation considered here. A simple extension of the theory shows that it holds equally for other methods of stressing, e.g. by dead-weight loading.

We can now understand the difference between weak brittle materials and strong brittle ones. A glass window-pane and a steel chisel are both brittle, for example, yet the first is weak and the second is strong. In weak brittle fractures there is very little plastic deformation and γ is then given mainly by the surface energy. For example, if $\gamma = 1$ J m^{-2} and $E = 10^{11}$ N m^{-2}, a crack 10^{-2} m long gives a breaking strength of only about 3×10^6 N m^{-2} $\simeq$ 450 psi in a material which could have an ideal strength of order 10^{10} N m^{-2}. In strong brittle fractures γ is greatly increased by the contribution of the plastic work. For example, in Fig. 21.16(a) the spacing of the plastic bridges might typically be of the order of the grain size, or the spacing of foreign inclusions. Fig. 21.17 now becomes the plastic stress-elongation curve of the bridges. Typically, this curve might rise to a stress of order $E/300$ over a plastic displacement of order 10^{-5} m, giving $\gamma \simeq 3000$ J m^{-2} and $\sigma \simeq 1{\cdot}7 \times 10^8$ N m^{-2} $\simeq$ 25 000 psi for a 10^{-2} m crack. These two examples bring out vividly the distinction that must be made between *microscopic* and *macroscopic strength*. In the first, we have $\sigma_m \simeq E/10$ and $\sigma \simeq E/30\,000$; in the second, $\sigma_m \simeq E/300$ and $\sigma \simeq E/600$.

There are of course many different ways in which atoms may pull apart or slide over one another in the highly stressed region round the tip of the crack. Subject to the effect of orientation of the stresses at the crack tip, σ_m is the failure stress of the weakest mode of failure in the tip. The failure will take place in this microscopically weakest mode and the alternative modes will remain dormant so long as their own failure stress is not reached. If this weakest mode has a large failure displacement, however, its fracture toughness will be high and the body will be macroscopically strong even when notches are present. This is probably the most crucial feature of a metal for heavy engineering use, since it enables the metal to remain strong and tough even when notched and grooved.

In highly ductile metals the plastic region cannot be confined to a narrow failure zone. Slip spreads out from the tip of the crack into the metal above and below, so that general yielding spreads throughout a large volume and the crack opens up by plastic deformation into a wide notch, giving a ductile rupture.

21.11   Brittle fracture of steel

Structural steel is usually ductile at atmospheric temperatures. However, some of the most disastrous and tragic of engineering failures, such as the complete break-up of ships, bridges and pressure vessels, have been caused by the brittle fracture of steel. Such fractures come without warning and take place very rapidly. A welded steel ship, for example, may break in two in a fraction of a second, due to a brittle crack which runs round its hull and deck at a speed of about 2000 m s^{-1}.

Temperature plays a major part in the effect. Except for F.C.C. metals, virtually all solids become brittle at low temperatures. The transition from ductile behaviour to brittle behaviour generally occurs in a narrow range, a few tens of degrees only, so that it is possible to characterize a material by a certain *transition temperature*. In mild steel there is a *crack-arrest temperature* which commonly occurs in the range 250–350 K. Below the crack-arrest temperature the metal will allow a crack, once started, to run as a rapid brittle fracture at a stress below 10^8 N m^{-2}, which is in the range of ordinary working stresses for steel in structural engineering. Above this temperature the metal will stop such a crack by extensive plastic deformation at the tip.

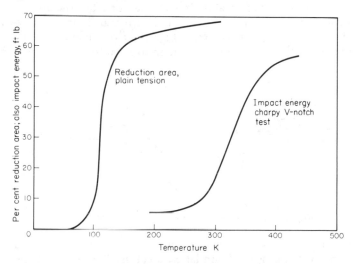

Fig. 21.19 Ductile-brittle transition in a Bessemer rimming steel, as indicated by reduction of area in a plain tensile test and by energy absorbed in a Charpy V-notch impact test

This running brittle crack has of course first to be started. It is because brittle cracking is difficult to initiate that structural steel is normally ductile at atmospheric temperatures. Plain tensile tests give a ductile cup-and-cone fracture with some 30 to 60 per cent reduction of area at room temperature (unless the metal is embrittled by some constituent such as iron oxide, carbide or phosphide along its grain boundaries, which gives a weak intercrystalline fracture). This ductile behaviour is repeated, with little change, at lower testing temperatures down to about 100 K. The *nil-ductility temperature* is then reached, below which the metal appears brittle even in plain tensile tests and the reduction of area drops sharply to almost zero (cf. Fig. 21.19).

There are two ways of producing brittle fractures in structural steel at atmospheric temperature in the laboratory. The first is to use a *crack-propagation test* in which a large plate of the metal, under load, is severely cooled at one end so that a brittle crack can be started and then run into rest of the plate at a higher temperature. This test allows the crack-arrest temperature to be measured but is expensive because it requires large specimens. The second method is to use a *notch test*. This has many forms but one of the simplest and most widely used is the *Charpy* or *Izod* V-notch impact test in which a small bar of standard dimensions, across which is a machined sharp notch, is struck with a swinging heavy pendulum to bend and break it open at the notch. The ductility of the specimen is then determined either from the energy absorbed, which is found from the amplitude of the pendulum before and after impact, or from the appearance of the fracture. Fig. 21.19 shows the ductile-brittle transition of a low-carbon steel as determined by a Charpy test. This transition is much more closely related to the crack-arrest temperature than to the nil-ductility temperature.

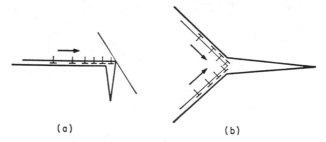

(a)   (b)

Fig. 21.20 Formation of cracks from dislocations

Experiments near the nil-ductility temperature have shown that, unless severely embrittled at grain boundaries, steel usually does not break until it begins to yield plastically. This has been shown by comparing brittle fracture stresses, in tension, with ductile yield stresses, in compression. Another indication is that, if the metal is strained part way through its yield elongation, small brittle cracks (*micro-cracks*) may be found in the Lüders bands but not in the unyielded regions. The fracture occurs by cleavage on (100) planes in the ferrite grains, approximately perpendicular to the main tensile axis, and the evidence suggests that these cracks may form by the coalescence of dislocations piled-up at grain boundaries or intersecting slip or twin bands, as suggested in Fig. 21.20.

It appears that the general condition for brittle cleavage in steel is that yielding should occur in the presence of a tensile stress greater than a certain value, typically about $E/200$. The yield stress of iron rises rapidly at low temperatures, probably due to the Peierls-Nabarro force (§ 17.4), and reaches this level at about 100 K. In the nil-ductility range deformation twins form and sometimes produce cleavage cracks at their points of intersection. The metal is then very brittle.

There are various ways of reaching the critical stress level at higher temperatures. Under impact loading the high rate of strain will raise the yield stress for slip and so raise the brittle temperature. At temperatures slightly above the nil-ductility range work hardening can raise the stress to the required level and cleavage may then occur after some plastic deformation. At much higher temperatures necking and ductile rupture usually start before this stress level is reached. Finally, there are the effects of a notch, which may be present *ab initio* in a notched specimen, or may be formed by ductile rupture.

The first effect is that the notch produces a large hydrostatic tensile stress near its root. In Fig. 21.21 we consider a long sharp notch the tip of which runs along the z axis (perpendicular to the diagram) through a thick plate. The x and y axes are as shown. The diagram shows qualitatively the

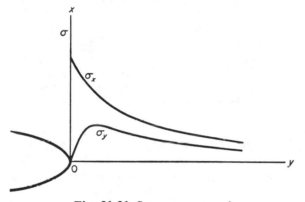

Fig. 21.21 Stresses at a notch

values of the principal tensile stresses σ_x and σ_y at various points along the y axis directly ahead of the tip, when the metal is elastic. The stress σ_x across the root rises to a maximum at the tip, as expected. The tensile strain associated with this stress tends to produce corresponding contractions along the y and z axes but these contractions cannot occur except to a limited extent because the metal is constrained to remain *in situ* by its surroundings. Lateral tensile stresses σ_y and σ_z are therefore also set up in the root. The stress σ_y rises, with σ_x, as the tip is approached but then falls off to zero at the surface because there is no tensile force acting through this surface. The stress along the notch, σ_z, is given by $\nu(\sigma_x + \sigma_y)$, where ν is Poisson's ratio.

A fairly similar stress distribution holds when the root of the notch becomes plastic, the main difference being that σ_x falls off somewhat, near the tip of the notch. The reason is that the value of $\frac{1}{2}(\sigma_x - \sigma_y)$, i.e. the maximum shear stress in the material, is fixed by the plastic shear strength of the metal. Calculations show that the maximum value of σ_x in this case can be some two to three times larger than the tensile yield stress of the metal, for the same shear yield stress. This enhanced tensile stress enables the stress condition for brittle fracture to be satisfied at a much lower yield stress than in plain tension, and the nil-ductility temperature is correspondingly raised, typically to about 200 K.

The second effect of the notch, which carries the notch transition temperature up into the range of the crack-arrest temperature, is to produce large plastic strains highly concentrated in the root. These strains give work hardening and ductile ruptures that can convert into brittle cracks, but without the expenditure of much plastic work because the volume of worked metal is so small. In this way the V-notch impact test is able to produce brittle fractures which start from small regions of ductile rupture at the root of the notch.

The fact that high *hydrostatic* tensile stress raises the brittle temperature is most important. Hydrostatic stress cannot increase the number of dislocations in a pile-up, or squeeze them more closely together, but it can help to expand a crack once this is formed. It follows that the conversion of glide dislocations into cracks cannot be the *hardest* stage in starting the fracture; the early growth of such a micro-crack, after it has formed, must require a higher stress. The reason appears to be that grain boundaries are powerful obstacles to small cleavage cracks. Many such cracks, which have stopped at grain boundaries, can in fact be observed in steel after deformation just above the nil-ductility temperature. Grain refinement in fact is amongst the most effective of techniques for improving the toughness of steel by lowering the transition temperature. Aluminium-killing, to produce a fine-grained metal, is one of the main steel-making techniques used for producing structural steels with greater toughness at temperatures down to about 250 K. To achieve toughness at much lower temperatures than this, a high-alloy steel (e.g. with a large nickel content) or an austenitic steel must be used.

The effect of grain boundaries can also sometimes be seen in brittle fractures in service. Small disconnected cleavage cracks may be observed in the grains immediately ahead of the main crack front (Plate 8B). These cracks do not penetrate the grain boundary regions and the process of fracture is completed by the ductile tearing and rupture of the bridges between the cleavage cracks, rather in the manner of Fig. 21.16(a). Most of the observed fracture toughness of the metal is contributed by the plastic work absorbed in these bridges.

Service brittle fractures in large steel structures generally start from large notches created by poor design and manufacture procedures. Badly made welds, for example, may leave unwelded regions in the joints which act as deep notches. Furthermore, high self-stresses may exist in the adjoining regions, due to the thermal strains produced by welding, and these self-stresses may produce large, local, plastic strains at the roots of the notches and so set off the fracture. The remedy is to use a careful inspection procedure, to eliminate welding defects, and to anneal the assembled structure to relieve the self-stresses. Where utmost safety is required, a steel should be used which has a crack-arrest temperature below the service temperature.

21.12   Chemical embrittlement

There are innumerable examples from metallurgical practice of normally strong and tough metals being made extremely weak and brittle by various chemical effects. The hot shortness and cold shortness of steel are two such examples. There are many ways in which chemical effects can ruin the mechanical properties of industrial metals and much scientific effort and process control is involved in protecting metals from them.

One way in which metals are embrittled is through the precipitation of *gaseous phases* inside them. Very high pressures can thus build up in small internal gas pockets, sufficient to break the surrounding metal. For example, from the formula $P = \exp (\Delta G^{\ominus}/RT)$, a standard free energy of gas precipitation of, say, 80 000 J mol^{-1} is capable of creating a gas pressure of order 10^4 atm $\simeq 10^9$ N m^{-2} at 1000 K. The embrittlement of steel due to the precipitation of hydrogen gas, from solution, in fine *hair-line* cracks is an important practical problem. It is overcome in steel castings by vacuum degassing (cf. § 13.9) but there are many ways in which hydrogen can enter and embrittle the metal in subsequent processing and use, e.g. by heating in reducing atmospheres and by acid pickling and electrolytic plating. There are many other examples of embrittlement by gaseous phases. Copper can be embrittled by annealing in reducing atmospheres. Hydrogen diffuses into the metal and reduces cuprous oxide particles, particularly along grain boundaries, to steam; the pressure of this then disrupts the metal along these boundaries. Another example is provided by the inert gases created in metals by nuclear radiation, such as fission gases in uranium fuel elements, which form gas pockets.

There are many metallurgical effects that can lead to brittleness. In high-chromium heat-resistant alloys the brittle intermetallic *sigma phase* (cf. § 14.4) tends to form along grain boundaries. In Al-killed structural steel the carbide tends to decompose slowly to *graphite*, which can embrittle grain boundaries. In high-strength precipitation-hardened aluminium alloys *denudation* (cf. § 20.2) produces narrow zones of soft metal along the boundaries. When such a microstructure is stressed, intense plastic shear occurs locally along these soft zones. This produces a situation geometrically similar to that of Fig. 21.20(b) and cracks are produced at the triple junctions of the grain boundary network.

Metals are frequently embrittled by the segregation to boundaries and surfaces of substances that lower the surface energy. The embrittling effect of hydrogen in steel is believed to be partly due to the adsorption of hydrogen on the crack faces, so lowering their surface energy. Another example is provided by *liquid-metal embrittlement*. When a polycrystalline metal is wetted by a liquid metal that forms a solid-liquid interface of low energy, the liquid may penetrate along the grain boundaries. If the *dihedral angle* is zero (cf. § 19.5), i.e. if $2\gamma_S \leqslant \gamma_{GB}$ in eqn. 19.31, where γ_S is the solid-liquid interfacial energy and γ_{GB} is the grain boundary energy, the liquid can penetrate completely along the grain boundaries and disintegrate the polycrystalline solid. Even if the dihedral angle is not zero the presence of the liquid reduces the applied stress needed to part the metal along its grain boundaries. Normally ductile and strong metals can be severely embrittled by exposing them while under tensile stress to certain molten metals; e.g. brass by mercury; copper by bismuth; and steel by solders and brazing alloys.

In the embrittlement of ductile metals by reduction of the surface energy, it seems probable that a major effect is that lower stresses are developed in the vicinity of the crack, with the result that less plastic deformation is produced in the adjoining grains by the presence of the crack. In other words, a reduction in surface energy may bring about a far greater reduction in the total work of fracture.

Still more striking effects are produced when metals are stressed in environments in which some chemical attack takes place. The free energy released by chemical dissolution can then contribute to the driving force for fracture. We shall discuss these *stress-corrosion* effects in § 23.9.

21.13 Creep rupture

A creep-resistant material is required to support stresses for long periods at high temperatures without deforming unduly or breaking. Full information on both of these is provided by determining the creep curve over a range of steady stresses and temperatures. In some applications, however, fracture is the main limiting factor, not excessive deformation. It is then possible to dispense with the full creep test and, instead, to make *stress-rupture* tests in which the time to fracture at a given stress and temperature is measured.

What happens, as regards fracture, when a specimen is held under constant tensile stress for a long time at constant temperature? At the low stresses used in typical creep tests the grains remain almost rigid, for example straining plastically by only about 0·01 in some 100–10 000 hours. The normal ductile rupture processes, involving large plastic strains and extensive necking, cannot occur. In this sense creep-rupture is a brittle fracture (although it develops very slowly). Although the stress is too low to operate the ordinary dislocation processes to any great extent, there are nevertheless certain thermally activated processes of deformation and fracture which can operate slowly at high temperatures, even at very low stresses.

The two most important of these processes are *grain boundary sliding* and *vacancy creep*. The atoms in a non-coherent grain boundary have two crystal surfaces on which to crystallize, but they are prevented from crystallizing on either by the forces from the other. They thus have a choice of alternative possible positions, between which they can move to some extent independently of their neighbours. Normally such movements occur randomly to and fro across the boundary, according as random thermal fluctuations throw them over the energy barriers between the positions. A small shear stress applied along the interface, although quite unable by itself to produce such movements, nevertheless slightly promotes those that cause the boundary to shear with the stress and slightly opposes those in the opposite direction. Provided the temperature is high enough, the boundary can thus slide even at very low stresses.

The temperature above which grain boundary sliding becomes strongly developed, which is called the *equicohesive temperature*, depends of course on the time of loading but is commonly about $0.5T_m$. Below it, ductile metals fail by ductile rupture with a large ductility; above it, rupture becomes localized in the grain boundaries and occurs after a strain of order 0·01 to 0·1.

Sliding produces large strains at the triple junctions where differently oriented grain boundaries meet. If the grains themselves are fairly soft, these strains may be accommodated by plastic shear of the material near the junctions. The effect of grain boundary sliding in this case is simply to increase the overall rate of creep. However, if the grains are strong the stresses built up at the triple junctions may become large enough to open up cracks along the grain boundaries there, in a manner geometrically similar to that of Fig. 21.20(b). This *triple-point cracking* is one of the major causes of creep rupture.

There are various ways to reduce triple-point cracking. One is to soften the grains, provided that the ensuing increased creep rate is acceptable. Another is to use a textured structure in which the grains have a fibrous shape, aligned along the main tensile axis, to avoid large components of shear stress along the boundaries. It is also possible to *pin* the boundaries by foreign particles that act as obstacles in them, although it is essential that such particles should adhere strongly to the adjoining grains, since otherwise cracks may nucleate alongside them. The nucleation of triple-point cracks is still not fully understood, but it appears that adsorbed impurity atoms in the boundaries which can lower the energies of crack surfaces play an important part in this process. There are considerable opportunities for further research here.

At high temperatures and low stresses triple-point cracking gives way to *cavitation fracture*, in which small spheroidal holes form all along certain boundaries and then slowly grow and coalesce. Grain boundary sliding is usually necessary to nucleate these cavities, probably at non-adherent foreign particles or at small ledges in the boundaries. Once formed, however, these holes appear to grow by vacancy creep, which can happen at extremely low stresses at high temperatures. Vacancy creep removes atoms from the surfaces of the holes and deposits them on nearby parts of the grain boundary where they become attached to the adjoining grains. The specimen thus grows larger and this causes the applied (tensile) stress to do work. If a spheroidal hole on a boundary perpendicular to the tensile stress σ grows to a radius r, the work done is $\frac{4}{3}\pi r^3\sigma$. For the values of r ($\simeq 10^{-6}$ m) and σ involved in creep cavitation, the work of *elastic* deformation round such a hole is negligible, as is shown by the fact that such holes in practice remain spheroidal and do not distort into the shapes of Griffith cracks. The surface energy of the hole is of course $4\pi r^2\gamma$. The condition for growth by vacancy creep at the stress σ is then given by

$$\frac{d}{dr}\left(\tfrac{4}{3}\pi r^3\sigma - 4\pi r^2\gamma\right) = 0 \qquad\qquad 21.17$$

i.e.

$$\sigma = \frac{2\gamma}{r} \qquad\qquad 21.18$$

For example, if $\gamma = 1$ J m^{-2}, a hole of radius $r = 10^{-6}$ m can grow at a stress $\sigma = 2 \times 10^6$ N m$^{-2} \simeq 300$ psi. The general problem of preventing or reducing creep cavitation in polycrystalline metals is similar to that of dealing with triple-point cracking. Careful control of impurity content seems the best approach but much more needs to be known about adsorption and wetting at grain boundaries. A more radical solution is to use creep-resistant alloys in the form of *single crystals*.

21.14 Metal fatigue

Engineering metals are often subjected to fluctuating loads in service. Thus, aircraft wings are buffeted in turbulent air, connecting rods are pushed and pulled in piston engines and leaf springs bent to and fro. In all such cases the metal is liable to fracture by *fatigue*, the most common of all causes of engineering failure. In most fatigue tests a specimen is subjected to a regular sinusoidally oscillating stress, at a frequency in the range 1000 to 10 000 cycles min^{-1}, with a mean stress at or near zero. Either a tensile bar is used and subjected to push-pull loading, or a rotating shaft is used, on which a bending moment is applied through loaded bearings. For each *stress amplitude S* used, where S is half the difference between the extreme stresses of the cycle, the *fatigue life or endurance* is measured, i.e. the number of cycles N to fracture, and an S, N curve is plotted as in Fig. 21.22.

Fatigue failure begins by the formation, fairly early in the life of the specimen, of a small crack, almost always at a point on the external surface. This crack then slowly grows into the material, roughly perpendicularly to the main tensile axis, as a sharp narrow fissure until the remaining un-broken cross-section is so small that it can no longer support the applied load and undergoes a simple tensile failure. Experiments have proved that the fatigue crack front advances a small distance during each stress cycle and each stage of advance is marked on the fracture surface by a small undulation or *ripple line*. These lines form a series of roughly circular and concentric arcs, which mark successive positions of the crack front and focus back to the origin of fracture. Under ideal conditions they are too fine and evenly spaced to be seen, except by special metallographic methods, but under practical conditions fluctuations in the amplitude of the stress cycle generally produce a few ripples much larger and more prominent than the rest, easily visible on the fracture surface with the unaided eye. A fatigue failure is therefore easily recognizable since the fracture surface consists of two distinct parts, one broken by fatigue and marked with concentric ripple lines, the other broken by simple tensile failure, i.e. ductile rupture or brittle fracture.

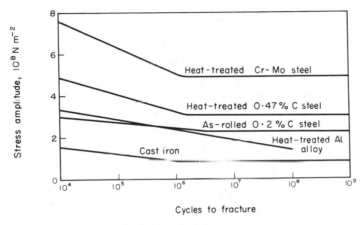

Fig. 21.22   Typical fatigue curves

Because fatigue cracks generally start at the surface, the fatigue strength of a metal can be greatly improved by inducing *compressive self-stresses* in the surface layers. This can be done by plastically working the surface by *surface-hammering* or by *shot-peening* (bombarding the surface with fine shot); or alternatively by diffusing into the surface layers some alloy element that expands the structure there. The fact that fatigue cracks grow slowly enables precautionary measures to be taken against fatigue failure in service. Critical components can be inspected for cracks periodically and taken out of service when badly damaged. Being fine and narrow, fatigue cracks are not easily seen but various sensitive techniques for crack detection, involving fluorescent penetrating inks or ultrasonic or electromagnetic probes, have been developed.

Fatigue cracks appear to be brittle but in fact are produced by plastic deformation. The important thing about *cyclic* stressing, at a stress amplitude comparable with the yield strength of the metal, is that it *continually* produces plastic deformation, alternately positive and negative in each cycle. Work hardening fails to stop this process. This is because sources of high localized plastic deformation, such as favourably oriented grains or roots of notches, which plastically deform in one half of the stress cycle, are then pushed back by the elastic strains in surrounding regions and so also plastically deform in the opposite direction during the other half of the cycle. This is shown by the *Bauschinger effect*, in which plastic yielding resumes, in the reverse direction, at a stress (σ_b) numerically smaller than that reached in the forward direction (σ_a), as in Fig. 21.23.

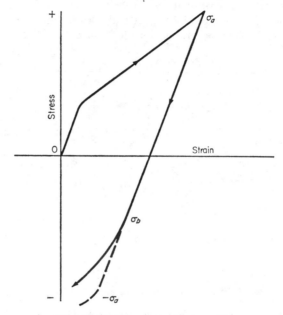

Fig. 21.23 The Bauschinger effect

It is this continual to-and-fro plastic deformation in localized regions that eventually produces and spreads the fatigue crack. The failure commonly starts in thick, plastically active, slip bands which run in from the surface of the metal. The repeated plastic oscillation of these slip bands, in and out of the surface, causes some of the metal in them to become gradually squeezed out through the surface, forming so-called *extrusions*; and at the same time narrow fissures gradually advance down these slip bands. When they have penetrated through a few grains, which may take some thousands or millions of cycles to achieve, depending on the stress amplitude, these fissures then change their direction of advance and begin to grow perpendicularly to the main tensile stress axis. They now advance much more rapidly by a process that produces the characteristic ripple lines. Each half-cycle of tensile stress pulls the faces of such a crack apart by plastically tearing the material at the root of the crack. During the compressive half-cycle the faces are pushed together again, including the plastically torn region. The process repeats exactly during the next tensile phase, except that the crack is now slightly longer than before because the plastically torn region has become part of the general crack surface.

One way to improve fatigue strength, then, is to make plastic deformation difficult. This leads to various practical measures. The surface should be smooth and gently contoured, without strain concentrators such as notches and sharp steps. The fatigue strength, i.e. stress amplitude for a given life, increases roughly in proportion to the tensile strength and any method of hardening the metal will usually correspondingly improve the fatigue strength. Since failure generally begins in the surface, methods of surface hardening, e.g. by shot peening, surface hammering, carburizing and nitriding of steel, etc. are important for improving the fatigue properties. Conversely, the formation of soft, ductile surface layers, for example by surface decarburization of steel, lowers the fatigue strength.

Since it causes dislocations to become pinned and immobile, strain ageing is effective in improving fatigue strength. So also is precipitation hardening although there is a complication about this in hard aluminium alloys at room temperature. The finely precipitated structure is unstable in the presence of alternating plastic strain; softening occurs along active slip bands which then become very susceptible to fatigue failure. It seems likely that the oscillatory movements of the dislocations in active slip bands cause vacancies to be generated and these, by diffusion, lead to dissolution and overageing of the precipitates nearby. High fatigue strength can then be obtained only at temperatures where vacancies are insufficiently mobile, i.e. below about $0.25 T_m$.

21.15 Fibre strengthening

The theory of interatomic forces shows that the ideal tensile strength of a crystalline solid is typically about $0.1E$. There is thus a great challenge to produce practical materials with strengths approaching this ideal value. The two things that weaken a solid are dislocations and cracks. Both have to be eliminated or made ineffective if the material is to be strong against both yielding and fracture. In general, the elimination of dislocations from crystalline materials, though not impossible, is difficult and expensive. It is not at all difficult to immobilize dislocations by more drastic methods than we have so far considered, for example by using intermetallic compounds or non-metallic crystals in which the dislocations have very high Peierls-Nabarro forces, but this merely means that we have produced brittle materials with very low resistance to crack propagation. This is the central dilemma facing the designer of strong materials.

In strong alloys it is solved by dispersing fine strong particles of such a substance in a softer ductile matrix. Shear rather than fracture remains the preferred mode of failure in this matrix and so the mixture combines fairly high strength with ductility. This solution to the problem is of course a compromise and cannot lead to strengths much above about $0.01E$. In addition, metals do not have outstandingly high Young's moduli, they are often heavy, chemically reactive and unstable at high temperatures.

The alternative approach is to use thin strong fibres, bundled together and bonded by an adhesive matrix. This is the basis of commercial *fibreglass*, in which thin glass threads are held together by a resin. A bulk strength of about $0.02E$ is obtained but the disadvantage is that E is rather small for glass. However, many solids such as alumina, graphite, boron,

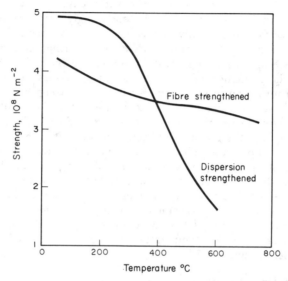

Fig. 21.24 Tensile strength of silver at various temperatures, showing effect of dispersion strengthening with alumina particles and of fibre strengthening with a small volume fraction of alumina whiskers (after Sutton and Chorné, 1963, *Metals Engineering*, **3**, 44)

silicon nitride and silicon carbide can be prepared in the form of thin strong fibres, with diameters about 10^{-6} to 10^{-5} m. Dislocations are immobile in them except at high temperatures; they have smooth surfaces free from deep notches and steps; and are light, refractory and have high Young's moduli. Some of these are now forming the basis of new composite materials of high strength.

One possibility is to add such fibres to creep-resistant alloys to increase the high-temperature strength. Consider a ductile metal rod with an elastic fibre along its axis. If the rod is stretched it will tend to flow plastically along the fibre and, in so doing, will subject this to a 'tug-of-war'. Clearly, if the fibre is sufficiently long, the accumulated force from this tug-of-war on the central length of the fibre can amount to a very large stress even if the metal is very weak plastically or adheres very weakly to the fibre. A sufficiently long fibre can in this way always be stressed up to its intrinsic strength by a weak matrix. Given a large number of such fibres, held together as a parallel bundle by such a matrix, the strength of the composite along the fibre axis becomes almost independent of the mechanical properties of the matrix, even though the latter is responsible for transferring the applied load on to the fibres. In this way it is expected that a metal can be strengthened against creep practically up to its melting point. Fig. 21.24 shows that this appears to be possible in practice.

Fibre-strengthening can also provide great toughness. If the fibre-matrix interface is fairly weak, a crack attempting to cut through a bundle of such fibres will tend to turn along the fibre axis rather than continue in the transverse direction. A simple illustration of this is provided by a bamboo rod which remains highly resistant to fracture even when sharply notched with a razor blade. In fibre-strengthened alloys, where there is likely to be fairly good adhesion between the fibres and the matrix, the greatest toughness is to be expected when the fibres have just such a length that they will tend to pull out of their sockets rather than break. The work done in pulling out all the fibres lying across the path of fracture, against the frictional forces provided by the adherent matrix, will then contribute a large factor to the work of fracture in eqn. 21.14.

FURTHER READING

AMERICAN SOCIETY FOR METALS (1962). *Strengthening Mechanisms in Solids.*

AVERBACH, B. L. (1959). *Fracture.* John Wiley, New York and London.

BIGGS, W. D. (1960). *The Brittle Fracture of Steel.* MacDonald and Evans, London.

COTTRELL, A. H. (1953). *Dislocations and Plastic Flow in Crystals.* Oxford University Press, London.

FRIEDEL, J. (1964). *Dislocations.* Pergamon Press, Oxford.

HONEYCOMBE, R. W. K. (1968). *The Plastic Deformation of Metals.* Edward Arnold, London.

KELLY, A. (1966). *Strong Solids.* Clarendon Press, Oxford.

KENNEDY, A. J. (1962). *Processes of Creep and Fatigue in Metals.* Oliver and Boyd, London.

KNOTT, J. F. (1973). *Fundamentals of Fracture Mechanics.* Butterworths, London.

KOVACS, I. and ZSOLDAS, L. (1973). *Dislocations and Plastic Deformation.* Pergamon, Oxford.

MCLEAN, D. (1962). *Mechanical Properties of Metals.* John Wiley, New York and London.

NATIONAL PHYSICAL LABORATORY (1963). *Symposium on the Relations between the Structure and Properties of Metals.*

OSBORNE, C. J. (1965). *Fracture.* University of Melbourne: Butterworth, London.

READ, W. T. (1953). *Dislocations in Crystals.* McGraw-Hill, New York and Maidenhead.

THE ROYAL SOCIETY (1964). 'Discussion on New Materials', *Proc. R. Soc.* **A282.**

TIPPER, C. F. (1962). *The Brittle Fracture Story.* Cambridge University Press, London.

ZACKAY, V. (1965). *High-Strength Materials.* John Wiley, New York and London.

Chapter 22

Plastic Working

22.1 Introduction

The ability to deform plastically is a supremely important property of a metal for general engineering use. Not only does it enable the metal to be both strong and tough in service; it also enables it to be shaped in the solid state by various processes of mechanical working.

There are two reasons for the mechanical working of metals: first, to produce shapes that would be difficult or expensive to produce by other means. These may range from shapes such as thin sheet or wire, which are simple in themselves but difficult to produce cheaply in any other way, to more complex shapes such as I-beams, steel rails, embossed coins, crankshafts, screw threads and motor car bodies. All these are possible by various processes of rolling, forging, extrusion, stamping, drawing and pressing.

The second reason is that the mechanical properties of metals and alloys are usually improved by mechanical working. The most obvious effect here is work hardening, commonly used as a means of strengthening materials such as copper and alpha brass. Working also enables tougher microstructures to be developed; for example, by closing and welding up cavities; by elongating slag particles and other inclusions into parallel strings (aligned, by careful design of the process, along directions that will bear the greatest tensile stress in the finished product); and by refining the grain size by recrystallization, homogenizing cored structures and breaking up and spheroidizing films of phases round the as-cast grain boundaries, all by hot working or by cold working and annealing.

The yield stress of a metal in a cold working operation rises to a high level through work hardening, e.g. 2×10^9 N m^{-2} in stainless steel, and a relatively massive plant is required to handle the work-piece. In hot working, the yield stress rarely exceeds about 7×10^7 N m^{-2}, so that a plant of the same size is then able to handle much larger work-pieces and give heavy deformations.

22.2 Simple elongation

In many processes such as wire drawing, extrusion and rolling of bar between grooved rolls, the metal is essentially elongated along one axis and reduced in its transverse dimensions.

The simplest example of such a process is, of course, provided by the tensile test itself. Consider a tensile bar which is being stretched uniformly by a force on its ends and let its current length, area and yield strength be l, A and $Y(l)$ respectively. When the length increases from l to $l + dl$ the increment of *plastic work* done is $Y(l)A\,dl$. The total work to stretch the bar from l_1 to l_2, noting that the volume $= V = Al$ is constant, is then given by

$$W = \int_{l_1}^{l_2} Y(l)A\,dl = V \int_{l_1}^{l_2} Y(l)\,d(\ln l) \qquad 22.1$$

The *work done in the metal per unit volume*, $w_i = W/V$, is thus

$$w_i = \int_{l_1}^{l_2} Y(l)\,d(\ln l) = \int_{A_2}^{A_1} Y(A)\,d(\ln A) = \int_{\epsilon_1}^{\epsilon_2} Y(\epsilon)\,d\epsilon \qquad 22.2$$

where A_1, A_2 and ϵ_1 and ϵ_2 are the corresponding cross-sectional areas and true strains. Thus w_i is simply the area under the true stress-strain curve

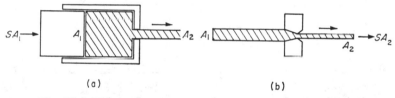

(a)   (b)

Fig. 22.1 Elongation by (a) extrusion and (b) wire drawing

between ϵ_1 and ϵ_2. In problems of mechanical working $Y(\epsilon)$ is often replaced by its average value $\bar{Y}$ in the range from ϵ_1 to ϵ_2, so simplifying the above expressions to

$$w_i = \bar{Y} \ln\left(\frac{l_2}{l_1}\right) = \bar{Y} \ln\left(\frac{A_1}{A_2}\right) = \bar{Y}(\epsilon_2 - \epsilon_1) \qquad 22.3$$

This simplification is reasonable in hot working, where there is no work hardening, and also in cold working if the metal is already heavily work hardened when it begins the deformation process under consideration.

We now consider the applied force which, by moving as the work-piece deforms, actually does the mechanical work on the material. Let it act on the area A with intensity S per unit area. In a simple tensile bar, pulled on its ends, we have, of course, $S = Y(l)$.

In several metal-working processes, however, S acts on a *constant area* A; e.g. $A = A_1$ in diagram (a) of Fig. 22.1 and $A = A_2$ in diagram (b). In moving a distance l the applied force does the work SAl, i.e. SV, and thereby works a volume V of the metal. Hence, equating this to the work W done in the metal,

$$S = w_i \qquad\qquad\qquad 22.4$$

and we see from eqns. 22.2 or 22.3 that in general $S \neq Y$. This fact is vital to the wire-drawing process since it allows us, by keeping the drawing reduction $(A_1 - A_2)/A_2$ sufficiently small, to pull the metal through the die by an applied tensile stress S that is *smaller* than the tensile yield strength Y of the metal. This crucial feature of the wire-drawing process enables plastic stretching of the exit wire, with all its problems of necking and tensile fracture, to be completely avoided. All the plastic deformation is done within the die, where the forces (from the die itself) are predominantly compressive, even though all the mechanical work is done by the tensile force pulling on the end of the exit wire.

Eqn. 22.4 applies only to *ideal* plastic working in which no energy is wasted in processes that do not contribute directly to the required change

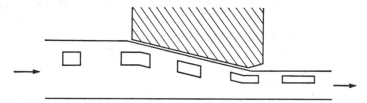

Fig. 22.2 Deformation in a die

of shape. Such conditions prevail in the central length of a tensile bar stretched uniformly by forces at its ends. Most other forms of mechanical working, however, are less efficient than this and a correspondingly larger S is needed to achieve the change of shape. The two main sources of energy loss are *friction*, between the workpiece and the dies or tools acting on it, and *geometrically redundant plastic deformation*. Fig. 22.2 shows what happens typically to a square element of the metal as it passes through a die. As well as changing to a rectangle, which conforms to the overall deformation, the element is sheared, first one way and then the other, and these shears contribute nothing to the overall deformation although they consume plastic work.

Let w_f and w_r be the work consumed by friction and redundant deformation, respectively, per unit volume of metal worked. We then define the *efficiency* of the process as

$$\eta = \frac{w_i}{w_i + w_f + w_r} \qquad\qquad 22.5$$

and hence

$$S = \frac{w_i}{\eta} \qquad\qquad\qquad 22.6$$

In practice, efficiencies commonly range from about 50 per cent in hot working to about 80 per cent in cold rolling.

22.3 Forging

Most mechanical working processes employ compressive forces, so that tensile necking and fracture are avoided. *Forging* is an example. Beginning with the blacksmith's hammer and anvil, the process has developed in various ways. Large pieces are forged under machine driven hammers and very large ones under slow hydraulic presses. In *closed die* forging a small component such as an engine crankshaft is shaped in a *die*, the two halves of which completely enclose it when brought together. The die blocks are driven together by a machine hammer, as in *drop forging* or *stamping*, or by a press. *Coining* is another example of closed die forging. *Upsetting* is a

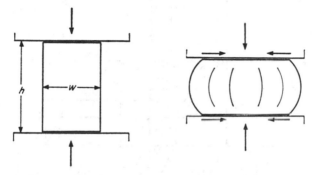

Fig. 22.3 Barrelling in simple compression

forging process in which the end of a rod is splayed out to form a bolt head. In *open die forging*, or *bar forging* or *press cogging*, an open-ended die is used so that the metal is free to elongate along it. This method is used for forging bars, shafts and rings. In *swaging* a small cylindrical rod is forged by passing it through a rotary machine in which it is hammered by converging dies. This process is particularly useful for working the more brittle metals without fracture.

As a simple example of forging, consider the compression of a regular slab between flat anvils, as in Fig. 22.3. In uniform deformation the true strain is given by $\ln (h_1/h_2)$ where h_1 and h_2 are the initial and final values of the height h. There is no tensile instability, of course, but there are other problems. An instability due to plastic buckling and sideways shear may occur unless the ratio h/w of height to width is below about 1·5. The other problem is *barrelling*, as shown in the diagram. This is caused by the force of friction acting inwards along the faces between the anvils and the work-piece, which results from the outward spreading of the metal along these faces. To prevent barrelling it is necessary to interrupt the process at frequent intervals, unload and lubricate the faces of the anvils.

If the slab in Fig. 22.3 is a circular cylinder the metal spreads out equally along all radial directions from the centre of the circular area of contact with the die face. Consider now the configuration of Fig. 22.4 in which the plastic zone between the die faces is long and thin. When squeezed by the dies it is constrained, against spreading outwards along the z axis (except

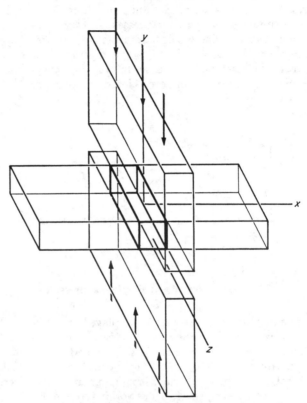

Fig. 22.4 Plane strain compression

slightly at its ends), by the forces exerted on its top and bottom faces, due to friction, and on its side faces by the adjoining parts of the work-piece outside the plastic zone. In the limit of a long, thin zone these forces of friction and *plastic constraint* completely prevent deformation along the z axis; the inward movement of particles in the plastic zone along the y axis is then exactly compensated by outward movement along the x axis. This type of *two-dimensional* deformation is known as *plane strain* and occurs, at least approximately, in several metal-working operations. It is important in the rolling of sheet for example, which is geometrically rather similar to Fig. 22.4, since it prevents the sheet from widening out along the line of contact with the rolls.

22.4 Mechanics of plane strain deformation

In Fig. 22.5 we show an elementary cube of the material in the plastic zone of Fig. 22.4 and see that the plane strain deformation can be produced by shear in the directions shown, on diagonal planes at 45° to the x and y axes. These are of course not crystallographic slip planes (although are commonly referred to as *slip lines*) but indicate the average shear systems of

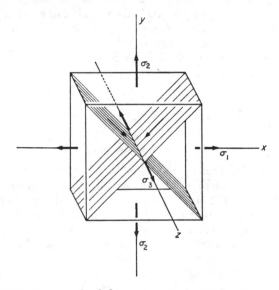

Fig. 22.5 Plane strain deformation of a cubical volume element

regions involving large numbers of crystal grains. The *positive directions* of the principal stresses, σ_1, σ_2, σ_3, along the x, y, z, axes respectively, are indicated in the diagram. Compressive stresses of course act in the negative direction.

What is the *yield criterion*, i.e. the values of σ_1, σ_2, σ_3, for this plane strain deformation? The simplest idea is that plastic shear occurs when the maximum shear stress in the material acts on the shear system shown and reaches a characteristic value k, the *shear yield stress* of the material. This is the *Tresca criterion*. Since a tensile stress σ produces a maximum shear stress $\sigma/2$ on a plane at 45° to it, and since positive stresses σ_1 and σ_2 produce shear stresses in opposite directions on the diagonal planes of Fig. 22.5, the yield criterion becomes

$$\sigma_1 - \sigma_2 = 2k \qquad\qquad 22.7$$

Is this reasonable? Suppose that $\sigma_1 = \sigma_3 = 0$. Then, with a negative σ_2, the cube of Fig. 22.5 is subjected to simple compression, so that in this case $\sigma_2 = -Y$, where Y is the uniaxial tensile yield stress. Substituting in eqn. 22.7 we see that

$$Y = 2k \qquad\qquad 22.8$$

on the Tresca criterion. However, in *uniaxial compression* the metal is free to use *any* shear plane at 45° to the y axis. In *plane strain compression* it is limited to only those planes which are *also* at 45° to the x axis. Does this restriction increase the resistance of the material to the compression? According to the Tresca criterion it does not. Experiment however shows that in practice there is usually a small effect. This can be interpreted as an influence of the third principal stress σ_3 on the yielding which is brought about mainly by the stresses σ_1 and σ_2.

A yield criterion involving all three principal stresses is needed. This is provided by the *von Mises* criterion

$$(\sigma_1 - \sigma_2)^2 + (\sigma_2 - \sigma_3)^2 + (\sigma_3 - \sigma_1)^2 = 6k^2 \qquad 22.9$$

Applied to uniaxial tension or compression, where $\sigma_1 = \sigma_3 = 0$ and $\sigma_2 = \pm Y$, it gives

$$Y = \sqrt{3}k \qquad\qquad 22.10$$

which differs from the Tresca conclusion. To apply it to plane strain deformation we need the value of σ_3. This principal stress is clearly intermediate in magnitude between the other two; it plays a part in the plastic deformation of the material, yet there is no deformation along its own axis. These conditions are clearly satisfied if σ_3 is equidistant between σ_1 and σ_2, so that the tendency of the one pair of principal stresses to extend the metal along the z axis is opposed by the equal tendency of the other pair to contract it along this axis. We thus have

$$\sigma_1 + \sigma_2 = 2\sigma_3 \qquad\qquad 22.11$$

in plane strain. Using this to eliminate σ_3 from eqn. 22.9, we find that the von Mises criterion reduces to eqn. 22.7 for plane strain. We thus take eqn. 22.7 as our yield criterion but use eqn. 22.10 in place of eqn. 22.8. The *plane strain yield stress*, i.e. the value of $-\sigma_2$ when $\sigma_1 = 0$, is then

$$-\sigma_2 = 2k = (2/\sqrt{3})Y = Y' \simeq 1\cdot 15\,Y \qquad 22.12$$

so that, compared with the uniaxial yield stress Y, the material is 15 per cent stronger. This represents the effect of constraining the deformation to only one family of diagonal planes.

22.5 The friction hill

The yield pressure $1\cdot 15\,Y$ which must be exerted by the faces of the die to produce plane strain compression is valid only under ideal conditions. We now consider how much the pressure is increased by friction.

In Fig. 22.6 we show a slab of height h between dies of width $2a$ and suppose that the thickness normal to the diagram is unity. Consider the forces acting on the volume element shown, of thickness dx at a distance x from the centre line. The net horizontal force due to the stresses σ_1 and $\sigma_1 + d\sigma_1$ is $h\,d\sigma_1$. Balancing this is $2\mu P\,dx$ due to the two frictional forces on the top and bottom faces, resulting from the pressure P and the coefficient of friction μ. Hence $2\mu P\,dx = h\,d\sigma_1$. Since $P = -\sigma_2$, eqn. 22.7 gives $d\sigma_1 + dP = 0$, so that $dP/P = -d\sigma_1/P = -(2\mu/h)\,dx$, which integrates to

$$P = Y'e^{(2\mu/h)(a-x)} \qquad\qquad 22.13$$

since $P = Y'$ at $x = a$. The pressure P exerted between the die and the metal thus increases towards the centre line as shown in Fig. 22.7, known as the *friction hill*.

Clearly, when $a \gg h$ this can lead to an enormous increase of pressure in the central regions, which raises great difficulties in certain processes such as the rolling of thin strip. There is however a limit to the frictional force because μP cannot exceed k, the shear yield strength of the metal. This limiting condition is reached when the metal *sticks* to the dies and shears plastically along a narrow boundary layer next to the die faces. The friction hill still continues to rise, towards the centre, but *linearly* instead of exponentially because μP is then replaced by the constant k (neglecting work hardening).

Dies and rolls used in hot working are usually rough and unlubricated, so that the coefficient of friction is then high, e.g. $\mu = 0\cdot3$ to $0\cdot6$. However, hot working is generally concerned with the heavy deformation of thick slabs, so that $2a \simeq h$. The exponential factor then does not rise very high and it is a fair approximation to take the *mean pressure* over the die faces as

$$\bar{P} \simeq Y'\left(1 + \frac{\mu a}{h}\right) \qquad\qquad 22.14$$

so giving an increase, due to friction, of order 50 per cent in typical hot working processes.

Friction is important in practically all metal fabrication processes. In cold rolling, highly polished rolls flooded with lubricating oil are used to keep μ below $0\cdot05$. In wire drawing, the dies (usually tungsten carbide) are similarly lubricated, e.g. with soap powder or tallow, to keep $\mu \simeq 0\cdot05$. In extrusion, the great pressure developed between the metal and the die can lead to sticking, due to pressure welding; an important advance in the extrusion of steel has been the development of a surface phosphate film on the billet which is able to hold in itself an oily constituent sufficiently strongly to prevent pressure welding in the dies, so making possible the *cold extrusion* of steel. For *hot steel extrusion* a viscous glass lubricant has been developed (*Sejournet process*). In the *machining* of metals the cutting tool parts the chip from the work piece by a process of plastic shear. The friction tends to be very high ($\mu \simeq 0\cdot5$ to $1\cdot0$) because the chip surface bearing against the

tool is freshly formed and hence free from adsorbed films which could reduce the adhesion. Lubrication is then essential. The cutting tool is flooded
with a solution of a water-soluble oil to reduce friction and remove heat.
For easy machining a *free-cutting* metal is used, which contains a built-in
lubricant in the form of dispersed particles of a suitable second phase, e.g.

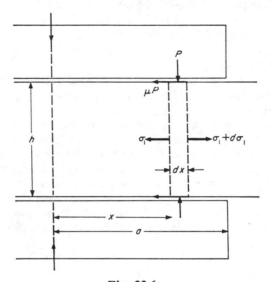

Fig. 22.6

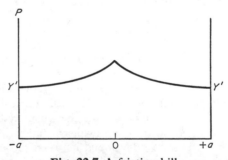

Fig. 22.7 A friction hill

lead in brass, manganese sulphide in steel. In the pressing of powdered
metals to form *powder compacts* by the plastic deformation and pressure
welding of the adjoining particles, it proves to be almost impossible to
produce a long cylindrical bar by pushing with a piston on one end of a
column of the powder in a cylindrical mould. The friction on the walls
shields the powder at the far end from the pressure exerted by the piston at

the near end. This problem is overcome by packing the powder in a sealed rubber tube which is then compressed by an external hydrostatic pressure.

22.6 Surface indentation

Let us now try to use the method of Fig. 22.4 to make a bar of approximately square cross-section. The thickness h of the metal is now about the same along the y and z axes, which allows it to spread considerably along the z axis and so partly defeats our purpose. If, to avoid this, we try to regain plane strain conditions by making the die very narrow, i.e. $2a \ll h$, we meet another difficulty. The plasticity does not spread to the centre of the slab. Instead, the metal immediately beneath the dies tends to slip sideways and so allow the dies to *indent* by heaping up at the surfaces alongside them.

Fig. 22.8 shows the effect. The metal next to the indenter is displaced by plastic flow, as shown, while the remaining part of the work piece remains undeformed. The pressure P required for this indentation is higher than

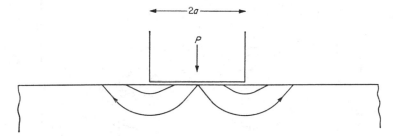

Fig. 22.8 Local deformation under a frictionless indenter

Y', however. The work done by P, as the indenter sinks in, is converted to plastic work by shear along the slip lines against the shear stress k. For a unit displacement this work is a product of k and the length of the slip lines (for unit thickness perpendicular to the diagram). Because they have to curve beneath the indenter to reach the free surface nearby, they are disproportionately long in relation to the width $2a$ of the indenter. (They must approach the free surface, and also the surface of the indenter when there is no friction, at 45°, since the normal to such a surface is necessarily an axis of principal stress.) Because they are so long, the work done through the shear stress k on them is correspondingly high and so also is the pressure P which provides the energy. We can see intuitively that a factor of about 2 or 3 is involved here. A calculation of the effect, which lies beyond the scope of the present treatment, shows in fact that

$$P = (2 + \pi)k \simeq 2 \cdot 82 Y \qquad\qquad 22.15$$

Because of this high *plastic constraint* factor, the surface indentation mode is replaced by the general deformation mode of § 22.3 and § 22.4 well

15 + I.T.M.

before $2a$ is increased to h. In practice it is found that $2a > \frac{1}{3}h$ is sufficient to ensure that plasticity spreads completely between the dies.

Resistance to plastic indentation is determined by an *indentation hardness test*, in which a small hard indenter is pressed into the surface by a standard load and the size of indentation produced by it is then measured. In most tests a *diamond* indenter is used, in the shape of a pyramid (*Vickers test*), a cone (*Rockwell test*), or an elongated pyramid (*Knoop test*), but in the *Brinell test* a sphere of hard steel or tungsten carbide is used. The *hardness number H* is defined as the load on the indenter divided by the area of contact between the indenter and the material, usually given in kg mm^{-2}. Although the plastic indentation is not plane strain, the plastic constraint factor remains at about the same value, so that

$$H \simeq 3Y \qquad\qquad 22.16$$

to a good approximation. This relation enables the hardness test to be used

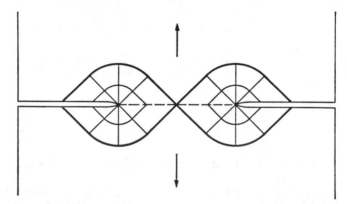

Fig. 22.9 Plastic constraint at the root of a notch

as a simple, cheap and non-destructive means of evaluating the yield strengths of metals, particularly for quality control during production and for diagnosis of failure in service. The advantage of the sharply pointed diamond indenter is that it always produces a geometrically similar indentation, irrespective of the hardness of the material (assuming this to be at least an order of magnitude softer than diamond) or the depth of the indentation. The geometry of the Vickers diamond, for example, is such that a plastic strain equivalent to a tensile elongation of about 8 per cent is produced in the indentation. In this case Y in eqn. 22.16 refers to the yield strength after a tensile strain of this magnitude. A spherical indenter has several disadvantages: the geometry of the indentation depends on the depth of penetration, so that Y is no longer measured at a single well-defined plastic strain; because the sphere first touches the surface tangentially, the earliest stages of penetration can be accommodated partly *elastically*, the material springing back again when the load is removed; and of course a steel ball is likely itself to deform when used on hard materials.

An important application of eqn. 22.15 to the behaviour of plastic metals when notched was made by *Orowan*. Fig. 22.9 shows such a notch in a tensile bar. Allowing for the reversal in the sign of the stress, each half of the bar acts on the other half, through the ligament by which they are joined, in exactly the same way as the punch acts on the specimen in Fig. 22.8. The same pattern of slip lines is applicable and the effective tensile yield stress of the ligament is thus given by eqn. 22.15. This is the effect referred to in § 21.11, which produces a tensile stress across the notch that is mainly hydrostatic tension and is some two to three times larger than the uniaxial tensile yield stress, for the same shear yield stress.

22.7 Rolling

An obvious technical development of the forging process in Fig. 22.4 is to turn the dies into rotating *rollers* and pass the metal continuously between them. Rolling in fact is much more economical than forging; it is quicker and uses less power. It is generally preferred except where large complex shapes are to be made and is the main process by which ingots, as cast after extraction and refining, are reduced in section to billets, blooms, slabs, bars and sheets. In steel-making, for example, the ingots are taken from their moulds while still hot and then stored in a *soaking pit* or furnace ready for hot-rolling at 1000–1200°C. The first rough rolling of such an ingot is done in a *cogging mill* between large cast iron rolls. This is usually a *two-high reversing mill*, i.e. two horizontal rolls one above the other, the direction of rotation of which can be reversed to pass the metal forwards and backwards. Sometimes a *three-high* non-reversing mill is used, the central roll rotating oppositely to those above and below, and the metal passes forward through the lower pair and back again through the upper pair. For the first few passes the *draught* (i.e. reduction of cross-section) is light, until the as-cast grain structure has been broken down and toughened. Reductions of 10–50 per cent per pass are then used. Usually more than one mill is used, the first being for rough shaping or 'cogging'; the later ones, which have hard and smooth rolls, being for 'finishing'. Plain cylindrical rolls are used for producing slabs, plates and sheets; grooved rolls for bars, blooms, billets and more complex sections such as rail track.

The mechanics of the rolling process is not fundamentally different from that of the forging process of Fig. 22.4. It might be thought that there would be no friction, since the contacting part of the roll moves along with the metal. However, because the metal is plastically elongated in the rolling direction, as it passes between the rolls, it speeds up. It thus moves slower than the roll on the entry side and faster on the exit side. Friction then acts along the roll-metal interface towards a central point where the speeds are equal and a friction hill is formed, like that of Fig. 22.7, but somewhat distorted, as in Fig. 22.10, due to the more complex geometry and also to work hardening, if any, as the metal passes through the rolls.

Consider now the *roll force*, per unit length of roll, which pushes the rolls apart; F in Fig. 22.10. This determines the size and hence cost of the mill needed for a given operation. The product of the roll force and its lever arm length to the centre of a roll ($\simeq \frac{1}{2}BC$ in Fig. 22.10) determines the *torque* on the roll and hence the power needed. The roll force cannot easily be accurately calculated. For a simple estimate we shall take it as the product of the mean pressure, given by eqn. 22.14, and the length BC in Fig. 22.10. We have $BC = \sqrt{BD \times AB}$. Assuming the angle BDC to be small, $BD \simeq 2R$ where R is the radius of the roll. Also $AB = \frac{1}{2}(h_1 - h_2)$.

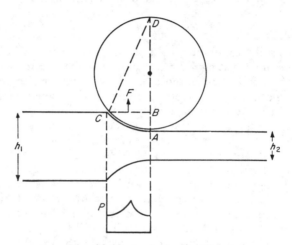

Fig. 22.10 Analysis of rolling

Hence $BC \simeq \sqrt{R(h_1 - h_2)}$. In eqn. 22.14 we take $a = \frac{1}{2}BC$ and $h = \frac{1}{2}(h_1 + h_2)$. Hence

$$F \simeq Y'\left[\{R(h_1 - h_2)\}^{1/2} + \frac{\mu R(h_1 - h_2)}{(h_1 + h_2)}\right] \qquad \textbf{22.17}$$

This formula is too approximate for detailed quantitative use, although it brings out most of the main factors involved. The most serious simplification is neglect of the *elastic flattening* of the rolls, which greatly increases the arc of contact when $R \gg BC$, and has a large effect on the roll force in cold strip rolling. Refined versions of the theory of rolling take this and other effects into account.

To keep the roll force small we use hot rolling, so that Y' is low; or, in cold rolling, use polished lubricated rolls to keep μ small. There are special difficulties in the rolling of thin sheet and strip, due to the tendency for the arc of contact to become very large compared with the thickness. Further measures are then often necessary. One is to apply *forward* and *back tension* to the strip, as in Fig. 22.11. In *Steckel rolling*, for example, the rolls rotate freely and the metal is pulled through between driven *coiling drums*.

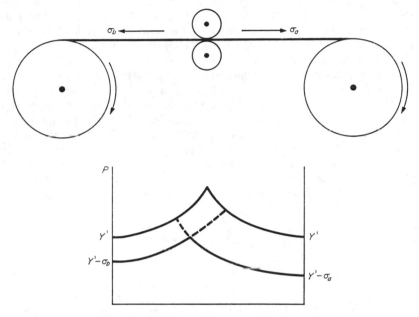

Fig. 22.11 Effect of front and back tension

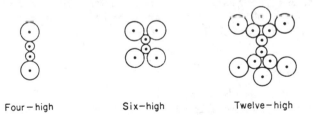

Fig. 22.12 Backing rolls

The front and back tension lower the friction hill, as shown in Fig. 22.11. This follows directly from the analysis in § 22.5. On the exit side of the roll we have, for example, $P = Y' - \sigma_a$ when a tension σ_a is applied.

The other special measure is to use rolls of *small* diameter, to avoid $R \gg h$ and to reduce roll flattening. However, long slender rolls are too elastically flexible to compress the metal unless they are supported by heavy *backing rolls*, as in Fig. 22.12. This principle has been particularly developed in *Rohn* and *Sendzimir* mills for rolling thin foil. In its extreme form the working rolls are like knitting needles backed by up to 18 heavier rolls.

The other main development in modern rolling practice is the *automated continuous mill*, sometimes fed by continuous casting and usually producing large coils of thin strip. The metal passes through a sequence of several mills, its thickness at each stage being continuously measured. The settings and speeds of each mill are controlled and adjusted to those of all the other mills by an on-line computer.

Plastic constraint is not a major factor in rolling because usually $R \gg h$. However, when $R < h$ the process changes to one of indentation, as discussed in § 22.6. This is *surface rolling*. Only the surface layers are plastically deformed and, because these are constrained by the purely elastic central part of the sandwich, the deformation produced in them as they enter the rolls is almost exactly cancelled by that produced as they leave. This technique is sometimes used for working the surface layers of a metal and for putting compressive self-stresses in them. Its main application is in *temper rolling* for overcoming the yield point in steel sheets before pressing.

Fig. 22.13 Leveller rolling

In the alternative process, *leveller rolling*, the metal is bent to and fro between alternate single rolls, as in Fig. 22.13. Temper rolling is generally more effective because the self-stresses in the surface layers help to delay the reappearance of the yield drop on ageing after rolling, since they assist the externally applied stress to produce localized yielding while the bulk of the cross-section is still lightly stressed.

22.8 Extrusion

Extrusion has the advantage of developing the shear forces for plastic deformation by compression in a confined chamber. The compressive forces are very large, requiring strong presses, but the absence of tensile forces enables metals to be heavily deformed without fracture. Extrusion was originally a hot-working process but with the development of strong presses and high-pressure lubricants, cold extrusion has also become an important process. Another advantage of extrusion is that it can produce long lengths with complex sections, which is very difficult by any other process.

There are many kinds of extrusion processes. Fig. 22.1(a) shows the simplest type of *direct* extrusion process. The same result can be obtained with rather less friction by *inverse* extrusion, as in Fig. 22.14(a). This diagram also shows *impact* extrusion, in which a small disc is hit with a die of slightly smaller size and thereby forced up into the shape of a tube. This method is used for the mass production of toothpaste tubes and similar small articles.

Ideally, the pressure required to produce an extrusion is $Y \ln (A_1/A_2)$ where A_1 and A_2 are the entry and exit cross-sectional areas. In practice

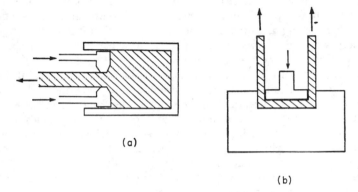

(a)

(b)

Fig. 22.14 Extrusion (a) inverse, (b) impact

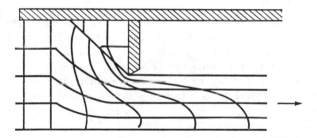

Fig. 22.15 Flow of metal in unlubricated extrusion

much higher pressures are needed because the deformation is very non-uniform, with a large redundant component, and because the friction is large. Fig. 22.15 shows the typical distortion of the metal. The geometry of deformation changes progressively during the extrusion. The first metal to emerge is practically undeformed and, still having the unrefined as-cast structure, is usually discarded. The metal then tends to flow down *shear lines* which form a converging cone between the main part of the billet and an annular pad of 'dead metal' behind the die. As extrusion proceeds, more and more heavily deformed metal emerges from the die. At a late stage an *extrusion defect* occurs. Oxidized metal from the surface of the billet, near the back, flows into the billet towards the axis and passes through the die, forming a cylinder of mechanically weak metal between a sound core and a sound outer cylinder. This defect can be avoided by using a piston with a diameter slightly smaller than that of the billet. The piston then punches through the billet, leaving the oxidized surface behind on a thin shell of metal which is discarded.

In hot extrusion, the *heat* created in the billet by the work of plastic deformation sometimes raises problems. If the extrusion is done slowly, so that the heat can be conducted away, the temperature does not rise appreciably during the process and the upper limiting temperature for extrusion is then the *solidus* of the metal or alloy. At more practical rates, a lower starting temperature must be used.

Most of the recent improvements in extrusion have aimed at lubricating the billet so that it can slide along its container and down the face of the die to the orifice, thereby obtaining more uniform deformation and eliminating the extrusion defect. An important recent process, which enables normally brittle metals to be extruded and eliminates the extrusion defect, is *hydrostatic extrusion* in which the billet is surrounded by a fluid. A conically tapering die is used, which the leading end of the billet is shaped to fit. The pressure is applied through the fluid and the metal remains separated from the wall of its container.

22.9   Wire drawing

The maximum deformation permissible in a single wire drawing operation is limited by the condition that the tensile stress pulling the wire through must be less than the yield strength of the emergent wire. We thus have $Y \leqslant S = Y \ln (A_1/A_2)$ from eqns. 22.3 and 22.4, which gives $A_1/A_2 \leqslant 2 \cdot 7$, i.e. a *maximum drawability* of 63 per cent reduction of area under ideal conditions. In practice, various factors affect this value. There is redundant deformation, as in Fig. 22.2, and there is friction even though the die is lubricated. A sharp taper in the entry cone of the die gives inhomogeneous deformation and a large redundant deformation. A gentle taper gives a large friction, due to the increased length of the die. The optimum semi-angle of the cone depends on the drawing reduction and generally lies between 4 and 12°.

The efficiency η is usually about $0 \cdot 5$ but the effect of this on the maximum drawability is to some extent offset by work hardening which may occur as the metal passes through the die and so enable the emergent wire to support a higher tensile stress. The reduction of area allowed in one pass is usually limited to about 30 per cent, so that the conversion of a rod to a fine wire has to be accomplished in a long series of drawing operations. Tungsten carbide is widely used as a die material, although the very finest wires are drawn through diamond dies.

A recent innovation is to immerse the wire and dies in an ultrasonically agitated fluid. The vibrations scrub the wire clean of foreign particles, which would otherwise score the surface of the wire as it passes through the dies. In this way very smooth surfaces are produced.

22.10   Deep drawing and sheet metal working

Metal sheets are shaped into bowls, cups, motor car panels, etc. by various processes. In *deep drawing* a circular blank is pressed to the shape of a cup by a punch, as in Fig. 22.16(a). The radius of the blank decreases during the operation as the metal flows radially inwards and then turns the corner to become the wall of the cup. In *stretch forming*, by contrast, the outer part of the blank is prevented from moving by a clamp and the metal is deformed by tensile stretching.

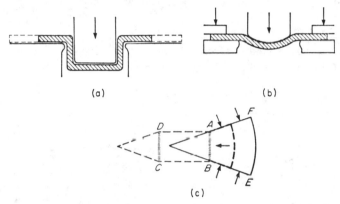

Fig. 22.16 Sheet metal working, (a) deep drawing, (b) stretch forming, (c) deformation of a radial element of the blank in deep drawing

Deep drawing is a true drawing operation. We see this from Fig. 22.16(c) which shows (ignoring the corners turned) the deformation of a radial sector of the blank to form the wall of the cup *ABCD*. The 'drawing die' in this case is provided by the neighbouring sectors of the blank, along *AF* and *BE*, as they all move in and converge together. The tensile stresses in the material are fairly low except in the walls of the cup. There is a danger of a tensile necking failure here unless the blank is well lubricated to allow it to slide in freely. There is of course no friction in the 'die' since all parts of the blank move inwards at the same rate. Buckling of the metal round the rim of the blank, to avoid the plastic compression as it moves inward, is a problem and it is usual to apply a light pressure through an annular 'blankholder', similar to the stretch forming clamp, to prevent this. If this pressure is too large, the process turns into stretch forming and there is then a danger of tensile necking failure. Careful control is needed to avoid both buckling and tensile necking.

There are many metallurgical problems in sheet metal working. We have already mentioned *stretcher strains*. Metal for deep drawing should be soft, ductile and without a great work hardening capacity, e.g. very low carbon steel. For stretch forming, on the other hand, work hardening is needed to suppress tensile necking. Austenitic stainless steels are thus particularly suitable for stretch forming.

Plastic anisotropy, due to textures developed in the rolled sheet from which the blanks are cut, is troublesome because it leads to *earing*, i.e. some sectors of the blank deform much more than others and are pulled preferentially inwards leaving the other sectors behind as 'ears'. Coarse grain size is another source of problems. Local differences of deformation from one grain to another produce a rough *orange-peel* appearance on the sheet. There is also the problem of excessive grain growth in lightly deformed regions when the sheet is annealed and recrystallized.

22.11 Seamless tube making

When forging a round bar from a billet, between flat dies, it is usual first to forge a square section, then reduce this to the required cross-sectional

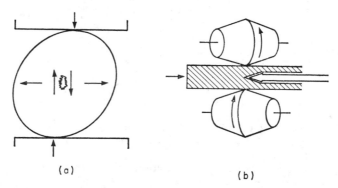

(a)   (b)

Fig. 22.17 (a) Formation of a central cavity when forging a round bar, (b) tube making by the Mannesmann process

size by compressing alternate sides of the square, and finally compress the corners to make an octagon and so on to a circular section. Why not simply use a circular section throughout, turning the bar repeatedly through small angles between successive squeezes? The reason is that this produces a hollow centre along the bar. The circular section is made slightly oval by each squeeze. This oval is slightly rotated between successive squeezes, with the result that the perpendicular line of action of the force from each die does not pass through the axis of the bar. As Fig. 22.17 shows, intense shear is then produced across the axis which causes internal necking by the processes discussed in § 21.1. The outward sideways movements of the left and right halves of the bar during the squeeze then expands and coalesces these localized necks into a continuous axial hole.

This effect is used deliberately in the *Mannesmann process* for making seamless tubes by rotary piercing. As shown in Fig. 22.17(b), a solid rod is fed between two obliquely aligned, double-cone shaped rolls which rotate it about its axis and continuously act on it in the manner of Fig. 22.17(a), so producing an axial hole. This hole is then expanded and smoothed over a mandrel, as shown. Further shaping and expansion of the tube is then done over a mandrel in various other types of special rolling mills (*Pilger mill, plug rolling mill, rotary expanding mill*).

22.12 Metal working at high speeds

Metals are often worked plastically at high speeds. For example, in fast wire drawing the metal may be pulled through at about 30 m s^{-1}, giving strain rates in the die of order 10^3 to 10^4 s^{-1}. In *explosive forming* even higher rates are reached, e.g. 10^5 s^{-1}. Under such conditions metals generally behave in a manner different from that under static loading conditions. They follow a different stress-strain curve. The plastic strain may become intensely localized in certain regions. Twinning may occur where ordinarily there is only slip. Phase changes may occur. Normally ductile materials may appear brittle and shatter.

We must distinguish between effects caused by changes in the *mechanics* of the loading and those which depend on changes in the properties of the material. Effects of the first type are due mainly to *inertial forces* and for their analysis we may assume that the properties of the material remain unchanged.

Any sudden displacement of part of the material is resisted by the force of inertia and this impulsive force creates *stress waves* which run through the material. These waves can be reflected at free surfaces, can pass through one another and can be focussed to give high stresses at points of convergence. The energy in a stress wave can be very high by elastic standards. If the kinetic energy $\frac{1}{2}\rho v^2$ per unit volume of a material of density ρ moving at speed v is suddenly converted, e.g. by impact, into elastic energy $\sigma^2/2E$, where E is Young's modulus, then

$$\frac{\sigma}{E} = \frac{v}{c_0} \qquad\qquad 22.18$$

where

$$c_0 = \sqrt{\frac{E}{\rho}} \qquad\qquad 22.19$$

is the velocity of elastic waves. Typically, $c_0 \simeq 6000$ m s^{-1} so that an impact velocity of order 300 m s^{-1} (670 mph) gives stresses of order of the ideal strength in an elastic solid. If the energy were to go completely into surface energy of fracture ($\simeq 1$ J m^{-2}) the average size of fragments would be only

10^{-7} m. The possibility of avoiding this shatter depends on consuming the energy in other ways, e.g. turning it into heat through plastic deformation.

The theory of plastic stress waves has been developed for waves in thin rods. In Fig. 22.18 we consider, as in high-speed wire drawing, a thin rod which is suddenly snatched at one end, extending it rapidly. A wave of tensile strain, accompanied by a stress wave, propagates along the wire to accommodate this motion. For simplicity, assume that the stress-strain curve is as shown. As regards the mechanics of wave propagation we can regard the plastic part of the curve, above the yield stress σ_y, as another 'elastic' part but with a lower modulus E_1 ($= d\sigma/d\epsilon$). If the stress at the end exceeds σ_y we get two waves, an elastic one of height σ_y and speed $\sqrt{E_0/\rho}$; and a plastic one of height σ and speed $\sqrt{E_1/\rho}$. At the plastic wave front there is an increase of stress from σ_y to σ. In practice, of course, the stress-strain curve is usually rounded so that a smoothly rising plastic stress wave is propagated.

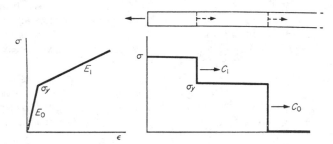

Fig. 22.18 Stress waves in a rod

If $d\sigma/d\epsilon$ becomes zero or negative, the wave speed is zero. This happens at the yield drop in mild steel. Once the yield elongation is completed, a positive $d\sigma/d\epsilon$ and a finite wave velocity are regained. The plastic wave in this case is led by a sharp discontinuity in strain, equal to the yield elongation. This *plastic shock wave* is of course a Lüders band. The nominal stress-strain curve also reaches zero slope at the ultimate tensile stress. It does not rise again after this point so that the wave velocity remains at zero once this stress has been reached and failure occurs by localized tensile necking. This is a problem in high-speed wire drawing and also in high-speed extrusion. Rod may emerge from the extrusion die at high speed. If the extruding force is suddenly cut off, the inertial forces in this rod due to deceleration can cause tensile necking and rupture.

There are many effects which can alter the mechanical properties of the metal under conditions of dynamical loading. A large Peierls-Nabarro force gives a yield stress that varies sensitively with strain rate, although in the ductile metals, particularly those of F.C.C. structure, the dislocations can usually move fast enough and in sufficient numbers to accomplish the applied strain rate at stresses not much higher than those required for slow deformation. At high temperatures, of course, very large changes can occur when the strain rate is increased into the range where recrystallization can no longer keep pace with the rate of work hardening. The other important effect is *adiabatic softening*. If the heat generated in a rapidly sheared zone cannot escape fast enough the metal in that zone becomes very hot and soft, so that further deformation continues preferentially there. Advantage can sometimes be taken of this effect. For example, in punching rivet holes in steel plate a neater hole with less plastic distortion and less consumption of energy can be achieved if the operation is done so quickly that the punched-out plug slides along an adiabatically softened cylindrical surface through the plate.

The process of *explosive forming* has become important in recent years. In this a metal blank is forced against a shaping die by the shock wave from an explosion. Usually the process is done in a tank of water and the force from a small charge of explosive is conveyed to the metal as a pressure wave through the water. Pulse pressures of about 5×10^{10} N m^{-2} are achieved. The main advantages of the technique are that strong steels and other hard alloys can be successfully forged to complex shapes without the need for large presses and that fairly cheap dies can be used, e.g. reinforced concrete lined with metal sheet. Also there is not much 'spring-back' when the pressure is removed and a good surface finish with a detailed impression of the die is obtained. There are many variants of the process. For example, as alternatives to chemical explosives, electrical spark discharges in liquids, electrically exploded wires and electromagnetic forces are all used to produce the shock pressure.

Very strong compressive shocks, approaching 10^{11} N m^{-2}, can be produced by detonating flat charges of explosive against a material. The pressure wave produces uniaxial compression in the direction of propagation and so this wave involves both shear and volume change. The plastic shear strength of the metal is so far exceeded that the pressure wave can be regarded as propagating effectively in a 'fluid', resisted by the bulk modulus of elasticity only. This bulk modulus increases, at very high pressures, so that the later parts of the pressure wave travel faster than the earlier parts and a shock wave is formed. In B.C.C. iron and steel the effects are complicated by a transformation to the more dense F.C.C. structure at pressures of the order of 2×10^{10} N m^{-2}, which gives an additional shock wave. These high stress shocks produce great hardening of the metal with little plastic deformation because fine deformation twins and plastic shear bands are formed. The notch-impact properties of the metal after this treatment are poor, possibly because the twins nucleate micro-cracks. Freshly cold-worked iron does not so readily form twins under shock loading. This may be because less time is required to start slip when the metal contains large numbers of unpinned dislocations.

22.13　Superplasticity

An important feature of processes such as vacancy creep, already discussed in § 21.13, is that they can operate, at high temperatures, even at very low stresses, giving a type of viscous flow in which the strain rate is proportional to stress. When deformation is produced by viscous flow there is no tensile necking, as is demonstrated by the fact that hot glass can be drawn into fine threads by simply pulling its ends apart. It follows that, if conditions are arranged so that a metal is deformed by vacancy creep, at stresses too low for the usual plastic processes, it can be stretched enormous amounts in tension, without necking. This is *superplasticity* and it opens up the possibility of being able to work metals and alloys by processes directly analogous to glass-blowing. Superplasticity becomes feasible, practically, when the microstructure of a multi-phase alloy is extremely fine, so that the migration distance for the vacancies is correspondingly short (§ 19.9).

FURTHER READING

ALEXANDER, J. M. and BREWER, R. C. (1963). *Manufacturing Properties of Materials* Van Nostrand, New York and London.

BACKOFEN, W. A. (1964). *Fundamentals of Deformation Processing*. Syracuse University Press, New York.

BRIDGMAN, P. W. (1952). *Studies in Large Plastic Flow and Fracture*. McGraw-Hill, New York and Maidenhead.

CRANE, F. A. A. (1964). *Mechanical Working of Metals*. Macmillan, London.

DIETER, G. E. (1961). *Mechanical Metallurgy*. McGraw-Hill, New York and Maidenhead.

FISHLOCK, D. and HARDY, K. W. (1965). *New Ways of Working Metals*. George Newnes, London.

FORD, H. (1963). *Advanced Mechanics of Materials*. Longmans Green, London.

HILL, R. (1950). *The Mathematical Theory of Plasticity*. Oxford University Press, London.

INSTITUTION OF METALLURGISTS (1964). *Recent Progress in Metal Working*. Iliffe Press, London.

NADAI, A. (1950). *Theory of Flow and Fracture of Solids*. McGraw-Hill, New York and Maidenhead.

PARKINS, R. N. (1968). *Mechanical Treatment of Metals*. George Allen and Unwin, London.

PHILLIPS, A. (1956). *Introduction to Plasticity*. Ronald Press, Cardiff.

PRAGER, W. (1959). *An Introduction to Plasticity*. Addison-Wesley, London.

SACHS, G. (1954). *Fundamentals of the Working of Metals*. Interscience, London.

TABOR, D. (1951). *The Hardness of Metals*. Oxford University Press, London.

WOLDMAN, N. E. and GIBBONS, R. C. (1951). *Machinability and Machining of Metals*. McGraw-Hill, New York and Maidenhead.

Chapter 23

Oxidation and Corrosion

23.1 Introduction

The free energy values in Chapter 7 showed us that most metallic elements exist naturally in the atmosphere in an oxidized condition. The main problem of extraction metallurgy is to convert them to the metallic state against this natural tendency. In this chapter we shall now consider the processes by which such metals revert to an oxidized state, i.e. processes of *oxidation* and *corrosion*. Although oxidation in the sense of removing electrons to convert atoms into positive ions occurs in most forms of metallic corrosion, it is usual to limit the term *oxidation* to the chemical reaction of a dry metal surface with an oxidizing gas, sometimes also called *dry corrosion* or *tarnishing*, and to refer separately to *wet corrosion* or *aqueous corrosion* when the chemical attack takes place through the medium of water.

Both types of corrosion are of course extremely important. Oxidation limits the use of metals in high temperature power plant and is responsible for the mill scale which gives hot-rolled steel such a rough surface. Wet corrosion, particularly the rusting of steel in moist air and water, is one of the major technological problems of modern society. It costs the United Kingdom well over £1000 million a year in metal losses and preventive measures. There are a few practical benefits from oxidation and corrosion: the green patina on copper roofs pleases the eye; the lead-acid battery depends on corrosion reactions; chemical etching, electrochemical machining and oxygen flame cutting are useful for shaping metals; and anodic oxidation is the basis of electrolytic polishing; but these are insignificant in comparison with the harmful aspects of corrosion.

23.2 Oxidation

Suppose that the surface of a reactive metal is exposed to an oxidizing gas such as dry air or oxygen, or hot dry steam. At high temperatures some metals form molten or volatile oxides, e.g. V_2O_5 on vanadium and MoO_3 on molybdenum; or the oxygen dissolves in the metal itself, e.g. in titanium;

oxidation then proceeds catastrophically, as fast as the oxygen supply will allow. More commonly, however, a thin film or thick scale of solid oxide forms over the exposed surface and this slows down the rate of further oxidation by separating the metal from the gas.

Thick scales are clearly visible. The thin films can also be seen, when about 100–1000 Å thick, from the interference colours produced by light rays reflected from their top and bottom surfaces. The tempering of quenched steel in fact is sometimes controlled by observation of the *temper colours* (straw at 200°C, blue at 300°C) produced on the clean metal surface when heated. Very thin oxide films, e.g. 20 Å produced on aluminium at room temperature, remain invisible however.

The rate of oxidation is usually measured by weighing a sample periodically to determine the amount of oxygen taken up. The average film thickness x can then be calculated from known properties of the oxidation

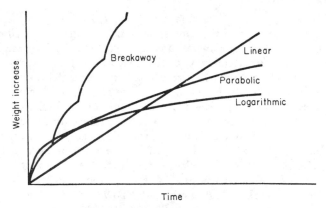

Fig. 23.1 Shapes of oxidation curves

product. The oxidation usually develops with time t according to one of the curves shown in Fig. 23.1, which are described approximately as *logarithmic* ($x \propto \ln t$), *parabolic* ($x \propto t^{1/2}$), *linear* ($x \propto t$), or as *break-away* (repeated rapid parabolic steps, giving an overall linear behaviour). The logarithmic law applies mainly to highly protective thin films, formed at low temperatures, e.g. on Fe below 200°C. Fig. 23.2 shows the effect of temperature on the oxidation of copper. The parabolic law is widely obeyed at intermediate temperatures, e.g. Fe between 200 and 900°C. The linear law applies to the initial stages of oxidation before the film is thick enough to be protective and also wherever the protection no longer increases as oxidation continues, e.g. Mo at temperatures where MoO_3 is volatile. Break-away effects are observed when something happens to the film, such as cracking or flaking off, which reduces its thickness. Repeated break-away on a fine scale can prevent the protective part of a film from increasing beyond a certain thickness and so give linear oxidation. The *plasticity* of oxide at high temperatures is important because it helps prevent cracking and break-away, thus improving the protectiveness of the film. For example, copper gives smooth parabolic oxidation curves at 800°C, where the oxide is plastic, and discontinuous break-away curves at 500°C, where the oxide is brittle.

The first oxygen molecules to land on a clean metal surface are usually *chemisorbed*; they dissociate and their atoms then bond chemically to metal atoms in the surface. A monolayer of oxygen forms rapidly over the surface in this way and is followed, more slowly, by the build up of a few adsorbed layers on top of it. Oxide is then nucleated, slowly at low temperatures because of an activation energy for nucleation, at certain favourable sites on the surface, e.g. on impurity particles, surface steps, ends of dislocations, etc. Small mounds grow from these nuclei and a thin oxide film forms over the whole surface.

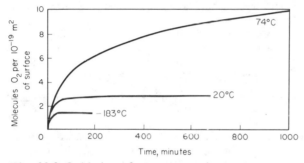

Fig. 23.2 Oxidation of copper at various temperatures

Further oxidation requires the passage of ions and electrons through this film. Usually the migration of ions is the slow, rate controlling step. The electrons can pass readily through a thin film ($\simeq 20$ Å) by the tunnel effect (cf. § 3.4) and also through a thick semi-conducting film by the processes discussed below.

An ion can move through an oxide film by the ordinary processes of lattice diffusion but the activation energy is generally too large for significant movement at temperatures where a thin film grows logarithmically. However, very strong electrostatic forces act through the film. This is because the oxygen atoms on the outer surface become negative ions, by extracting electrons from the metal underneath the film, and so exert an electrostatic attraction on the positive ions in the metal. While the film is thin, this force is so strong that the work done by it on an ion which moves through the film can provide a significant fraction of the activation energy, sufficient to enable the ion to move. The ions are thus pulled through the film by the force. As the film thickens, the force becomes weaker since the distance between the positive and negative ions is increased. The activation energy is then less affected by the electrostatic work and the rate of migration—and hence oxidation—slows down drastically, giving a type of logarithmic law.

Logarithmic growth is sometimes observed on much thicker films. This is believed to be due to an effect of *porosity* in the film. There is initially a rapid passage of ions along the pores but later, as the pores become congested and blocked, the rate slows down, giving a roughly logarithmic curve.

Parabolic growth sets in at temperatures where the ions can pass through the film by thermal movement alone. It was originally assumed that oxygen ions passed through the film and formed oxide at the metal-oxide interface. Experiments have proved, however, that the metal ions frequently migrate through the film to the outer surface. The method is to use inert markers, as in the Kirkendall experiment (cf. Fig. 19.18). For example, a small patch of chromic oxide may be placed on a clean surface of iron. After oxidation it is found that this marker substance lies at the metal-oxide interface, *not* the outer surface of the film, showing that metal has migrated upwards through the film, forming new layers of oxide at the outer surface.

$$Cu^+ \quad Cu^+ \quad Cu^+ \quad Cu^+ \quad Cu^+ \quad Cu^{2+}$$
$$Cu^+ \quad \square \quad Cu^{2+} \quad Cu^+ \quad Cu^+ \quad Cu^+$$
$$Cu^+ \quad Cu^+ \quad Cu^+ \quad Cu^+ \quad \square \quad Cu^+$$
$$Cu^+ \quad Cu^+ \quad Cu^+ \quad Cu^+ \quad Cu^+ \quad Cu^+$$

Fig. 23.3 Copper sub-lattice in cuprous oxide, containing vacant sites and cupric ions

The nature and speed of such migration processes depend on the *defect structure* of the oxide lattice in such a film. Consider the Cu_2O film on oxidized copper. We can think of the crystal structure as consisting of three interpenetrating sub-lattices, two for the copper and one for the oxygen. Ideally, all lattice sites are correctly and completely filled, with no supernumary atoms in interstitial sites. In practice, however, cuprous oxide contains slightly less copper than the exact formula Cu_2O requires. This leaves the oxygen sub-lattice still perfectly complete but leads to a small number of vacant sites in the copper sub-lattices, as shown in Fig. 23.3.

There is an essential difference between these vacancies and those in a metal. These are vacant *ionic* sites. The oxide as a whole must be electrically neutral. Therefore, to each vacant cuprous ion site there must be a corresponding *missing electron* in the crystal. This is achieved by removal of a second electron from one of the nearby copper ions, converting it into a *cupric* Cu^{2+} ion. The defect structure thus consists of a number of vacant cuprous sites and an equal number of cupric ions, as shown in Fig. 23.3. The missing electrons cause the oxide to be a *p-type intrinsic semi-conductor* (p = positive), since each missing electron is equivalent to a *positive hole* in the electron band structure of the crystal (cf. § 4.5). Electronic conduction occurs by the transfer of an electron from a cuprous ion A to a neighbouring cupric ion B, so making A cupric and B cuprous; in effect, by the migration of positive holes through copper lattice. Ionic conduction occurs by the migration of the vacant sites in the copper lattice. This is therefore a case in which oxidation occurs by the migration of the metal through the film to the outer surface. The conductivity of the film—and hence the rate of oxidation—increases with increasing deviation of cuprous oxide from the exact Cu_2O formula. This deviation can be produced by increasing the oxygen pressure of the surrounding atmosphere.

Oxides such as NiO, FeO and Cr_2O_3 are similar to Cu_2O in having vacant metallic ion sites. There are, however, other types of defect ionic lattices. Zinc oxide contains more zinc than the formula ZnO warrants. The two sub-lattices are themselves complete and the excess zinc ions are accommodated interstitially, together with loosely attached electrons to maintain electrical neutrality. These extra electrons provide electronic conductivity by their movements in the conduction band (cf. § 4.5) and so the oxide is an *n-type* intrinsic semi-conductor (n = negative). The film grows by the interstitial migration of zinc ions from the metal to the outer surface.

The oxides Fe_2O_3, TiO_2 and ZrO_2 all contain excess metal and so are *n*-type semi-conductors, but in these cases the deviation from stoichiometry is achieved by having vacant sites in the oxygen sub-lattice. Oxygen therefore migrates through the film to the metal and the film grows downwards into the metal, instead of at the outer surface. This inward movement and oxidation at the metal-oxide interface tends to produce compressive stresses in the oxide film. This is often an advantage, since it produces a dense film, with pores closed up. However, when the film becomes thick it may break up to relieve these stresses, so leading to break-away oxidation.

The law of parabolic film grow is very easily derived. Let the film thickness be x at time t. As in the analysis leading to eqn. 12.30, we have diffusion through an interfacial layer and we assume that the concentrations of the migrating species are constant at the two surfaces of this layer, theoc constant values being set by the local thermodynamical equilibrium between the oxide and metal, at the one surface, and between the oxide and atmosphere at the other. There is thus a constant difference Δc in concentration across the diffusion zone. From Fick's law, the rate of transport through unit area is $D\Delta c/x$. This is proportional to the rate of growth dx/dt. Hence $dx/dt \propto x^{-1}$, i.e.

$$x^2 = kt \qquad\qquad 23.1$$

The constant k can be calculated in terms of the diffusion coefficient D, the free energy of oxidation, which determines Δc, and the structural parameters of the oxide crystal.

23.3   Prevention of oxidation

It follows from the above considerations that if the oxide is highly resistant to the passage of either electrons or ions, a high resistance to oxidation is possible. Aluminium, being a metal of fixed valency with a high affinity for oxygen forms an oxide, alumina, which is a good electrical insulator and gives a highly protective film. A high free energy of oxidation tends to produce a dense, almost stoichiometric, chemically stable oxide film with good protective properties, e.g. Al_2O_3, Cr_2O_3, TiO_2, BeO. An additional requirement, however, is that the specific volume of the oxide should be *slightly* greater than that of the metal, so as to form a compact continuous film (*Pilling-Bedworth rule*). If smaller than the metal, it tends to provide a porous and non-protective film. This appears to be why the films on metals with large atomic spacings, e.g. Na, K, Ca, Mg, are poorly protective. If the oxide is *too* bulky, however, the stresses set up by its formation may cause it to spall off, leading to break-away oxidation.

Clearly, one important method of obtaining oxidation resistance is to provide the surface of the object in question with a protective oxide film. This can be done in many ways. One is to add a suitable alloy element to the metal. *Hauffe's valency rule* is often useful here, although not absolutely reliable; if the oxide is deficient in metal atoms, add an alloy element of lower valency to reduce the number of vacancies present; conversely, if the oxide contains an excess of metal, add an alloy element of higher valency.

A fairly noble metal such as silver or copper can be protected by *selective oxidation*. In this, about 1 per cent of a highly reactive metal such as aluminium is dissolved in the basis metal, which is then carefully oxidized in an atmosphere too weak to oxidize more than the alloy metal. As the atoms of this diffuse up to the surface of the metal they are oxidized and a stable film is formed; e.g. Al_2O_3 formed on a sterling silver, containing 92·5 per cent Ag, 6·5 per cent Cu, 1 per cent Al, to give a *tarnish-resistant silver*.

Larger alloy additions are made to iron and steel to gain oxidation resistance. Aluminium and silicon both give very protective oxide films, but in the amounts required they embrittle the metal and this restricts their use in practice. The most common addition is chromium; 12 per cent in stainless cutlery, more in austenitic stainless steel (e.g. 18 per cent Cr, 8 per cent Ni) and heat-resistant steels. When ordinary iron is heated at temperatures above 600°C a thick scale is formed, consisting of FeO next to the metal, with Fe_2O_3 on the outside and Fe_3O_4 in an intermediate layer. When the above amounts of chromium are added, the inner layers are replaced by a layer of protective Cr_2O_3, beneath a thin outer layer of Fe_2O_3. The outstanding protective qualities of Cr_2O_3 at high temperatures make chromium an essential alloy addition for most heat-resistant alloys, such as *nichrome* electric heater wire (80 per cent Ni, 20 per cent Cr) and various alloys for gas turbines.

Protection is also obtained by coating the object with a suitable material. In addition to gold plating and chromium plating, there are various processes of cladding metals, by *rolling* on thin surface layers (e.g. pure Al on Al-alloys), by *dipping* in molten metal baths (e.g. Sn), by *calorizing* (i.e. heating the object in a powdered $Al-Al_2O_3$ mixture), by *chromizing* (i.e. heating steel in chromous chloride vapour), and by *metal spraying* (e.g. Al on steel). It is also possible to apply ceramic layers, e.g. silicides, borides, Al_2O_3, Cr_2O_3, by spraying and other means, although it is essential in such cases to provide only a thin layer, otherwise there is a risk of cracking and flaking off. Great efforts have been made to develop oxidation-resistant coatings for high-temperature gas turbine blades, particularly on the refractory metals such as molybdenum and niobium. Silicide and aluminide coatings have been mainly explored, e.g. $MoSi_2$ on Mo and $NbAl_3$ on Nb. Severe difficulties have been experienced, particularly due to cracking of the coatings caused by thermal strains on heating and cooling, but some of the more complex coatings that have been developed, e.g. Cr—Ti—Si on Nb alloys, are moderately successful.

23.4 Corrosion in acids and alkalies

A simple form of corrosion is the solution of a metal such as zinc or iron in dilute acid. The metal goes into solution as ions; and hydrogen ions from the solution are converted to gas bubbles on the metal. The hydrogen ions in the solution, by their conversion to neutral atoms, absorb the excess electrons from the metal, the electrical charge of which would otherwise bring the corrosion reaction to a halt (cf. § 9.4). The metal continues to dissolve fairly evenly over its surface until most of the hydrogen ions are removed and the hydrogen potential, given by eqn. 9.5, is reduced to that of the metal.

The attack of steel by dilute acids is prevented by *tin plating*. Tin lies near hydrogen in the electrochemical table (cf. Table 9.2), has a large hydrogen overvoltage (cf. Table 9.3) and so provides a protective coating. However, if there is a hole in the coat the iron beneath is severely attacked by dilute inorganic acids through the electrochemical effect discussed in § 23.5. On the other hand, tin forms complex ions with citric and other acid juices in fruit cans, which shifts its electrode potential so strongly in the negative direction that it dissolves in preference to the iron, so that this remains protected even at holes in the tin plating.

Consider now the effect of hydrogen overvoltage. Since the corrosion occurs by the flow of positive metallic ions from the metal into the solution we can interpret it as a flow of electrical current I driven by an electrical potential E. Only when the current is negligible does E have the value given by eqn. 9.5. At finite rates *electrode polarization* occurs, i.e. E is changed by the voltage drop across the surface of the metal necessary to overcome the activation energy barrier opposing the electrode reactions there (cf. § 9.4). Fig. 23.4 shows the half-cell potentials for metal dissolution and hydrogen evolution at various rates, as indicated by the corrosion current. The vertical distance between each pair of curves indicates the potential E available to drive the corrosion reaction at a given current I. Corrosion can thus develop up to the rate that makes this E zero, i.e. up to the value of I at which the curves cross. We see that zinc, because of its large hydrogen overvoltage, is corroded more slowly than iron, i.e. $I_{Zn} < I_{Fe}$, even though zinc lies further than iron from hydrogen in the electrochemical table (cf. Table 9.2).

Tenacious oxide films can protect metals from simple acid corrosion. Thus aluminium is attacked very slowly until the alumina film is dissolved. Nitric acid and hot concentrated sulphuric acid are oxidizing agents. If they are used, or if oxidizing agents such as chromates or chlorates are added to acids such as hydrochloric acid, corrosion may be aided by the extraction of electrons from the metal to take part in reactions which create nitrogen oxides, or sulphur dioxide, or hydroxides. However, when the oxidizing agent is very strong the acid can oxidize the surface of the metal

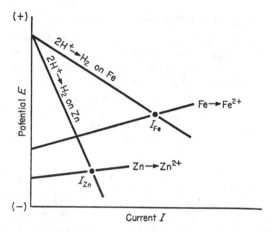

Fig. 23.4 Effect of polarization on corrosion currents developable on zinc and iron

directly and a tenacious oxide film may then be formed. The metal then becomes immune to further attack and behaves like a noble metal. It is said to be *passivated*. Hot concentrated nitric acid passivates iron by forming an Fe_2O_3 film on its surface. In this condition the metal is remarkably resistant to corrosive attack. Unfortunately this film is brittle and fragile and so is not a practical substitute for the Cr_2O_3 film on stainless steel.

Lead has good resistance to many acids. It lies close to hydrogen in the electrochemical table and has a large hydrogen overvoltage (Table 9.3). Against sulphuric acid it is protected by a film of insoluble lead sulphate which forms over its surface.

Many metals are attacked by alkaline solutions. The sensitivity of aluminium kitchen ware to washing soda is familiar. Caustic soda solutions form compounds with many metals, e.g. sodium ferroate (Na_2FeO_2) on iron and similar compounds (e.g. zincate, aluminate, stannite, plumbite) on other metals. Hydrogen is evolved as the metal dissolves, through a reaction such as

$$Zn + 2NaOH \rightarrow Na_2ZnO_2 + H_2 \qquad\qquad 23.2$$

even though the H^+ concentration in the alkali is very low. The reason is that the formation of, for example, zincate ions keeps the concentration of Zn^{2+} ions in solution very low. This reduces the electrode potential so far below that of hydrogen that the metal continues to dissolve and release hydrogen.

23.5   Electrochemical corrosion

Suppose that a piece of iron is partly immersed in a neutral salt solution, e.g. NaCl, open to the air. The iron begins to dissolve as Fe^{2+} ions and the metal becomes negatively charged due to the removal of positive ions from it.

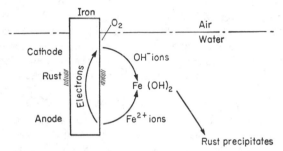

Fig. 23.5 Electrical circuit in electrochemical corrosion

The corrosion reaction is then halted unless this negative charge is also removed. The displacement reaction $Na^+ \rightarrow Na$ is not possible because sodium is more reactive than iron. Hydrogen ions can absorb electrons and become neutral atoms and molecules on the surface of the metal but the H^+ concentration is too low in neutral solutions for this to be other than a minor process. In fact, if the iron and salt solution are sealed off from the air, very little rusting takes place.

We have to consider the system open to the air, however, as in Fig. 23.5. Corrosion then continues without cease on the lower part of the metal, labelled *anode* in the diagram. Here Fe^{2+} ions are liberated into the solution and electrons into the metal. These electrons are now consumed in a a *cathodic reaction* which occurs on the metal near the water line. Essentially this reaction is the conversion of oxygen, dissolved in the water from the air, to negative ions, the reaction product being hydroxyl ions, i.e.

$$O_2 + 2H_2O + 4e \rightarrow 4OH^- \qquad\qquad 23.3$$

This is one form of *electrochemical corrosion* in which an electrical circuit is created, with an anode, a cathode, an electrolytically conducting path between them through the solution and an electronically conducting path between them through the metal. The anodic and cathodic reactions necessarily occur at the same rate and it has been proved that a current flows round the circuit at a rate equivalent, in the Faraday sense (cf. § 9.1), to the amount of metal corroded.

We notice that in this *differential aeration cell* the corrosion occurs where the oxygen concentration is *low*. Also that, although a thin oxide film forms on the cathodic areas and a certain amount of rusting occurs in the region between anode and cathode, no corrosion product is deposited on the corroding region. At the anode, Fe^{2+} ions pass into solution where they meet OH^- ions from the cathode, also in solution, and there form ferrous hydroxide, $Fe(OH)_2$. Depending on the state of aeration in the solution, this may then oxidize to $Fe(OH)_3$ or to red-brown rust, i.e. $Fe_2O_3.H_2O$, or to black magnetite, Fe_3O_4.

Electrochemical corrosion by differential aeration is very important in practice. Drops of water which collect in crevices or in gaps between riveted or bolted plates, or beneath specks of grit sitting on the surface, can all act as differential aeration cells for localized corrosion, since oxygen can enter the water from the air at the surface but cannot readily penetrate to the ends of the crevices. Smooth, continuous, clean surfaces are resistant to this form of corrosion.

What determines the rate of this corrosion? We have a set of interlinked processes—penetration of oxygen from the air to the cathode, reactions at anode and cathode, electronic and electrolytic conduction—and they are all controlled by the slowest one. Electrolytic conduction can be the rate-controlling process in pure water but the conductivity of salt solutions is usually more than adequate and in such cases the supply of oxygen to the cathodic area is usually the rate-controlling process. This has several important consequences. For example, if the area of metal near the water-line is increased, the rate of corrosion increases proportionately, even when concentrated on an anode of unchanged area. Small anodic regions in contact with large cathodic ones, e.g. ends of crevices in the surfaces of large sheets, can thus be very intensively attacked. It also follows that, when the conductivity processes are not rate-controlling, the anodic attack need not spread itself evenly over the entire anodic area of the metal. It can concentrate locally and intensively on the most highly anodic spots, such as pits, scratches, cracks in surface oxide films, plastically deformed and highly stressed regions and foreign inclusions.

When different metals are joined and exposed to a conducting solution, *galvanic corrosion* can occur. Fig. 23.6 shows an example. Iron passes into solution as Fe^{2+} ions and the electrons left behind are conducted through to the copper, where they take part in the reaction 23.3, which produces OH^- ions at the surface. This form of corrosion is very damaging because it concentrates sharply on the less noble metal at the metal-metal junction, where deep attack occurs. At the junction a large corrosion current can pass because the electrical resistance of the short path through the electrolyte is low.

Many common forms of metal joining, e.g. brazing, welding, soldering, bolting, provide junctions at which galvanic corrosion can develop. Because of its extremely damaging effects, such joints between electrochemically dissimilar metals are avoided as far as possible in components exposed to water, particularly sea water, and moist atmospheres. Galvanic corrosion on a microscopic scale can also occur between the constituents of multiphase alloys and of impure metals which contain foreign particles and intermetallic compounds.

It is not always possible to predict from the electrochemical table which metals will be anodes or cathodes, when joined, or what the cell voltage will be. For example, if a highly reactive metal is covered by a protective oxide

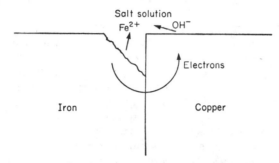

Fig. 23.6 Galvanic corrosion between iron and copper

film, its ions cannot easily leave it and go into solution even though the metal is readily ionizable. Thus titanium behaves like a noble metal in a galvanic cell and oxide-coated aluminium acts as the cathode to a zinc anode, despite the positions of these metals in Table 9.2. Stainless 18-8 steel usually exists in the noble, passive state but if its oxide film is destroyed it becomes reactive and much more susceptible to galvanic attack. By taking pairs of metals and alloys in a given medium such as sea-water and observing which become the anodes, we can prepare a *galvanic series*, similar to but not identical with the electrochemical series, which gives them in order of increasing reactivity, as in Table 23.1. For example, stainless steel or brass fittings enhance the corrosion of mild steel in sea water.

Galvanic corrosion is put to practical advantage in methods of *sacrificial protection* and *cathodic protection*, which pump electrons in the protected metal, to charge it negatively and oppose the escape of positive ions. *Humphrey Davy* in 1824 suggested that copper fastenings on wooden ships could be protected from seawater corrosion by placing pieces of zinc or iron in contact with them at positions round the hulls. These metals are

TABLE 23.1.   GALVANIC SERIES IN SEA-WATER

increasing reactivity	Ti, Monel (70 Ni, 30 Cu), Passive 18-8 steel, Ag, Inconel (80 Ni, 13 Cr, 6·5 Fe), Ni, Cu, 70-30 brass, 60-40 brass, Sn, Pb, Active 18-8 steel, Cast iron, Mild steel, Al, Zn, Mg.

preferentially dissolved and the copper absorbs their electrons and is protected. A familiar modern example is *galvanized iron sheet*. If holes occur in the zinc coating and electrolytic cells form there when the sheet is moist, the iron is cathodically protected by sacrificial corrosion of the zinc (at temperatures below 70°C). Cathodic protection is important for ships and for pipelines in moist soils. Auxiliary anodes are placed in the corrosive medium at suitable intervals, in contact with the metal to be protected. They may consist of a reactive metal such as magnesium, or a material such as graphite or iron which is then positively charged by an electrical connection to the protected metal through a D.C. generator, which ensures a flow of electrons from the sacrificial anodes to the protected metal.

23.6 Atmospheric corrosion

The rusting of iron and steel in the atmosphere depends very much on the humidity. If the air is so wet that a film of moisture forms on the surface of the metal, electrochemical corrosion can occur at weak spots in the oxide film. An important point in this case is that the large surface is very accessible to oxygen because the electrolytic film is thin, so that the rate of corrosion is not limited by the rate of supply of oxygen.

In less humid air, where the surface remains dry, corrosion can still occur because iron oxide is hygroscopic. Below about 60 per cent relative humidity there is no rusting, only an air-formed oxide film; rusting occurs and develops severely in humidities above 80 per cent.

This corrosion is greatly enhanced by the presence of foreign particles in the air which settle on the metal, particularly particles of hygroscopic salts, e.g. ammonium sulphate or sodium chloride. Sulphur dioxide is especially harmful and is mainly responsible for the high rates of rusting experienced in industrial areas. Its effect is enhanced by the presence of soot particles in the atmosphere, which absorb it and carry it in concentrated form to the surfaces on which they settle.

If such air is carefully filtered to remove all dust particles and impurities, it no longer rusts iron. Instead the air-formed film slowly grows thicker and more protective. Thus, if clean iron is wrapped in, for example, clean muslin for long periods it will become fairly resistant to atmospheric corrosion.

23.7   Intergranular corrosion and dezincification

Because of abnormalities of structure and composition, grain boundaries are often the targets for localized anodic attack. This tendency exists to some extent in all polycrystalline metals but is particularly severe in certain alloys. The famous example is stainless steel which, in a certain metallurgical condition, becomes so sensitive that its grain boundaries can be completely corroded away and its otherwise unattacked grains can simply fall apart. The metal is normally protected by the chromium in solid solution, which allows a passivating oxide film to form over its surface. However, heating to 500–800°C can cause the chromium near the grain boundaries to be removed from solid solution to form chromium carbide precipitates. The thin layers alongside the grain boundaries, being depleted of chromium, are then no longer protected by the passivating film and become anodic relative to the rest of the surface. Being so narrow, they are strongly attacked by the corrosion current generated from the cathodic reactions over the rest of the metal surface. This highly specific form of corrosion is sometimes referred to as *weld decay* because the heat treatment given by fusion welding can produce zones of highly susceptible metal alongside the welds. The effect is overcome by *stabilizing* the stainless steel through the addition of a small amount (0·5 per cent) of a strong carbide-forming element such as titanium or niobium which captures the carbon and leaves the chromium free to remain in solution.

There are many other special types of corrosion. One which is sometimes important in brass is *dezincification*, the selective removal of zinc from the alloy, leaving behind a spongy mass of copper. This was once a major problem in ships' condenser tubes. It seems likely that the zinc is removed anodically, leaving vacant sites behind, particularly near grain boundaries, although there may also be some solution and deposition of copper ions. The effect is overcome by the addition of about 0·05 per cent of arsenic to the brass.

23.8   Prevention or restraint of corrosion

Most methods of preventing corrosion depend on separating the metal from the corrosion agent by some kind of protective layer. Before discussing these we shall consider methods that do not depend on such layers. We have already mentioned some; for example, the important technique of cathodic protection by the use of sacrificial anodes; also the diminution of atmospheric corrosion by reducing air pollution or, indoors, by filtering and drying the air.

There are several other methods, which follow directly from the general principles outlined above. Electrochemical corrosion depends on conduction through the electrolyte; hence it is an advantage, where possible, to keep the conductivity of the electrolyte as low as possible, for example by storing ships in clean river water rather than sea water. Electrochemical corrosion also depends on metallic conduction from the anode to the cathode; hence, wherever possible, an insulating washer should be used to separate two different metals in a joint, although the effect of this can be nullified if the cathode metal dissolves in the electrolyte and then deposits on part of the anode metal. Electrochemical corrosion also depends on supply of oxygen to the cathode; hence *de-aeration* of the electrolyte to remove oxygen is a useful method of inhibiting this corrosion. This can be done by heating the solution, or by holding the solution under reduced pressure or under an inert gas such as nitrogen; or chemically by the addition of sodium sulphite or hydrazine.

Corrosion in acids depends on the electrode potential and overvoltage. Nickel has a high overvoltage and so, in large concentrations, is a useful alloying element for giving resistance to alloys under conditions where the oxygen content is low. *Monel metal* (70 Ni, 30 Cu) is an example of such an alloy. Where utmost resistance is essential, as in dental fittings and laboratory ware, noble metals such as platinum and gold are of course used.

We consider now methods which depend on protective layers. Corrosion in neutral solutions is sometimes overcome by adding *cathodic* or *anodic inhibitors* to the solvent. With iron in salt solution, for example, corrosion normally continues unabated at weak spots in the oxide film because both the anodic and cathodic products (Fe^{2+} and OH^- respectively) dissolve in the solution and no precipitate is formed near the site of the electrode reactions. Cathodic and anodic inhibitors provide such precipitates at the cathode or anode respectively, and so stifle the corrosion reaction.

An important cathodic inhibitor is calcium bicarbonate, present naturally in hard water, which forms calcium carbonate on the cathodic parts of the metal and blocks the reaction there. Soluble magnesium or zinc salts ($MgSO_4$ or $ZnSO_4$) are sometimes added to neutral solutions because they form hydroxide precipitates on the cathode. All such precipitates must be electronic insulators if they are to be effective, since otherwise the cathodic reaction could continue on the outer surface of the precipitate.

Anodic inhibitors, on iron and steel, convert the Fe^{2+} ions as they leave the metal into insoluble precipitates which block the anodic reaction, usually at holes in the oxide film. The main precipitate is ferric oxide. Typical substances used as anodic inhibitors are potassium chromate and sodium phosphate; also various alkalies, although these are effective only when there is oxygen present in solution. Since anodic inhibitors have no direct effect on the cathodic reaction, there is a considerable risk of an intensive anodic reaction at localized weak spots, leading to pitting, if the inhibitor fails to seal off the anode completely. It is essential therefore that the concentration of anodic inhibitor should exceed the minimum amount needed for a complete film. Fortunately, some substances such as sodium benzoate and cinnamate appear not to produce pitting, whatever the amount, and these are used as anodic inhibitors in motor car radiators.

Hot-worked steel is usually *pickled* in hot sulphuric acid solution to remove the mill scale, for example before tinning or galvanizing. It is important to minimize the attack of the acid on the metal itself while the scale is being removed and for this purpose *restrainers* are added to the pickling solution. These are organic substances such as thiourea, quinoline and pyridine which contain sulphur and nitrogen. These polar atoms enable the molecules to attach themselves to the surface of the metal and so form an adsorbed layer which inhibits the corrosion process. An important requirement of pickling restrainers is that they should limit the uptake into the metal of hydrogen, released by the acid attack, which can cause blistering and hydrogen embrittlement.

There are many circumstances where it is impossible or undesirable to form protective layers by adding substances to the corrosive medium. It is then necessary to modify the metal itself, either by adding an alloy element which enables a protective film to be formed (e.g. Cr in stainless steel; also 14 per cent Si in cast iron which gives an acid-resistant SiO_2 layer) or to coat the surface of the metal directly with a protective layer.

Many kinds of metallic coatings are used. Inert metal coats, e.g. gold, copper and nickel, must be perfectly continuous to protect a metal such as iron; otherwise, as in tin plate, iron exposed at gaps can become anodic relative to the coating metal and so be intensely attacked. Base metal coats (e.g. zinc or iron) provide sacrificial protection of the metal beneath, as discussed in § 23.5. Coats of chromium and aluminium provide protective oxide films. The most widely used method of applying most coats is electrodeposition, although cladding by rolling, spraying, calorizing and dipping are also important.

Amongst the non-metallic coatings we note the various *glassy enamels*, which are excellent so long as they remain uncracked; miscellaneous coating materials such as organic plastics, rubber, pitch and grease; also *phosphate* coatings, which are formed on steel by dipping the metal into a solution of phosphates and phosphoric acid; and the process of *anodizing*, particularly effective on aluminium, in which a thick tenacious oxide film is built up by electrolysing the metal, as anode, in a bath of chromic, sulphuric or oxalic acid.

The most familiar method of applying a protective coat is, of course, by *painting*. Oil paints consist of pigments suspended in linseed oil to which is added a thinner (white spirit or turpentine) and a drier (a metal linoleate or naphthenate). The main pigments used are red lead, zinc oxide and chromate, iron oxide and powdered aluminium. Paint films are electronic insulators and so the cathodic reaction cannot occur except at the metal surface. This reaction is not inhibited, however, because the film is usually porous and allows oxygen and water to migrate through it fairly readily. The effectiveness of a paint depends upon the high resistance it provides in the electrolytic path between anodic and cathodic areas of the metal and also, when certain pigments are present, on the anodic inhibition it can provide. Slightly soluble chromates in the paint can slowly dissolve out when water is present and passivate the underlying metal surface. Red lead reacts with linseed oil to form soaps which then break down into lead salts of various fatty acids (e.g. azelaic, suberic, pelargonic acids) which are good anodic inhibitors.

As regards world economy, by far the most important corrosion problem is the rusting of structural steel in moist air. Anodic attack occurs in crevices, where local pockets of ferrous sulphate collect, formed from sulphur dioxide adsorbed from the air and then converted to the sulphate, and it leads to the formation of rust layers on the cathodic areas of the metal surface. Because this rust cannot easily be brushed completely off, paint is frequently applied over it, giving only short-lived protection. This adherent rust could, however, form a stable substrate for a permanent paint film. The difficulty lies in the pockets of ferrous sulphate, which catalyse additional rusting beneath the paint. The swelling which occurs there, due to the volume of rust created, throws up small bulges which break through the paint and open the way for a resumption of atmospheric attack. To extract the sulphate ions from the crevices a paint containing metallic zinc can be used. Zinc particles in contact with the steel then become anodes and the direction of current flow in the crevices is reversed, so that the anions flow outwards. It is necessary to include powdered cadmium in the paint so that, after the zinc particles next to the steel have been sacrificially corroded, metallic contact is still maintained between the steel and the zinc particles further out in the paint film.

23.9 Stress-corrosion

We now consider the disintegration of a metal under the combined action of chemical corrosion and mechanical stress. The first recognition of *stress-corrosion cracking* came with the problem of the *season-cracking* of brass cartridge cases, in the nineteenth century, a type of brittle intergranular fracture which occurred when cold-worked brass, containing large self-stresses, was exposed to atmospheres containing traces of ammonia, and which could be prevented by giving the metal a mild annealing treatment after working. Since then many other examples have been identified; for example, *caustic embrittlement* in boilers, due to the joint action of high stress and large hydroxyl ion concentrations; intergranular cracking of strong aluminium alloys; transgranular cracking of magnesium alloys; and transgranular cracking of austenitic stainless steels in solutions containing chloride or hydroxyl ions.

In stress-corrosion cracking there is usually very little overall corrosion, but the metal breaks by the passage of a macroscopically sharp and narrow crack across the line of principal tensile stress; for example, at a speed of order 10^{-3} m hr^{-1} in austenitic stainless steel stressed in hot MgCl$_2$ solution.

The mechanism of stress-corrosion cracking is not yet fully established. There are several possible effects—for example, embrittlement by uptake into the metal of hydrogen released through corrosive attack at the tip of the crack, or dissolution in the corrosive medium of material in stacking faults in slip bands—but it seems likely that one general cause of transgranular cracking is anodic dissolution of plastically deformed metal at the tip of the crack. Fig. 23.7 illustrates this. The yield stress is locally exceeded in the region ahead of the crack tip (A). The metal there deforms plastically,

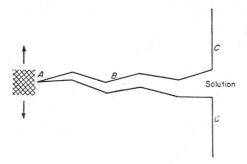

Fig. 23.7 Stress-corrosion crack

slightly opening the tip, and a highly anodic metal surface, where dissolution can occur readily at dislocations, is thereby exposed at the tip. Most of the anodic reaction, which balances the cathodic conversion of oxygen and water to hydroxyl ions on the outer surface C, then occurs at the tip, the weakly anodic sides of the crack (B) remaining fairly inactive. Concentration polarization in the solution at the tip, which would oppose this effect, might be overcome by the gradual opening of the crack due to the plastic yielding at the tip, so that fresh liquid is continually sucked in.

Electron microscope observations have suggested that the metals and alloys most susceptible to transgranular stress-corrosion cracking have low stacking fault energies. Whether this means that stacking faults themselves are the targets for localized anodic attack, or whether piled-up groups of dislocations are attacked, is not clear. It implies, however, that resistance to this type of stress-corrosion cracking may be increased by adjusting the alloy composition to increase the stacking fault energy.

Intergranular stress-corrosion cracking may have several causes. The season cracking of α-brass in ammoniacal environments may be due to local

dezincification along the grain boundaries (cf. § 23.7). The stress opens up
such a boundary into an intergranular crack, the tip of which is thus exposed
for further dezincification. High strength aluminium alloys are sometimes
susceptible to intergranular cracking in chloride solutions. The effect here
seems to be a consequence of that discussed in § 20.2 and § 21.12. Localized
plastic deformation occurs in the narrow soft zones adjoining the grain
boundaries. Presumably, the highly dislocated metal formed and exposed
by this deformation at the tip of an intergranular crack dissolves anodically
in the manner of Fig. 23.7.

23.10 Corrosion fatigue

The fatigue strengths of metals and alloys are often greatly reduced by the
presence of a corrosive environment. Fig. 23.8 shows a typical example of

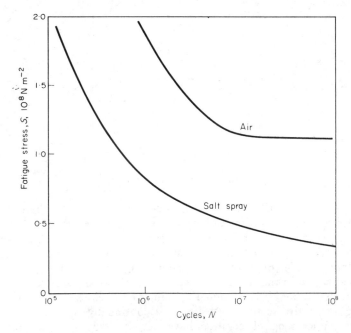

Fig. 23.8 Effect of corrosive environment on the fatigue S-N curve of an Al-Cu
alloy

this *corrosion fatigue*. The *damage ratio* (i.e. corrosion fatigue strength
divided by air fatigue strength) due, for example, to a salt water environment is about 0·4 for Al alloys, 0·5 for stainless steels, 0·2 for plain carbon
steels and 1·0 for copper. It is accentuated by rapid movement of the
corroding medium over the surface of the metal, which washes away any

protecting layers that might form. Since this *erosion-corrosion* often occurs in practice, it is necessary to simulate it in laboratory tests such as the *salt-spray test*.

In a sense, almost all practical fatigue tests are conducted under corrosion conditions because even the air can reduce the fatigue life of a metal by about an order of magnitude compared with fatigue in a vacuum. The most striking effects occur, however, in environments such as sea water and salt spray.

We do not yet have an established, detailed, theory of corrosion fatigue. The form of the fatigue crack is usually not much altered by the presence of the corrosive agent. The crack merely grows faster. Possibly, a process similar to that of Fig. 23.7 operates. During the tensile phase of the fatigue cycle the tip of the crack is stretched plastically and this exposes plastically strained metal there to the corrosive agent which is simultaneously sucked into the fatigue crack. Most of the methods developed for protection from simple corrosion, e.g. protective coatings, inhibitors, cathodic protection using sacrificial anodes, and those developed for protection from simple fatigue, e.g. shot-peening and surface alloying to promote compressive stresses in the surface, are also suitable for improving the corrosion-fatigue strengths of metals.

FURTHER READING

CHILTON, J. P. (1961). *Principles of Metallic Corrosion.* Royal Institute of Chemistry.

DEPARTMENT OF TRADE AND INDUSTRY (1971). *Report of the Committee on Corrosion and Protection* (Chairman: Dr T. P. Hoar). H.M. Stationery Office, London.

EVANS, U. R. (1960). *The Corrosion and Oxidation of Metals.* Edward Arnold, London.

— (1968). *The Corrosion and Oxidation of Metals: First Supplementary Volume.* Edward Arnold, London.

— (1975). *The Corrosion and Oxidation of Metals: Second Supplementary Volume.* Edward Arnold, London.

— (1963). *An Introduction to Metallic Corrosion,* 2nd Ed. Edward Arnold, London.

KUBASCHEWSKI, O. and HOPKINS, B. E. (1962). *Oxidation of Metals and Alloys,* Butterworth, London.

UHLIG, H. H. (1948). *The Corrosion Handbook.* Chapman and Hall, London.

Chapter 24

Electronic Structure and Properties

24.1 Electrons in a lattice

In § 4.5 and § 4.6 we sketched an elementary picture of the electronic structure of solids and in § 19.1 we used the idea of free electrons to discuss cohesion in metals. In this chapter we shall go further into the electron theory of solids, partly for general interest, partly to see what it can say about the structures of alloys and partly to discuss the electrical and magnetic properties of solids.

In § 19.1 we used the *standing wave* solutions of Schrödinger's equation given in § 3.4. These enabled us to find the energy levels of free electrons, but have disadvantages. As in Fig. 3.2, they suggest that the electron density varies *periodically* through the metal, even though in the free electron theory the potential inside the metal is assumed *constant* so that there ought to be an equal chance of finding an electron anywhere inside the metal. This difficulty comes from the standing waves, which do not move along with time. In terms of electrons a standing wave means that we do not know whether an electron is moving up or down a line through the metal.

To deal with a current of electrons which we know is going one way, *running waves* that move along in time have to be used. We might consider, for example, $\sin 2\pi[(x/\lambda) - vt]$ for a wave of wavelength λ and frequency v. Then, as the time t increases, x must also increase to keep up with a given point on the wave, i.e. the wave moves in the $+x$ direction with speed $c = v\lambda$. Any linear sum of such sine and cosine terms is a solution of Schrödinger's equation and it turns out that the most convenient sum for representing moving electrons is of the form

$$\psi = A(\cos \theta + i \sin \theta) = A e^{i\theta} \qquad \textbf{24.1}$$

where A is a constant, $i = \sqrt{-1}$ and $\theta = 2\pi[(x/\lambda) - vt]$. The use of $\sqrt{-1}$ is allowable because we have to take the *square* of ψ to obtain a physical quantity (i.e. the probability of where the electron is) and $\sqrt{-1}$ does not appear in this square. As mentioned at the end of § 3.3, the 'square of

ψ' for a complex function is $\psi\psi^*$ where ψ^* is ψ with $+i$ replaced by $-i$. We then have

$$\psi\psi^* = A^2(\cos\theta + i\sin\theta)(\cos\theta - i\sin\theta) = A^2(\cos^2\theta + \sin^2\theta) = A^2$$
                                                                                    24.2

with the welcome result that the electron density is *constant* through the metal.

We shall not need to refer explicitly to time in the running wave function and so shall regard the $\exp(2\pi i\nu t)$ part of it as absorbed into the factor A. Our wave function for moving electrons is thus

$$\psi = Ae^{ikx} \quad \text{or} \quad Ae^{-ikx} \qquad 24.3$$

according as the movement is up or down the x axis. The symbol

$$k = \pm\frac{2\pi}{\lambda} \qquad 24.4$$

is called the *wave number* and is the number of wave crests in a distance 2π along the x axis. In three-dimensions k is a vector, the *wave vector*, and

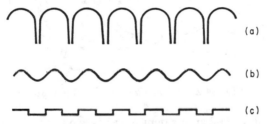

Fig. 24.1 Periodic potentials through a lattice

gives the direction of propagation as well as the wavelength, i.e. it has components k_x, k_y, k_z, along the three axes x, y, z. From eqn. 3.4 we have

$$k = \frac{2\pi m v}{h} \qquad 24.5$$

so that k is a measure of the momentum mv of the particle. (This simple interpretation of k breaks down, however, when the electron moves in a periodic lattice.) The energy E of a free electron increases parabolically with k, i.e.

$$E = \tfrac{1}{2}mv^2 = \frac{h^2k^2}{8\pi^2 m} \qquad 24.6$$

We now consider the periodic lattice field of the metal by taking a potential energy V in eqn. 3.10 that varies periodically with x, as in Fig. 24.1 (diagram (a) is taken through the centres of atoms, diagram (b) not). For simplicity we consider *nearly* free electrons, for which the variation of V is small, as in diagrams (b) and (c), since these ought to behave almost like free electrons. We can take the average value of V as zero.

We shall find that for most values of k the electron is very like a free electron. However, *at critical values of k this similarity to free electrons totally breaks down, even when the periodic variation of V is very small.* Many important properties of crystals depend on this effect. The diffraction of electrons by crystals (cf. § 17.8) shows that something of this kind must happen. An electron shot into a crystal from outside experiences the same periodic field as an internal one once inside; and experiment shows that such electrons are rejected by the crystal if they satisfy Bragg's law. Evidently, reflection off lattice planes must prevent the native electrons, also, from moving in modes that satisfy Bragg's law.

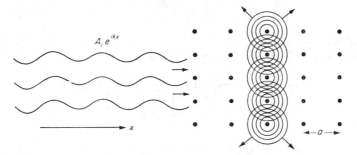

Fig. 24.2 Scattering of a wave by a lattice

Consider in Fig. 24.2 an incident wave $A_0 \exp(ikx)$ advancing in the $+x$ direction perpendicularly through a set of lattice planes. The moving electrical field represented by this wave interacts with the atoms and these interactions generate wavelets which spread out equally and spherically from each atom, as shown for the middle lattice row. These wavelets correspond to the Huygens wavelets in optics that spread out from the lines of a diffraction grating. All waves from one row of atoms are in phase, since they are all created simultaneously by the same crest of the incident wave. Because of this, they interfere constructively and build up into two waves similar to the incident wave (i.e. *plane waves* moving along the x axis). One of these goes forward with the incident wave. The other goes backward, in the $-x$ direction, and is represented by the wave function $A_1 \exp(-ikx)$. This backward wave indicates the possibility that the electrons may be *reflected* backwards by the row of atoms. The amplitude A_1 depends on how strongly the atoms can reflect the electrons and it increases with increase in the amplitude of V.

We must consider next the superposition of the reflected waves from successive rows of atoms. For arbitrary values of k the backward waves from different rows are usually out of phase with each other and do not add up to a significant reflected wave. We then have $A_1 = 0$ and the electrons with such a k continue moving in their incident motion without disturbance from the periodic lattice field. Since the wave function $A_0 \exp(ikx)$ remains unchanged, these electrons behave like free electrons. The fact that the lattice does not affect their uniform motion implies that *infinite* electrical conductivity should be possible in perfect crystals. We shall discuss this in § 24.4.

Consider now the critical values of k, where Bragg's law is satisfied. If k is such that the path difference between waves reflected off neighbouring rows of atoms is equal to a whole number $n\lambda$ of wavelengths, successive reflected waves interfere constructively and add up to make a strong reflected beam. In terms of k this Bragg reflection condition is

$$ k = \pm \frac{\pi n}{a} \qquad\qquad 24.7 $$

where $n = 1, 2, 3 \ldots$. These are the critical k values. Even though the individual wavelets are very weak when V is nearly constant, a *large* number of them can always add up to give a strong reflected beam, so that the electron is reflected when it penetrates the crystal sufficiently deeply. Thus, when k satisfies eqn. 24.7, the total wave function is not even approximately like the function $\exp(ikx)$ which represents an electron going in one single direction. It must contain equally strong incident and reflected waves. Since the reflected wave is $\exp(-ikx)$ the total wave function must be of the form

$$ e^{ikx} + e^{-ikx} \qquad (= 2\cos kx) $$

or   24.8

$$ e^{ikx} - e^{-ikx} \qquad (= 2\,i \sin kx) $$

However, as we have already seen, these sine and cosine functions represent *standing waves*. This means that at the critical k values the electron has zero average velocity through the lattice, because it is repeatedly shuttled forwards and backwards by the reflections from the periodic lattice field; and also that the electron density varies periodically through the lattice. Since this density goes as the square of the wave function, its periodicity when $n = 1$ is simply the periodicity a of the lattice itself. The sine function has its nodes at the places where the cosine has its crests and vice versa. Because of this, one of the standing wave functions places the maxima of the electron density at the points where V is lowest, in which case the electron energy is then *lower* than that of a free electron, and the other places the maxima at the highest points of V, in which case the electron energy is higher; cf. Fig. 24.3. The free electron parabolic E, k relation, eqn. 24.6, is thus interrupted by breaks at the critical k values, since at each of these

points there are two E values, one higher and one lower than the free electron value.

This is shown in Fig. 24.4. The E, k relation follows the free electron
parabola for most of the way, but at the critical values $\pm k_1$, $\pm k_2$, etc. it

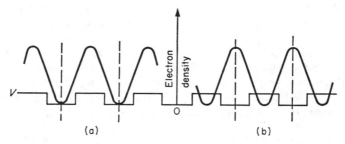

Fig. 24.3 Periodic electron densities associated with critical wave functions k
giving maxima (a) in high-V regions, (b) in low-V regions

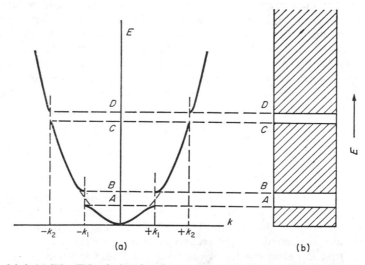

Fig. 24.4 (a) The E,k relation for electrons moving in a certain direction through
a crystal and (b) the energy bands for this direction

deviates and flattens to produce the discontinuities from A to B, and
C to D, etc. The energy levels available to an electron in the crystal are thus
broken up into *bands*, separated by *gaps AB, CD*, etc. Thus, if we gradually
increased the wave number of an electron, starting from the bottom of the
curve, the energy would increase steadily until it reached the level A. At
this point the smallest further increase in k would require the electron

energy to be increased to B, correspondingly to the transfer of the maxima in the periodic electron density from the troughs to the crests of V. This particular energy band diagram refers of course to one particular direction through a crystal. Similar diagrams, with other critical k values, apply to other directions. When presented collectively for all directions, in a diagram discussed in § 24.2, the ranges of k between the energy gaps are called *Brillouin zones*. In each such zone the energy levels are closely spaced, forming an almost continuous spectrum of allowed values. The *first zone* contains all levels up to the first discontinuity, the *second zone* all those between the first and second discontinuities, and so on.

Consider now electrons with k values *near* a critical value, say k_1. The scattered wavelets from neighbouring lattice rows are in this case *almost* in phase with one another and interfere constructively. Only those waves from distant rows are so much out of phase as to interfere destructively. The critical wavelength is $2a$ (eqns. 24.4 and 24.7, with $n = 1$). The path difference for reflections from lattice rows m spacings away is $2ma$. The number of wavelengths over this length of path is $k(2ma/2\pi)$, which is simply equal to m when $k = k_1$. The interference certainly becomes destructive when k differs from k_1 sufficiently to make the number of wavelengths differ from m by $\pm\frac{1}{2}$, i.e. when $kma/\pi = m \pm \frac{1}{2}$, or

$$k = \frac{\pi}{a}\left(1 \pm \frac{1}{2m}\right) = k_1\left(1 \pm \frac{1}{2m}\right) \qquad 24.9$$

This roughly estimates the number of rows, m, that can reflect constructively for a given value of k. This number increases, as k approaches k_1, and lattice reflections become increasingly probable, so that the wave function of the electron gradually changes from the $\exp(ikx)$ type towards the $\cos(kx)$ type. The number of lattice rows needed to give a strong reflected beam depends on the scattering power of the individual atoms, i.e. on the amplitude of variation of V. An increase in this amplitude thus not only widens the energy gap AB but also widens the range of values over which the E,k relation deviates from the free electron parabola.

24.2 Brillouin zones

We now apply the above theory to a two-dimensional square lattice, of lattice constant a, as shown in Fig. 24.5. Consider electrons moving at angle θ to the atomic rows parallel to the Y axis. Then, from Bragg's law, reflection occurs at the wave numbers

$$k = \pm\frac{n\pi}{a \sin \theta} \qquad 24.10$$

The *boundary of the first Brillouin zone*, i.e. the position of the first discontinuity in the E,k relation, for this direction of motion, is then given by setting $n = 1$ in this relation. The result is simplified if we split the vector

k into its components, k_x along X and k_y along Y. Since k_x is perpendicular to the reflecting rows we are considering, and since $k_x = k \sin \theta$, we simply have $k_x = \pm \pi/a$; i.e. the E,k discontinuity due to this reflection occurs

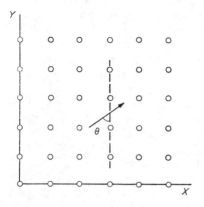

Fig. 24.5 Electrons in a square lattice

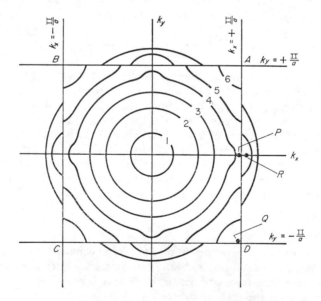

Fig. 24.6 The first Brillouin zone for the two-dimensional square lattice

for all electrons with this value of k_x, irrespective of their k_y value. Similarly, reflection from the atomic rows parallel to the X axis occurs whenever $k_y = \pm \pi/a$.

These results are best expressed by a diagram drawn in *k-space*. We construct axes, parallel to the X and Y axes of Fig. 24.5, along which k_x and k_y are measured, as shown in Fig. 24.6. Then any point in this diagram has one value each of k_x and k_y and so represents a particular state of motion for the electron in the two-dimensional crystal of Fig. 24.5. The discontinuities are then shown by marking in the values of k_x and k_y at which they occur. The first discontinuity occurs wherever any one of the following four conditions is satisfied, $k_x = \pm\pi/a$, $k_y = \pm\pi/a$, and so occurs at all points on the square $ABCD$ in Fig. 24.6. This square is the boundary of the first Brillouin zone for this lattice. Boundaries of higher zones can be found by considering Bragg reflections off other lattice rows and by taking higher values of n in Bragg's law. We notice that the zone boundaries in Fig. 24.6 are parallel to the lattice rows (*planes* in three dimensions) that produce them. This is a general property of zone boundaries which follows directly from Bragg's law.

The usefulness of such diagrams is increased by marking in *energy contours*, i.e. those k values at which the electrons have the same energy. By constructing E,k diagrams for various directions through the crystal, we can find the k values for any given energy and then plot them in the zone diagram. Typical energy contours for nearly free electrons are shown in Fig. 24.6. The low-energy contours 1, 2, 3, are circles centred on the origin. This follows because electrons with k well away from the critical values resemble free electrons and these behave the same whatever direction they move since there is no periodic field to give them directionality. Further out from the origin (e.g. contour 4) the contours bulge out towards the nearest part of the zone boundaries, because the E,k relation breaks away from the free electron parabola near such a boundary and E rises more slowly with k. High-energy contours, e.g. 5 and 6, intercept the zone boundaries, showing that the only allowed electron states at such energies lie in the corners of the zone.

The energy contours do not continue smoothly across a zone boundary because of the energy discontinuity here. Thinking of Fig. 24.6 geographically as a contour map of a basin-like hollow (height corresponding to energy) we see the boundary of the zone as a vertical cliff. The energy contours of the second zone, a few of which are shown, depend on the height of this cliff. It is important to know whether the two zones *overlap*, i.e. whether the *lowest* energy levels in the second zone have *lower energy* than the *highest* levels in the first zone. Consider the two cases:

(1) *Large discontinuity; no overlap.* For example, suppose in Fig. 24.6 that $P = 4.5$, $Q = 6.5$, and the height of the cliff from P to R is 4, so that $R = 8.5$. The lowest level, R, in the second zone is then higher than the highest level, Q, in the first zone, and the *energy band* structure of the crystal is as in Fig. 24.7(a).

(2) *Small discontinuity; overlap.* For example, suppose that the cliff from P to R is only 1, so that $R = 5.5$ and lies below the highest level ($Q = 6.5$) in the first zone, as in Fig. 24.7(b). We see in this case that, although there

16*

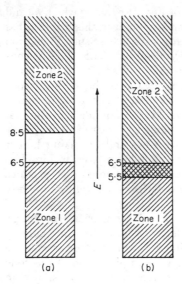

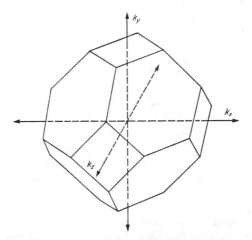

Fig. 24.7 Energy bands for (a) large discontinuities and (b) small discontinuities

Fig. 24.8 The first Brillouin zone for the F.C.C. lattice

are *always* forbidden ranges in single E,k curves for particular directions through the crystal, the total energy spectrum for all directions of motion does not have forbidden ranges.

For actual metals the Brillouin zones are polyhedra, the plane surfaces of which are parallel to the corresponding reflecting planes in the crystal. These planes are the same as those that give x-ray reflections (cf. § 17.7) and the strength of an x-ray reflection is a guide to the reflecting power of the planes from which it comes; and hence to the height of the energy cliff at the zone boundary caused by these planes. Fig. 24.8 shows the first Brillouin zone for the F.C.C. lattice. X-ray analysis shows that reflections in this lattice occur first from {111} and {200} planes, so that the boundaries of the first zone are parallel to these.

Which quantum states in a Brillouin zone are occupied by electrons? To answer this, we first consider how many states there are in a zone. The increase in the value of k from one state to the next, in a one-dimensional crystal of length L, is obtained from the relation $k = \pm 2\pi/\lambda$ and then by fitting successively shorter wavelengths into this length L. There is a rather subtle point here. We cannot simply use eqn. 3.8 to connect λ and L because this was based on *standing* wave functions whereas k refers to *running* waves. For each value of λ there are two such waves, going up and down the wire respectively. We can no longer fix nodes of these waves at the ends of the wire. We need a different kind of boundary condition at these ends. A complete theory of how an electron is reflected from an end is complicated. However, for the mere purpose of *counting* the number of quantum states this theory is unnecessary. It proves to be sufficient simply to assume a *periodic* boundary condition; i.e. to assume that λ has the same phase at each end of the wire. The allowed values of λ are then given by $n\lambda = L$, where $n = 1, 2, 3 \ldots$ etc. This replaces eqn. 3.8. The corresponding allowed values of k are then $\pm 2\pi n/L$ and the increase in k from one state to the next is $2\pi/L$. Halving the number of allowed values exactly compensates for doubling the number of quantum states belonging to a given value of λ, associated with the fact that each λ now represents two running waves, $k = \pm 2\pi/\lambda$. It can be proved in fact that the distribution of quantum levels is independent of the boundary conditions or the shape of the boundary.

In three dimensions the volume given to each quantum state in k-space is simply $8\pi^3/V$, where V is the volume of the crystal. The volume of the cube which represents the first Brillouin zone in a simple cubic lattice (cf. Fig. 24.6) is $(2\pi/a)^3$, where a is the lattice constant. The number of quantum states in the zone is thus V/a^3, i.e. *exactly equal to the number of lattice points in the crystal*. Since each state can hold two electrons of opposite spins, the capacity of the zone is two electrons per lattice point.

A similar analysis shows that, in B.C.C. and F.C.C. lattices also, the first zone holds one quantum state per lattice point. In C.P. Hex. crystals the zones do not hold a simple integral number of quantum states. The precise value depends on the axial ratio. The first zone of zinc, for example, holds 1·792 electrons per atomic site.

The usual rules apply to the occupation of the quantum states by electrons. At 0 K the electrons fill the states of lowest energy, with two of opposite spins in each. Thus, for monovalent cubic metals (e.g. Na, Cu) the first zone is half filled. With higher valencies it matters whether or not the zones overlap. Fig. 24.9 shows what happens in the two cases as more and more electrons are added. In the non-overlapping case there is a state, at two electrons per lattice point where the first zone is completely full and all others are completely empty. This is impossible in the overlapping case; there are always *partly-filled* zones, since a new one begins to fill before the preceding one is completely filled. The boundary of the occupied region of k-space, called the *Fermi surface*, is of course always a contour through states of the same energy.

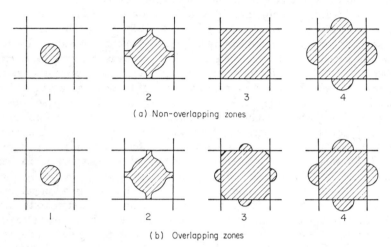

(a) Non-overlapping zones

(b) Overlapping zones

Fig. 24.9 Various stages (1, 2, 3, 4) in the occupation of Brillouin zones by electrons

24.3   Metals and insulators

The effect just described has long been used to explain the difference between metals and insulators. Consider a direction $+X$ through a crystal along which an electric field will be applied. Beforehand, no current flows along this direction. For every electron moving with a certain speed in this direction there is another moving with the same speed in the reverse direction, $-X$. This is because the occupied states in k-space are distributed symmetrically about the origin. Fig. 24.10 (full line) shows this. Here we give the probability p of occupation of a state, at 0 K, as a function of the wave number k. All states up to the Fermi surface, at $\pm k_m$, are full and all others empty.

When the field is applied, the electrons accelerate in the $+X$ direction and the whole distribution curve shifts sideways, to the position of the broken line. Most electrons remain compensated by the motion of others in the reverse direction. However, those electrons in the states near $+k_m$ are uncompensated since there are none in the corresponding states near $-k_m$.

Thus, while all electrons are affected by the field, only those near the Fermi surface produce a current.

Suppose now that a Brillouin zone is completely filled and that there is no overlapping, as in Fig. 24.11(a). The Fermi distribution *cannot* now be

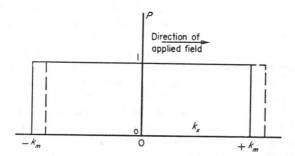

Fig. 24.10 Displacement of electron distribution from a symmetrical to an unsymmetrical position by the application of a field

displaced unless some electrons are lifted up the energy cliff into the next zone. In general, the electrons cannot acquire the energy to do this unless the field is thousands of volts strong. But while the electrons all stay in the first zone there is no electrical current since the distribution remains symmetrical about $k = 0$. Substances of this type are therefore *insulators*, not metals.

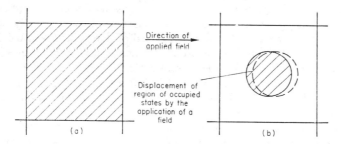

Fig. 24.11 Effect of an applied field on filled and partly filled zones

When zones are partly filled, as in Fig. 24.11(b) (also Fig. 24.9, except diagram (a), 3), the distribution can be shifted sideways by moving electrons into nearby states *in the same zone*. Because of the close spacing of the adjoining energy levels, the lifting of electrons into them can be accomplished by arbitrarily small applied voltages. Unsymmetrical distributions, which give a flow of current, can thus be produced by small fields. Such substances are *metals*.

Semi-conductors have a weak electronic conductivity which increases with increasing temperature. As explained in § 4.5, they may be either *intrinsic* semi-conductors or *impurity* semi-conductors. In the first, the zones do not overlap but the energy gap is small, so that thermal agitation can lift a few electrons from the top of the filled zone into the bottom of the empty one. Both zones can then conduct. In the second type, impurity atoms of different valencies either donate a few electrons to the empty zone or remove a few from the filled zone, again with the help of thermal agitation.

There is an obvious similarity between this aspect of the zone theory of solids and the molecular bond picture outlined in § 4.5. We can pursue this further by considering what happens to atomic energy levels when a number of separate atoms are gradually brought together. We saw in § 4.3 that when two identical atoms, e.g. hydrogen, are brought together, their atomic energy level splits into two, a low one which corresponds to a *bonding* wave function and a high one which corresponds to an *anti-bonding* function. When a large number of atoms are brought together the same thing happens except that the splitting now produces a whole band of closely spaced levels, as in Fig. 24.12. The splitting and broadening of the energy band occurs first for the valency electrons, since these lie mainly in outer regions of the atoms, and the levels of the inner electrons do not split until the atoms are much closer.

This approach to the energy band structure is particularly applicable to *transition metals*, which owe many of their special properties to the partly filled *d* shell below the valency electrons. Fig. 24.13 shows in a highly simplified way the broadening of the bands in a metal such as iron.

The 4s band is broad and, being derived from an atomic *s* state which holds only two electrons per atom, it has a low *density of states* (cf. § 19.1). This is shown by the $N(E)$ curves of diagrams (b) and (c). By contrast, the 3d band is narrow and, being composed of five superposed bands from each of the five 3d states of the atom and hence able to hold ten electrons per atom, it has a very high density of states. To fill it completely, nearly eleven electrons per atom are needed because the overlapping part of the 4s band has also to be filled, to the same level. In nickel and copper this part of the 4s band holds about 0·6 electrons per atom. Thus, in nickel and the preceding transition metals the 3d band is only partly filled in the solid, whereas in copper this band is full and there is one electron per atom in the 4s band, as in diagrams (b) and (c) of Fig. 24.13.

The partly filled *d* band is responsible for many properties of transition metals, e.g. high cohesion, poor electrical conductivity, strong paramagnetism, ferromagnetism, large electronic specific heat. As regards the last of these, we saw in § 19.2 that only those electrons within an energy range of about kT from the Fermi level E_F can contribute to the specific heat of a metal at a temperature T. The high density of states in the *d* band means that, in a transition metal, there are many more electrons than usual, near the Fermi level, and the electronic specific heat is correspondingly high.

Returning to the comparison of the zone and atomic theories of energy bands, we notice a fundamental difference in that the zone theory requires a periodic lattice structure whereas the atomic theory does not. In fact we used a basically different argument in § 4.5 to explain the difference between metals and insulators; one that depended on the distribution of the electrons

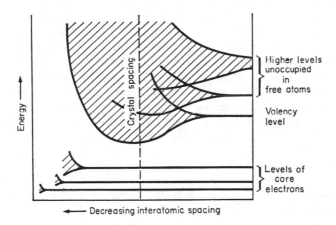

Fig. 24.12 Broadening of atomic energy levels when atoms approach to form a crystal

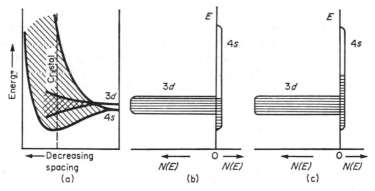

Fig. 24.13 (a) broadening of 3d and 4s bands in metals such as iron; (b) filling of levels in a transition metal and (c) filling in copper

in *real* space, i.e. in the material itself. We regard this as the more fundamental explanation since, if a periodic crystal structure leading to filled zones were the *only* thing that prevented a material from being a metal, we would expect vitreous glass to be a metal. Similarly, in reverse, there are crystalline solids such as nickel oxide which have partly filled zones but which are nevertheless insulators, not metals, presumably because the electrons must remain equally distributed amongst all their molecules.

It is at first sight surprising that the nearly-free electron model of § 24.1, on which the zone theory of metals is built, works so well. We have ignored the fact that the actual lattice field felt by an electron varies violently, the potential (cf. eqn. 3.2) going to $-\infty$ at each atomic nucleus. We have also ignored the fact that the wave function of a nearly-free electron must become an atomic-like wave function inside each atom, i.e. it must oscillate to and fro in the atom, according to its ns, np ... etc. character (cf. § 3.5 and § 3.6), and the minimum number of oscillations is determined by the fact that all the simpler atomic wave functions in the atom are already filled by the core electrons. Because of these oscillations, the kinetic energy of the nearly-free electron becomes high inside each atom, sufficiently so that it practically balances the drop in potential energy near the nucleus. This necessary rise in kinetic energy towards the centre of the atom can in fact be thought of as a 'pseudo-potential' which very nearly cancels out the actual potential of the nucleus. The difference between them is a rather weak potential which can be put into the Schrödinger equation to generate the nearly-free electron theory with its simple sinusoidal wave solutions which go smoothly through the atoms with no atomic-like oscillations.

24.4 Conductivity of metals

When an electric field is applied to a metal the free electrons are accelerated towards the positive end. Because a steady flow of current is produced by a constant voltage, we deduce that this acceleration is opposed by some resistance which the electrons meet during their motion. To make a start on this, we suppose that the resistance is due to the electrons colliding with the lattice ions and assume that the extra velocity acquired by an electron from the accelerating field during the interval between successive collisions is lost at each collision. It can then be shown that the *resistivity* R of the metal is given by

$$R = \frac{mv}{ne^2l} \qquad \textbf{24.11}$$

where n = number of conduction electrons per unit volume, v = average electron velocity, m = electron mass, e = electron charge and l = *mean free path*, i.e. average distance travelled between successive collisions.

The resistance of a pure metal usually depends on absolute temperature T as in Fig. 24.14, varying as T at temperatures above about 100 K and as T^5 at low temperatures, reaching zero at 0 K. How does this behaviour fit in with eqn. 24.11? We know that only the electrons near E_F contribute to the conduction, so that n and v refer to these only. But the Fermi distribution is hardly altered by temperature (cf. § 19.2) so that n and v should be nearly constant. Since m and e are constant, practically the whole effect in Fig. 24.14 must be due to a variation of mean free path with temperature.

Calculations in fact show that, for pure copper or silver, l is about 100 atomic spacings at room temperature and increases to infinity at 0 K. The mean free path is so long at low temperatures that the thickness of the metal noticeably affects the resistivity, because the electrons are scattered off the inside of the free surface.

The collisions of the electrons with the lattice ions cannot be pictured in a simple classical way as the bumping of particles against each other, for if this were the case the mean free path would be similar to the atomic spacing and would hardly change with temperature. We have to consider the scattering of the ψ waves and deduce from it the movements of the electrons. We saw in § 24.1 that scattered wavelets from a perfectly periodic lattice reproduce the complete wave again so that there is no disturbance to the motion of the electron through the lattice. The resistance of a perfect

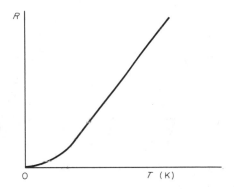

Fig. 24.14 Variation of resistance with temperature

lattice should thus be zero. Further analysis shows that this conclusion also holds if the lattice is in its lowest state of natural vibration (zero-point oscillation). The wave-mechanical picture thus explains why $R = 0$ at $T = 0$ in Fig. 24.14.

In certain metals, e.g. Pb, Sn, Hg, Nb, the resistance remains at zero for a few degrees above absolute zero and then, at a critical temperature, sharply rises to join the curve in Fig. 24.14. Vanishing resistivity at *finite* low temperatures is *superconductivity*. We shall discuss it later.

Electrical resistance is due to incoherent scattering of the electron waves which occurs at places where the periodicity of the lattice is disturbed. Irregularities on an atomic scale of size are particularly effective whereas long-range features such as gradually varying elastic strains have little effect. Thermal vibrations throw the atoms out of alignment and this causes the scattering responsible for the effect in Fig. 24.14. Irregularities such as impurity or alloy atoms, vacancies and dislocations also cause resistance. The wave mechanical nature of the scattering process is particularly manifest in the large difference between the resistivities of ordered and disordered solid solutions. Alloy atoms, through their different electrical charges and sizes, are strong scatterers of electrons. If the atoms are distributed randomly in the solution the scattering is incoherent and a high resistance results. If they are ordered, the periodicity of the superlattice enables the wavelets from different atoms to be scattered coherently and the resistance is then not much greater than that of a pure metal. The resistivity of an alloy of superlattice composition can thus vary sensitively with the heat treatment according to the state of order produced, as shown in Fig. 24.15.

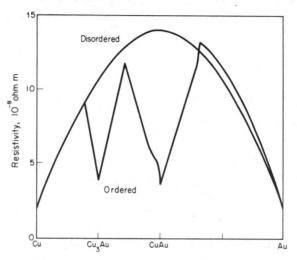

Fig. 24.15 Electrical resistivity of copper-gold alloys with ordered and disordered structures

We now consider why some metals are better conductors than others. In Table 24.1 the conductivities at 0°C are given. There is little regularity, apart from the notably high values of the copper, silver, gold, group. We can get a simpler picture by taking out the effects on the resistivity of the different amplitudes of thermal vibration in the various metals. From the theory of lattice specific heats (cf. § 19.2) it is possible to deduce correction factors which allow the conductivities to be compared at constant amplitude of thermal vibration. These modified values are given in the third column of the Table.

We notice that the monovalent non-transition metals are in general better conductors than either the alkaline earths or the transition metals. These effects can be explained by the zone theory. A nearly empty zone is a poor conductor because there are few electrons to carry the current. A nearly full

TABLE 24.1. ELECTRICAL CONDUCTIVITY OF METALS

Metal	Conductivity at 0°C (10^6 ohm^{-1} m^{-1})	Corrected for Thermal Vibrations
Li	11·8	12·9
Na	23	24
K	15·9	15·3
Cu	64·5	9·1
Ag	66·7	12·4
Au	49	8·1
Be	18	2·0
Mg	25	8·1
Ca	23·5	11·1
Ba	1·7	1·0
Zn	18·1	6·1
Cd	15	4·5
Ti	1·2	0·21
Cr	6·5	0·51
Fe	11·2	1·14
Co	16	1·7
Ni	16	1·9
Al	40	9·5
Sn	10	1·2
Pb	5·2	3·4
Bi	1·0	0·5

(Mott and Jones, 1936, *Theory of the Properties of Metals and Alloys*, Oxford University Press, London, pp. 246–7).

zone is also poor, because the number of empty states in the zone, into which the electrons can be shifted to give the necessary asymmetry (cf. Fig. 24.11), is then small. Furthermore, electrons near a zone boundary are liable to undergo Bragg reflections during their motion and this reduces their acceleration in the applied field; they behave as if they had a large 'effective mass'.

The monovalent non-transition metals are good conductors, then, because they have half-filled zones which give plenty of electrons near the Fermi surface that behave very like free electrons. The alkaline earth metals have lower conductivities because they have a nearly full zone and a nearly empty zone, neither of which is as good as a half-filled zone. The low conductivity of bismuth is particularly due to a zone effect and it is noteworthy that, when bismuth melts, its conductivity *increases* despite the increased scattering from the disordered atoms. The zone structure is evidently partly destroyed, along with the lattice structure, when the crystal melts.

The low conductivities of the transition metals are due to small mean free paths. When an electron is deflected by an irregularity it goes into some different quantum state. The more empty states there are available, at about the same energy, the greater the chance that it will be so deflected by a given irregularity. The numerous empty states at the top of the d band in transition metals greatly enhances the scattering power of irregularities in them.

The high conductivity of aluminium should be noted. Weight for weight, aluminium wire has twice the conductivity of copper wire and so, reinforced by a steel core for strength, is now widely used for electrical power lines. In electrical generators and motors, however, where compactness is essential, copper wire is used since it has two-thirds higher conductivity for the same cross-sectional area, as well as being stronger at the operating temperature.

Some metals and alloys find uses based on their *low* electrical conductivities. For example, *nichrome* (80 Ni, 20 Cr) resistance heater wire has only 1·5 per cent of the conductivity of copper. The associated low thermal conductivities of such materials are also useful in cryogenic and other thermal apparatus; for example, 18/8 stainless steels, *constantan* (60 Cu, 40 Ni) and *German silver* (52 Cu, 26 Zn, 22 Ni) all have thermal conductivities of order 20 W m^{-1} K^{-1}, whereas that of copper is about 400. On the other hand, low thermal conductivity is a disadvantage in the *machining* of metals, since it leads to excessive heating and welding between the work piece and the cutting tool. This is partly responsible for the poor machinability of titanium and austenitic stainless steel.

24.5 Semi-conductors

The ease with which the distribution of electrons between filled and empty energy bands can be modified by heat, light and impurities in semi-conductors and the large changes of conductivity which can result from this have led to the development of many semi-conducting devices of great practical value: for example, rectifiers, transistors, photo-cells, thermo-electric generators, solar batteries and light-producing devices based on phosphors. The starting point for some of the most important of these is the *p-n junction*. We have seen in § 4.5 that an *n*-type impurity semi-conductor is produced by adding a trace of a higher valency element such as phosphorus to a crystal of, for example, silicon or germanium; the extra electrons introduced by this element go into the conduction band and act as negatively charged current carriers. Similarly, a *p*-type impurity semi-conductor is produced when an element of lower valency such as indium is added instead. The impurity atoms in this case abstract electrons from the top of the filled band so leaving empty states there which act as if they were positively charged current carriers, i.e. *positive holes*. The *p-n* junction is formed when two such regions, one *p*-type and one *n*-type, exist side by side in the same crystal.

Initially, all the positive holes lie in the *p*-type region and all the free electrons in the *n*-type. Each region is electrically neutral at this stage, the total charge of its holes or electrons being precisely equal and opposite to that of its ionized impurity atoms. The holes and electrons are mobile, however, and in their random migrations some will cross over into the other side of the junction. Both the transfer of positive holes to the *n*-side and of electrons to the *p*-side cause the *n*-side to become *positively* charged relative to the *p*-side, so that a voltage drop V then exists across the junction, as shown in Fig. 24.16.

This voltage drop V brings the two sides into equilibrium. If a hole now crosses from *p* to *n*, or an electron from *n* to *p*, its potential energy increases

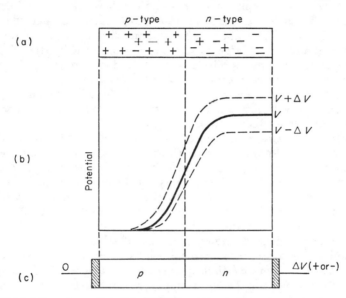

Fig. 24.16 Migration of positive holes and negative electrons across a *p–n* junction (a) produces a voltage drop V from the *n*-side to the *p*-side (b). This voltage drop is modified by the application of a forward bias $-\Delta V$ or reverse bias $+\Delta V$ through terminals (c).

by eV, where e is the magnitude of its charge. We now have an example of the dynamical equilibrium described at the end of § 6.6. There is a large number of holes in the *p*-side. In their migrations, these holes approach the junction but most of them are turned back by the potential energy barrier. Only a fraction $\exp(-eV/kT)$ of them has the necessary thermal energy to cross over into the high energy side of the junction. There are few holes in the *n*-side, but any one of these that approaches the junction has no difficulty in passing through to the low energy region on the other side. In equilibrium these two counter-currents are equal. Let c_0 and c_1 be the concentrations of holes on the *p*-side and *n*-side, respectively. Then

$$c_1 = c_0 e^{-(eV/kT)} \qquad 24.12$$

Exactly the same argument applies to the equilibrium of the electrons, except that n is now the low energy (high concentration) side and p is the high energy side.

Let us call the flow of either type of current carrier 'downhill' when it goes from its high to low energy region and 'uphill' in the opposite direction. In equilibrium, the downhill current is equal and opposite to the uphill flow, for both current carriers. We now apply a *biassing voltage* $+\Delta V$ or $-\Delta V$ to the junction, as shown in Fig. 24.16(c). If ΔV is small, the downhill flows are not appreciably altered. However, the uphill flows are strongly altered because V is replaced by $V + \Delta V$ or $V - \Delta V$ in the Maxwell-Boltzmann formula. If ΔV is negative (*forward bias*) the uphill flows are strongly increased. They can thus become much larger than the downhill flows, so that a relatively large current flows through the junction. If ΔV is positive (*reverse bias*) the uphill flows can be reduced to negligible magnitude, leaving only the relatively small and constant downhill flows to carry the current. We thus have a *rectifier*. The p-n junction conducts a large current when the voltage acts in one direction and a very small current when it acts in the reverse direction, as shown in Fig. 24.17.

The *junction transistor* consists of two p-n junctions back to back. Fig. 24.18 shows the p-n-p transistor, in which a thin layer of n-type material is sandwiched between two p-type regions. The main current flow in this case is provided by the positive holes but it is equally possible to make an n-p-n transistor in which the electrons carry the current. In the absence of an applied voltage each of the p-n junctions comes to equilibrium, in the manner discussed above, so that the central n layer becomes positively charged relative to the p regions, as shown in Fig. 24.18(a).

We now connect three terminals, $e = emitter$, $b = base$ and $c = collector$, to the transistor and make b slightly negative and c more strongly negative, relative to e, so changing the electrical potential in the transistor, as shown in Fig. 24.18(b). As regards the e-b rectifier junction, we now have a forward bias and so a strong current of holes will flow from the emitter (hence its name) into the n layer. Some of these holes will combine with electrons in this layer and, since b is a source of electrons relative to e, this constitutes a flow of current through this part of the transistor, between the e and b terminals. However, because the n layer is thin (e.g. 10^{-5} m), most of the holes, entering it from the e region, reach its far side without meeting any electrons. There they fall down the large potential drop into the collector region, thus constituting a large flow of current between the e and c terminals. This gives the *amplifying* action of the transistor. Because the central layer is so thin, which prevents a large current flowing through the base terminal, a relatively large voltage drop must be applied between e and b to pass a small I_1 through this part of the transistor circuit. This same large voltage drop then ensures that there is a large flow of holes from the e to c regions, so giving a large current I_2 through this part of the transistor circuit. The voltage drop from e to b varies with the current I_1; hence small changes

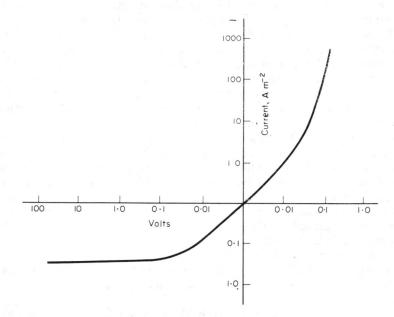

Fig. 24.17 Current across a *p–n* rectifier junction

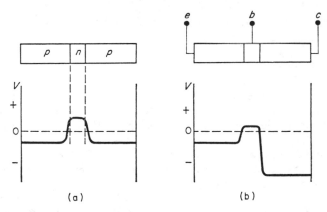

Fig. 24.18 The junction transistor showing (a) the variation of potential (b) the effect of applying a small negative voltage to *b* and a large one to *c*

in the current passed through the base electrode give large changes (e.g. × 100) in the current passed through the collector electrode.

Many other types of semi-conductor and transistor devices have been developed. Semi-conductor technology raises a number of interesting and challenging metallurgical problems. Junctions are made from single crystals, because the current-carrier paths must not be interrupted by grain boundaries. Zone refining and crystal growing (called crystal 'pulling' from a melt) are essential for producing crystals with high degrees of purity and perfection. Usually a crystal is grown of n-type or p-type material by 'doping' the melt with the necessary impurity (e.g. As or Sb for p-type; In, Ga or Al for n-type). Most crystals are made from germanium or silicon, although intermetallic compounds (e.g. indium antimonide and gallium arsenide), metallic oxides and other semi-conducting materials are being developed.

A p-n-p germanium transistor is typically made by taking a small thin slice of n-type crystal, to which the base connection is made. Small specks of indium or indium-gallium alloy are placed on the top and bottom faces of the disc and melted, so producing small p-type regions by diffusion into the germanium. These form collector and emitter spots on the crystal. Indium electrodes are fixed to these spots and electrical leads soldered to them.

There are innumerable variations in these techniques. Many transistors are made by forming the necessary regions and connections on one face of a substrate crystal. Electro-etching, electro-plating, evaporation, deposition, growth of thin deposited crystal layers ('epitaxial' growth) and masking to treat selective areas by deposition or acid etching, are some of the many special metallurgical processes developed for preparing semi-conductor devices. In *semi-conductor integrated circuits*, used in micro-miniaturized equipment, complete electrical circuits are prepared by vapour deposition, through masks, on to a single silicon slice, the transistor parts of the circuits then being formed by diffusion of some of the deposits into the underlying crystal.

24.6   Zone theory of alloys

We saw in § 14.3 and § 14.4 that many alloy phases, or phase boundaries, occur at certain values of electron concentration. Thus, when metals of higher valency are added to copper, silver or gold, similar phases usually appear at similar electron concentrations; the α (F.C.C.) phase boundary is reached at an electron concentration of 1·4, a β phase (often B.C.C.) appears at 1·5, and so on. H. Jones (1934, *Proc. R. Soc.*, **144**, 225; 1937, Proc. Phys. Soc., **49**, 250) proposed that these effects come from the Brillouin zone structure of the alloys.

We saw in § 19.1 that the density of states $N(E)$ increases as $E^{1/2}$ for free electrons (cf. eqn. 19.6). Consider the effect of a Brillouin zone on this free electron parabola, as shown in Fig. 24.19. Suppose we are adding electrons to an initially empty zone. At first, while only low energy states are being filled, the free electron relations are followed and the $N(E)$ curve (OA) follows the free electron parabola. As the expanding Fermi surface approaches the zone boundaries, the energy E rises more slowly with the wave number k (cf. Fig. 24.4) so that the density of states climbs high in this energy range. Later, as the corners of the zone are reached the density of states drops, eventually to zero as the zone becomes full. When Brillouin zones overlap, the total $N(E)$ curve is obtained by superposing and adding their $N(E)$ curves, as shown in Fig. 24.20.

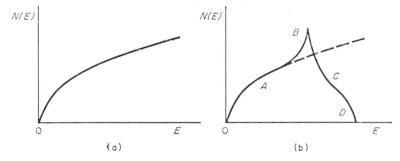

Fig. 24.19 Density of states (a) for free electrons and (b) as modified by the zone structure

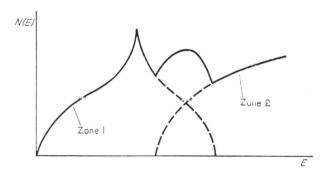

Fig. 24.20 The total $N(E)$ curve obtained from the sum of individual $N(E)$ curves of overlapping zones

Consider an alloy system with two possible phases, 1 and 2, with $N(E)$ curves as in Fig. 24.21. Suppose that the solvent metal is monovalent and that phase 1 is its primary solution. Its electron concentration is too low for the Fermi level to reach the point A in the diagram. As we increasingly add a second metal of higher valency, the electron concentration and Fermi level increase. At first, while the free electron parabola is still followed, both phases are equally favoured, so far as the Fermi energy of the electrons is concerned.

Beyond A, the $N(E)$ curve for phase 1 rises to a peak at B, as the zone boundary of this phase is reached, whereas that for phase 2 is still, at this stage, following the free electron parabola. In this range of electron concentration the Fermi energy of the electrons is lower in phase 1 than phase 2, which favours the thermodynamical stability of phase 1. Beyond B, however, the position reverses sharply because $N(E)$ drops steeply for phase 1 whereas that of phase 2 climbs higher. At alloy compositions that take the Fermi level to just beyond B, therefore, the Fermi energy rapidly becomes unfavourable to phase 1 and, other things being equal, we can expect to find the phase boundary of phase 1 in this composition range. Similarly, the best composition for phase 2, so far as Fermi energy is concerned, is that which puts the Fermi level at C. If we now identify phase 1 with the F.C.C. α-phase, based on Cu, Ag, or Au, as solvent, and phase 2 as the

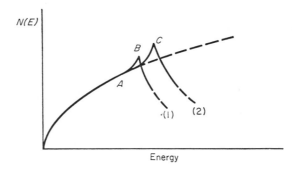

Fig. 24.21 $N(E)$ curves for alloy phases

corresponding β-phase, then point B should occur at an electron concentration just below 1·4 and point C should occur at 1·5.

Let us assume that the electrons are so nearly free that the Fermi surface is practically a sphere when it first touches the zone boundary. Since $E = h^2/2m\lambda^2$ (cf. § 3.4), the electron wavelength at the Fermi level is, from eqn. 19.4, given by

$$\lambda = 2\left(\frac{\pi V}{3N}\right)^{1/3} \qquad\qquad \textbf{24.13}$$

When the Fermi sphere just touches the zone boundary at the nearest points, this wavelength is the Bragg reflection wavelength $\lambda = 2d$ for waves normal to the reflecting planes responsible for this zone boundary. In F.C.C. crystals the most widely spaced planes that give Bragg reflections are the $\{111\}$, for which $d = a/\sqrt{3}$ where a is the lattice parameter. We also have $a = (4V/N_0)^{1/3}$ where N_0 is the number of atoms in a volume V of the crystal. Hence

$$\lambda = \frac{2}{\sqrt{3}}\left(\frac{4V}{N_0}\right)^{1/3} \qquad\qquad \textbf{24.14}$$

Eliminating λ between these two expressions gives the electron concentration N/N_0, beyond which the F.C.C. phase is expected to become unstable, as

$$\frac{N}{N_0} = \frac{\pi\sqrt{3}}{4} = 1\cdot36 \qquad\qquad 24.15$$

The most widely spaced planes that give Bragg reflections in B.C.C. crystals are the {110}. A repetition of the above calculation for this case then gives $N/N_0 = 1\cdot48$ as the ideal electron concentration for the β-phase.

Similar arguments can be developed for other electron compounds. The Jones theory can thus successfully account for the compositions of this class of alloy phases. A serious difficulty exists, however. Experimental methods have been developed to determine the positions of Fermi surfaces for some metals and these have shown that in, for example, pure copper *the Fermi surface already touches the zone boundary*. This would seem to destroy the whole basis of the argument, yet there is no doubt that large numbers of alloy phases do form where the Jones theory predicts they should. It seems likely that when solute is added to the pure metal the Fermi surface becomes *more spherical* and so pulls away from the zone boundary at low concentrations; and then expands, more or less spherically, with increasing electron concentration in the manner described above. Reasons have been given for expecting the Fermi surface to become more spherical in dilute alloys but the theory of this effect is still not very firm.

24.7   Magnetic properties

When an object is placed in a magnetic field, it becomes magnetized and a force is exerted on it. Its *intensity of magnetization J* is indicated by the strength of this force and is determined by the *magnetic field strength H* and the *magnetic susceptibility* κ of the material in it, i.e.

$$J = \kappa H \qquad\qquad 24.16$$

Diamagnetic materials (e.g. Cu, Ag, Au, Bi) have a small negative κ, i.e. are weakly repelled by the field. *Paramagnetic* materials (e.g. alkali and alkaline earth metals; transition metals except when ferromagnetic) have a small positive κ, i.e. are weakly attracted. *Ferromagnetic* materials (e.g. Fe, Co, Ni, Gd, below the Curie temperature) have a large positive κ and can also retain magnetization after the external field is removed, i.e. can be permanent magnets.

Experiment has shown that ferromagnetism is caused by the spin of the electron. Each electron behaves like a small bar magnet which can point its north pole either up or down the applied magnetic field, according to the direction of its spin. Very crudely, we might picture the electron as a small electrically charged sphere spinning about its axis, the rotation of the charge producing its own magnetic field centred about this axis. For a piece of material to be magnetized, more of its electrons must spin one way than the other, otherwise the magnetic effects of the opposed spins cancel out. In a paramagnetic material an applied field pulls some spins round to its own direction, so upsetting the balance of opposed spins and hence magnetizing the material. When the applied field is removed, the spins revert to a balanced distribution with no resultant magnetization. In a ferromagnetic material the alignment of parallel spins can continue permanently without an external field, so that in these there is some internal interaction which aligns the spins.

What is this interaction? Since the electrons behave like bar magnets there is of course a magnetic force between them. But calculation shows that this is far too small to align the spins. The *electrostatic force* between electrons is strong enough, however, and the way in which this can align the spins to give ferromagnetism was first realized by *Heisenberg*. We saw in § 4.3 that when two wave functions of the same sign overlap they reinforce each other and, because the electron density goes as the square of their sum, they represent a heaping up of the electron charge density in the overlapping region. If the electrons have opposite spins they can go into such a quantum state. There is then, however, a strong electrostatic repulsion between them because of their closeness. In the hydrogen molecule this repulsion is more than balanced by their electrostatic attraction to the two nuclei, so that this quantum state is energetically preferred. But, given a different atomic configuration, this conflict of the repulsive and attractive forces can go the other way. It does so within a single atom, so that the *p* states, for example, fill up in the order described in § 3.6 (Hund's rule). The sign and magnitude of this so-called *exchange interaction* depend, in fact, sensitively upon the different distributions of the electrons, according as these occupy states with parallel or opposed spins. If the exchange interaction energy is positive a state of lower energy is reached when the electrons have parallel spins and so occupy spatially different quantum states and keep away from one another. This condition is favoured, between different atoms, when the atomic radius is somewhat larger than the ionic radius and the atomic wave functions concerned are rather small near the nuclei. Electrons in *d* and *f* states are particularly suitable.

Even if the exchange interaction favours parallel spins, there is still the effect of the Pauli principle to be considered. If the spins are all parallel, only one electron can go into each quantum state, so that some electrons may be forced up into states of high kinetic energy. It follows that the electrons responsible for ferromagnetism must come from partly filled atomic shells, so that the energy bands formed from these shells have vacant quantum states available when the electrons align their spins; and also that the density of states in these bands must be high, so that the increase in kinetic energy, when spins align, is smaller than the exchange interaction energy.

Electrons in the atomic cores cannot give ferromagnetism because they are in filled shells. The outer valency electrons also cannot because they are in states similar to the bonding state of hydrogen and because the density of these states is low. The partly-filled *d* states of the transition metals are favourable, however, and these metals are either ferromagnetic or strongly paramagnetic. The atomic spacing is critical. If the atoms are too far apart, the exchange interaction is too weak to resist thermal agitation, which throws the spins out of alignment. If they are too close, the exchange interaction changes sign and the kinetic energy factor increases. Conditions are most favourable when the atomic radius is 1·5 to 2·0 times greater than the radius of the atomic d shell. In the iron group of metals we have the following ratios of atomic to shell radii:

	Ti	Cr	Mn	Fe	Co	Ni
	1·12	1·18	1·47	1·63	1·82	1·98

It is interesting to note that if the interatomic spacing of manganese is increased by alloying with interstitial nitrogen, or in a *Heusler alloy* (e.g. CuAlMn), a ferromagnetic compound is produced.

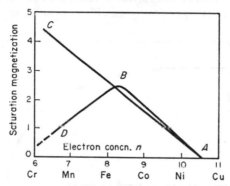

Fig. 24.22 The saturation magnetization per atom of ferromagnetic metals and alloys

The maximum, i.e. *saturation*, magnetization is obtained when as many electrons as possible spin in parallel in the d band. There are five quantum states per atom in the $3d$ band, each of which can hold two electrons of opposite spins. Picture, then, the band as split vertically down the energy scale into two half-bands, side-by-side, each holding all states with one spin direction. Let the total number of $3d$ and $4s$ electrons per atom be n, distributed so that there are x and $n - x$ per atom in the $4s$ and $3d$ bands respectively. Consider substances in which $(n - x) > 5$, so that saturation magnetization is obtained by completely filling one of these half-bands. The saturation magnetization, measured in electron units per atom, is then $5 - (n - x - 5)$, i.e. $10 + x - n$. Experiment shows that $x = 0·6$ in nickel and we can assume that this value also holds approximately for the elements immediately preceding it, i.e. in this part of the periodic table we deduce the saturation magnetization to be $10·6 - n$. Experimental values can be obtained by varying n through a series of ferromagnetic alloys. Fig. 24.22 shows the theoretical relation ABC and the experimental one

ABD. There is agreement from *A* to *B*, but when *n* falls below about 8·3 the saturation magnetization decreases steadily in practice.

Evidently it is impossible for one half-band to remain full when the other is more than half empty. As *n* falls below 8·3 the full band begins to empty and continues to do so until both half-bands have equal numbers of electrons at about $n = 5·6$. L. Pauling (1938, *Phys. Rev.*, **54**, 899; 1947, *J. Am. chem. Soc.*, **69**, 542; 1949, *Proc. R. Soc.*, **A196**, 343) has correlated this with the build up of *cohesion* in the first long period of the periodic table. As indicated by interatomic spacings and melting points, the cohesion builds up rapidly in the early transition metals and then remains roughly constant through the remaining ones, as far as nickel. This suggests that the *d* band can be divided horizontally, on the energy scale, into an upper and lower band. The lower band holds about 5·2 electrons per atom and gives cohesion, i.e. the exchange interaction favours states of the bonding type, containing electrons of opposite spins. The upper band holds about 4·8

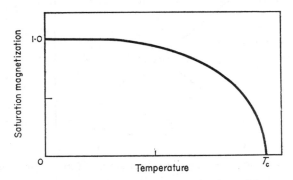

Fig. 24.23 Variation of saturation magnetization with temperature

electrons per atom and gives ferromagnetism instead of cohesion. Going from potassium towards copper the bonding states fill up first. Beyond chromium, the extra electrons then go into the ferromagnetic band and build up to a maximum magnetization between iron and cobalt (26 per cent Co in Fe) when one half of the ferromagnetic band is full and the other half is empty. This second half then fills up, reducing the saturation magnetization, which eventually reaches zero between nickel and copper (60 per cent Cu). In the second and third long periods it seems that the entire *d* band is involved in cohesion, since these metals are not ferromagnetic (apart from the rare earth gadolinium) and since the cohesion reaches a high peak when the band is about half full (cf. Mo and W).

We have of course so far been considering conditions at low temperatures, where the system seeks the state of lowest internal energy. Experiment shows that the saturation magnetization of a ferromagnetic substance decreases at higher temperatures, becoming zero at the *Curie point* T_c

(780, 1075, 365°C, respectively, for Fe, Co, Ni). Above this temperature the substance is paramagnetic.

Evidently, as the Curie point is approached, an increasing number of electrons set their spins in opposition to the main orientation. This raises the internal energy but also raises the entropy since, if M of the N electrons available in the substance for ferromagnetism reverse their spins, there is an additional entropy, $S = k \ln [N!/M!(N - M)!]$ relative to the fully magnetized state. Fig. 24.23 is very similar to Fig. 14.19 and in fact has the same thermodynamical basis. We have here another example of a co-operative effect. When the spins are all aligned a large increment of internal energy is required to reverse the spin of any one electron against the combined forces from all the others. As more and more reversals occur, this opposition is progressively weakened and the way is made easier for still more reversals, until the system eventually loses control of the situation completely.

The coefficient of thermal expansion of a ferromagnetic material often shows deviations in the neighbourhood of the Curie point. This is because the exchange interaction favours a certain interatomic spacing and so pulls the actual spacing slightly in this direction. On heating to the Curie point this effect is relaxed and the spacing reverts to the uninfluenced value. In iron, this produces a very low coefficient of expansion just below the Curie point. *Invar* (36 per cent Ni in Fe) has its Curie point at about 200°C and, in consequence, a very low coefficient of expansion at room temperature which makes it a valuable material for length standards. Higher nickel alloys are also designed with coefficients that match that of glass and so can be used for making metal-glass seals.

The slight change in lattice constants which results from magnetization produces the change in length known as *magnetostriction*. In small applied fields, iron expands in the direction of the field and nickel contracts. This effect is utilized in *magnetostrictive oscillators* for converting electrical vibrations into mechanical vibrations, e.g. in ships' echo-sounding devices.

24.8 Magnetic domains

If the electron spins are spontaneously aligned at temperatures below the Curie point, why is it that a bar of iron can be demagnetized at room temperature? The answer lies in the *domain structure* of a ferromagnetic material. The material is divided up into small domains, in each of which the spins are aligned. The direction of alignment varies from one domain to another, however, and the mutual cancellation of the opposite magnetic fields of oppositely aligned domains makes the material as a whole appear unmagnetized. When an external magnetic field is applied the magnetic field energy of domains aligned with the field is lowered and that of those aligned against it is raised. The favourably oriented domains then grow at the expense of the others by the movement of the boundaries (called *Bloch walls*) between them. These boundaries are not to be confused with grain boundaries and can exist within a single grain. At high field strengths the boundaries often move in sudden jerks, as they pull away from anchoring imperfections in the crystal, giving small sharp increases in the overall magnetization known as *Barkhausen jumps*.

It is important to realize that a crystal exists naturally in the form of a domain structure. Consider two bar magnets. If placed side by side, north pole against south, they attract. The magnetic field energy is lowest when they have this arrangement in which there are no *free poles*, i.e. each pole is attached to one of opposite sign. The same effect occurs in a single crystal, the magnetic energy of which is reduced when there is a domain structure in which each pole of every domain is joined to opposite poles of neighbouring domains. Fig. 24.24 shows a simple domain pattern of this type. Such patterns, called *Bitter patterns*, can be revealed by spreading colloidal magnetite on the surface, which acts like sub-microscopic 'iron filings'.

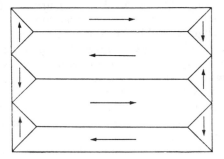

Fig. 24.24 A simple domain structure of the type found in silicon-iron

In 4 per cent silicon-iron the domains are typically about 10^{-4} m across. A lower limit to the domain size is set by the surface energy of the Bloch walls. This energy is positive and so favours a structure of coarse domains. Its origin is explained by reference to Fig. 24.25 which shows how the direction of electron spins varies across a '180° Bloch wall'. The exchange interaction energy does not reach its lowest value in the Bloch wall, because the electron spins there are not in parallel. They become more parallel as the boundary thickens, but opposing this tendency is the effect of *magnetic anisotropy*. Magnetization occurs more readily in some crystal directions than others. The spontaneous magnetization within the domains occurs along these *directions of easy magnetization*, e.g. ⟨100⟩ in iron and ⟨111⟩ in nickel. Across a domain boundary the magnetic axis is forced to rotate away from the direction of easy magnetization and the thicker the boundary the more atoms there are with unfavourable directions. This effect opposes the tendency of the exchange interaction to widen the boundary and a balance is struck at an equilibrium thickness which is about 300 atoms in iron. The surface energy of a 180° wall parallel to {100} in iron is 0·0018 J m⁻².

The existence of directions of easy magnetization is utilized in transformer steels, the laminae of which are prepared, by rolling and critical annealing, in a state of preferred crystallographic orientation so arranged that a direction of easy magnetization is aligned with the direction of the magnetic field. In this way magnetic 'softness' is obtained and hysteresis losses are reduced. Elaborate procedures have been developed to produce the required recrystallization textures in silicon-iron (cf. § 21.7).

The ease of magnetization also depends on the state of internal strain in the material and on the presence of impurities; these can anchor the domain boundaries to fixed positions in the crystal. For this reason it is common to

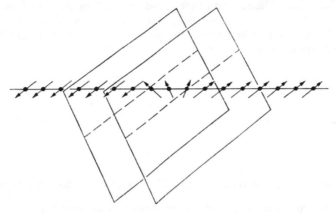

Fig. 24.25 Change in direction of magnetization across a 180° Bloch wall

remove non-metallic impurities from transformer steel by annealing in hydrogen. Internal strains affect the magnetic 'hardness' through magnetostriction; thus *permalloys* (e.g. 50 to 80 per cent Ni, balance mainly Fe) have been developed which have no magnetostriction and so have high magnetic permeability and low hysteresis losses at low magnetizing forces. These are widely used in electrical apparatus for communications, where the currents are usually small. For heavy power equipment, e.g. transformers, motors, generators, silicon steels (e.g. 4 per cent Si) are used. Here the main need is to minimize eddy-current losses. The silicon helps in this by raising the electrical resistivity of the material to about four times that of pure iron. It also improves the magnetic permeability by deoxidizing the metal and reducing carbon to the magnetically more innocuous graphitic form.

Magnetic coercivity is due mainly to the local magnetic fields which form on non-magnetic inclusions or holes in the metal. Free poles must form on the surfaces of these and magnetic field energy belongs to them. But these free poles can be eliminated and the magnetic energy reduced if the inclusions are contained in a Bloch wall, since a domain structure such as that in Fig. 24.26 can then develop round them. The Bloch walls thus become anchored to the inclusions and holes. They are then hard to move. It is difficult to magnetize the material but, once magnetized, the material is equally difficult to demagnetize and so is a permanent magnet.

High magnetic coercivity can also be obtained, not by anchoring the domain walls, but by using particles *too fine to contain domain walls*, i.e. of thickness smaller than the natural domain wall thickness of the material. Such particles are spontaneously magnetized and their direction of magnetization cannot easily be changed since the domain walls (which are to magnetism what glide dislocations are to yield strength) are absent. For example, the coercive force of elongated iron particles 150 Å diameter is about 10 000 times larger than that of bulk iron. A new class of permanent magnets has now been developed, based on synthetic microstructures prepared from aligned fine needle-shaped particles of iron embedded in a matrix of lead or organic plastic.

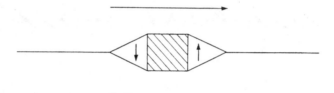

Fig. 24.26 Domains round a non-metallic inclusion (shaded) in a Bloch wall

The most common class of permanent magnet alloys is the *Alnico* series. These are complex alloys and contain some 6–12% Al, 14–25% Ni, 0–35% Co, 0–8% Ti and 0–6% Cu in 40–70% iron. They are hard, brittle materials and are shaped by either casting or sintering, followed by grinding. They are precipitation-hardening alloys and are heat-treated to produce rod-like precipitates some 300 Å diameter and 1000 Å long, lying along ⟨100⟩ axes in the B.C.C. matrix. It seems likely that these precipitates are single domains and that the coercivity is due to the same effect as in the artificial fine-particle magnets. An important feature of the alnicos is their response to *magnetic annealing*. The magnetic remanence and coercivity can be strikingly increased by cooling the alloy, from about 1300°C to room temperature, in a strong magnetic field, before ageing at about 550°C. This magnetic annealing causes all the precipitated rods to form along the ⟨100⟩ axis nearest to the direction of the field.

24.9 Superconductivity

Some metals become *superconductors* at low temperatures. When cooled below a critical temperature T_c their electrical resistance drops sharply to zero. One striking way to demonstrate this is to induce a current in a closed ring of a superconducting metal. Provided the temperature remains below T_c this current continues to circulate round the ring as a 'persistent current' without any external aid, more or less indefinitely. Such a current has run for 2·5 years without perceptible decay, which implies a resistivity of less than 10^{-23} ohm m.

Superconductivity has its origins in a small attraction between some of the conduction electrons. We usually think of two electrons as repelling one another electrostatically but their repulsive field is weakened at large separations by the screening effect of other, nearby, electrons (cf. § 4.5). One general source of an attraction between them has been identified. This is a second kind of screening due to the positive ions of the lattice. As an electron moves through the lattice it pulls the nearby ions towards itself, so creating a slightly higher density of positive charge in its neighbourhood. Under certain conditions, this positive charge due to the distortion of the ionic lattice more than compensates the electronically screened negative charge of the electron, so that another electron may then be attracted by the net positive charge.

The effect is very weak, which is why superconductivity is destroyed by thermal agitation at temperatures above the order of 1 K in most metals. The element with the highest T_c is technetium (11·2 K). More familiar metals with high T_c (K) values are Nb (9·46), Pb (7·18), V (5·3), Hg (4·15) and Sn (3·72). The alloy with the highest known T_c is Nb_3AlGe (20·75).

The importance of the lattice ions in superconductivity is shown by the *isotope effect*. Different isotopes of mercury (also tin) are found to have different critical temperatures, proportional to $M^{-1/2}$, where M is the atomic mass of the isotope concerned. Since the velocity of elastic waves and the frequency of atomic vibrations also vary as $M^{-1/2}$ (cf. eqn. 19.14), this is good evidence that superconductivity in such elements is caused by an interaction between electrons and lattice vibrations.

The theory of how the weak electron-electron attraction leads to superconductivity is very complicated. It turns out that the attraction is strongest between electrons, in pairs, with opposite momenta and spins. The electrons at the Fermi surface correlate their motions, relative to one another, so as to take maximum advantage of this effect. When in this organized state of motion their total energy (i.e. kinetic energy plus interaction energy which is negative) is reduced and there is a finite energy gap separating this organized state from any more excited state of motion. This gap can be visualized as a thin shell round the Fermi surface but it does not produce an insulator or semi-conductor because, when an electric field is applied to the material, the whole Fermi distribution together with its gap can drift to an unsymmetrical position, as discussed in § 24.3 (cf. Fig. 24.10), so causing a current to flow. The difference between this and ordinary conductivity is that now the electrons cannot be scattered off lattice irregularities. There are no quantum states at the same energy into which they can be scattered.

To change their state they would have to jump up the energy gap and a lattice irregularity such as a phonon cannot supply an electron with such an amount of energy. Once the Fermi distribution is displaced from the origin in k-space, it will therefore stay in this displaced position even when the applied field is removed, since the scattering which would eventually bring it back to the symmetrical position is now suppressed. The displaced distribution thus produces a persistent current.

Several practical uses for superconductivity have now been found. Small superconducting devices are used as switches (*cryotrons*) and memory storage elements in electronic computers. Superconducting wires are now wound

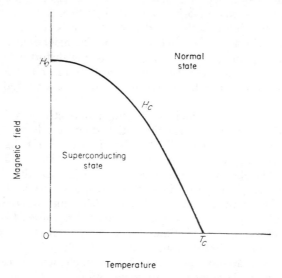

Fig. 24.27 Effect of a magnetic field on the superconducting transition

into solenoid coils to make high-field electromagnets and it is possible that these will find large-scale use in heavy-current electric engineering, even though they have to be cooled in liquid helium.

To understand such devices and the superconducting materials used in them, we must consider the relation between superconductivity and magnetism. Fig. 24.27 shows what happens when a superconductor is operated in a magnetic field. Above a critical field H_c the superconductivity is destroyed and the material then behaves 'normally'. The value of H_c rises parabolically, from zero at T_c to a maximum H_0 at 0 K. The critical fields of pure elements are not particularly high, e.g. $H_0 \simeq 10^{-2}$ to 10^{-1} tesla, whereas values of 1 tesla ($= 10^4$ gauss) and above are required for high-field magnets. The superconductivity is also destroyed, even in the absence of an external magnetic field, when the current flowing in the superconductor reaches the value at which its own associated magnetic field at the surface of the superconductor exceeds the critical value.

An important related effect is the *Meissner effect* which appears in *ideal* superconductors (e.g. soft, annealed pure metals). As shown in Fig. 24.28, if a specimen is placed in a magnetic field H ($< H_c$) while in the normal state and is then cooled down into the superconducting range, the magnetic lines of force are *pushed out* when it becomes superconducting. The superconducting state is a definite thermodynamical state with a free energy that is lower than that of the normal state at these low temperatures. Work has to be done in expelling the magnetic flux and it can be shown that this is equal to $H^2/8\pi$, per unit volume of superconductor. The critical

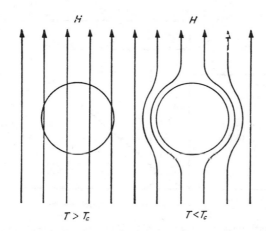

Fig. 24.28 The Meissner effect, i.e. expulsion of magnetic flux from a super-conductor

field H_c at any given temperature below T_c is then determined by the relation

$$G_n = G_s + \frac{H_c^2}{8\pi} \qquad\qquad 24.17$$

where G_n and G_s are the free energies of the normal and superconducting states, per unit volume, at this temperature.

When the flux expulsion is looked at more closely it is found that the external magnetic field H does not drop sharply to zero at the surface of the superconductor. It penetrates slightly into the superconductor, falling rapidly to zero through a thickness of order $\lambda (\simeq 10^{-7} \text{ m})$, called the *penetration depth*. A simple argument for this is that the supercurrent is carried by a finite number of electrons and therefore needs a finite, though small, thickness of material at the surface for its passage.

Because of this penetration depth, extremely thin fibres can remain super-conducting to much higher magnetic field strengths, since the flux expulsion, which adds the magnetic work to the free energy of the superconducting state and so makes the latter thermodynamically unstable relative to the normal state, is greatly reduced. A superconductor composed of fine filaments can thus continue to operate in a strong magnetic field, which opens the way towards strong superconductive magnets.

The question now arises, in connection with the Meissner effect, of why a thick superconductor in a magnetic field H does not divide up internally into alternating filaments or lamellae of superconducting and normal regions, so as to avoid having to do the work $H^2/8\pi$ involved in the expulsion of the magnetic flux. Admittedly, some flux would be expelled locally

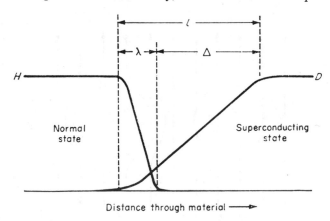

Fig. 24.29 Variation of the magnetic field H and the degree of superconductivity D through an interface between normal and superconducting regions

from the superconducting strands to become *trapped flux* in the neighbour-ing normal ones, but if the alternations were on a sufficiently fine scale the work associated with this could be very small.

The answer lies in a property of the material known as the *coherence length l*. It is perfectly possible and quite common for superconducting and normal regions to exist side by side in the same material but the interface between them cannot be infinitely thin because the superconducting state depends on the correlated motions of paired electrons and the spacing of the electrons in such a pair is typically of order 10^{-6} m; this spacing largely determines the coherence length.

As we go through an interface from a normal to a superconducting region, the magnetic field H and degree of superconductivity D vary as shown in Fig. 24.29. The first falls off rapidly, in a distance of order λ, and the second builds up more gradually, in a distance of order l. The *effective* thickness from which flux is expelled is thus increased by an amount of order

$\Delta \simeq l - \lambda$ and the additional magnetic work associated with this appears as a *positive surface energy* of the interface. It is this surface energy that prevents the material from dividing up into fine superconducting and normal regions.

Two possibilities for producing the desired filamentary structure still remain, however. First, the material may be made heterogeneous. This can be done by alloying for example, to produce a two-phase mixture of superconducting and normal alloy phases. Another method is by cold working. It is known that superconductivity is sensitive to mechanical strains and bundles of dislocation lines seem to provide just the required filamentary structure. In this way we can create *hard* superconductors that withstand strong magnetic fields without becoming normal.

The second possibility has to do with modifying the surface energy. If the electronic mean free path between scattering centres is reduced by alloying (cf. § 24.4), the effect is to increase the penetration depth and decrease the coherence length. By sufficient alloying it is possible in fact to make Δ *negative*. We then have, instead of a *type I* superconductor with a positive surface energy, a *type II* in which the surface energy is *negative*. Because this energy is negative the superconductor exists *naturally* in a state of finely divided superconducting and normal regions.

It seems probable that the excellent high-field superconductivity of alloys such as Nb_3Sn is due to a combination of this effect with the action of dislocation bundles in stabilizing the filamentary state by pinning the lines of trapped magnetic flux. The critical fields reached in the cold-worked state can rise to 30 times greater than those in the annealed state.

Superconducting solenoids made from Nb_3Sn wire have achieved fields of over 6 tesla. This alloy is extremely brittle, however, and greater progress has been made in preparing hard drawn wires of Nb_2Zr, which is more ductile. Superconducting magnets are now made commercially.

FURTHER READING

AMERICAN SOCIETY FOR METALS (1959). *Magnetic Properties of Metals and Alloys.*
BROCK, G. E. (1963). *Metallurgy of Advanced Electronic Materials.* Interscience, London.
HUME-ROTHERY, W. and COLES, B. R. (1969). *Atomic Theory for Students of Metallurgy.* Institute of Metals.
KITTEL, C. (1953). *Introduction to Solid State Physics.* John Wiley, New York and London.
LONDON, F. (1960). *Superfluids: Macroscopic Theory of Superconductivity.* Dover Publications, London.
LYNTON, E. A. (1962). *Superconductivity.* Methuen, London.
MOTT, N. F. and JONES, H. (1936). *The Theory of the Properties of Metals and Alloys.* Oxford University Press, London.
PEIERLS, R. E. (1955). *Quantum Theory of Solids.* Oxford University Press, London.
SCHROEDER, J. (1962). *Metallurgy of Semi-Conducting Materials.* Interscience, London.

SEITZ, F. (1940). *Modern Theory of Solids*. McGraw-Hill, New York and Maidenhead.

SHASHKOV, YU. M. (1961). *The Metallurgy of Semi-Conductors*. Pitman, London.

SHOCKLEY, W. (1950). *Electrons and Holes in Semi-Conductors*. Van Nostrand, New York and London.

SHOENBERG, D. (1965). *Superconductivity*. Cambridge University Press, London.

SMITH, R. A. (1959). *Semi-Conductors*. Cambridge University Press, London.

WYATT, O. H. and DEW-HUGHES, D. (1972). *Metals, Ceramics and Polymers*. Cambridge University Press.

ZIMAN, J. M. (1960). *Electrons and Phonons*. Oxford University Press, London.

— (1964). *The Theory of Solids*. Cambridge University Press, London.

Chapter 25

Properties and Uses

25.1 Strength and cost

The choice of a material for a given practical use depends on the *effectiveness* of the material in that use and on its *cost*.

Effectiveness depends on many properties, which is why commercial alloys are often complex materials; it is usually necessary to develop some properties to an outstanding degree while maintaining several others at an acceptable level. Even the simplest household hardwares require some solidity and mechanical strength, chemical stability and resistance to chemical attack and temperature changes, in addition to manufacturing properties such as ease of casting, forging, pressing or machining. Most materials for engineering use must possess certain properties to an outstanding degree; for example, tensile strength and ductility in steel girders; elastic resilience in springs; plastic toughness in motor car bumpers; or others such as lightness, corrosion resistance, heat resistance, high or low melting point, electrical conductivity, magnetic susceptibility, optical reflectivity or transparency, high or low neutron cross-section, etc.

Cost refers to the whole completed article in service. It thus includes not only the intrinsic cost of the raw material, which is determined by factors such as the abundance, accessibility and ease of extraction of the ore, and the degree of chemical refinement and alloying needed; it also includes manufacturing costs, which bring in additional properties of the material such as ability to be shaped by casting, working, welding and machining; and also operating costs, e.g. cost of repeated painting to prevent corrosion. The fabrication costs of engineering structures are often so high that the cost of the basic metal or alloy is small by comparison. In such cases a more expensive material may be used if it makes fabrication easier and cheaper.

The property which is required of materials used in really large amounts is *mechanical strength*. The common structural materials, steel, cement and timber, all have very high strength/cost ratios and are used in amounts some 20 to 50 times greater than the next most common materials, plastics and aluminium. Table 25.1 gives some typical average values of strength and cost. To derive a strength/cost ratio we divide the ultimate tensile stress of the material, in 10^6N m^{-2}, by the cost of unit volume in £ per m^3.

TABLE 25.1.   ILLUSTRATIVE AVERAGE VALUES OF STRENGTH AND COST OF STRUCTURAL MATERIALS

Material	Tensile strength 10^6 N m^{-2}	Density 10^3kg m^{-3}	Cost (weight) £ per 10^3kg	Cost (volume) £ per m^3	Strength/ cost (volume)
Low alloy steel bar	970	7·8	90	700	1·38
Mild steel sheet and bar	380	7·8	50	330	1·15
Reinforced concrete	14	2·4	8	19	0·74
Timber	7	0·5	33	16	0·44
Grey iron castings	190	7·3	90	660	0·29
Aluminium castings	210	2·7	290	780	0·27
Magnesium castings	190	1·7	400	680	0·28
Polyvinyl chloride	60	1·4	200	280	0·21
Brass strip	460	8·6	260	2240	0·20
Fibreglass	760	1·87	2000	3740	0·20
Zinc die castings	280	6·6	225	1485	0·19
Copper strip	280	8·9	320	2450	0·11
Polythene	14	0·9	200	180	0·08

(Based on W. O. Alexander, 1965, *Metallurgical Achievements*, Pergamon Press, Oxford, p. 179)

We notice that the density of a material is important in the strength/cost ratio. For example, in terms of cost per unit weight, aluminium and magnesium castings are amongst the most expensive of the materials listed but their low density brings down their cost per unit volume, which substantially increases the load which they can carry for a given expenditure.

25.2   Plain carbon steels

Carbon is the cheapest and one of the most effective alloying elements for hardening iron. Fig. 25.1 shows the general effect of carbon content on the strength and ductility of plain carbon steels in the normalized and quench-hardened conditions. In normalized (or hot-rolled or annealed) steel the carbon contributes to the hardness mainly through the formation of *pearlite*, up to 100 per cent at 0·8 per cent C. We see that 1 per cent C nearly quadruples the strength but drastically reduces the ductility. Quenching from the austenitic range produces *martensite* in the higher-carbon steels and this leads to a further increase of strength, up to about $1·7 \times 10^9$ N m^{-2}, in high-carbon steels.

Plain carbon steels are conveniently divided into three groups; *low carbon* (< 0·3% C), *medium carbon* (0·3 to 0·7% C) and *high carbon* (0·7 to 1·7% C).

The low-carbon steels combine fair strength with high ductility and excellent fabrication properties (rolling, drawing, welding), and are generally used in the hot-rolled or annealed or normalized condition. These are the structural steels, used in very large amounts (90% of all steel) for bridges, buildings, ships, vehicles and boilers, particularly in large pieces which cannot readily be heat-treated. Medium carbon steels are quenched and tempered to develop great toughness with good strength, the properties required of engineering components such as shafts, gears, connecting rods, rails, rail axles and drop forging dies. High carbon steels are quench hardened and lightly tempered to develop great hardness with a little toughness for springs, dies and cutting tools.

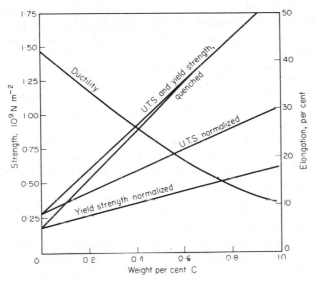

Fig. 25.1 Effect of carbon on the strength and ductility of plain carbon steel

The effects of carbon content in low carbon steels are important. *Dead mild steels*, which contain 0·05 to 0·15% C, have excellent ductility, which is essential for the deep drawing of sheet but their strength is low, although a hard strong surface can of course always be produced by case hardening. Their high ductility makes them difficult to machine unless their manganese sulphide content is increased to make them free-cutting.

As the carbon content is increased above 0·2 per cent the strength rises into the range required for structural steels used for beams, tubes, ship plates, etc. but the ductility falls below the ideal level for deep drawing. Problems of brittle fracture in thick, coarse-grained sections and of welding due to the formation of brittle, quench-hardened zones adjoining the welds also appear. Structural steel is used in such large amounts that if its strength could be increased by some 10 to 20%, without losing notch toughness and weldability and without increasing its cost, there would be immense economic benefits. Great efforts have thus been made to develop improved structural steels.

The most attractive approach has been to refine the ferrite grain size, since this both raises the yield strength (cf. Fig. 21.8) and lowers the ductile-brittle transition temperature (cf. § 21.11). Aluminium-killing is one method of achieving this and is sometimes used in steels for critical components such as nuclear reactor pressure vessels. Aluminium also fixes nitrogen and so greatly diminishes strain ageing and strain-age embrittlement. However, aluminium-killed steels are expensive because of ingot pipes and have poor creep properties. Silicon-killed steels are more commonly used; both for boilers because they have better creep strength and also for ship plates, because they can be produced in the semi-killed form with high ingot productivity, but their generally coarser grain size can lead to brittleness at low atmospheric temperatures, particularly in thick plates. The main improvements here have been to reduce the carbon content and to add small amounts of other alloy elements to refine the grain size and to raise the strength. Manganese is helpful and appears to reduce the ferrite grain size by lowering the transformation temperature. Steels containing about 0·15% C and 1·0 to 1·5% Mn have been developed, with good notch toughness at 0°C. A more recent tendency is to add about 0·01% or more of niobium to the steel and to reduce the carbon content still further. The formation of fine NbC particles restricts grain growth and also produces a useful precipitation hardening within the ferrite grains. Niobium is preferred to other strong carbide-forming elements, such as titanium, because it does not deoxidize and so allows a semi-killed steel to be made.

The toughness and strength of medium carbon steels are achieved by quenching, to form martensite, and then tempering at 350 to 550°C to produce dispersed *spheroidal carbide*. The hard brittle cementite then strengthens the metal but, existing as spheroids in a soft yielding matrix, it is not subjected to such high stresses as would cause it to fracture. *Ausforming* (cf. § 21.8) is an important new process for producing very high strengths with good ductility in medium carbon steels.

Great hardness can be easily produced in high carbon steels by quenching and lightly tempering, e.g. at 250°C. Such steels are thus useful for small cheap cutting tools. Their main limitations are poor hardenability, which restricts their use to thin sections and also necessitates water quenching, with dangers of distortion and quench cracking; and their rapid softening at temperatures above about 350°C, which limits their use for high-speed metal cutting. These difficulties are overcome by alloying.

25.3 Alloy steels

There are numerous ways in which the properties of steel can be usefully developed by alloying. Leaving aside special purpose alloys such as the magnet steels and invars, the development of alloy steels rests on six principal features, (1) hardenability, (2) formation of carbides, (3) solubility, (4) stabilization of austenite or ferrite, (5) corrosion resistance, and (6) precipitation hardening.

To get full hardenability during a quench the metal must pass the 550°C range without transforming. Nucleation at these temperatures occurs at austenitic grain boundaries, so that a coarse grain size increases the hardenability, but this leads to brittle coarse-grained quenched structures. The alternative is to shift the nose of the *TTT* curve to the right by adding suitable alloy elements, such as Mo, Mn, Cr, W, Si and Ni. (When in solution Ti and V are very effective, but usually they exist instead as carbides in steel.) Numerous medium alloy steels have been developed, containing a total of up to about 5% of such elements, together with 0·3 to 0·5% C, and are used in the oil quenched and tempered condition for engineering components such as shafts, gears, connecting rods, brackets and bolts. Manganese is a cheap addition for the simpler types of alloy steel. Chromium gives deep hardening and is often added together with nickel, which goes into solid solution, lowers the M_S temperature, and improves the toughness of the steel. Many important *nickel-chromium* steels contain 0·5 to 5% Ni with up to about one-half this amount of Cr. Molybdenum is also added in small amounts to prevent temper brittleness and in larger amounts to increase hardenability and resistance to softening on tempering. At higher alloy contents, such steels become *air-hardening*, i.e. quench harden by air cooling.

Alloy elements differ greatly in their tendency to form alloy carbides in steel. The main carbide-forming elements are Ti, Nb, V, Mo, W and Cr. This is not simply related to their thermodynamic affinities for carbon, because affinity for iron provides a counter-tendency. Thus, Al has a fair affinity for carbon but is not a carbide former (in fact, it promotes the decomposition of cementite to graphite) because it has a high affinity for iron. In alloy steels which contain relatively large amounts of carbide-forming elements it is practically impossible to retain large amounts of carbon in solution in austenite, except at extremely high temperatures, and so a very hard martensite cannot be made. The extremely strong carbide-forming elements, Ti and Nb, extract carbon, not only from cementite, but also from other carbides and force their alloy elements, e.g. Cr, Mo, W, into solution in the iron. Only small amounts of Ti and Nb are usually added.

The carbide-forming elements promote finer grain size, and hence improve the toughness, by providing some undissolved carbide particles at the quenching temperature. The hard carbide particles give the metal a good cutting edge, which is important in tool steels, and their stability allows the metal to be used at dull red heat without softening unduly. The most common type of high speed tool steel contains 18% W, 4% Cr, 1% V and 0·7% C. It is heated to 1150–1300°C, to bring sufficient carbon and alloy elements into solution to give the required martensitic hardness, then oil or air cooled to above room temperature (to avoid quench cracking) and immediately tempered at about 550°C to toughen and harden the metal by precipitating fine complex carbides (*secondary hardening*). The wear resistance provided by a large content of hard carbides has other uses; for example, ball bearings are made from a steel containing 1·4% Cr and 1% C.

The extent to which alloy elements dissolve in iron is broadly governed by the size factor rule. Solid solution hardening has some practical uses. It has enabled high-strength, low-alloy, structural steels to be developed which are used in the hot-rolled condition and which are weldable, the carbon content being reduced to about 0·1%. Small amounts of various elements, e.g. Cr, Ni, P, Si, Mn, Mo and Cu, are added. Phosphorus is

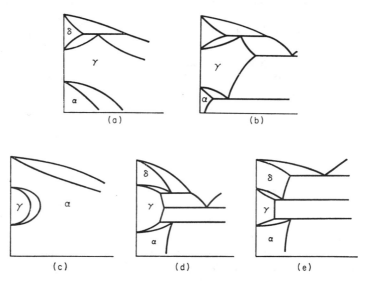

Fig. 25.2 Types of iron alloy phase diagrams: (a) open γ field; (b) expanded γ field; (c) closed γ field; (d) and (e) contracted γ field

particularly effective in strengthening ferrite by solid solution hardening, but the amount that can be added is limited by its embrittling effect. Copper usefully increases the corrosion resistance. An entirely different example of solid solution strengthening is provided by some creep-resistant and tool steels which are strengthened by large amounts of dissolved cobalt.

Alloy elements in solid solution considerably change the relative free energies of the B.C.C. and F.C.C. phases, producing various types of phase diagram of which those in Fig. 25.2 are typical. The substitutional alloy elements that produce open or expanded γ fields lie near the end of each transition series in the periodic table. These include the 'full' metals (cf. § 19.1) whose atoms behave rather like hard spheres that tend to favour close-packed crystal structures. The elements that produce contracted or closed γ

fields are mainly the stable B.C.C. transition metals. Interstitial carbon and nitrogen expand the γ field, probably because there are larger interstices for them in F.C.C. iron than B.C.C. The classification of alloys according to their effect on the phase diagram is as follows:

Open γ field: Mn, Co, Ni, Ru, Rh, Pd, Os, Ir, Pt
Expanded γ field: C, N, Zn, Au
Closed γ field: Be, Al, Si, Ti, V, Cr, Ge, As, Mo, Sn, Sb, W
Contracted γ field: B, Nb, Ta, Re

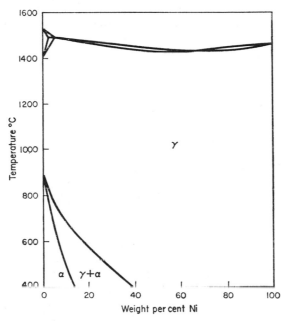

Fig. 25.3 Iron-nickel phase diagram

When carbide-forming elements such as Ti or Cr are added to carbon steels, the single-phase γ field shrinks drastically because the carbon can lower its free energy substantially by going out of solution from the austenite to form an alloy carbide. A wide γ + *carbide* two-phase field is then produced. Many tool steels have this type of phase diagram.

The opening up of the γ field makes possible the development of austenitic steels. Fig. 25.3 shows the phase diagram of the iron-nickel system. It is unnecessary in fact to add enough nickel to reach the γ phase boundary at room temperature, because a small addition can depress the γ/α transformation temperature range enough to prevent the transformation from occurring and so make austenite *metastable* at room temperature. In an 18% Cr

8% Ni stainless steel the phase boundary is reached at 650°C. The diffusion of alloy elements, which is necessary to produce equilibrium ferrite at this temperature, is slow and does not occur significantly during air cooling from this temperature. The austenite could transform to ferrite without change of composition at 340°C, but nickel depresses the M_S temperature and the transformation is prevented from occurring by nucleation difficulties.

Austenitic types of steel can be made more cheaply by substituting Mn (and in some cases N) for part of the Ni, although a complete replacement leads to hot working difficulties. *Hadfield's steel*, which contains 12% Mn and 1% C, is remarkable for its toughness, which is a result of the metastable condition of its austenite. This steel is water quenched from 1000°C to make it austenitic at room temperature. It is then only moderately hard but, being F.C.C., is very resistant to fracture. Any attempt to cut or abrade its surface leads immediately to intense local hardening round the region of contact, due to transformation of the austenite by strain. It is thus immensely tough and resistant to wear, properties which are exploited in rock-crushing equipment and railway cross-overs.

Several alloy additions can make steel resistant to oxidation and corrosion, Al and Si for example, but the outstanding one is of course chromium. The amount needed to give resistance to scaling at high temperatures increases with temperature, but generally must exceed about 12% Cr for protection in oxidizing environments. Chromium produces a closed γ field and single phase austenite cannot be produced in iron-chromium alloys which contain more than 12% Cr. This is not a problem in *stainless irons*, which are virtually carbon-free irons that contain about 13% Cr and are used in the ferritic condition for furnace components. In stainless steel cutlery, however, martensite is needed to give a sharp hard cutting edge. Fortunately, the addition of carbon expands the γ loop, for example out to 18% Cr at 0·6% C. Medium carbon steels containing sufficient chromium for good stainless qualities (e.g. 15% Cr) can thus be quench hardened for cutlery use. In austenitic 18% Cr 8% Ni stainless steel, the addition of the nickel further improves the corrosion resistance, as well as converting the alloy to the metastable F.C.C. structure and so giving it ductility, toughness and excellent cold working properties, which have led to its widespread use for kitchenware, surgical instruments and in chemical plant.

Requirements mainly for high speed aircraft have led to the development of *high-strength transformable stainless steels*, for gas turbine components, engine casings and parts of air frames. These steels have good high-temperature strength, are stainless and weldable. There are two main types: *fully transformable steels* and *controlled transformation steels*.

The first type undergo 100% transformation from γ to α on cooling from the austenitic range to room temperature. This defines their chromium content at 12%; this must be as high as possible for corrosion and oxidation resistance but cannot exceed 12% if α is to be avoided in the initial γ

structure. The carbon content is kept to 0·10 to 0·15 for weldability and the steel is tempered at 650°C to produce secondary hardening by precipitation of alloy carbides. Small amounts of Mo and V are added to ensure that the precipitates are stable carbides such as Mo_2C which do not overage rapidly. However, Mo and V are ferrite-forming elements and prevent the attainment of a wholly austenitic structure at high temperatures. This effect is countered by adding a small amount of nickel or manganese. Thus a typical composition is 0·1% C, 12% Cr, 1·5% Mo, 0·3% V, 2% Ni; in the air-quenched and tempered condition it has a yield stress of about 8 $\times$ 10^8 N m^{-2}, a tensile strength of about 10^9 N m^{-2}, with about 20% elongation and a Charpy V-notch impact energy greater than 50 ft lb at room temperature.

Controlled transformation steels are austenitic at room temperature, which allows them to be fabricated by cold working. They are then transformed by cooling to about $-80°C$ and finally tempered at 700°C to precipitate carbides and complete the transformation. A small amount of α in the initial γ structure is more tolerable in these steels and so the chromium content is raised to about 16% to improve the corrosion and oxidation resistance, the carbon being held at about 0·06% C for weldability and ductility. A delicate balance of other alloy elements is necessary to produce carbide precipitates resistant to tempering without unduly increasing the α in the initial $\gamma + \alpha$ structure and without depressing the $\gamma \rightarrow \alpha$ transformation temperature below the range attainable by fairly cheap refrigeration methods. A typical composition is 0·06% C, 16·5% Cr, 5% Ni, 2% Mn, 1·5% Mo, 0·9% Al, 2% Co, with a strength in the range 1·2 to 1·5 $\times$ 10^9 N m^{-2} and 20–25% elongation.

These steels provide fairly simple examples of how modern sophisticated alloys are developed. Their complexity of composition is in no sense accidental but is a logical and inevitable consequence of the close interplay of the basic factors, such as corrosion resistance, γ/α equilibrium, transformation temperatures and carbide stability, that govern the properties of the metal. With so many inter-related variables as these, a large number of alloy elements is needed to achieve the degrees of freedom necessary to bring them all under fairly independent control.

As we have seen, *precipitation hardening* in steels is mainly a matter of precipitating carbides. Other precipitates are, however, sometimes useful; e.g. nitrides in aged Bessemer steel for rail track; also nitrided case-hardened steel. A more recent development is the use of intermetallic compounds for precipitation hardening in austenitic steels. The addition of Ti and Al, for example, to the basic Fe-Cr-Ni alloy allows compounds such as Ni_3Ti, NiAl, Ni_3(Al, Ti) and Ni(Al, Ti) to precipitate at temperatures in the range 700–750°C, giving strengths in the range 0·7 to 1·1 $\times$ 10^9 N m^{-2} with good ductility and toughness. Another example is provided by the *maraging* steels, used mainly in rockets and high-speed aircraft. Basically, these are iron-nickel alloys containing very little carbon but with enough nickel

($\simeq 20$ to 25%) to form martensite on air cooling to room temperature. Alloys are added to produce hardening by precipitation of Mo-rich or Ti-Ni intermetallic compounds, which raises the yield strength into the range $1\cdot4$ to $1\cdot8 \times 10^9$ N m^{-2} with good fracture toughness. Subject to certain precautions these alloys are also weldable. Their advantage here is that they remain soft when welded, and so can be straightened out afterwards.

25.4 Cast iron

On a strength/cost basis cast iron is second only to structural steel amongst the metals and alloys and it has the great advantages of excellent casting and, in the grey form, machining properties, which allow it to be fashioned into intricate shapes. Steel forgings and stampings are about twice as expensive as plain sheet and bar, so that in strength/cost value they are not markedly superior to cast iron; and of course they cannot easily be produced in shapes as complex as those possible in iron sand castings.

Many types of cast iron can be made, according to the pig iron used, the melting conditions and alloying and heat-treatment procedures, but they all contain $2\cdot4$ to $4\cdot0\%$ C. In *white* cast irons, which are usually made by limiting the content of graphite-forming elements such as silicon to low levels (e.g. below $1\cdot3\%$ Si), all the carbon exists as cementite and the name white refers to the bright fracture produced by this brittle constituent. In *grey* cast irons, which generally contain 2 to 5% Si, most of the carbon exists as flakes of graphite embedded in a ferrite-pearlite matrix and these flakes give the fracture a dull grey appearance. An intermediate type is *mottled* iron which contains both graphite and eutectic cementite.

In iron, carbon is thermodynamically more stable as graphite than as cementite. Fig. 25.4 compares the metastable iron-cementite and the stable iron-graphite phase diagrams. The greater stability of the graphite phase is evident from the fact that the carbon-rich boundary of the γ phase is shifted to the left when this phase comes into equilibrium with graphite instead of cementite. The fact that cementite commonly forms in iron and steel is due to the sluggishness of the reaction to graphite, when carbon precipitates out from solution in iron. At the low carbon contents of typical steels graphite is rarely formed, except sometimes locally in grain boundaries, causing embrittlement, when aluminium-killed steels are held for long periods at elevated temperatures.

At the higher carbon contents typical of cast irons the carbon may separate as either graphite or cementite when the casting cools through the eutectic range and solidifies. Cementite is favoured by rapid cooling (*chilling*) and by the presence of carbide-forming alloys such as chromium. In practice, white cast irons are usually made simply by keeping the silicon content low. When the silicon is raised to about 3% a grey cast iron is formed, even when rapidly cooled. Nickel, another graphite-forming element, is sometimes also added to grey cast irons.

Grey iron is used in the as-cast form for innumerable purposes, e.g.

engine cylinder blocks, drain pipes, domestic cookers and lamp posts. The graphite flakes, which act mechanically like small cracks in the structure, give the metal good machinability, since the chips break off easily at the flakes, and a high damping capacity, which is important for absorbing machine vibrations, but they reduce the strength and ductility of the metal. White iron is too hard to be machined and must be ground to shape. This and its brittleness limit the direct use of this type of iron although the hardness of cementite is important for resistance to wear and abrasion. White iron is thus sometimes used for brake shoes and tips of plowshares.

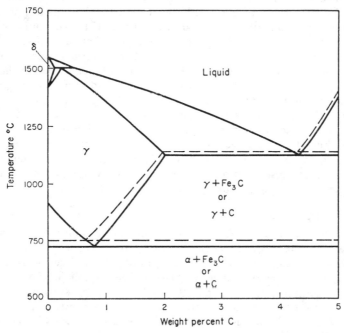

Fig. 25.4 The iron-cementite (metastable; full lines) and the iron-graphite (stable; broken lines) phase diagrams

Chilled iron castings also have some uses. These are basically grey iron sand castings, certain parts of which are transformed to white iron at the surface by rapid cooling brought about by metal chills fitted in the mould.

The main use for white iron however is as the starting material for *malleable cast irons*, in which the cementite is eliminated by heat-treatment, after casting. These contain about 0·6 to 1·0% Si, to promote the decomposition of the cementite during the heat-treatment but not enough to produce graphite flakes during casting. *Whiteheart* malleable iron is made by heating the casting in an oxidizing environment (either an iron oxide packing or a controlled atmosphere) to decarburize the surface layers.

Blackheart malleable iron is made by heating the casting in a neutral environment (iron silicate scale or slag) for many hours at about 900°C to decompose the cementite to rosette-shaped graphite nodules, which do not have the deleterious crack-like effect of graphite flakes. Blackheart iron is an important engineering material since it combines the casting and machining properties of grey iron, which enable intricate shapes to be made easily, with mechanical properties akin to those of structural steel, e.g. tensile strength 3·5 to 4·5 $\times$ 10^8 N m^{-2} with 5–20% elongation. It is widely used in agricultural and engineering machinery.

The structure of graphite nodules is sensitive to the composition of cast iron and several types of improved irons have been developed in which the structure is controlled by various alloy additions. Under certain conditions *spherulitic nodules* are formed. These are roughly spherical in shape and each is built up from a number of graphite crystals which grow radially outwards from a common centre with their basal planes perpendicular to the radial axis of growth. This structure is promoted by additions of magnesium, cerium or calcium, but opposed by sulphur in the form of MnS. In *nodular cast irons* the structure is developed directly, in an as-cast grey iron, by adding a small amount of magnesium or cerium alloy to the metal in the ladle. Other alloys such as silicon and nickel are also added and tensile strengths of order 6×10^8 N m^{-2} with 10% elongation are obtained. The combination of these good mechanical properties with excellent casting and machining properties is bringing these irons into wide use.

25.5 Strength and weight

Wherever gravitational or inertial forces play a dominating part in an engineering system, the *strength/weight* ratio of its constructional materials is more important than their strength alone. It has been estimated, for example, that it is worth paying £100 to reduce the weight of a commercial supersonic aircraft by 1 kg; and £40 to reduce that of a subsonic aircraft by the same amount. It is the operating cost factor that opens the way to the use of expensive lightweight structural materials that would otherwise be excluded by the simple strength/cost criterion of § 25.1. Aircraft and rockets are of course the obvious examples of this but we can also see the importance of strength/weight ratio in such things as, for example, bridges, motor cars, rail coaches, buildings and overhead cables. The historical development of bridges from early stone construction, with their massive arches of short span, to suspension bridges of slender structure with enormous span, provides a clear example of the influence of materials of higher strength/ weight ratio on engineering design.

The heavy non-ferrous metals and alloys are never used for strength alone, since steel is stronger and cheaper, but the low density of the light metals aluminium and magnesium makes them competitive with steel in applications where the strength/weight ratio is important. For example, a given weight of a high-strength aluminium alloy is far stronger than the same weight of structural carbon steel. Table 25.2 gives examples of the *specific strengths* of various materials, i.e. tensile strengths divided by specific gravities.

TABLE 25.2. SPECIFIC STRENGTHS AND YOUNG'S MODULI

Material	Specific gravity	Tensile strength, 10^6 N m^{-2}		Young's modulus, 10^9 N m^{-2}	
		Actual	Specific	Actual	Specific
(bulk)					
Aircraft structural steel	7·8	1 310	170	207	26·5
Mild steel	7·8	380	49	207	26·5
Duralumin	2·8	550	200	72·4	25·9
Mg alloy	1·8	345	190	44·8	24·9
Ti alloy	4·5	1 240	280	120	26·5
Beryllium	1·8	310	170	300	170
Fibreglass	1·87	760	410	41	22·1
Methyl methacrylate	1·2	48	40	2·8	2·3
Timber	0·5	7	14	13·1	26·2
(Fibres)					
Carbon whisker	2·2	24 000	11 000	965	440
Alumina whisker	4·0	15 000	3 800	345	86
Asbestos fibre	2·4	1 500	620	183	76
Flax	1·5	700	470	96	64
Nylon	1·07	500	470	4·8	4·5
Hard-drawn steel wire	7·8	4150	530	207	26·5

In light slender constructions *elastic stiffness* is of great importance. The failure of a compressed column by elastic buckling is minimized by expanding the cross-section in the form of an I-beam, a hollow tube or honeycombe but the walls of this cannot be made too thin. Aluminium, despite its low Young's modulus, has an advantage here since its favourable strength/weight ratio allows thicker walls to be used. The same principle applies to stressed aircraft skins, which may wrinkle under compressive stresses if too thin. Thus an aluminium alloy skin $1·5 \times 10^{-3}$ m thick is much more resistant to wrinkling than a steel skin 5×10^{-4} m thick with the same weight and tensile strength.

By exploiting the methods discussed in Chapter 21 it is possible to increase the strengths of materials up towards their ideal values. In practice, however, difficulties due to excessive elastic deformation must generally limit the maximum working stress to about $E/100$ (E = Young's modulus). The limiting factor in well-designed structures with well-designed strong light materials is thus the *specific Young's modulus*, i.e. E/ρ or E divided by specific gravity, values of which are given in Table 25.2. The similarity between steel, aluminium, magnesium, titanium and wood in respect of this property partly accounts for the close competition between these materials for aircraft structures. For a decisive advance in this aspect of aircraft development, materials of much higher specific modulus must be brought into use. Beryllium is an obvious possibility but is held back at present by its expensiveness and brittleness. Materials reinforced with strong, light, stiff fibres offer another line of advance.

Amongst existing lightweight materials, the aluminium alloys are of course the outstanding examples and we shall discuss them in § 25.6. The other main ones are based on magnesium and titanium.

Magnesium is a soft metal but its strength can be raised to the order of 3.5×10^8 N m^{-2} by alloying and other metallurgical treatments. As mentioned in §14.3, the electrochemical and size factors limit the capabilities of magnesium for forming solid solution alloys. However, magnesium dissolves aluminium at high temperatures sufficiently to give Mg-Al alloys good age-hardening properties. Typical casting alloys of this type contain about 9% Al together with about 0.5% Zn and 0.25% Mn to improve corrosion resistance. Small additions of zirconium greatly improve the strength and ductility of magnesium by refining the grain size; and additions of cerium and thorium improve the creep resistance at elevated temperatures. Several such alloys have been developed (e.g. casting alloys containing about 0.7% Zr, 3.5% Zn and 2.5% Ce; and wrought alloys containing 0.7% Zr and 3% Zn) and used in aircraft, but their susceptibility to corrosion, particularly in marine environments, has restricted their use. The great chemical reactivity of magnesium is of course exploited in the use of sacrificial magnesium anodes to protect steel from corrosion and in magnesium ribbon and powder for pyrotechnics. There is no great fire risk in the general use of magnesium, however, so long as the metal is not finely divided, although about 1% Al and 0.015% Be are added to magnesium for nuclear fuel element cans to ensure that the metal maintains a fully protective oxide film even when molten. (Magnesium has a low neutron absorption cross-section and is non-reactive with uranium.)

Titanium combines the high strength of a transition metal with a density nearer to that of a light metal. Because of its excellent strength/weight ratio it is becoming increasingly used in aircraft engines and structures. It is also extremely resistant to oxidation (except at high temperatures) and marine corrosion, and so is used in chemical plant and marine components. However, its high cost, low Young's modulus and indifferent creep strength have restricted its applications. The metallurgy of titanium alloys is dominated by the phase change which occurs in the metal from α-Ti (C.P. Hex), below 885°C, to β-Ti (B.C.C.) above. The temperature of the α/β phase boundary is raised by additions of aluminium and tin, and useful alloys are made, in which the α phase is stabilized and hardened by dissolved solute, by adding about 5% Al and 2.5% Sn. Transition metals are more soluble in β than α and so lower the α/β temperature, generally producing a eutectoid transformation similar to that produced by carbon in steel, although the 'martensite' formed by quenching titanium alloys through this eutectoid lacks the hardness of carbon steel martensite, because the alloy solute in this case is dissolved substitutionally, not interstitially. The most useful β-stabilized alloys are those which contain some α phase, which may either be present when the alloy is hot-worked (in the $\alpha + \beta$ phase field) or may be formed on ageing, after the alloy has been worked at high temperature and then cooled down while mainly in the β structure. A typical alloy of this type contains about 3% Cu, 1·5% Fe, with a tensile strength of 10^9 N m^{-2} and 15% elongation. Much stronger alloys are possible, based mainly on the β structure; one example contains 16% V and 2·5% Al, with a tensile strength of $1·9 \times 10^9$ N m^{-2}. Traces of interstitially dissolved nitrogen, oxygen, carbon and hydrogen greatly affect the mechanical properties of titanium and must be carefully controlled. Nitrogen and oxygen greatly harden the metal but also embrittle it when allowed to increase beyond a few hundred parts per million. Carbon in excess of 0·3% forms titanium carbide. Hydrogen embrittles the metal by forming brittle titanium hydride precipitates when the metal is cooled after heat treatment.

25.6   Aluminium and its alloys

In contrast with the 'traditional' metals such as iron and copper, the tonnage use of which has remained fairly steady in recent years, aluminium has increased steadily in amount used by about 7 per cent a year for the past 25 years and is now second only to steel in amount produced. Amongst the common materials, only the plastics have a faster rate of growth of production and use than aluminium.

The reasons for the success of aluminium are clear. It has excellent corrosion resistance, which allows it to maintain an untarnished and attractive natural surface in the atmosphere with little or no treatment. It is light, moderately cheap and has excellent fabricating properties, e.g. it can be rolled down to thin foil, and has high thermal conductivity. These properties commend its use in every household. For a given weight it conducts electricity better than any other metal and so is used in overhead electrical cables. Its strong alloys have a higher strength/weight ratio than steel and this, together with the stiffness that comes from the thickness of section that is allowable because of its lightness, has made it a major structural material for aircraft, transport, and other engineering uses.

Aluminium has successfully competed with copper for overhead cables and its alloys are competitive with copper alloys (e.g. brasses) for general engineering use. Its competition with steel is intense. The successful replacement of tin plate by aluminium sheet for beer cans has caused the steel industry to introduce thinner cans in an effort to hold on to this traditional market. Aluminium competed successfully with stainless steel for curtain walling on skyscrapers, but not with structural steel and reinforced concrete for bridges (except portable bridges). The immunity to brittle cleavage fracture at low temperatures, which comes from its F.C.C. crystal structure, has enabled aluminium (as Al-Mg alloy) to be used for large tanks for storing liquid methane at $-165°C$. The competition between aluminium, steel, plastics, wood, glass and ceramics for domestic uses is self-evident.

Pure aluminium is widely used for cooking utensils and electrical cables because of its high conductivity. It is also used for wrapping foil, for mirrors and reflectors and for cladding on aluminium alloys to protect them from corrosion. Its good corrosion resistance can be further improved by electrolytic *anodizing* to build up a thick $(2·5 \times 10^{-5}\,m)$ dense oxide film which can be coloured with dyes if desired.

The main alloying elements used in aluminium alloys are silicon, magnesium, copper and zinc, although more limited use is made of many other elements. The simple eutectic which aluminium forms with silicon at 12% Si is the basis of important casting alloys that contain 11–13% Si. These have excellent casting fluidity and so can be sand and die cast into intricate shapes for engine cylinder blocks, gearbox cases, crankcases, etc. When 'modified' by the addition of about 0·05% sodium to refine their structure they have good strength, toughness and ductility.

A disadvantage of some alloy additions such as copper (as also impurity elements such as iron) is that weak spots are then formed in the oxide film, at the sites of second phase particles, and the corrosion resistance is then greatly reduced. Fortunately, silicon additions do not have this effect and the Al-Si casting alloys have good corrosion resistance. Another alloy element which maintains good corrosion resistance is magnesium. This enters into solid solution in the aluminium and additions of 7% form the basis of useful wrought alloys which are very ductile and can be raised to a fair level of strength (e.g. $3 \times 10^8\,N\,m^{-2}$ psi with 18% elongation) by cold working. Their good corrosion resistance, particularly in marine environments, and welding properties make these Al-Mg alloys useful in shipping, aircraft and storage tanks.

For higher strengths, heat-treatable alloys are used. We discussed the principles underlying these precipitation hardening alloys in Chapter 20. As regards binary alloy additions, only copper is really effective. Silicon, magnesium and zinc, added singly, give poor age hardening properties because stable, fine-scale, intermediate precipitates are not formed. Satisfactory properties are however obtained from ternary and more complex alloys, based on Mg_2Si in the Al—Mg—Si system and $MgZn_2$ in Al—Zn—Mg. The 4% copper addition forms the basis of the series of duralumin alloys, a modern type of which typically contains 3·5 to 4·8% Cu 0·85% Mg, 0·9% Si, 1·0% Fe, 1·2% Mn and 0·2% Zn and in the fully heat-treated condition has a tensile strength of about $4 \times 10^8\,N\,m^{-2}$ psi with 8% elongation. For strength at elevated temperatures, as needed for example in aluminium pistons, nickel is also added. A typical alloy of this type contains some 4% Cu, 2% Ni and 1·5% Mg.

The reduced corrosion resistance due to the copper content is a disadvantage of the duralumin alloys. It can be overcome by cladding or spraying with pure aluminium. Alternatively an alloy hardened with Mg_2Si can be used, e.g. 1·5% Mg, 1·3% Si, 0·6% Fe, 0·7% Mn, 0·1% Cu, 0·5% Cr. This has a lower strength in the fully heat-treated condition, e.g. $3·1 \times 10^8$ N m^{-2} with 10% elongation, but its good corrosion resistance allows it to be used unprotected in the atmosphere.

The highest tensile strengths are obtained from fully heat-treated Al—Zn—Mg—Cu alloys, e.g. 6% Zn, 2·5% Mg, 1% Cu, which can reach $5·5 \times 10^8$ N m^{-2}. However, their ductility is low, about 5%, which gives them a low fracture toughness, and they are prone to intercrystalline cracking and stress-corrosion cracking due to localized overageing in the grain boundaries. The addition of a trace of silver improves the grain boundary structure and properties of these alloys. At a lower strength, alloys containing 4.5% Zn and 1% Mg are useful because they can be welded without losing their heat-treated strength. This is possible because they age at room temperature in about one month after welding and the weld develops a strength equal to that of the parent metal. Having a heat-treated strength of about $2·8 \times 10^8$ N m^{-2} and 15–20% elongation, they are good alloys for on-site constructional work.

25.7 Chemical inertness

Chemical properties, particularly corrosion resistance, greatly influence the value and use of metals and alloys. We have already seen this in the development of magnesium and aluminium alloys. Steel loses some of its supremacy because of the cost of the various measures needed to protect it from corrosion, e.g. painting, tinning, galvanizing, electroplating and alloying. Aluminium and titanium, by contrast, enjoy outstanding corrosion resistance; and a metal such as chromium owes almost its entire use to its ability to resist corrosion and to confer this resistance on metals to which it is added.

Corrosion resistance can come of course either from great chemical reactivity, leading to the formation of stable oxide films, as on Al, Ti and Cr, or from *chemical inertness*, which allows the almost bare metal surface to exist unattacked. We shall now consider metals which mainly owe their practical value to this chemical inertness.

The ability of a *noble* metal to remain uncorroded with a bare metal surface is valuable, not only for corrosion resistance itself but also in many other ways; for example, it allows the metal to make good electrical contacts, to be soldered, brazed and welded readily, and to preserve a bright metallic surface of attractive appearance. The value of the noble metals is further enhanced by their ductility for mechanical shaping and by their good alloying capabilities.

Gold is of course mainly used in jewellery and dental fillings. The pure metal (24 carat) is very soft and malleable, which allows it to be beaten out into thin gold leaf, but for most practical uses gold is hardened by alloying (e.g. 18 carat, 75% Au). Copper and silver are usually added in balanced proportions to preserve the colour of gold, although nickel, zinc and cadmium are also sometimes added.

Some coins are still made from *silver*. The fine colour and high polish shown by this metal have maintained its pre-eminence for tableware. Silver resists many corrosive environments and is used to line brewing vats, although it is of course tarnished by sulphur compounds. It is particularly suitable for electrical make-and-break contacts because its weak affinity for oxygen and sulphur enables thin layers to form, which prevent its surfaces from sticking together, without appreciably increasing the electrical contact resistance. Amongst the silver alloys, we may note *silver solder* (30% Cu and 10% Zn in Ag) which, with a melting point of about 775°C and good strength, is useful for joining metals both in jewellery and fine engineering work.

The exceptional corrosion resistance of *platinum* has led to its use in vessels for holding highly corrosive reagents, particularly at high temperatures. There are advantages also in the greater hardness of the platinum metals; for example, platinum is used for lightly-loaded electrical contacts and for nozzles from which glass and rayon fibres are extruded, and *iridium* is used on pen nib tips. *Palladium* now competes with platinum for jewellery and both metals are used in dentistry. *Rhodium* is used with platinum for thermocouples at temperatures up to 1700°C, for example in the steel industry, and is also sometimes coated thinly on silverware to prevent tarnishing.

25.8 Copper and its alloys

Copper owes its competitive position to the fact that it provides properties near to those of a noble metal at a price near to that of a cheap structural metal. Certainly, it cannot compete with steel or aluminium in applications where the strength/cost or strength/weight ratio is the critical factor, but its chemical inertness, high conductivity, good working and alloying properties, and ease of soldering and brazing nevertheless ensures a wide range of uses.

The advantages of its chemical inertness and good fabrication properties are evident in its use as sheet for roofs (where the green patina produced by slight atmospheric corrosion enhances its attractive appearance) and for domestic water pipes. The combination of high thermal conductivity with these properties leads to its wide-scale use in motor car radiators, cooking utensils, distillery and chemical plant, engine cylinder heads and fireboxes. For some of these applications its strength can be increased without much loss of conductivity by the addition of 0·5% As (which pins dislocations) or 0·5% Cr (which can treble the strength by precipitation hardening).

It has a higher electrical conductivity, per unit cross-sectional area, than any other metal except silver and so is widely used in electrical apparatus, sometimes with 0·1% Ag added to prevent softening at slightly elevated temperatures. Its ease of soldering and low contact resistance, due to its chemical inertness, give it additional decisive advantages for the wiring of electrical circuits and for brushes and overhead contact wires for electric trains. Copper for overhead wires is usually strengthened by cold working and alloying with 0·9% cadmium; in this way the strength can be raised to over $5·5 \times 10^8$ N m^{-2} with only 14% loss of conductivity.

The main alloying elements for copper alloys are zinc, tin, aluminium, nickel and silicon, although many others are also often used in small amounts (e.g. P, Be, Mn, Cd, Pb, Cr). Copper alloys are mostly strengthened by solid solution hardening and work hardening (exceptions are the Cu-Be and Cu-Cr precipitation-hardening alloys). Most alloy elements are added for solid solution strengthening and also, below certain limits of composition, to increase ductility, e.g. for deep drawing. Additional reasons for alloying are cheapness (Zn), corrosion resistance (Ni, Al, Sn), machinability (Pb), weldability (Si), deoxidation (P; cf. § 21.12).

The most common alloying element, zinc, increases the strength of copper, e.g. by 50% at 30% Zn, increases the ductility (up to 30% Zn), reduces both cost and density, lowers the melting point (which is an advantage for casting alloys), but reduces the corrosion resistance (dezincification and season cracking). Copper dissolves up to 36% Zn in primary solid solution and the *alpha brasses*, formed within this composition range, are notable for their ductility in cold working operations. *Gilding brass* (5% Zn) has good corrosion resistance; so also has *red brass* (15% Zn) which is used for plumbing and imitation gold jewellery, but the most important alloys are *cartridge brass* (30% Zn), the great ductility of which makes it particularly suitable for deep drawing and cold working generally, and *yellow brass* (34%), the cheapest brass that can be cold worked, which is used for screws, rivets and tubes. At these high zinc contents corrosion becomes a problem and other alloys are added to rectify this. *Admiralty brass* (29% Zn, 1% Sn, with 0·05% As which reduces dezincification) has fair resistance to corrosion by sea water. A more recent alloy, developed to have high resistance to corrosion in marine condenser tubes and power stations, is *aluminium brass* (22% Zn, 2% Al, 0·4% As).

Beyond 36% zinc, the B.C.C. *beta phase* appears with the alpha phase and makes cold working difficult, although hot working is relatively easy and brasses such as *Muntz metal* (40%) are widely used for hot extrusion into tubes and other more irregular shapes. A small amount of tin may be added to give resistance to corrosion by sea water, as in *Naval brass* (40% Zn, 0·75% Sn). More complex shapes, e.g. for plumbing fitments, are made by casting. To meet the need in this case for good casting and machining properties other alloys are also added, as in *cast yellow brass* (38% Zn, 1% Sn, 1% Pb). Lead, which forms dispersed insoluble soft particles in the matrix, is widely used in copper alloys to give free-machining properties. It also acts as a lubricant when these alloys are used for sliding bearings.

A 40% Zn brass is frequently used as a hard solder for brazing steel components together; the name *brazing* in fact derives from the original use of brass for this purpose.

While the name *brass* has remained closely linked to the copper-zinc alloys, the name *bronze* is now applied not only to the copper-tin alloys but to many other copper alloys as well. Tin strongly increases the elastic limit of copper, giving hard, strong alloys, and the ductility is also improved, at least up to 10% Sn. Higher tin contents give much lower ductilities but the strength continues to rise and useful casting alloys can be made up to 19% Sn, although the traditional *gun metal* contains about 9% Sn. *Coinage bronze*, used for 'copper' coins contains some 0·5% Sn and 2·5% Zn, and *phosphor bronze*, the high elastic resilience and corrosion resistance of which leads to its use as a material for springs, contains some 5% Sn and 0·2% P. *Leaded bronzes* contain up to 4% Pb, for machinability and much larger amounts (e.g. 30% Pb) when used for sliding bearings. *Porous bronzes* are also made, from sintered powders, for use as oil-impregnated bearings.

Apart from strength and hardness, the advantage of the tin bronzes lies in their good corrosion resistance. The high cost of tin however has led in recent years to its substantial replacement by other metals such as aluminium, silicon and manganese. The effect of aluminium on strength and ductility is broadly similar to that of tin and *aluminium bronzes* are made with 4 to 7·5% Al, for working into strip, wire, rod and tube, and 7 to 9% Al for casting. The aluminium gives the alloy an excellent corrosion resistance and this, together with the fine golden colour achieved at about 7% Al, has led to the wide use of aluminium bronze for imitation gold jewellery and ornamental architecture. Because of its high strength and corrosion resistance, aluminium bronze is widely used for gears and other moving parts in machinery, condenser tubes and ships' fittings. *Silicon* has broadly similar effects to those of aluminium, but its lower solubility restricts the amount added. A typical composition is 3% Si and 1% Mn. *Silicon bronze* is particularly suitable for welding because the silicon forms a protective silica flux. *Manganese bronze* is really a type of Muntz metal (40% Zn) containing a small amount of manganese, which acts as a deoxidizer and grain refiner, together with small amounts of tin, iron and aluminium for strength and corrosion resistance.

Beryllium bronze (1.8% Be in Cu) is one of the few strongly precipitation-hardening copper alloys. The very high strength (e.g. 9×10^8 N m^{-2} yield stress and $1·2 \times 10^9$ N m^{-2} tensile strength) attainable from it in the solution treated, cold worked and temper hardened condition, together with its good corrosion resistance, makes it competitive with steel for springs, particularly in electrical apparatus where high-conductivity, non-magnetic springs are required. Because of its high resistance to wear it is also sometimes used in place of phosphor bronze for highly loaded bushes and seats in machinery.

Nickel forms an important series of alloys with copper, the *cupro-nickels*.

The very good corrosion resistance obtained at about 30% Ni has led to the large-scale use of this alloy for condenser tubes and other applications where corrosive conditions are severe. At 40% Ni an alloy is obtained with a high electrical resistance that varies little with temperature, useful for electrical resistors in control switchgear. *Nickel silver* (18% Ni, 18% Zn, in Cu) with its silvery colour and good corrosion resistance, is widely used for tableware, particularly when electroplated with silver (EPNS ware).

25.9 Fusibility

There are many uses for metals which have low melting points. Their fusibility may be put to direct use, as in liquid metal coolants, dental amalgams, automatic sprinklers and solders; or exploited for the mass production of cheap castings in permanent metal moulds; or used indirectly, through the mechanical softness and malleability which usually results when a solid metal is fabricated or used at a temperature near its melting point, as in the manufacture of 'tin foil' or the use of white metal bearings. The fusible metals include zinc, lead, tin, cadmium, bismuth, mercury and the alkali metals (cf. Table 13.2).

Zinc is one of the cheapest metals and the ease with which it can be pressure die cast into a permanent metal mould (cf. § 13.10) enables it to be fashioned cheaply into articles of complex shape. Many small parts of motor cars, e.g. door handles, locks, carburettors and petrol pumps are made from zinc die castings, sometimes electroplated with chromium. The zinc metal used in these castings is of high purity, to avoid corrosion and disintegration along impure grain boundaries, and it is alloyed with about 4% Al and 0·02% Mg. The aluminium strengthens the zinc and also prevents the molten metal from attacking the steel casting dies. Zinc has of course several other uses; for example, for galvanizing steel sheet; for making brass; and as zinc sheet for roofs, dry batteries, food cupboards and organ pipes.

Lead can be cast easily into complex shapes, is very soft and malleable, has good corrosion resistance, particularly to sulphuric acid, is dense, and is not unduly expensive. These properties determine its main uses; e.g. in storage batteries, cable sheathing, solder, bearing metal, type metal, x-ray shields, ammunition shot, and sheet and pipe for the chemical and building industries. The pure metal recrystallizes and creeps readily at room temperature. It can be mechanically shaped, cut and welded easily and its softness (strength 1·5 to 3×10^7 N m^{-2}) enables it to be extruded readily on to electrical cable as a sheathing material with good corrosion resistance. To harden the metal and improve its corrosion resistance further, antimony is added, increasing from about 1% Sb for cable sheathing, to 9% for storage batteries and 10 to 15% for die-casting alloys. In these higher amounts, antimony improves the casting properties of lead and allows the metal to be age hardened, so increasing its tensile strength to about 8×10^7 N m^{-2}. The other main alloying element is tin and the lead-tin system is the basis of many bearing metals, solders and type metals, discussed below. More complex alloys, containing lead, bismuth, tin and cadmium, sometimes in eutectic proportions, are used in fusible plugs and printing type (cf. § 15.3).

About one-half of the *tin* produced is used in tinplate. The traditional method of dipping the steel sheet into molten tin is now becoming replaced by an electroplating process, to economize in the use of this rather scarce and expensive metal by producing a thinner coating. A recent use for molten tin has appeared in the *float process* for making plate glass by floating molten glass on the surface of a tin bath. Tin bronzes consume about 7% of the tin produced. In tin-based alloys the main alloy elements are lead and antimony. *Britannia metal*, used for tableware, consists of 7% Sb and 2% Cu in tin. There are many important tin-lead alloys, e.g. tinman's and plumber's *solder* (cf. § 15.3) and *pewter* (similar to Britannia metal but with about 10% Pb included). *Tin foil* ('silver paper') is made by rolling a sandwich of lead sheet in tin sheets down to a foil about 10^{-5} m thick; in recent years this has of course been largely replaced by the cheaper aluminium foil.

Tin and lead are the main constituents of most *white metal bearing alloys* or *Babbitt metals*, which are used as thick pads enclosed in steel journal boxes for supporting rotating shafts and other moving parts. It is difficult to align a long shaft so accurately that it will run perfectly through several fixed bearings. The advantage here of the soft white metal bearings is that the bearing metal yields to the pressures on it and so aligns itself to the shaft. This softness also prevents high stresses developing at asperities between the bearing and shaft, and so allows the shaft to run in dirty oil without becoming scored by particles of grit which get into the bearing interface. Although such a bearing metal must be soft, a certain stiffening is also necessary to enable it to support the bearing pressure and to prevent it becoming squeezed out at the ends. For this reason some antimony and sometimes copper is added. For supporting the heaviest loads the preferred alloy contains about 8% Sb and 8% Cu in Sn, but the high tin content makes it expensive. For supporting lighter loads a cheaper lead-based bearing is usually used, e.g. 10 to 15% Sb and 5 to 10% Sn in Pb.

Turning now to other fusible metals, *cadmium* is used as a bearing metal (3% Ni, or 1% Ag, or 1% Cu, in Cd) in aircraft engines; it is more resistant to fatigue than the tin and lead bearings, but expensive. *Indium* is sometimes used as a coating for bearings to improve their resistance to corrosion and wear. *Mercury* is familiar from its use in thermometers and other measuring instruments, and also as the basis of dental amalgams. A pasty alloy is made by mixing a powdered alloy of Ag_3Sn (sometimes with Cu replacing part of the Ag) with mercury and immediately inserted into the waiting dental cavity, where it 'sets' as inter-diffusion takes place and new solid phases form. The tin content has to be carefully controlled to avoid large changes of volume during setting. *Sodium* is used as a liquid metal coolant in the shanks of engine valves and also, because of its low neutron absorption cross-section, in fast neutron nuclear reactors, where immense amounts of heat have to be removed from a very small reactor core. *Potassium* is sometimes added to the sodium to reduce the melting point; the Na—K system has a eutectic at $-12°C$ and 23% Na.

25.10   High-temperature properties

The ability to withstand high temperatures without melting or corroding, and to support stress at those temperatures, is crucially important in certain uses of materials. The most familiar example is the coiled-coil tungsten filament in the gas-filled lamp. In this application sheer high melting point is needed, since the intensity of light emitted increases very steeply with temperature. Along with its extremely high melting point, tungsten also has the advantage of low evaporation losses (about 10^{-5} kg m^{-2} s^{-1} at 3000°C). The coiled-coil form reduces heat losses from the filament and the gas filling increases the life by reducing evaporation. Grain boundary sliding is a problem because it reduces the cross-section and so leads to local overheating and early failure. An early solution was to grow the filaments as single crystals but the present-day method is to prevent the grain boundaries from sliding by keying them with insoluble particles, e.g. of thorium oxide.

Other applications where high temperature properties are important are in high speed cutting tools, thermocouple elements, resistance heaters, heat engines and boilers, rocket combustion chambers and nozzles, and kinetically-heated high-speed aircraft and space vehicles. In high-speed cutting tools the requirement for great hardness at red heat has led to the development of high-speed tool steel (cf. § 25.3) and also to tungsten carbide and other refractory metal carbide tools in which the carbide grains are cemented together by a small amount of cobalt. High melting point, oxidation resistance and a high thermoelectric e.m.f. are the main properties required for thermocouple elements and are satisfied by the combinations platinum-platinum plus 13% rhodium; and chromel-alumel (i.e. 10% Cr in Ni and 2% Al in Ni); molybdenum-tungsten thermocouples are sometimes used in protective atmospheres up to 2500°C. The main material used for resistance heating elements, on account of its good oxidation resistance, ductility and electrical resistance, is nichrome (20% Cr in Ni). Others are kanthal (25% Cr, 5% Al, 3% Co, in Fe), platinum, silicon carbide, molybdenum (protected) and carbon (protected).

The important of high temperature materials in heat engines and power plant generally stems directly from the thermodynamic fact that the efficiency of converting heat into work increases with the temperature of the hottest part of the engine. For example, the specific thrust (i.e. kg thrust per kg of air per sec taken in) of an aircraft engine could be almost doubled if the turbine inlet temperature were increased from 1000°C to 1400°C.

The systematic development of high-temperature creep-resistant materials began in the 1920s with the need for greater efficiency in steam power plants. Development in this field has continued steadily ever since, although it has been rather overshadowed by the more spectacular development of 'superalloys' for jet engines and gas turbines. Early high-temperature engineering design was dominated by the desire to minimize creep deformation, but the very thick sections which resulted from this suffered in some cases from distortion and thermal cycling fatigue effects due to the large temperature gradients set up in them. The more recent trend has been towards thinner, more flexible sections, using stronger materials and accepting some creep deformation. The other trend has been the increasing recognition of the importance of rupture under creep conditions. The time to rupture, under a given load at a given temperature, is now often a more critical design parameter than the (small) creep deformation produced during this time.

We have already discussed some aspects of high temperature creep-resistant materials in § 20.9, 21.9, 21.13, 21.15, 23.3 and 25.3. The early developments were concerned with the improvement of boiler steels, leading to the use of silicon killing rather than aluminium killing for carbon steels, and to the addition of 0·5 to 1·0% Mo (together often with up to 2·25% Cr or 0·25% V) which enables stable fine carbides to be formed. Traces of boron are now sometimes added to gain additional creep strength. Because of their low chromium content, these steels have inadequate corrosion resistance above 550°C. For oxidation resistance at 600°C about 7% Cr is required but steels in this composition range tend to have duplex ferrite-austenite structures and unsatisfactory creep properties. There is then a choice between non-transformable alloy ferritic steels (e.g. 8% Cr, 3% Mo, 1·5% Ti, with low C) or fully austenitic steels such as those discussed in § 25.3. For high-temperature steam power plant, austenitic stainless steels of the 18% Cr, 12% Ni, 1% Nb type (sometimes with Mn replacing one-half of the Ni and with small additions of V and B) have been developed, which are hardened by precipitation of niobium carbide, particularly on dislocations. The difficulties here are a tendency towards brittle grain boundary ruptures due to excessive hardening in the grains, particularly near welds, and a susceptibility to corrosion by sulphates and chlorides in furnace flue gases.

The development of superalloys for gas turbines began with the attempt to strengthen the heat-resistant 80–20 Ni-Cr alloy by precipitation hardening and this work led to the discovery of the nimonic alloys. Nickel has proved a remarkable matrix metal for high-temperature alloys and is capable of maintaining good strength at temperatures up to about $0·7\,T_m$ (K), a temperature fraction which is only exceeded by materials such as $Al + Al_2O_3$ (S.A.P.). The nimonics are hardened mainly by precipitation of intermetallic compounds of the $Ni_3(Ti, Al)$ type. A typical example is *Nimonic 90*, which contains about 20% Cr, 2·5% Ti, 1·5% Al, 18% Co, small amounts of Fe, Mn and Si, and less than 0·1% C. This will support, without rupture in 75 hr, about $3 \times 10^8\,\mathrm{N\,m^{-2}}$ at 750°C or about $2 \times 10^8\,\mathrm{N\,m^{-2}}$ at 815°C. The best nickel alloys can support about $1·3 \times 10^8\,\mathrm{N\,m^{-2}}$

for 100 hr at 1000°C in the wrought condition, or at 1025°C in the cast condition. It seems unlikely that more than about 50°C further increase in the creep-resistant temperature can be gained by additional development of nickel alloys through conventional precipitation-hardening and solution-strengthening methods. Dispersion-strengthening, e.g. by ThO_2 in Ni, can enable higher temperatures to be reached but the strength levels attained may not be sufficient for turbine blades.

Cobalt-base alloys may perhaps resist oxidation better than the Ni-base alloys at high temperatures but are unlikely to have higher strengths. The best chromium-base alloys seem to be equal to the best cast nickel-base alloys in their high-temperature strength, and also have good oxidation resistance, but their brittleness at room temperature is a disadvantage which is slowly being overcome. The refractory metals such as molybdenum and niobium have good high temperature strength and offer a clear possibility of raising working temperatures to 1200°C but their poor resistance to high temperature oxidation is a great disadvantage. There is some promise that coatings can be developed for niobium alloys that will protect the metal even during thermal cycling (cf. § 23.3).

So far, the problem of developing a material with good strength at 1200°C or above, and with resistance to oxidation and to brittle fracture, has proved intractable. It now seems doubtful whether it can be solved by conventional metallurgical methods. Almost certainly ceramic materials will find applications here. These can be light, highly refractory and resistant to oxidation. Their brittleness is a problem which may be overcome by using them in a dispersed fibrous form, so that a crack does not have an uninterrupted path through the material. Probably the solution for a high-temperature creep-resistant material will be a metallic matrix, chosen for its oxidation resistance, which is strengthed by ceramic fibres or whiskers. Silicon nitride whiskers can now be made in bulk and the production of boron and carbon fibres and alumina whiskers is being intensively studied.

The frictional heating of high-speed aircraft raises problems of the creep resistance of light-weight structural materials. Aluminium alloys can be used up to speeds of about Mach 2, but at Mach 3 and above more refractory metals such as stainless steel and titanium have to be used. Again, there are opportunities for fibre-reinforced materials with lightness, strength and oxidation resistance, at temperatures up to about 400°C, to be used in aircraft structures. It should be possible, by fibre strengthening, to use a structural metal almost up to its melting point but there is also the possibility that fibre-reinforced plastics will be developed, stable up to these temperatures, to replace the structural metals.

25.11   Electromagnetic and nuclear properties

Although the amounts of metal used in electrical, magnetic and nuclear applications are minute compared with the tonnage use of metals as structural materials, the immense value of metals in these applications is demonstrated by the great size of the electrical and nuclear industries which depend critically upon them. So critical is this dependence that both halves of the cost-effectiveness criterion often run to striking extremes. When a material has really unique properties for which there is a strong practical need and no real alternative, it will usually be brought into use whatever its cost. There are many examples of this in both the electrical and nuclear engineering fields.

We have already discussed the electrical and magnetic properties, e.g. electrical and thermal conductivity, semi-conductivity, superconductivity and ferromagnetism, and the metals and alloys which have been developed for these properties, in Chapter 24 and § 25.6 and § 25.8.

In § 2.5 we noted some of the factors which govern the choice of materials for nuclear reactors. Metals are particularly suitable because, in addition to their usual engineering and manufacturing properties, they have high thermal conductivities and are immune to radiation damage by ionization (cf. § 4.6). The mobility of point defects also enables many metals, particularly F.C.C. metals, to be fairly insensitive to knock-on radiation damage (cf. § 19.10). Many severe difficulties remain, however. Often the materials which, for nuclear reasons, it is most essential to use have many difficult metallurgical and engineering features. They may be difficult to extract (e.g. Be), may contain undesirable and not easily removable impurities (e.g. Hf in Zr, B in stainless steel), may have awkward crystal structures or undergo crystallographic transformations (e.g. U and Pu), may oxidize readily (e.g. U and graphite), may be brittle, particularly after irradiation (e.g. U, Be and ferritic steel), and may be highly toxic (Pu and BeO).

The main fuel element material for the earlier types of gas-cooled reactors is *uranium*. We have discussed its problems of radiation growth, wrinkling, creep, and fission gas swelling, and how they are overcome, in § 19.10. For more advanced reactors, working at higher temperatures, the crystallographic transformations which set in above 660°C have largely ruled out the use of uranium in its metallic state and most recent reactor designs are based on the use of *ceramic* fuels, particularly oxides and carbides of U, Pu and Th. Uranium dioxide is now widely used in both water-cooled and advanced gas-cooled reactors because of its chemical stability and high melting point, although its low thermal conductivity is a disadvantage. Attempts have been made to develop liquid metal fuels (e.g. U in liquid Bi) but problems of liquid metal attack on their containers have proved difficult.

The main moderator materials in use are *graphite* and *heavy water*. Graphite suffers from effects of knock-on radiation damage (stored energy and radiation growth), although these become fairly unimportant at temperatures above about 250°C, where much of the radiation damage anneals out. At higher temperatures oxidation is a problem but this has been overcome in advanced gas-cooled reactors by adding a trace of methane to the carbon dioxide coolant.

Materials for fuel cans have to satisfy several stringent requirements.

The can must provide a complete and reliable envelope to prevent escape of radioactive fission products and to protect the fuel from chemical attack. It must itself be chemically compatible with the fuel and resistant to attack by the coolant. It must endure severe neutron and fission fragment irradiation. It must be made from a material with a low neutron absorption cross-section. It must have a good thermal conductivity. The can wall must be thin, both to conserve neutrons and also to minimize the temperature drop from the fuel to the coolant; and yet it must also be strong if the fuel itself is mechanically weak or if it is required to withstand a pressure of fission gases created by the fuel.

For gas-cooled reactors, using uranium metal fuel, *magnesium* is the main canning material. The two main contenders for UO_2 fuel cans in advanced gas-cooled reactors have been *beryllium* and *austenitic stainless steel*. Despite its very low neutron cross-section, beryllium has not so far been used, because of its cost and low ductility. Stainless steel cans are now used successfully. The main problem has been to produce very clean, inclusion-free steel and to fabricate it into very thin tubes (4×10^{-4} m wall thickness), this thinness being necessary to minimize neutron losses in this material of moderate neutron absorption cross-section. A remaining problem is some embrittlement which appears after prolonged irradiation at high temperatures. This appears to be caused by the formation of small helium gas bubbles on grain boundaries, created by the neutron-induced decomposition of boron, present in trace amounts in the steel.

In fast reactors, the problem of neutron absorption in the fuel can is less severe and metals such as *stainless steel* and *niobium* are used. Compatibility with the liquid sodium coolant is a problem. The oxygen content of the sodium has to be kept very low (< 10 parts per million) to avoid corrosion of the can.

In water and steam cooled reactors, corrosion in high-temperature water (e.g. 280°C at 70 atm. pressure) or steam at 500°C present special problems for canning materials. *Zirconium* (with hafnium removed) is the favoured material, usually in the form of *zircalloy* (1·5% Sn, 0·15% Fe, 0·05% Ni in Zr) or other alloys (e.g. 2·5% Nb in Zr). A recent trend is to add about 0·4% tellurium, which improves the resistance of zirconium to corrosion and absorption of hydrogen, in water at high temperatures and pressures.

FURTHER READING

ALEXANDER, W. O. (1965). *Metallurgical Achievements*. Pergamon Press, Oxford.
AMERICAN SOCIETY FOR METALS (1961). *Metals Handbook, Properties and Selection of Materials*.
FROST, B. R. T. and WALDRON, M. B. (1959). *Nuclear Reactor Materials*. Temple Press, London.
MARIN, J. (1952). *Engineering Materials*. Prentice-Hall, New Jersey.

SHARP, H. J. (1966). *Engineering Materials: Selection and Value Analysis.* Heywood, London.
SMITHELLS, C. J. (1962). *Metals Reference Book.* Butterworth, London.
YOUNG, J. F. (1954). *Materials and Processes.* John Wiley, New York and London.
ZACKAY, V. F. (1965). *High-Strength Materials.* John Wiley, New York and London.

Index